The World of
FLUORESCENT MINERALS

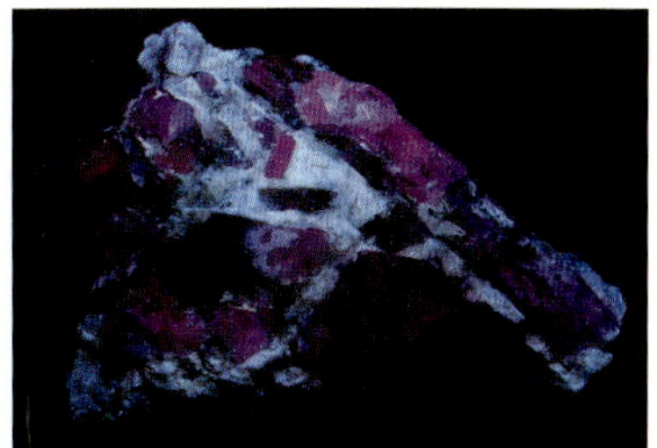
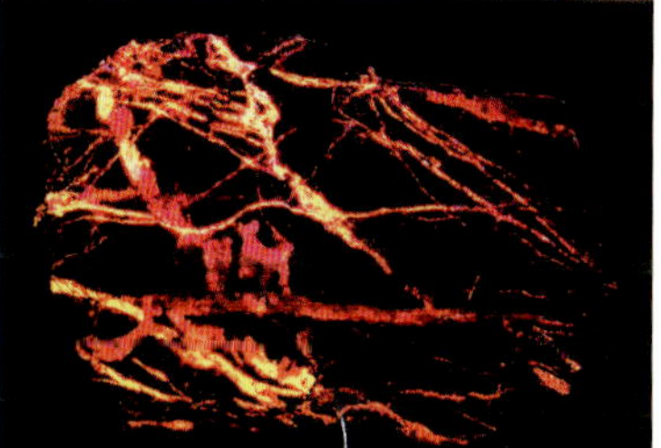
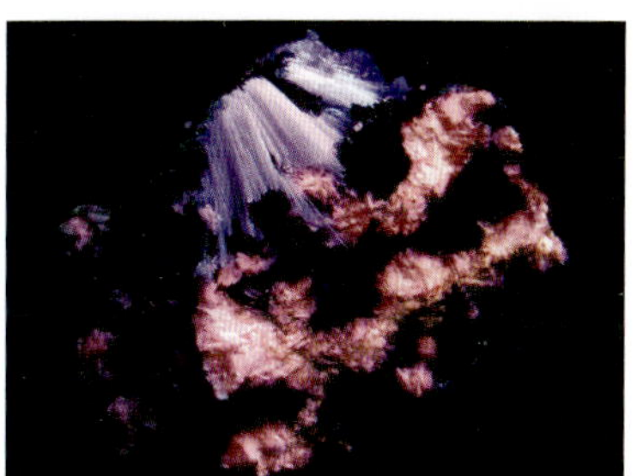
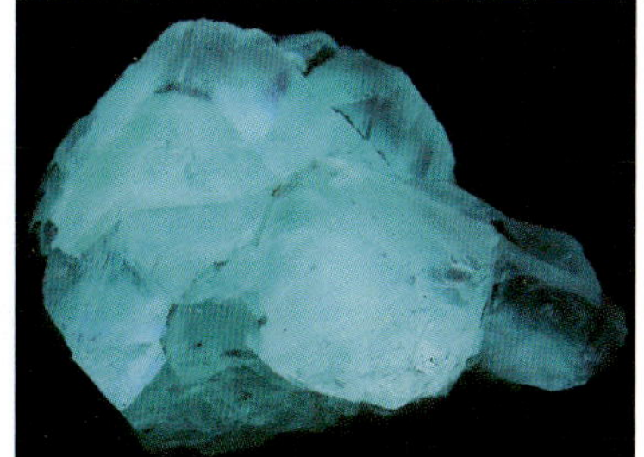
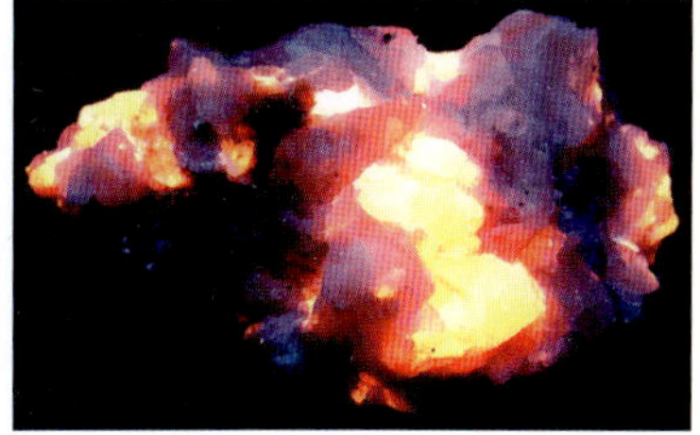
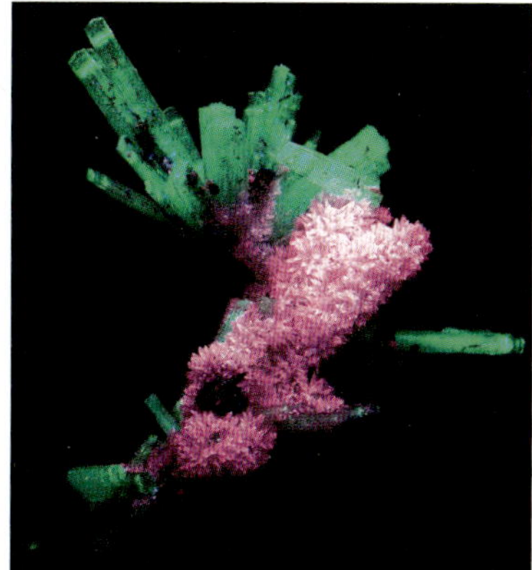
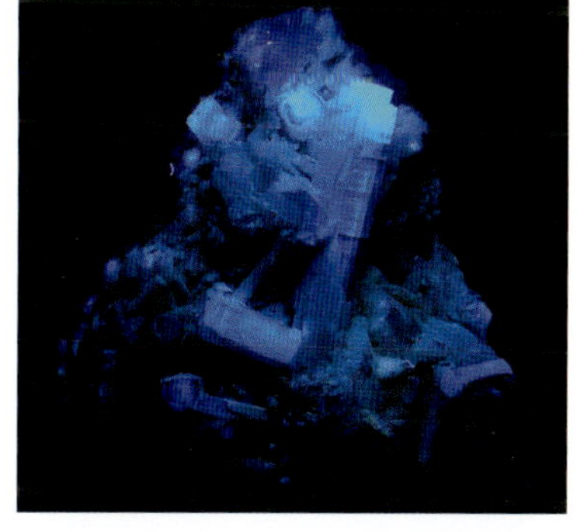
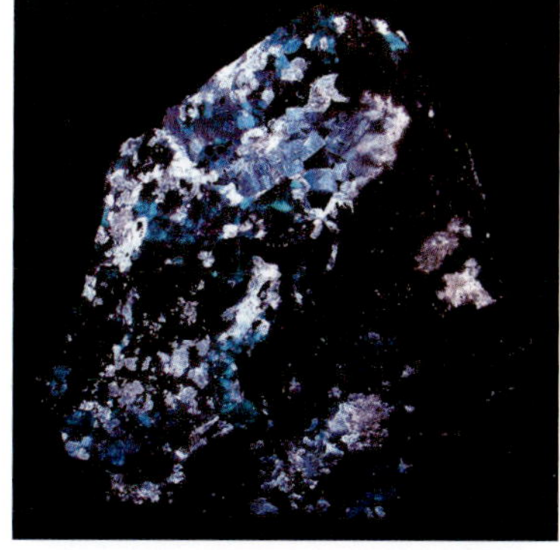
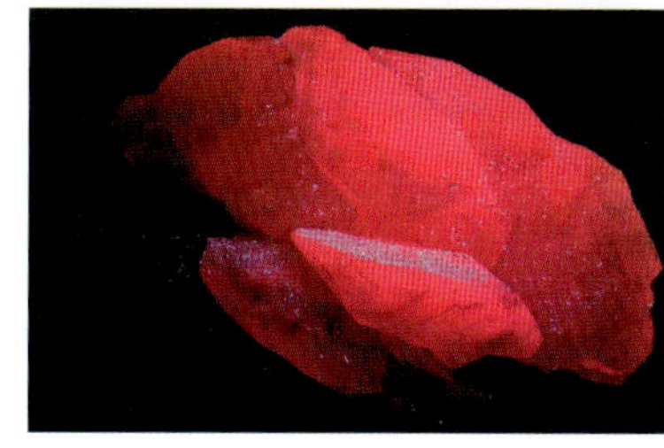
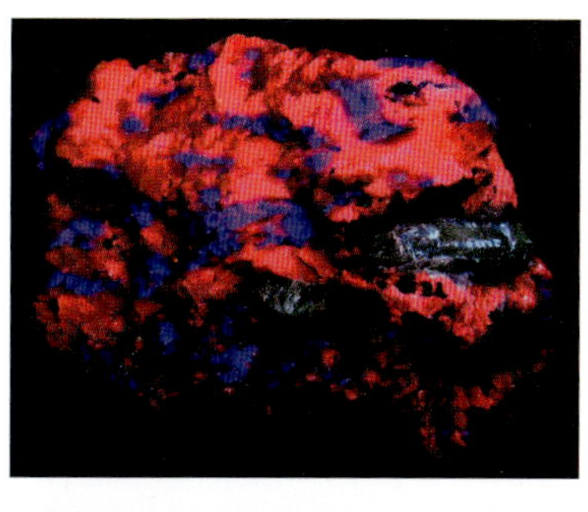
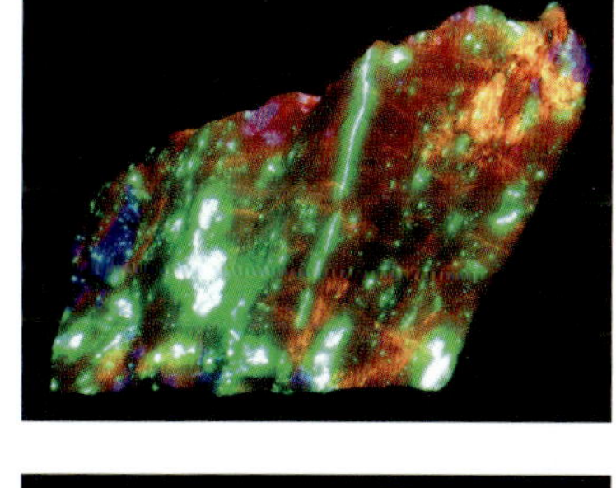
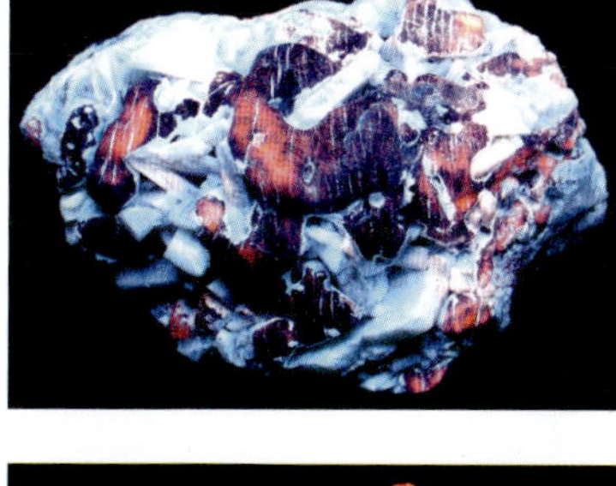
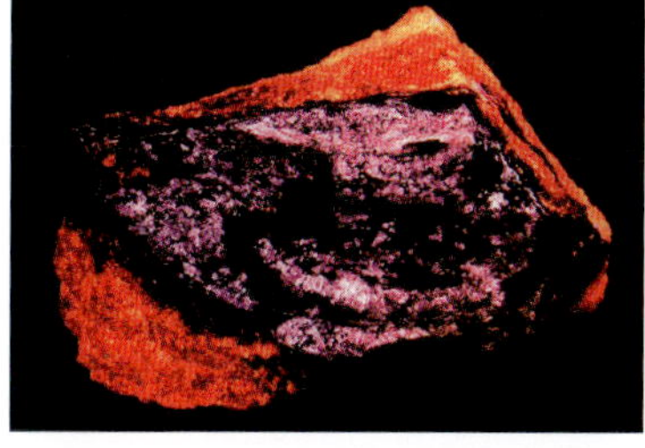
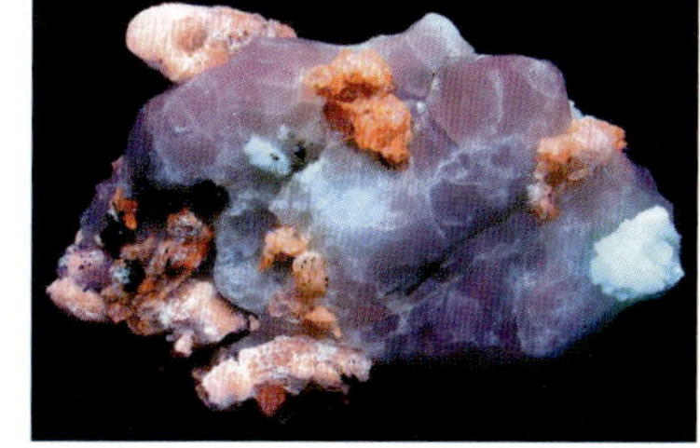
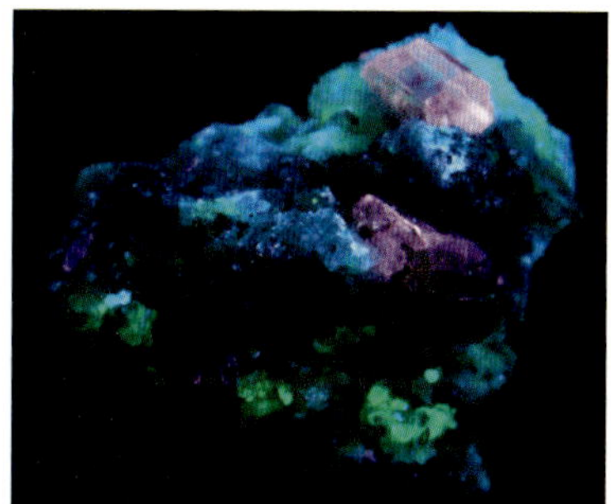
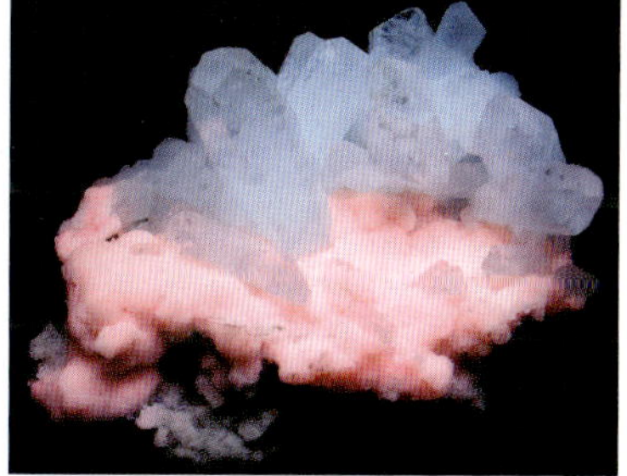
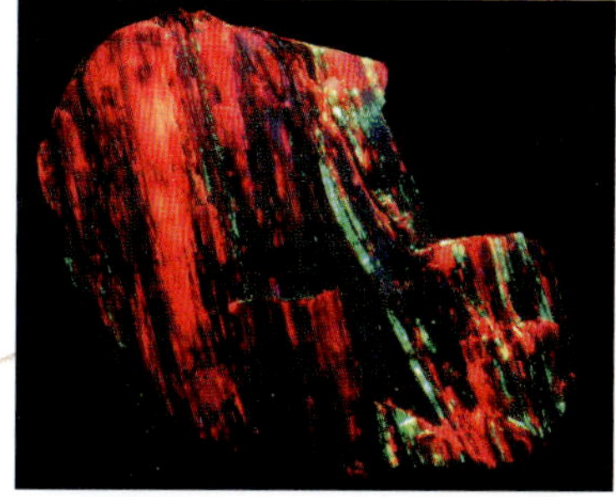
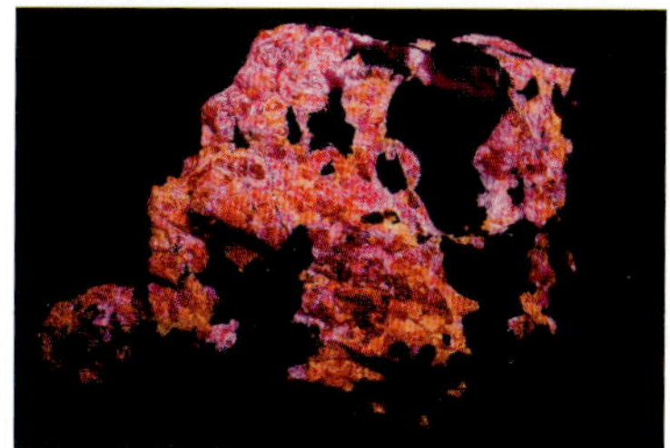
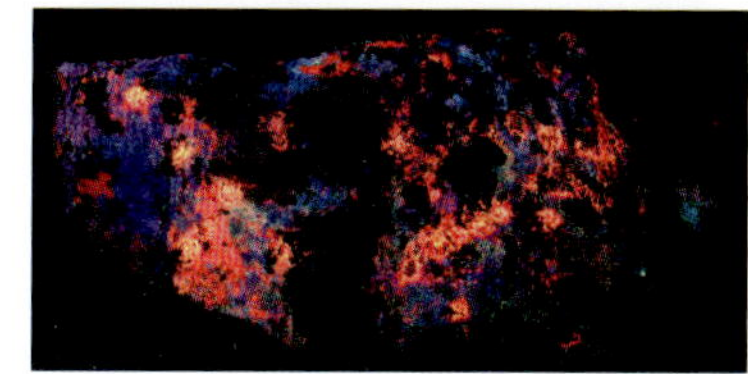

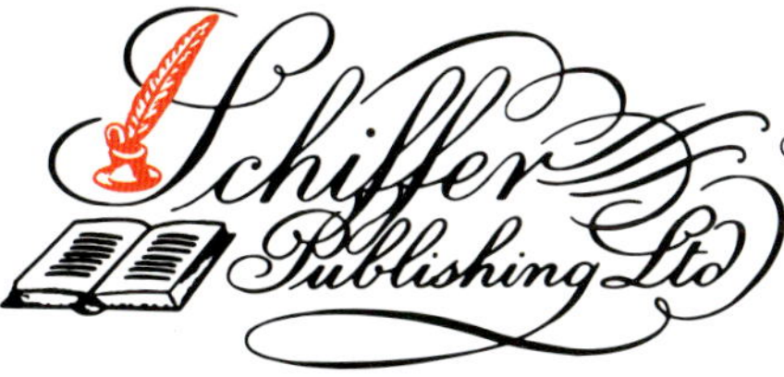

4880 Lower Valley Road · Atglen, PA 19310

Stuart Schneider

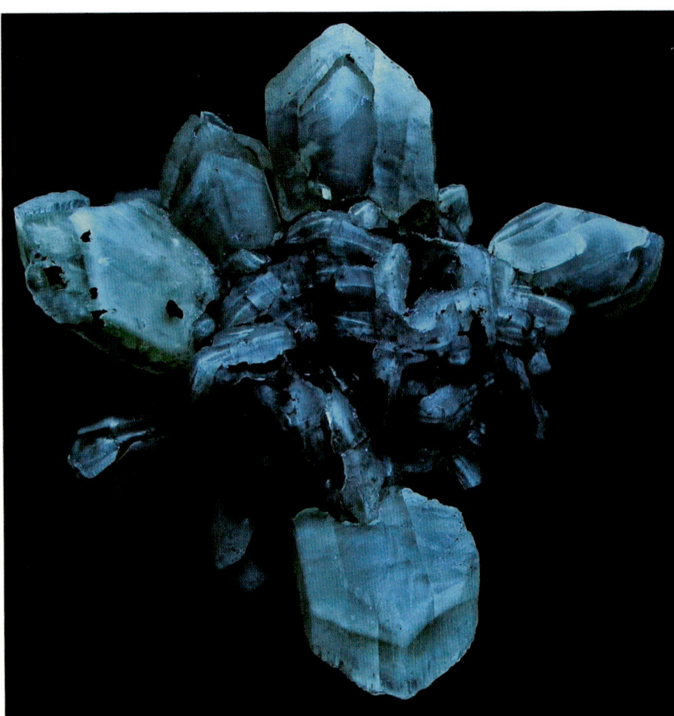

Library of Congress Cataloging-in-Publication Data:

Schneider, Stuart L.
The world of fluorescent minerals / Stuart Schneider.
p. cm.
ISBN 0-7643-2544-2 (pbk.)
1. Fluorescent minerals—Collection and preservation. I. Title.

QE364.2.F5S368 2006
549'.125—dc22

2006012981

Designed by John P. Cheek
Type set in Korinna BT

ISBN: 978-0-7643-2544-1
Printed in China
5 4 3 2

Other Books by this Author

The Book of Fountain Pens and Pencils, ISBN: 0-88740-394-8, $79.95 (with George Fischler)
Fountain Pens and Pencils: The Golden Age of Writing Instruments, ISBN: 0-7643-0491-7, $89.95 (with George Fischler)
The Illustrated Guide to Antique Writing Instruments, ISBN: 0-7643-0980-3, $19.99 (with George Fischler)
Collecting Picture and Photo Frames, ISBN: 0-7643-0610-3, $39.95
Collecting Flashlights, ISBN: 0-7643-0041-5, $29.95
Cigarette Lighters, ISBN: 0-88740-952-0, $39.95 (with George Fischler)
Handbook of Vintage Cigarette Lighters, ISBN: 0-7643-0932-3, $19.95 (with Ira Pilossof)
Halloween in America, ISBN: 0-88740-707-2, $29.95
Collecting Lincoln, ISBN: 0-7643-0270-1, $69.95
Collecting the Space Race, ISBN: 0-88740-535-5, $34.95
Ronson's Art Metal Works, ISBN: 0-7643-1194-8, $39.95
Halloween Costumes & Other Treats, ISBN: 0-7643-1410-6, $29.95 (with Bruce Zalkin)

And of particular interest to Mineral Collectors:

Collecting Fluorescent Minerals, ISBN: 0-7643-2091-2, $29.95

Published by Schiffer Publishing Ltd.
4880 Lower Valley Road
Atglen, PA 19310
Phone: (610) 593-1777; Fax: (610) 593-2002
E-mail: Info@schifferbooks.com

For the largest selection of fine reference books on this and related subjects, please visit our web site at
www.schifferbooks.com
We are always looking for people to write books on new and related subjects. If you have an idea for a book please contact us at the above address.

This book may be purchased from the publisher.
Please try your bookstore first.
You may write for a free catalog.

Contents

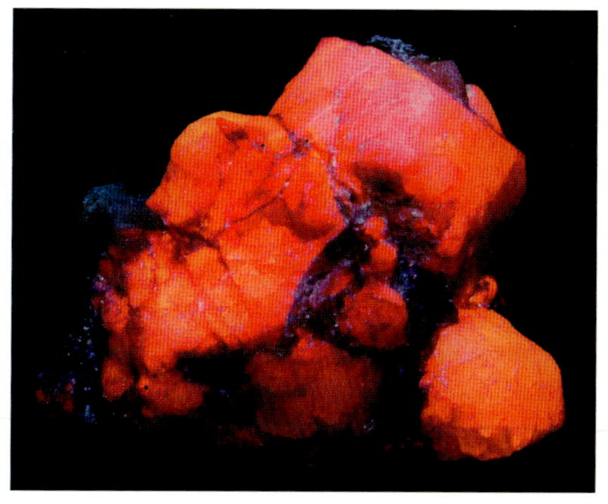

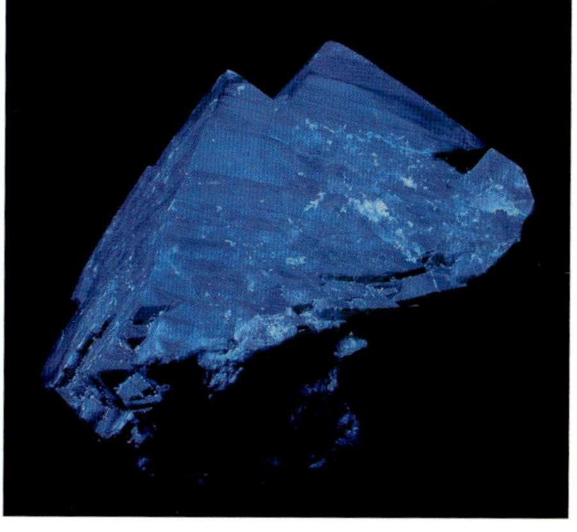

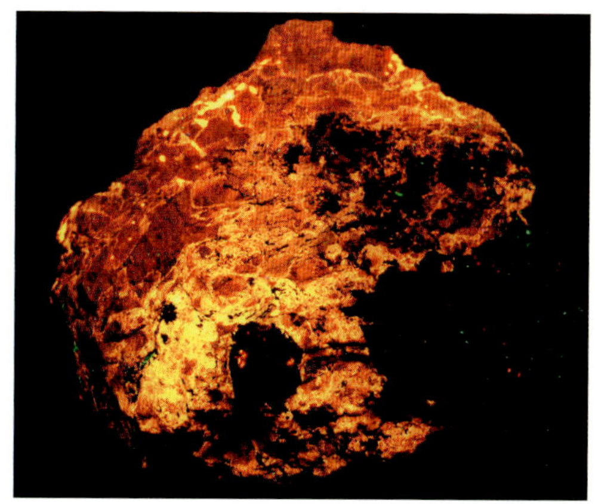

Acknowledgments

I would like to acknowledge the friends who advised me during the creation of this book. Thanks to Earl Verbeek, the curator of the Sterling Hill Mineral Museum, Richard Bostwick, author and updater of the fluorescent-mineral list for Franklin and Sterling Hill, and Jacques Poulin of Canada whose knowledge of Canadian minerals is exceptional. These individuals were willing to take the time to look over the manuscript, photos, and photo descriptions before publication, identify problems that needed the expert's eye and make suggestions that improved the book immeasurably.

Note: I added about 60 photographs and captions after my reviewers looked over the manuscript. Although I had help from many sources, if there are any errors in the book, I take full responsibility for them.

Thanks to the Franklin Ogdensburg Mineralogical Society (FOMS) for their help and field trips. I really look forward to their meetings and talking with the members. To Robert and Richard Hauck, of the Sterling Hill Mining Museum in Ogdensburg, New Jersey, who are always helpful. With all the things they have to do to keep the place running smoothly, they still look like they are having a good time. To Rebecca Schneider, who continues to accompany me on digging trips and to my wife, Peggy who often asked when the weather turned colder "When does the rock digging season actually end?"

I especially wish to acknowledge the contributions of (in alphabetical order) Scott Allen, C. Richard Bieling, Kevin Brady, Dave Bunk, Mark L. Cole of MinerShop for most of the Greenland material, Joseph Dorris, Alexandr G. Efimenko, George Elling, Jeffrey Fast, Donald Fife, Bill Gardner, Nicole Gariepy, Jamal-UL-Hasnain Gillani, Kurt Hennig, J. Hyrsl, Greg Jacobus, Aisha and M. Arif Jan, Jeanne's Rock & Jewelry of Texas, Raymond Klinger, Steve Kuitems, Greg Lesinski, Ed Letscher, Freddy Lubbers, Darryl MacFarlane, Veronica Matthews, Don Newsome of UV Systems, Inc., Philip Persson, George Polman, James Poole, Ernie Schlichter, Gennadiy Skublov, Gar Van Tassel, Yuri Tsygankov, Howard Van Iderstine, Charlie Ward, Dru Wilbur, and James Zigras. These people helped me locate great fluorescent minerals, offered information about the minerals, and aided my fluorescent-mineral education.

Once again, my thanks to those individuals for their time, their marvelous minerals, and above all their willingness to share so much of their knowledge with me.

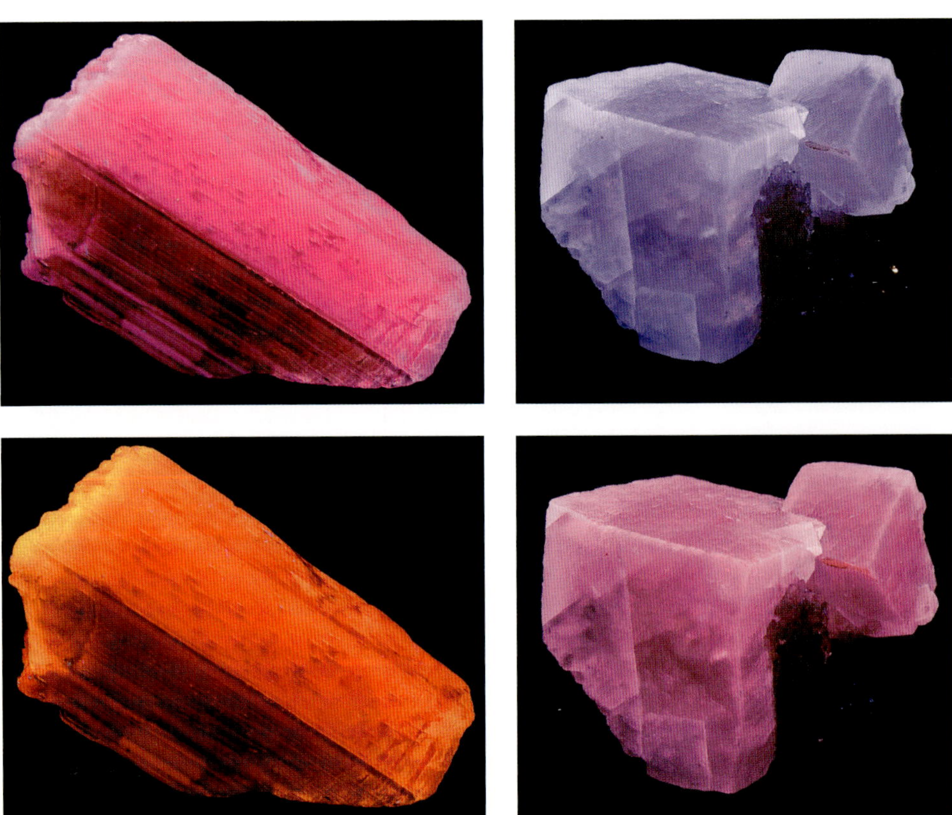

Introduction

My first book on fluorescent minerals was so well-received by mineral collectors that I knew a second volume would be a welcome addition to the first. Collectors and reviewers told me that they liked a book with loads of photographs that showed what the fluorescent minerals looked like in daylight and under the ultraviolet lamp. This book will again provide the best color photographs available of specimens that were not covered in the first book, plus new photos of interesting pieces. The goal was to include more worldwide minerals while still respecting the importance of New Jersey's superb collecting environment.

The first book contained some errors. A few minerals were misidentified. Others were given obsolete names that the mineral-collecting community still uses, but which mineralogists do not currently recognize as valid names of mineral species (e.g. zinc schefferite, beta-willemite, etc.). Some chemical formulas were taken from older literature and have been replaced by newer formulas. These corrections were provided to me by a group of readers who took the time to bring them to my attention. Those corrections will be incorporated in future editions (of which I hope there are many) of the first book. Although I write books for collectors of fluorescent minerals, many of whom are not trained mineralogists or geologists, it is important to provide the best information available while still writing in a easy-to-understand, non-technical manner.

The Colors in The Book

The colors in this book may not be exactly what you see in person. This is attributable to two things:

1) The colors are captured with a digital camera and then the photographs are "developed" on the computer. The monitor is color calibrated so that the colors are true to what is seen by this photographer. The photographs are saved as digital files, burned onto a disk, and then passed on to the publisher. The publisher uses a page layout program to typeset the book and once typeset, it is again burned to a disk. This disk goes to the color separators, who work on the photographs to make printing plates (four if you count black as a color). The printing plates are sent to the printer, where the book is printed with colored inks. Variations occur between the photographer's output and the final printer's product.

2) With many Franklin/Sterling minerals, the brightness of the willemite and the calcite are greater than the other less-bright minerals. This causes certain colors to be overexposed and others underexposed in a typical photograph. When looking at minerals under the ultraviolet lamp, the eye is much more sensitive to a wide range of intensities and hues than any camera or film. The camera is not able to balance color, brightness, and hue like the eye. By reducing saturation and brightness of the willemite and calcite and increasing saturation and brightness of the other minerals, I have attempted to show what they actually look like.

The book uses the following abbreviations: UV (ultraviolet), SW (shortwave), MW (midwave), and LW (longwave).

Value Guide

I have included a value guide to help collectors. A value guide is comparative. It allows you to evaluate two items to determine their relative value. It is useful in buying and trading and it can help give you a feel for the rarity of a piece.

The items in this book are valued by my experience buying and selling fluorescent minerals over the past seven years. Values are given in ranges and I have tried to give values that dealers can sell at and collectors will buy at. Generally these are the fair ranges that the minerals usually trade in. Note the size of the piece that is valued. Some are small pieces. It they were larger pieces, they would be valued higher.

What makes prices in the marketplace differ for the same mineral? 1) People who make their living selling minerals usually charge more than the hobbyist who collects and sells off his duplicates or excess material at a mineral show. One person is making a living while the other is often just covering gas expenses. 2) The person that you deal with may not have a clue as to how to value fluorescent minerals, but may just pick a value. If their pieces do not sell they may lower their price or they may raise their price if several people want the pieces. 3) You may be able to buy some minerals at the mine where they were discovered. The price at the mine may be $1.50 per pound. The same piece may sell for $45.00 a pound at a mineral show. The mineral collector may have to book a flight, rent a car, and stay in a hotel to visit the mine. These costs add to the few dollars a pound that the collector is paying. 4) Many fluorescent minerals are bought and sold at auction on eBay. If two people want the same piece, the price that is paid may exceed the values set in this book. If only one person wants the mineral the week that it is offered at auction, it may sell for a much lower price. 5) Often the pieces offered on eBay are smaller or not as high a quality as the pieces offered by dealers at shows or on the Internet. 6) Aesthetics or the way a mineral looks on display or appeals to the buyer will affect the value. A piece that displays well is worth more than a piece containing the same minerals that is unattractive.

Value ultimately is the price agreed upon by a willing seller and a willing buyer. The buyer must decide if the price of a mineral offered is a good deal for him. You will occasionally overpay for one specimen and occasionally you will underpay for another. In the end, it probably evens out.

Fluorescent Minerals

Fluorescent minerals are available from all over the world. New discoveries are being made each year and collectible specimens are coming out of Afghanistan, Pakistan, Russia, China, and Chile, to name just a few places. Some pieces in collections of worldwide material have never had ultraviolet lamps shined upon them. Many mineral dealers are becoming aware of the growing number of fluorescent-mineral collectors and try to find new minerals for these collectors.

Currently the best places to find fluorescent minerals are at rock and gem shows (my favorites are the Tucson Rock & Gem Show each January/February and the Franklin, New Jersey, mineral shows that take place two times a year), on the Internet, or at the mines or mine dumps. The world's largest online auction site, eBay, has a large section for "Fluorescent Minerals" with ever-changing offerings. Trading with other mineral collectors is a great way to improve your collection. The best way to find those collectors is to join a mineral collecting group like the Fluorescent Mineral Society or a local mineral group.

Fluorescence, Tenebrescence, Triboluminescence, and Thermoluminescence

While this was covered in my first book, I think it is worth repeating, as some of the people who read this book may not have the first book. The phenomenon of fluorescence has been observed for hundreds of years. The mineral fluorite lends its name to the word "fluorescence". Fluorescence is a physical (as opposed to a chemical) process. For a mineral to fluoresce, light (which is electromagnetic radiation) of one wavelength, strikes a fluorescent mineral. A reaction in the mineral causes light to come out of that mineral at another, longer wavelength. When those emitted wavelengths are within the visible spectrum, we can see the different wavelengths as different colors. The light emitted is sometimes confined to a narrow band and it may yield a pure intense color that is startlingly beautiful.

As you study your fluorescent minerals, you may notice that some of them have other interesting properties. Some minerals change color, usually turning darker — violet, red, purple, or raspberry color — when exposed to shortwave or longwave ultraviolet light. Over time, they will change back to their original color. This reversible color change is called tenebrescence, reversible photosensitivity, photochromism, etc. The best known example of tenebrescence is seen in hackmanite which is a color-changing variety of sodalite. Hackmanite can be found as masses, opaque crystals, and clear crystals. Most hackmanite turns a raspberry color when exposed to shortwave ultraviolet light. The color fades over time or in direct sunlight. There is a hackmanite from Afghanistan that will turn bright purple in seconds when exposed to shortwave ultraviolet light. Tenebrescence also occurs in some other minerals.

A partial explanation for the change is that, in tenebrescence the electrons in the specimen after exposure become unstable and the specimen darkens. As long as the specimen stays in the dark, the electrons stay the same. If the specimen is exposed to sunlight, energy is added to the electrons, they become stable, and the color disappears.

Triboluminescence is another interesting property in some minerals. It was observed in diamond, quartz, and sphalerite. When a metal object scratches sphalerite or it is crushed, orange flashes of light appear. Some examples have been found that are so sensitive that the scratch of a fingernail across them causes the flashes to appear. Triboluminescent minerals are not necessarily fluorescent minerals.

Thermoluminescence, another interesting property, is noticed in several minerals, but is most easily seen in chlorophane, a variety of fluorite. When chlorophane is heated, it gives off light and glows. Some chlorophane is so sensitive that the heat of a person's hand is enough to cause it to glow. Although the tenebrescence effect is usually able to be repeated many times, thermoluminescence is usually a one-shot affair. These are just a few of the amazing things that some minerals do in addition to fluorescing in intensely beautiful colors.

Fluorescence in Everyday Life and Into the Future

The question that many people ask about fluorescent minerals is "What are they good for, besides collecting?" The same question can be asked about fossils, Barbie™ dolls, and many other collectibles. The study of fluorescent minerals can be a steppingstone to the wider scientific application of fluorescence in our lives. Television devotees should know that the color television is a direct result of the study of fluorescent minerals. These minerals and how they fluoresce are the basis of the creation of phosphors (man-made fluorescent materials) that produce color in a television. Cathode ray tubes provide the power to excite the phosphors and make them fluoresce in color. Computer monitors and television screens are the fruits of the scientific study of why and how certain minerals fluoresce.

The fluorescent light that fills every office building and many homes started out with experiments by General Electric to make a more efficient light source. They needed fluorescing inorganic materials to stand up to the heat generated by electric lights. Scientists at GE hand-ground willemite powder. The willemite powder was applied to the inside of glass tubes. Mercury was vaporized in the tube, an electric current was passed through the vapor, and it gave off shortwave light. The willemite coating the inside of the bulb absorbed the shortwave light and gave off longer visible wavelengths of light. Willemite was replaced by phosphors and eventually magnesium tungstate and zinc beryllium silicate. These tests took place in 1935 to 1938 and fluorescent lighting was introduced at the World's Fair in 1939. Further work brought us the fluorescent lighting tubes that we know today. Phosphors can be mixed to emit many different spectra of visible light, tailored for specific uses such as cool white office light, warmer home lighting, daylight lighting tubes for plant growing, specialized lighting tubes for aquarium use, etc.

Fluorescent illumination and observation are the fastest growing fields in microscope design and technique in the medical and biological sciences. Fluorescent microscopy is used in DNA sequencing. Molecules have a fluorescent chemical tag or marker attached. The fluorescence of the tag allows detection of the molecule and identity of the DNA chain. The same tags are being used in studying immune responses in cells. The fluorescent chemical tags are attached to antibodies. This allows microscopic identification of the sites where the antibody attaches in a cell.

Cellular biologists can study the structure of DNA, proteins, and what is occurring in a cell by identifying where and how the fluorescent markers are moving or grouping. Fluorescence microscopes, devised in the early 1900s by August Kohler, Carl Reichert, and Heinrich Lehmann, were originally thought to have no valid scientific purpose. Now they are the research tool of the twenty-first century for cellular biologists.

Oceanographers use fluorescence to locate organic material dissolved in sea water. The fluorescence helps to detect pollutants and evaluate the health of an area of the sea.

Fluorescent dyes are used in laundry products. They cause your clothing to fluoresce in sunlight. Whites and pale colors look brighter. The fluorescent brighteners counteract the yellowing caused by incompletely cleaned fabrics.

Gemologists use the fluorescence of gems to help differentiate a genuine gem from imitators. Natural diamonds can fluoresce under longwave ultraviolet light in several colors — blue, orange, yellow, and yellow-green — or they can be dead under longwave ultraviolet light. Diamond look-alikes generally fluoresce more brightly under shortwave ultraviolet light — cubic zirconia (orange), YAG (weak red), white sapphire (blue), synthetic spinel (white), zircon (yellow-orange), corundum (blue) — and any "diamond" that phosphoresces after exposure to UV is synthetic. Also, if a diamond has a crack in it, clear fillers can make it look like a flawless gem. UV light can make the material in the crack fluoresce, exposing the repair.

Rubies fluoresce red under longwave ultraviolet light. Synthetic rubies generally fluoresce more brightly than natural rubies. Emeralds generally do not fluoresce, but synthetic emeralds will fluoresce red under longwave ultraviolet light. Tanzanite does not fluoresce, but some of its imitators do.

Natural blue sapphires generally do not fluoresce, but heat-treated blue sapphires will fluoresce blue-white (SW). Natural yellow sapphires generally fluoresce orange (LW) and orange-yellow (SW). Natural pink sapphires generally fluoresce red (LW & SW). Natural white sapphires generally fluoresce blue (SW). Synthetic blue sapphires fluoresce blue-white or green (SW). Other colors of synthetic sapphires generally fluoresce red (SW).

Kunzite, alexandrite, benitoite, scheelite, sodalite, scapolite, orthoclase (feldspar group), and moonstone (feldspar group) gems all fluoresce. The extent of some gems fluorescence can aid in identifying where the gem originated. The study of gems under UV light is interesting and can be informative.

The television show, CSI, introduced television viewers to the use of fluorescent lights to find semen stains, urine stains, and hidden fingerprints. A fluorescent dye is used to react with the proteins or oils in fingerprints and they show up under UV light. Semen and urine stains show up under LW light.

Modern stamps are imprinted with a fluorescent dye. Sorting machines in the post office use the fluorescing stamp to ascertain the correct positioning of an envelope so that the optical character reader can read the address and postal code for automatic sorting. As an anti-counterfeiting device, new currency has a fluorescent thread embedded in the paper. This thread fluoresces brightest under longwave ultraviolet light and less bright under shortwave ultraviolet light.

Paper makers are now able to examine the components of paper under the fluorescence microscope by treating the fibers with one fluorescent dye and the binder with another fluorescent dye. This enables them to study the structure of the paper and creates ways to examine specialty papers for properties that would normally not be visible under the microscope.

Manufacturers of explosives are required to dope their products with chemicals and fluorescent markers. If an explosive is used in a crime, the material can be studied and the maker of that material can be identified. This will hopefully enable the police to determine where that particular explosive was sold and follow its path from maker to user.

Several companies make fluorescent dyes to add to paints, makeup, and even water. Tom Warren was a pioneer in the production of fluorescent paints and was a friend of Walt Disney. Some of the earliest Disneyland "dark" rides used LW lamps and fluorescent paints made by Ultra-Violet Products, Inc. Using fluorescent paints, painters can make a painting look one way under daylight and another way under UV light. Theatrical productions can use ultraviolet lamps in their lighting setups to make the actors look different under UV light — think of a scene where a character is transported to heaven or hell and those around him now look angelic or devilish. Water artists are creating water sculptures that change color as UV lamps play across the piece.

This summary of fluorescence in everyday life just skims the surface. It is not meant to be exhaustive, but just show you how expansive the subject can be. The next time someone asks you what fluorescent minerals are good for, you now have a few ideas where the study of fluorescence is used to make our lives better.

Ultraviolet Lamps

Ultraviolet lamps are covered in the first book, but here are a few thoughts about them.

Ultraviolet lamps for mineral use are different from the ultraviolet lamps used in nail salons to help nail polish dry faster, or the fluorescent-tube poster lights sold in party stores. Ultraviolet lamps are specially filtered shortwave (SW), midwave (MW), or longwave (LW) UV lamps. These cost anywhere from $30.00 to $400.00 or more. The cost difference is in the power and construction of the light. The lower-end UV lamps have less output and the more expensive lamps have greater output. The UV light emitted by SW, MW, and LW lamps is invisible but the lamps also emit wavelengths of visible light.

The wavelength of visible and ultraviolet light can be measured in nanometers (nm). Shortwave ultraviolet lamps emit light at approximately 254nm. Shortwave UV is the most widely used light source for displaying fluorescent minerals since more minerals fluoresce under SW than under LW or MW.

Midwave lamps emit light at about 312nm which lies between SW and LW. Some minerals will fluoresce brighter under MW or will show a different color than either LW or SW.

Glass and plastic will block some but not all of the SW and MW light emitted by a UV lamp, so additional precautions must be taken. BE WARNED, SW and MW UV light can cause sunburn and burn the eyes. It is recommended that safety glasses offering 99.9% UV protection be worn at all times. Such safety glasses are available at UV mineral dealers and at many hardware stores. Fluorescent collectors also know that the UV lamp should be aimed away from your face and that, just like with visible light, UV light can be reflected back by various materials. If you have sensitive skin or use some of the newer high-powered ultraviolet lamps, their output can burn your skin like a sunburn. It pays to wear long-sleeved shirts when working under or near the ultraviolet lamps.

Longwave light is emitted in a wider spectrum than SW. The peaks of LW spectra are at 352 nm and 368 nm. Ultraviolet lamps can be purchased in one of two wavelengths; one at 352 nm (called 350 LW) or at 368 nm (called 370 LW). Most minerals react the same way to both wavelengths, but some minerals fluoresce better under one, while others fluoresce better under the other. If given a choice, most collectors go with the 350 LW.

LW UV is closest to visible light and will pass through most glass and plastic. LW lamps and their filters are less expensive than SW lamps and filters. The black light bulbs sold at party stores are a LW UV source, but they usually give out too much visible light to have much use to fluorescent-mineral collectors.

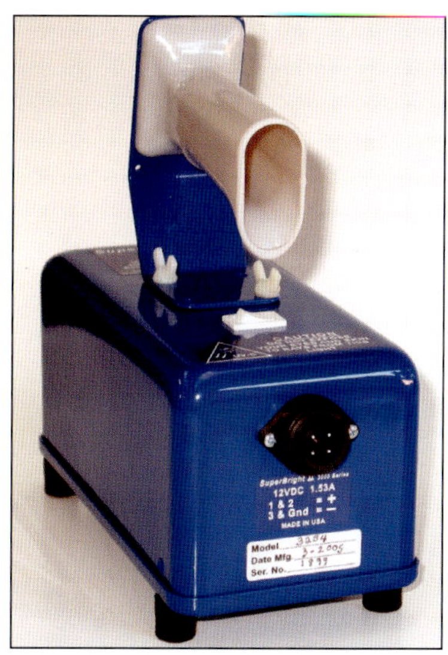

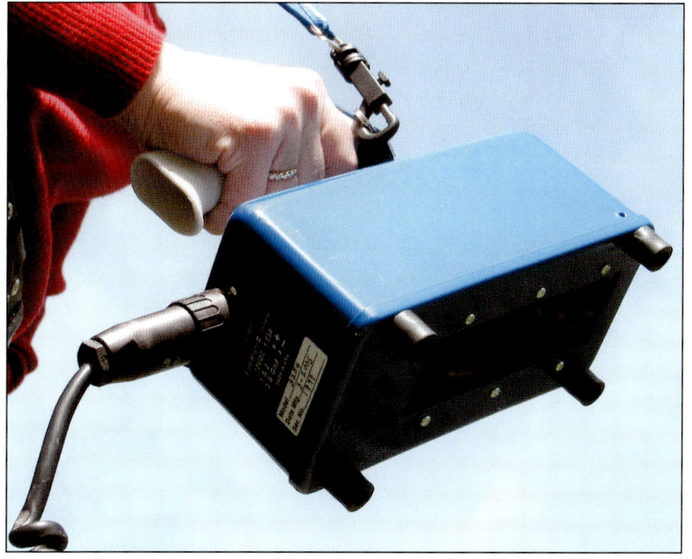

Two views of the new UV Systems' SuperBright II portable UV lamp. I used the first version of this light to take the photos for this book. This new version offers a hands-free feature, brighter output, and an improved electrical connector.

Activators

One of the first questions non-collectors ask about fluorescent minerals is "Do all rocks fluoresce under ultraviolet light?" Fluorescent minerals are obviously different from other minerals. The typical mineral does not fluoresce under UV light. The fluorescence is usually caused by small amounts of other elements called "activators." Some impurity has invaded the minerals in the specimen during its creation and that impurity activates the fluorescent quality of the mineral. Some minerals need two activators (co-activators) to cause fluorescence. Different activators may account for the different color responses found in certain minerals. Examples can be found by looking at the fluorapatites from New Jersey. Some fluoresce orange (SW), others fluoresces pale blue (SW), and another a plum color (MW). The brightness of the fluorescence also varies with the type or quantity of the activator. Too much of an activator in the mineral can cause the mineral not to fluoresce.

Some minerals are self-activated and fluoresce without activators. Examples are scheelite and autunite.

As there are activators, there are also "quenchers" of fluorescence. If they are present, in often the tiniest amount, the mineral will not fluoresce. Iron is a known quencher. However with an exception to every rule, a few minerals that contain iron, such as some albite and sphalerite, do fluoresce.

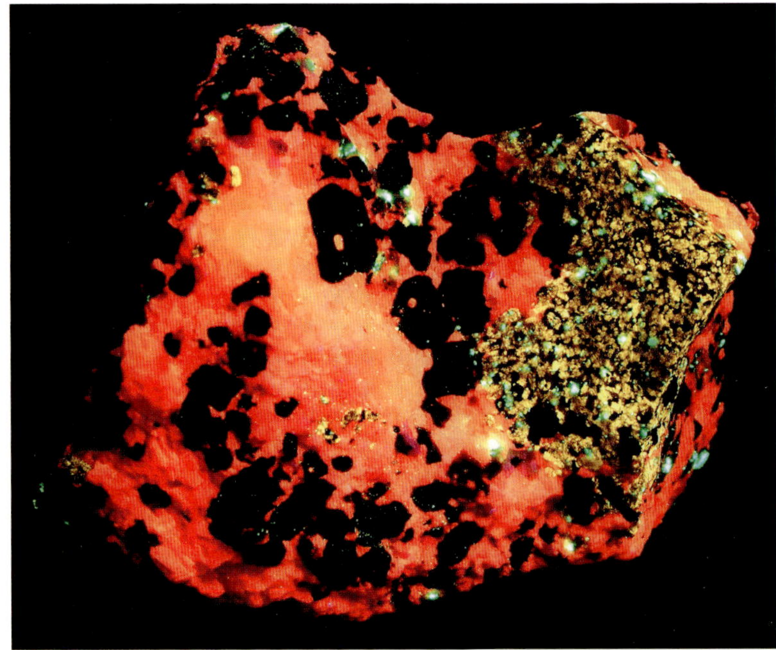

Finding Fluorescent Minerals

Sussex County, New Jersey contains more fluorescent minerals than any other place of comparable size in the world. Hopefully, this will encourage you to take a trip to the Franklin, New Jersey, area if you have an opportunity.

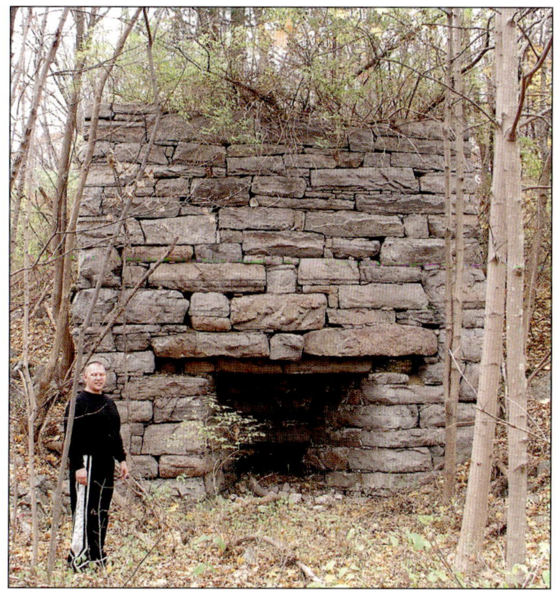

Franklin, New Jersey was originally called Franklin Furnace. There were numerous furnaces in the town. These particular remains are of a lime furnace that was used by the area's iron mills. Limestone was quarried and then burned to obtain a product that aided in the making of iron.

The Franklin & Sterling Hill Mines, New Jersey

There are several good books on the history of Franklin and surrounding mines and my first book has a more expanded section on the history of the mines. Franklin, located in Sussex County, New Jersey, is the home of the Franklin mine. The town was originally called Franklin Furnace. The town next to Franklin is Ogdensburg, the home of the Sterling Hill mine, which operated from 1848 until 1986. Through an amazing series of geological events, there are an incredible number of different minerals in the area. There are about 89 different fluorescing minerals that have come out of these two mines and surrounding quarries. The next nearest quantity of fluorescent minerals comes from Mont Saint-Hilaire, Québec, Canada, which has about 45 different fluorescent minerals.

Collectors of fluorescent minerals can still go to the two mines and for a fee, prospect on the mine dumps. A "dump" is where the rock that was mined, but not used, was dumped. Bring your UV lamp and a blanket to cover your head for daylight digs. Nighttime digs are also held for mineral society groups, so it is worthwhile to join a mineral society such as the Franklin-Ogdensburg Mineralogical Society or the Fluorescent Mineral Society, Inc. There are also special digs for members of the Franklin Mineral Museum and the Sterling Hill Mining Museum. (Mine contact information, collectors groups, Internet links, and places to buy lamps are listed in the Source section at the back of the book). The dumps are generally composed of low-grade zinc ore and other minerals that were removed from the mine and dumped in nearby areas.

Certain minerals are associated with the Parker shaft, some of which are unique to the area. The Parker shaft of the Franklin mine was used from 1895 to 1910 when it was closed. The Palmer shaft then became the main shaft.

Some of the material excavated from this area was discarded on the Parker dump and it is in this dump material that these rare minerals were first found and described. They became known as "Parker shaft minerals." Collectible examples are roeblingite, margarosanite, pectolite, prehnite, xonotlite, clinohedrite, esperite, hardystonite, cuspidine, minehillite, charlesite, and nasonite. Many times you will see these minerals offered as "from the Parker shaft." While that is inaccurate, unless they were mined before 1910, you will see it fairly often on descriptions when these minerals are sold. The real beauties of these minerals are associations of several minerals in one specimen.

At the Sterling Hill mine in Ogdensburg, you can take a tour through the upper mine shaft (the lower ones are flooded). Although the Sterling Hill and Franklin mines are just a few miles apart and mined the same veins of zinc, there are fluorescent minerals that are found at Sterling Hill and not Franklin and vice versa. Also there are occasional visual differences between the Franklin mineral and the Sterling Hill mineral. The photographs in this and my first book will hopefully give you a feel for some of the differences and similarities.

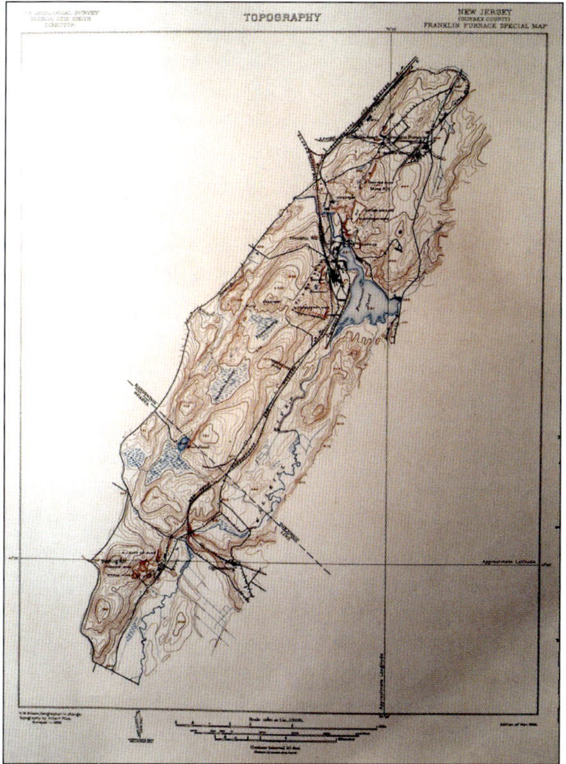

The Franklin Furnace area mines, roads, and railroads as they were in 1908. There are several different mine areas, most served by railroad tracks. There are close-ups of the Sterling Hill mine and the Franklin mine areas.

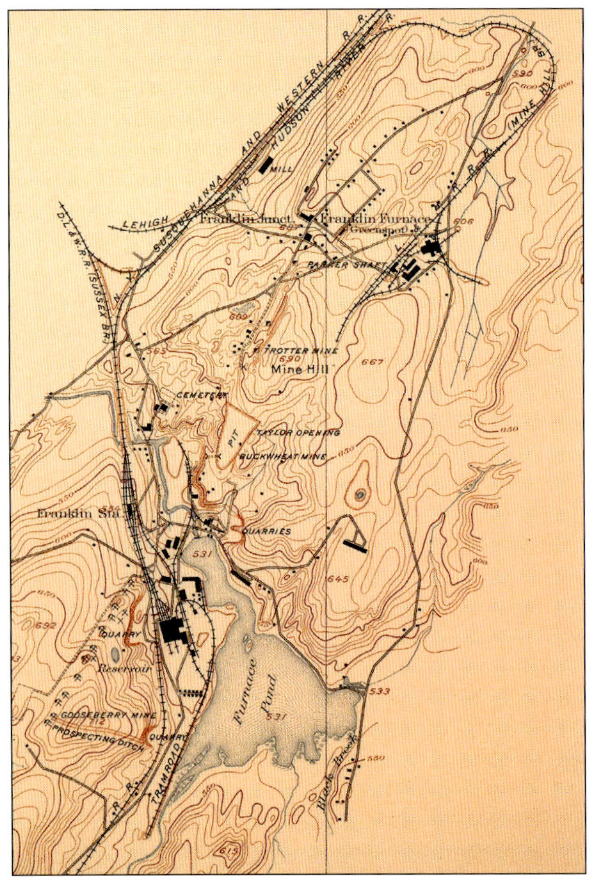

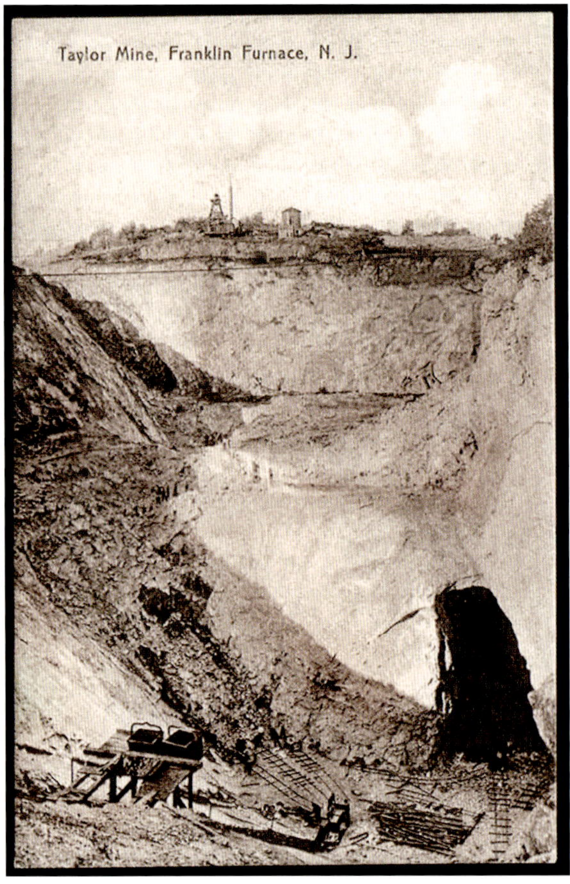

The Taylor mine in the Franklin mining district of New Jersey from a photo taken around 1896. There were many smaller zinc mines in the Franklin area until all the mines were brought under one "roof" in The Great Consolidation of 1897. The Taylor mine among others, was absorbed by the New Jersey Zinc Co.

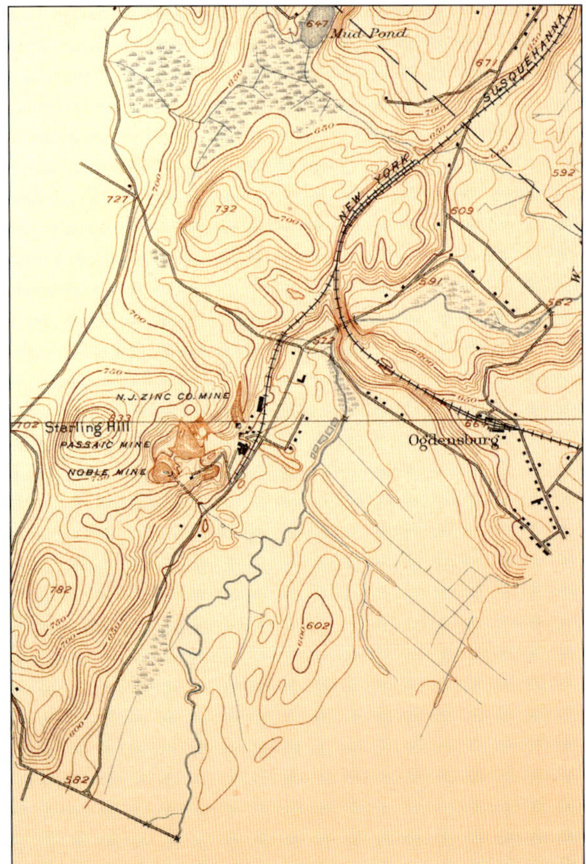

The Franklin Furnace area mines, roads, and railroads as they were in 1908. There are several different mine areas, most served by railroad tracks.

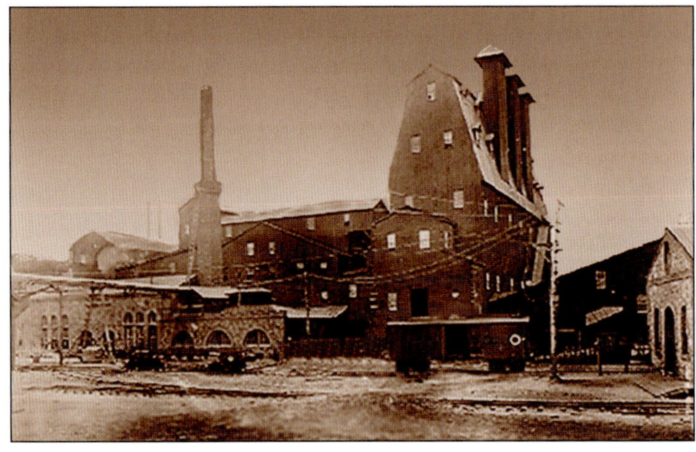

The Edison mine buildings from just before the turn of the century. Thomas Edison started an iron mine in Ogdensburg, New Jersey. Fluorescent fluorite and fluorapatite can be found in its dumps.

Fluorescent Franklin-Sterling Area Minerals and their Typical Fluorescent Colors

This list includes typical responses (e.g. calcite can be found that fluoresces blue, but it is not typical). Some are found just at Sterling Hill, some are just at Franklin, some in the local surrounding area marble, and some are found at several mines. This table has been adapted from the list created over the years by Richard Bostwick. His original checklist is updated annually and can be found in the annual program of the Franklin-Sterling Gem & Mineral Show. Note that beta-willemite is not a species. Many collectors use the colors below to try to tell the different minerals apart. It is a good place to start, but advanced collectors and mineralogists familiar with Franklin area minerals will tell you that this is not an acceptable method of differentiating or identifying the minerals. Professional analysis of a suspected mineral is the only accepted method of identification. For the collector, this is often an option that is not practical. Use all the resources available to you and enjoy the collecting experience. Use these charts as an aid to your identification.

Fluorescence Table — Franklin/Sterling Hill

Mineral	SW	LW
Albite	velvety purple-red	
Aragonite	white, yellow, green	cream
Barite	cream, yellow, or white	
Barylite	violet, red	
Bassanite	violet	
Beta-willemite	yellow, orange-yellow	
Bustamite	cherry red (in some specimens)	cherry red
Cahnite	cream	weak yellow
Calcite	orange-red	
Canavesite		violet
Celestine	violet	weak pale yellow
Cerussite		yellow
Chabazite	green	
Charlesite	pale blue	
Chlorophane (not a species)	teal blue	teal blue
Chondrodite	yellow, yellow-orange	
Clinochrysotile	orange-yellow	
Clinohedrite	pumpkin orange	
Corundum		cherry red
Cuspidine	orange-yellow	
Datolite	cream, yellow-green	weak yellow
Diopside	blue-white	
Dolomite	red	
Dundasite	weak yellow	weak yellow
Dypingite	grey-blue	blue
Epsomite		cream
Esperite	lemon yellow	weak yellow
Fluoborite	cream	
Fluorapatite	burnt orange, gold, or blue	
Fluorapophyllite	white	
Fluorite	blue-green	violet-blue, blue-green
Genthelvite	weak green	green
Guerinite	white	pale yellow
Gypsum	cream, pale blue, pale violet	
Hardystonite	purple-blue	weaker purple-blue
Hedyphane	cream, orange	weak red
Hemimorphite	pale yellow	white
Hexahydrile	white	white
Hodgkinsonite		cherry red
Humite	pale yellow	
Hyalophane	weak red-purple	
Hydrotalcite		cream
Hydroxyapophyllite	weak white	
Hydrozincite	bright sky blue	weak yellow
Johnbaumite	orange	
Junitoite		weak yellow
Magnesiohornblende	greenish blue	
Manganaxinite	red	
Margarite	weak white	weak white
Margarosanite	sky blue or pink-red	
Marialite	yellow	pink
Mcallisterite	cream	
Meionite	orange, pinkish red	no response or orange, yellow, cream
Meta-ankoleite	green	
Metalodevite	green	
Microcline	weak blue or red	weak blue
Minehillite	violet	weak yellow
Monohydrocalcite	green	green
Nasonite	pale yellow	
Newberyite	cream	
Norbergite	yellow	
Pargasite	greenish blue	
Pectolite	orange, purple	
Pharmacolite	white	
Phlogopite	yellow	
Picropharmacolite		white
Powellite	yellow	cream yellow
Prehnite	orange-pink, peach	
Quartz	yellow or green	
Roeblingite	red	
Samfowlerite	weak red	
Scheelite	pale yellow	cream yellow
Smithsonite	weak yellow	weak yellow
Sphalerite	weaker orange or blue	orange, pink, or blue
Spinel		cherry red
Starkeyite	white	white
Strontianite	violet	violet
Talc	cream yellow	
Thomsonite	pale yellow	
Tilasite	yellow	
Titanite	yellow-orange	
Tremolite	pale blue	yellow
Turneaureite	orange	
Uranospinite	green	
Uvite	yellow	
Willemite	green	weaker green (in some specimens)
Wollastonite	orange or yellow	
Xonotlite	violet	
Zincite	weaker pale yellow	pale yellow
Zircon	yellow-orange	
Znucalite	green	

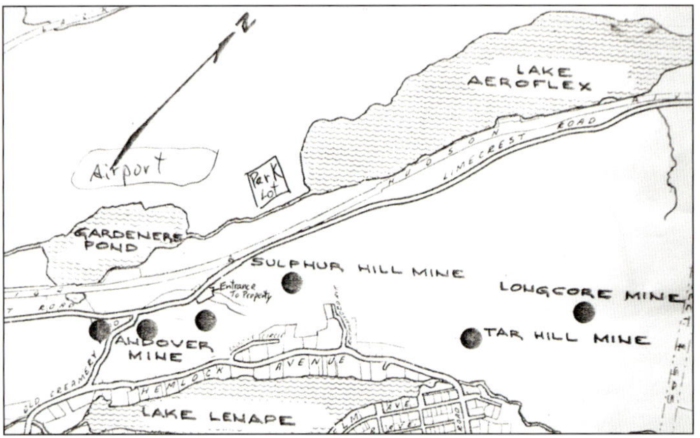

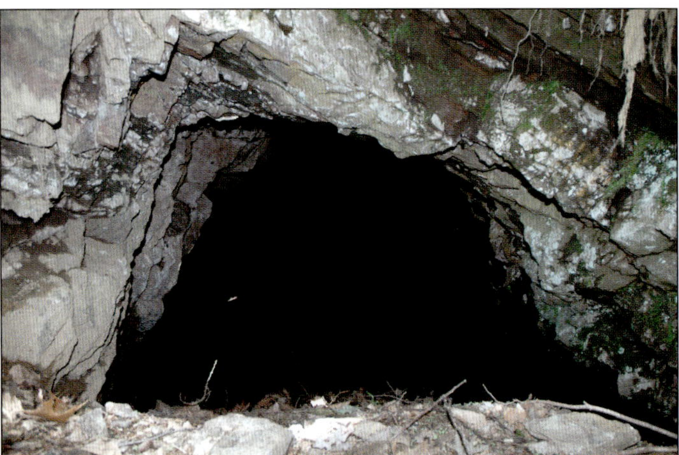

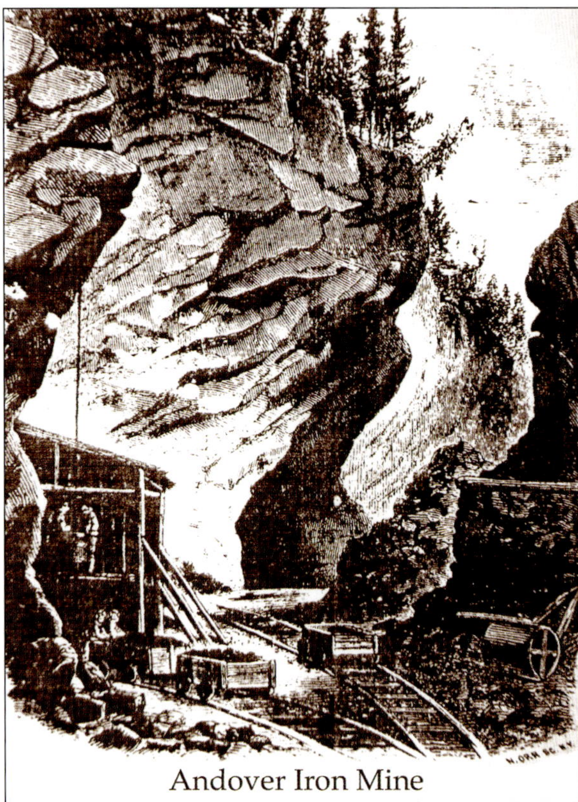

The Andover mine near Andover, New Jersey showing the entrance to the mine, a map of the mine area, and an early print of the mining activities.

Collecting At The Andover Area Mines In Sussex County

Northern New Jersey has a number of abandoned mines and quarries where fluorescent minerals can be found. The iron mines in the Andover area in northwestern New Jersey have been mined from the mid-1670s to the 1880s. They include the Andover mine, the Sulphur Hill mine, the Longcore mine, and the Tar Hill mine. A few fluorescent minerals have come out of these mines including a small find of yellow-fluorescing willemite. In a recent expedition back to the mines (they are on private property and permission was obtained in advance from the property owner) several fluorescent minerals — feldspar and calcite — were found.

These mines are about fifteen miles from the Franklin/Sterling Hill mines but the geology is different. Rich iron ore outcroppings in the form of magnetite and hematite were discovered in the mid-1600s and open pit mining began soon after. Iron furnaces and forges were built in the area. When the ore was smelted, the iron ingots were transported across land on wagons to local rivers where they were floated down to southern New Jersey. Almost all of the iron was then shipped to England by the English mine owners.

As the Revolutionary War began, the output of the mine was still being shipped to England. This continued for the first two years of the war, until, in 1778, the Continental Congress convinced New Jersey to take over the land from the English owners. After the change in ownership the iron was used to create armaments for the Continental army. The giant iron link chain that was laid across the Hudson River at West Point to stop the English from moving up river was forged at Andover.

The mines remained in operation until about 1790 when they were abandoned. The reason was that all of the area forests had been cut down to obtain fuel to transform the ore into iron and there was no more wood. The cost of carting in wood or coal and the cost of hauling out the iron were such that production became unprofitable.

The mining operation started up again in 1848. Within a few years, the mine was producing over 4,000 tons of iron ore per month. Mining had become profitable again due to the creation of the Sussex Mine Railroad and the use of the Morris canal system to transport the ore to Pennsylvania. Pennsylvania had great coal fields that could be used to turn the ore into iron. The increasing cost of transporting the ore to Pennsylvania proved to be the downfall of the mining operation. It closed in the 1880s.

The access to the mines is along Limecrest Road and just north of Old Creamery Road. There is no parking at the entrance and the signs along the road say that the state forest is closed. The state owns forty feet in from the road and the rest of the property, where the mines are, is owned by an Andover resident. It is private property and you will need permission to go there. A path leads into the forest and then splits after one hundred yards. Off to the right is the Andover mine. The remains of the mine are a large cut through the rock and three tunnels that followed the ore body. There are also dumps across from the tunnels. Each of the three mine entrances was flooded to about two or three feet of water when I was there so I could not go all the way into the tunnels. On the ceiling of the first tunnel, I found a small piece of rich iron ore (magnetite). In the third tunnel I was able to scan the walls with my SW ultraviolet lamp. There were areas of a reddish rock that fluoresced dark red. Some of this material was collected. Also on the wall was a tiny group of white calcite crystals that fluoresced soft pale green under shortwave ultraviolet light. I never found any yellow-fluorescing willemite.

If you take the left-hand fork when you enter the forest, you start to climb, and if you continue, you eventually come to the original 1700s Sulphur Hill mine pit that is about eighty feet deep. This may be where the yellow fluorescing willemite was found years ago. The pit is too steep to climb down the sides without a rope ladder. There is a small entrance tunnel to the pit that was also filled with water, too deep to enter. Southeast of the pit are the dumps. I covered myself with a large plastic tarp and lit the area with my ultraviolet lamp. I again found some deep red-fluorescing specimens. There are differences in the Sulphur Hill mine and the Andover mine. One is that Sulphur Hill had more sulfides and less hematite than the Andover mine.

The Longcore mine and the Tar Hill mine, each with a group of smaller pits are northeast of the Sulphur Hill mine. It was getting too late to explore them. When I went in June, the poison ivy was everywhere. There were loads of mosquitoes and I was warned about the presence of poisonous snakes. I covered myself with insect repellent and watched where I put my feet and hands. While this is not nearly as exciting as collecting at Sterling Hill, it did give me a chance to find a few fluorescent minerals and see where miners worked hundreds of years ago in New Jersey.

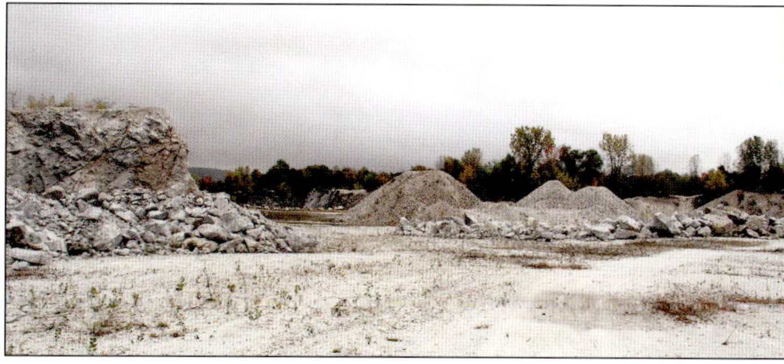

The Franklin quarry is a closed marble quarry that is located partway between the Franklin mine and the Sterling Hill mine. It is closed to the public but is available for digs with the Franklin-Ogdensburg Mineralogical Society once or twice a year. This photo was taken at a dig in October, 2005. The quarry is known as a place to find fluorescent minerals such as norbergite, diopside, tremolite, margarite, phlogopite, marialite and/or meionite, and some years ago, large uvite crystals. There are also nice nonfluorescent minerals found such as edenite, rutile, molybdenite, and also feldspars.

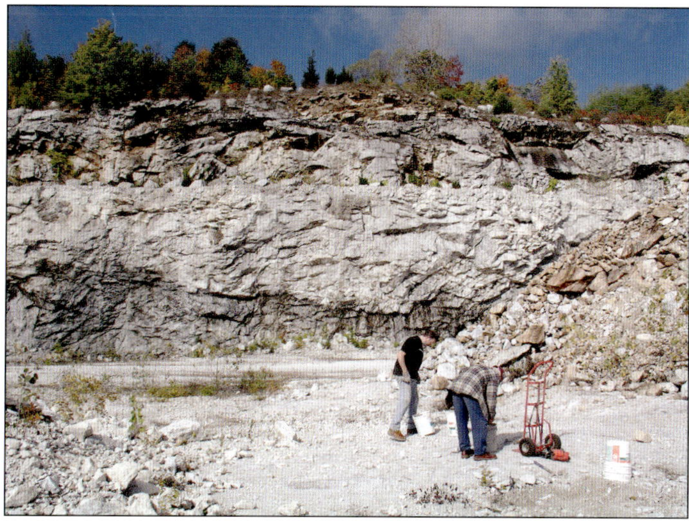

This is the area of the Franklin quarry where margarite was found. Some years ago, collectors say they found football-sized pieces of rock with beautiful radiating plates of margarite. It appears that the material found these days is leftover bits and pieces of rocks broken years ago.

Canada

Canada is a large and geologically diverse country. Collectors will find numerous places to dig for minerals. The Red River floodway in Winnipeg, Alberta, has produced an incredible number of fluorescent selenite roses. This area may no longer be available for public digging.

The Bancroft district in the province of Ontario encourages mineral collectors to visit, has a mineral museum, an annual mineral show, and the Chamber of Commerce provides written guides, a digging site (Bear Lake Diggings), and sample mineral kits to interested collectors. Fluorescent-mineral collectors can find feldspar and scapolite. They may also find calcite, zircon, sodalite, hackmanite (a variety of sodalite), fluorite, scheelite, and other fluorescent minerals. The Bancroft area has beautiful apatite crystals that may fluoresce under MW.

The Long Lake Mine

The Long Lake mine near Parham, Ontario, produced some spectacular fluorescent pieces that were shown in my first book. They look rather plain in daylight, but their real beauty comes out under SW and LW UV. They contain several fluorescent minerals that form bands, patches, and patterns that look stunning. The minerals are aragonite (peach & pale blue), a member of the humite group (probably chondrodite that fluoresces creamy yellow), calcite (red-orange), diopside (blue-green), and hydrozincite (pale blue).

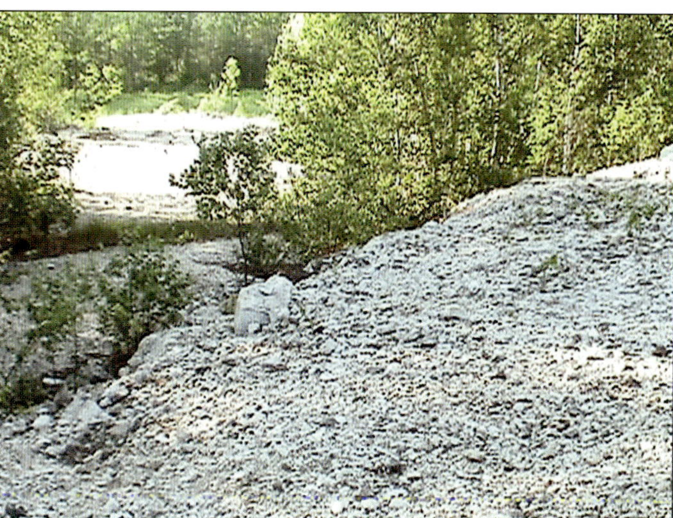

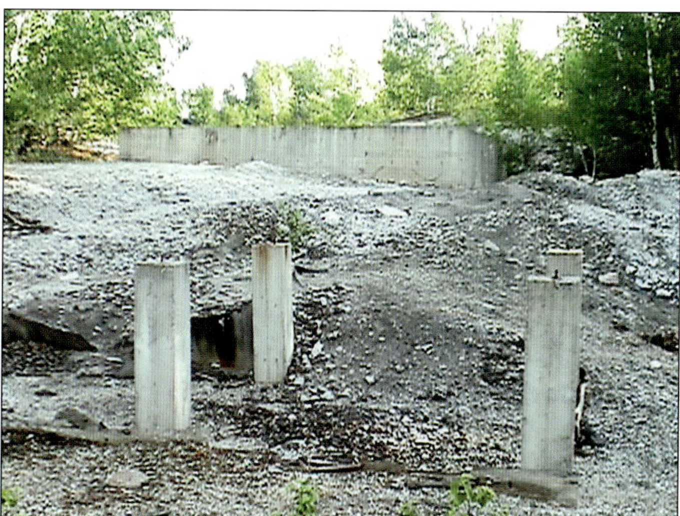

Long Lake

The mine is 8.8 km northwest of Parham, Frontenac County, Ontario, Canada. The zinc mine was started by a local resident, Ira Leslie Benn. Leslie, as he was known, studied mineralology at Queen's University and went on the first mineral survey of northern Ontario. He discovered zinc ore near Parham in 1901 and soon after began mining the area. The property, under Leslie's guidance produced over 100 tons of zinc ore. The mine was later transferred to James Richardson and Sons Ltd. of Kingston, Ontario, who worked the mine for a short period of time before ceasing operations. The area can be visited, but all that remains are just dumps containing the fluorescent material.

Scott Allen of New Jersey Mineral visited the Long Lake zinc mine area and writes as follows:

> We started the day by leaving Kingston, Ontario. We headed north for about an hour when we came to the town of Parham. The road that we were traveling, Route 38, turned right so we turned left onto Long Lake Road that heads out northwest of the town center. We stopped to ask for specific directions and spoke

to a local man. We learned that the site is owned by Elgin Young. He pointed us to the site, which was nearby, and we entered and parked the car. After exploring the site during the daylight hours the sun finally went down. What a sight when we turned the ultraviolet lights on, everything lit up. The variety of specimens was excellent (color wise). I believe that we spent five hours collecting, but we could have collected all night. The bugs were incredible, so bring your repellent. We ended up returning to the site a week later and collected about 500 to 600 pounds of material.

The province of Québec is important to fluorescent-mineral collectors. It contains the Mont Saint-Hilaire area that has produced many fluorescent minerals. Other areas in the province have produced vesuvianite, diopside, vlasovite, albite, weloganite, quartz, zircon, and yellow-fluorescing wernerite (a name used by collectors and dealers for a massive form of meionite, a species of the scapolite group).

Mont Saint-Hilaire

Mont Saint-Hilaire (MSH) is in the province of Québec, Canada, just a short drive south-east of Montreal. The mountain is an intrusive alkaline complex, the dominant rocks are nepheline syenite, sodalite syenite, and hornfels and it is rich in sodium and rare earth elements. Minerals of all sizes are found in cavities, vugs, and dikes in a quarry owned by the Poudrette family. Up to now about three hundered and sixty different minerals have been indentified, about fifty of them being type locality minerals, and many of them are fluorescent minerals. MSH is well known for its spectacular specimens of sérandite, carletonite and catapleiite, and for its large number of rare micros. Collecting is permitted only by joining organized field trips that are arranged by the Montreal Mineral Club (Club de minéralogie de Montréal). The fluorescent table has recently been revised by Jacques Poulin. His revision has more detail than the list below.

Fluorescence Table — Mont Saint-Hilaire

Mineral	SW	LW
Albite	cherry red	
Analcime	green, white	green
Calciohilarite	pale blue	pale yellow
Calcite	white, red, orange	white, red
Catapleiite	green	green
Cerussite	yellow (MW)	yellow
Cryolite	pale yellow, pale blue	
Datolite	white, violet-blue	yellow
Elpidite	yellow & green	green
Fluorapatite	blue, orange, yellow	grey-pink
Fluorapophyllite	blue, yellow-green	blue, green
Fluorite	violet blue, pale yellow	blue, pale yellow
Franconite	pale yellow	
Gaidonnayite	green	green
Gaultite	green	
Genthelvite	green	green (tan, orange - MW)
Gobbinsite	pale blue	
Gonnardite	white	white
Griceite	pale yellow	yellow
Helvite	deep red	red
Hemmimorphite	pale blue, pale yellow	white-yellow
Hibschite	pale yellow	pale blue
Kogarkoite	weak blue	white
Lalondeite	violet blue	violet blue
Leucophanite	pink-violet	pink
Leucosphenite	pale yellow	
Lintisite	pale blue	
Lorenzenite	yellow	
Magadiite	pale yellow	pale yellow
Makatite	white-blue	white-blue
Meionite	pale yellow	orange
Microcline	cherry-red	
Milarite	white-blue	
Monteregianite	green	green
Natrolite	green	dull green
Nenadkevichite	green	green
Parakeldyshite	white-orange	
Pectolite	weak pink	pink to violet
Polylithionite	yellow	
Quartz	yellow	yellow
Searlesite	blue-green	
Sheldrickite	pale blue	
Sodalite	orange, red	orange
Sodalite (hackmanite)	orange	orange
Sphalerite	orange	orange
Strontianite	white	white
Terskite	pale yellow	pale grey
Tetranatrolite (now gonnardite)		
Thaumasite	white-blue	white-blue
Thornasite	green	green
Tugtupite	red	orange
Villiaumite	red	
Vinogradovite	yellow	
Vuonnemite	green-yellow	weak yellow
Vuoriyarvite-Kite	orange	
Willemite	green	green, brown-yellow
Wohlerite	orange	
Wollastonite	white	white
Wurtzite		orange
Zepophyllite	pale yellow	pale yellow
Zircon	yellow	brown-yellow

Other Places

St. Lawrence County, New York

Two larger mining companies and many smaller ones were active in this central northern New York area - The larger ones were the Gouverneur Talc Company (GTC) and the Zinc Corporation of America (ZCA). Some of the mines have been closed and the land "reclaimed" in accordance with New York State law. The former dumps are now grassy park-like areas. There are differences to be found in the dumps associated with GTC (talc mining) and ZCA (zinc mining). If you go on a dig with members of the St. Lawrence County Mineral and Gem Society, you can still gain access to some of the dump areas. One of the most interesting fluorescent minerals from this area is sphalerite. It can fluoresce orange, sky blue, blue-green, gold, pink, or green under LW. Collectors can also expect to find nice examples of anthophyllite (red under LW & SW), calcite (bright sky blue SW and yellow LW), diopside (sky blue SW), fluorapatite (yellow SW), norbergite or chondrodite, (lemon to orange-yellow SW), parvowinchite (formerly called tirodite — red or hot pink under LW and less bright red under SW), phlogopite (yellow SW), talc (bright to dull yellow under SW), tremolite (tangerine orange under SW and less bright orange under LW), wernerite (a collector's name for meionite or marialite which fluoresces bright orange-yellow under LW), and willemite (bright green under SW & LW).

The Western United States

This is a very abbreviated discussion of what is available in this part of the country. There are thousands of mines in the western U.S. Many of these were gold, silver, borax, lead, copper, zinc, and uranium mines. They produced incredible non-fluorescent crystals and rare minerals (e.g. the red rhodochrosite crystals from the Sweet Home mine in Colorado). Fluorescent minerals can be found almost everywhere, but most mines do not have a great variety of fluorescent minerals, usually 5 or less per mine.

The mines of Arizona have been productive for fluorescent-mineral collectors. They can find andersonite (bright blue-green or green-white LW & SW), barite (orange LW), bayleyite (weak yellow-green LW & SW), calcite (orange red SW, also a variety that is white SW and green LW, and a variety that is green SW and white LW), cerussite (pale yellow LW & SW), crocoite (brown SW), eucryptite (cherry red SW), fluorite (violet-blue LW, less intense violet blue SW), mimetite (red-orange SW & LW), pyromorphite (white or yellow LW), scheelite (blue-white SW), schröckingerite (bright blue-green or green-white LW & SW), swartzite (bright yellow-green LW & SW), wickenburgite (dull orange to pink SW), willemite (green, yellow, cream SW), zunyite (cherry red SW), and others.

California is home to many mines, and minerals that fluoresce, such as aragonite (pale blue SW), benitoite (bright blue SW), calcite (orange, red, green, or white SW), caliche (caliche is a general term for any secondary calcium carbonate that forms in sediments or in voids in bedrock just below the surface in semiarid regions - peachy orange SW), calomel (yellow SW), chalcedony (a variety of quartz - green SW), colemanite (yellow-orange SW), dumortierite (blue SW), halite, (a sea salt - bright red-orange SW), hanksite (pale yellow SW & LW), hyalite opal (green LW & SW), hydrozincite (blue SW), scheelite (pale blue SW), trona (blue-gray, blue-green SW & LW), and others.

Without trying to shortchange the other western states, each has interesting fluorescent minerals as shown throughout the pages of this book, and many more that were not available at the time of photography. Local collecting club members know where they can find good specimens. Joining the Fluorescent Mineral Society will introduce you to the chapters closest to you. Meeting other collectors and sharing information is the best way to learn what is available and where you can legally dig.

Mexico

Some incredibly attractive fluorescent pieces are coming out of Mexico. Prices are usually reasonable. In Durango County, Mapimí is world-famous for its minerals (including adamite and calcite). The Ojuela mines near Mapimí are a group of mines that contains about 500 miles of tunnels. The area also has a suspension bridge built by Roebling & Co. to haul the ore across the mountain. Cerro del Mercado has great apatite crystals that fluoresce, some better than others.

Mexico produces: adamite (green SW & LW), aragonite (pale blue or green SW), Terlingua-type calcites (blue SW, pink LW), an unusual red calcite (red SW), geodes containing chalcedony (green SW), danburite (blue & green SW), fluorapatite (orange-yellow or pink SW), fluorite (red or violet LW), hyalite opal (green LW & SW), and hydrozincite (bright bluish-white SW and cream MW). There is great overview of collecting minerals in Mexico by the late Walt Bowser on the Internet located at: http://epe.lac-bac.gc.ca/100/201/300/cdn_rockhound/1997-2000/fall97/cr9701408_mexico.html.

Greenland

Greenland is a relatively new collecting area for a dedicated group of fluorescent-mineral collectors. The search for fluorescent minerals in Greenland is covered in my first book in a section written by Mark Cole. Trips to the area are arranged by Mark Cole and each year new material is being found. On these trips, most of the time is spent at the Ilimaussaq Complex. It is known as the most mineral-rich area in Greenland. There are about 200 minerals identified to date. Note that locations in Greenland can be spelled differently for the same location. Some use the Danish spelling, others use the old Greenlandic spelling, and others use the new Greenlandic spelling. Some examples: Kangerluarsuk (Kangerlussaq or Kangerdluarssuk), Ivigtut (Ivittuut), and Ilimmaasaq (Ilímaussaq) The following is adapted from a list created by Mark Cole.

Fluorescence Table — Greenland

Mineral	SW	LW
Albite	dark red	
Analcime	green	
Barylite	violet, red	
Beryllite	red	
Calcite	red	
Catapleiite	green	
Chkalovite	green	
Elpidite	green	
Fluorite	blue	blue
Leucophanite	pink-violet	pink
Microcline	blue/white, red	
Plagioclase	dark red	
Polylithionite	yellow/white	
Quartz	blue	
Sodalite	orange	bright orange
Sorensenite	yellow/white	
Steenstrupine	green	
Tugtupite	red	salmon, orange
Ussingite	orange green	
Villiaumite	red, orange	
Zircon	yellow	

Pakistan and Afghanistan

There are beautiful and interesting minerals currently coming out of Pakistan and Afghanistan, though the areas where they are found are not easily accessible. They are considered to be hideouts for insurgents and home to an illicit trade in opium. It is amazing that any material actually reaches the mineral-collector market. Pakistan and Afghanistan share a 1500 mile border. Most Afghan minerals come into Pakistan through this border. Peshawar in Pakistan serves as the main market for the minerals found in both countries. Gilgit, Pakistan, is also an important place to buy minerals.

Afghanistan

A very tenebrescent sodalite, var. hackmanite, that is appearing with more regularity on the market, comes from the Sar-e-Sang district. This district constitutes the east side of an approximately 6 mile stretch of the Kokcha River valley in the Badakhshan Province. The Sar-e-Sang district also produces fluorescent fluorapatite, phlogopite, calcite, and members of the feldspar and scapolite groups.

Spodumene, var. kunzite, crystals are found in pegmatites in the Laghman and Kunar Provinces in the Nuristan region. The provinces are situated in the extremely mountainous area of northeast Afghanistan. Nuristan is also a good source of fluorescent tourmaline, var. elbaite, fluorapatite, and several members of the feldspar group.

Pakistan

Geologically, Northern Pakistan, consisting of the North-West Frontier Province, Northern Areas, and Azad Kashmir, is a dynamic environment for the formation of fluorescent minerals. The Hindukush, Karakoram and Himalayan mountain ranges host almost all the gemstone and extraordinary mineral deposits. The Northern Areas have yielded excellent specimens of tourmaline and topaz. The Hunza Valley produces superb specimens of ruby and pargasite. Beautiful calcite, dravite, fluorpaptite, fluorite, herderite, spodumene, var. kunzite, zircon, and members of the tourmaline and feldspar groups can be collected near Gilgit.

Note: Specimens, when purchased, are often labeled with general area names such as "Gilgit," "Shengus," "Haromosh," etc. Gemstone and mineral mining operations are mostly small-scale and run by local, private operators. There are many middlemen between the actual mining and eventual sale. Locality information for many specimens is often inconsistent and inadequate.

Information about some Specific Types of Minerals

Scheelite

Scheelite is a popular mineral for fluorescent-mineral collectors. It is the primary ore of tungsten and is a calcium tungstate whose formula is $CaWO_4$. It can form perfect eight-sided dipyramidal crystals (dipyramid equals two four-sided pyramids — tetragons — joined base to base or a pseudo-octahedron, since it technically is not really eight sided, but looks like it is eight sided). Scheelite commonly occurs as compact or granular masses in contact-metamorphic deposits, especially limestone intruded by granites, in granite pegmatites, and in quartz veins. Its color is brown, tan, golden-yellow, or cream. Its luster is vitreous and occasionally it is gemmy and transparent. Scheelite's hardness is 4 .5 to 5, making it a soft mineral for lapidary use, although faceted gems have been cut from Chinese and Brazilian material.

The search for United States sources of scheelite contributed to improvements in the portable shortwave ultraviolet lamp. Once this portable ultraviolet lamp became available and affordable, and collectors found out what it could do, it was the catalyst that advanced the hobby of collecting fluorescent minerals.

Tungsten became commercially valuable in the 20th century when it began to be used to harden steel and was made into electric-light filaments. Thomas Edison originally used carbon filaments in his earliest light bulbs, but they burned out quickly. He needed a material that could glow brightly, but not melt. Tungsten, which could be drawn into a very thin wire, was the perfect choice. Pure tungsten is a steel-gray to tin-white metal and has the highest melting point of all metals: 3,410 degrees Celsius. At temperatures over 1650 degrees Celsius, it has the highest tensile strength of all metals and has excellent corrosion-resistance. It was the perfect material for light-bulb filaments and contributed greatly to the acceptance and success of Edison's invention.

Before our involvement in World War II, much of the tungsten used in the United States came from China. Without tungsten, metal working tools wore out quickly. One tungsten carbide drill bit would last twenty times longer than a molybdenum steel drill bit. Because of the war in Europe, the United States anticipated a tungsten shortage and encouraged the mining industry to find sources of scheelite in the United States. Ultra-Violet Products, Inc., of California, was one of the few companies that made portable ultraviolet lamps. Since scheelite fluoresces under shortwave ultraviolet light, use of one of these lamps would enable miners to locate scheelite in the field. It turned out that scheelite often occurs in conjunction with gold-bearing quartz. As miners fanned out across the western United States, they found that many exhausted gold mine dumps fluoresced bright pale blue and were full of scheelite. These old gold mines were once again valuable mining properties. United States sources of scheelite were found in abundance and during the war years, there was no shortage of tungsten and tungsten carbide tools.

Tungsten is known as "tungsten" in the Americas and as "wolfram" in Europe and surrounding countries. Its chemical symbol is W which was derived from the wolfram name. By the 1500s, Saxonian tin miners knew that there was a heavy unknown ore that, if some of it got mixed into the tin ores during refining, reduced the tin yield. Refiners said that it was as if something ate up the tin like a wolf eating a sheep. It eventually was named wolfram because it ate tin like the wolf and produced a dirt-like tin impurity, "Rahm", the German word for dirt.

In the year 1751 the Swedish chemist and mineralogist Axel Fredrik Cronstedt (1722-1765) was working with a heavy mineral, which he named "tungsten". This mineral was actually scheelite. The word tungsten is derived from the Swedish, "tung," meaning heavy, and "sten," meaning stone. Pure scheelite crystals are very heavy for their size. Cronstedt assumed this was a new element, but he could not separate the metal from the mineral.

Scheelite was named for one of its original discoverers, a Swedish chemist named Karl Wilhelm Scheele (1742-1786). Scheele was born in Pomerania, Germany, but spent most of his life in Sweden working as a pharmacist. He was a self-taught chemist who is also credited with discovering oxygen in 1772-73. In 1781, Scheele wrote that the heavy mineral was a compound of lime and an acid, that he called tungstic acid. Scheele suggested the possibility of obtaining a new metal by reducing this acid.

It was at Bergara, Spain's first research institute, that the metal was first isolated from tungstic acid. This occurred in 1783 by the brothers, Fausto d'Elhúyar (1755-1833) and Juan José d'Elhúyar (1754-1796). Scheelite kept its misnomer, tungsten, for many years. In 1821 the mineral was formally named scheelite by C. C. Leonhard, a scientist studying metals. He chose to name it in honor of Scheele.

At its purest, scheelite fluoresces bright pale blue. Scheelite is a "self-activated" mineral. Its fluorescence is due to the mineral itself, rather than an activator, which most other minerals need to be fluorescent. To illustrate the idea of an activator, think of calcite. Generally, it is not fluorescent. Add a tiny amount of manganese and an even smaller amount of lead and the calcite will fluoresce bright red-orange under SW UV. The manganese and lead are activators.

The best examples of large scheelite crystals currently come from China. The largest weigh several pounds. Some material from a different location in China is embedded in fluorite and has beryl crystals growing alongside the scheelite crystals.

Powellite

Powellite is closely related to scheelite. It is a calcium molybdate whose formula is $CaMoO_4$. Powellite forms crystals similar to those of scheelite. They share a comparable atomic structure. The two minerals form a series in which the molybdenum of powellite is substituted for the tungsten of scheelite . Molybdian scheelite will show a yellow fluorescence. Under LW UV, pure scheelite shows little fluorescence. As the molybdenum content increases, it will show a yellow response under LW UV.

Powellite was named after the American geologist, John Wesley Powell (1834-1902), a director of the U.S. Geological Survey. The first reported find of powellite was at the Peacock mine, Seven Devils District, Adams County, Idaho in 1891. Powellite fluoresces pale yellow to yellow (SW & sometimes LW). It has a hardness of 3.5 to 4 and its luster is adamantine or resinous.

Currently, the finest, large (one-half inch to two inches) powellite crystals are coming from India and are often found with stilbite or scolecite. They can be equated to diamonds, the bigger and more perfect, the higher the price. A three quarter inch wide crystal can sell for \$300 to \$2,000. Color and clarity also play a large role in determining price. The crystals with rich, root-beer-brown color, that contrast strongly against the dominantly pale-colored minerals of the matrix, generally command the highest prices. Fortunately for fluorescent-mineral collectors, the paler the daylight color of the crystal, the brighter the fluorescence.

Feldspars

The feldspar group of minerals seems to be an under-collected group among fluorescent-mineral collectors. Feldspars are a series of aluminum silicates with a hardness of about 6 on the Mohs hardness scale and their name is derived from the German "Feldt spat." Feldt means field and spat is a rock that does not contain ore and cleaves well. Feldspar was officially given its name by Johan Gottschalk Wallerius in 1747. It is important in glass making because it acts as a fluxing agent, reducing the melting temperature of glass components. It is also important in making ceramics since it melts gradually and lowers the melting temperature of clays. It also improves the strength, toughness, and durability of both.

Feldspars are divided into two categories: 1) plagioclase series, which contains calcium and/or sodium, and 2) alkali feldspars, which contains potassium and often sodium. Telling the two apart is not easy. Striations are often visible on plagioclase, and resemble fine parallel lines cut into the face of the rock. Plagioclase can be gray to grayish-white, but they also can be white, pink or pale yellow.

The feldspar group has at least 20 species, six are in the plagioclasse series: albite, oligoclase, andesine, labradorite, bytownite, and anorthite. These plagioclase feldspars form a series of minerals with two end members — albite and anorthite — which represent the sodium and calcium extremes of the plagioclase series. Albite is 90 to 100% albite (sodium) and 0 to 10% anorthite (calcium). Oligoclase is 70 to 90% albite and 10 to 30% anorthite. Andesine is less than 50 to 70% albite and up to 50% anorthite. Labradorite is 30 to 50% albite and 50 to 70% anorthite. Bytownite is 10 to 30% albite and 70 to 90% anorthite. Anorthite is 90 to 100% anorthite and 0 to 10% albite. All the intermediate compositions between pure albite and pure anorthite are possible. The series is arbitrary, but usefully divided into six species according to the albite:anorthite ratios of their compositions.

The alkali feldspars include microcline, sanidine, and orthoclase. Orthoclase feldspars are usually pale colored white, pink, yellow, or cream. There are varieties within some of the species and also intergrowth of species. Other feldspars include: anorthoclase, celsian, paracelsian, hyalophane, and perthite. Of interest to collectors of fluorescent minerals is hyalophane. Hyalophane is a potassium, barium, aluminum silicate and is recognized as a separate mineral species. Hyalophane from the Franklin, New Jersey, area fluoresces a weak red-purple.

Feldspars tend to twin easily and one crystal can even be multiply twinned on the same plane, producing parallel layers of twinned crystals. Crystals tend to be blocky and have two directions of cleavage at nearly right angles. The name orthoclase comes from Greek and means upright fracture, referring to its right-angle fracture. The name plagioclase also comes from Greek, meaning oblique fracture.

Albite is a very common and important rock-forming mineral, and some varieties fluoresce. Albite commonly forms from pegmatictic magmas, low-viscosity melts rich in volatile components like water, in which ion migration can be very rapid and crystallization correspondingly fast. The well-formed crystals of albite form in open cavities. Albite was named (from the Latin-"albus") for its white color. Albite will often fluoresce cherry red.

Microcline was named for its small inclined cleavage (slightly less than right angles) and can be white, pink, pale yellow, or sometimes blue-green. The blue-green or green variety is called amazonite.

Not all of the feldspar group minerals fluoresce, but there are enough that an interesting collection can be put together. One of the appealing things about the feldspars is that they can form beautiful crystals that make them attractive in daylight as well as under UV light. Any time you are at a rock show and see examples of albite, sanidine, amazonite, microcline, or hyalophane, be sure to examine them under the shortwave ultraviolet lamp. Most of the feldspars are not brightly fluorescent with a few exceptions such as Australian amazonite and some Afghanistan and Pakistan feldspars. Feldspars can fluoresce deep red, pale yellow, pale blue, grey, or pink.

Sphalerite

Sphalerite is usually yellow, brown, or gray to gray-black. Its luster is resinous. The pale-colored and red varieties have very little iron and are translucent. The darker, more opaque varieties contain more iron. Sphalerite has a yellow or light brown streak (usually a result of the iron. Without iron it is a very light color) and a hardness of 2.5 - 4. It is an important ore of zinc and alters to other zinc minerals such as hemimorphite, smithsonite, and willemite. It is usually found in association with galena, pyrite, andradite, franklinite, calcite, serpentine, amphiboles, and pyroxenes, along with dolomite, marcasite, pyrite, barite, and fluorite. Sphalerite is usually found in crystalline form and almost always contains iron. Sphalerite can be found worldwide, but most is not fluorescent due to its containing iron, a fluorescence quencher. The mineral has an interesting characteristic — it cleaves in a total of six directions.

Sphalerite was officially named in 1846, by E.F. Glocker, a metal physicist who also named eight other minerals including salt (halite) and meerschaum (sepiolite). Sphalerite is also known under other names such as: blende, zinc blende, mock lead, false galena, and black-jack. Blende is German for blind or deceiving. In some mines it was difficult for miners to tell sphalerite from galena, an ore of lead. This led to its being named *sphalerite* which is Greek for "treacherous rock." Sphalerite has a higher index of refraction than diamond, which gives it a high luster and sparkle when transparent crystals are cut as gems. Unfortunately, it is rather soft so, gemstones will chip or crack if they are banged around. It has the same chemical formula as wurtzite and matraite — zinc iron sulfide —- but has a different structure.

Sphalerite occurs in a wide range of hydrothermal environments. It can be found worldwide. Franklin, New Jersey, area sphalerite is low in iron and shows a nice fluorescence that can appear orange or blue, due to different activators — possibly silver for blue and manganese for orange. When the two are found together in one specimen the effect is as if the material fluoresces pink. Some of the Franklin-area material fluoresces a beautiful golden color, actually a red-orange or yellow-orange. The blue-fluorescing sphalerite is locally called cleiophane. The orange-fluorescing material is probably almost as low in iron as the blue-fluorescing material. Fluorescent sphalerite usually has a moderately long phosphorescence and is triboluminescent. There is a great deal of sphalerite from around the world that is not fluorescent, phosphorescent, nor triboluminescent.

The sphalerites from the talc and zinc mines in the Balmat-Edwards area of New York show an incredible variety of fluorescent colors: gold, orange, pink, green, and yellow. Those from the midwestern and southeastern United States fluoresce with a yellow to orange-yellow color.

There are German sphalerites that fluoresce orange LW. Australian sphalerites from the Broken Hill area and the Burragorang Valley have been found that fluoresce yellow LW.

Willemite

Willemite is a zinc silicate ore. It is one of the main fluorescent minerals at Franklin and Sterling Hill, New Jersey, and can be found in great abundance there. It is much rarer in the rest of the world. Franklin willemite was first described in 1822 and was referred to, for the next few years, as siliceous oxide or silicate of zinc. The Franklin discoverers did not give it a formal name. In 1830, in Belgium, the same mineral was discovered and formally named Willemite in honor of William I, king of the Netherlands from 1813 to 1840. When the Franklin, New Jersey, material was found to be the same mineral, it was thereafter called willemite.

In the Franklin area, willemite can be different colors: white, yellow, blue, gray, green, red, brown, or black. All of them fluoresce green, except beta-willemite. Beta-willemite is a name collectors have adopted for a micro-crystalline yellow to yellow-orange fluorescing form of willemite with a strong phosphorescence. The harder to find daylight colors of willemite are blue, yellow, red, and bright green. Brown willemite is locally called *troostite*.

Willemite can also found from: New York; Arizona; Utah; New Mexico; California; Minas Gerais, Brazil; Mont Saint-Hilaire, Canada; Beltana, South Australia; Tsumeb, Namibia (some of the best willemites from this area are gemmy, botryoidal blue willemite, gemmy yellow willemite, and pale green willemite); Broken Hill, Zambia; Sardegna, Italy; Vrancice, Czech Republic; Portugal; Sweden; Belgium; and Greenland.

Photographing Fluorescent Minerals

Photographing fluorescent minerals is not very difficult, but takes patience, time, and a bit of experimentation. All the rocks in the book were photographed with digital cameras. Not all digital cameras are good at capturing the color of fluorescent minerals under shortwave or longwave light. I can tell you what works well for me. You will need to experiment to find what works well for you.

A few words about color balance: Your brain automatically balances the color input it receives from your eyes. For example, if you look at a red shirt indoors under artificial light and then take it outside in bright sunlight, it still looks like a red shirt. Film cameras generally do not balance color. Digital cameras have a "white balance" which tries to turn all colored light into "daylight" colored light. Looking at photographs of the red shirt taken inside and outside, the color red will look different in each photo. The color of light is measured in "kelvin degrees."

- *Daylight* is about 5,600 kelvin (light at noon on a clear day).
- *Candlelight* is about 1,900 kelvin.
- *Tungsten bulbs* (regular incandescent light bulbs) are about 2,800-3,200 kelvin (more yellow light).
- *Halogen lights* are about 3,200 kelvin.
- *Fluorescent lights* are 3,000-3,500 kelvin (warm), 4,100-4,200 kelvin (cool), or 6,000-7,000 kelvin (daylight).
- *An overcast sky* is about 7,000 kelvin (indirect daylight).
- *A clear blue sky* is about 20,000 kelvin (this is also light in a shady area - more blue).
- *Shortwave UV light* is higher than 20,000 kelvin.
- *Longwave UV light* seems to be even higher kelvin and more blue than shortwave).

A good simple introduction to light color and how a camera sees it, can be found at the following web address: http://members.shaw.ca/jimht03/light.html.

Depending upon your light source, your camera will "see" the light differently. With a digital camera, depending upon its white balancing abilities, it will try to balance the light for daylight. This means that if you shoot a picture using regular light bulbs, the camera will try to convert the yellowish light given off by the bulb into daylight. Digital cameras give you the option of helping the white balance by setting the camera's white balance for the kind of light you are shooting under.

I used two different digital cameras for the photographs in this book — a Nikon D70 (six megapixel) with a 60 mm macro lens for the daylight photos and an Olympus 5050z (5 megapixel and no longer made, but available on eBay) for the UV photos.

I used the Olympus 5050z for the daylight and the UV photos in my first book. I have tried other cameras to shoot UV-lighted minerals and found that the 5050z seems to have the best white balance, best automatic light control, and good close-up ability to allow results that are easy to work with in Adobe Photoshop™. I tried some of the newer Olympus cameras - the E-1, the E-Volt 300, and the 8080z. Unfortunately, Olympus has changed the automatic light-metering parameters of these cameras, and they no longer automatically compensate for the color of the UV lamps. Simply, they work the same way as the Nikon D70 does. The Nikon could be used alone, but while I could produce decent shortwave ultraviolet light photos with the Nikon, using the white balance set to "shade" "+1," all longwave ultraviolet light photos taken with that camera came out blue with all other colors washed out. For daylight photos, the Nikon provides superbly crisp photos with good shadow detail, but I used the Olympus 5050z for the ultraviolet light shots. It was easier to use the Olympus for the ultraviolet light shots than to reset the white balance on the Nikon after each shot.

I use a computer program called Adobe Photoshop (version 6.0.1) (you can also use Photoshop Elements™ that costs about $100). It allows me to "rebuild" colors that are not easily captured by the camera. For example, the yellow of esperite under ultraviolet light is especially difficult to capture. It comes out a shade of green or sometimes white and needs to be recolored yellow in Photoshop.

I use a professional shooting table made of curved white lucite with a colorless curved lucite overlay to allow the specimen to seem as if it is floating in space. This helps to cut down on shadows under the piece in the daylight photos. A curved piece of white or off-white paper, as the shooting surface, should work quite well. I use two studio lights. One is mounted on a boom stand and the other on a light stand two or three times as far away as the boom-mounted light. This enables me to light the piece from above and create shadows in the crevices in the specimen. The shadows give the piece texture or a three-dimensional effect. The second light is in front of the specimen, but to one side. It softens the shadows and allows detail to appear in the shadow areas.

My boom stand is an old television microphone stand that I picked up at a photography flea market for under $100. It weighs fifty pounds and opens out to twelve feet. I mention this since new boom stands are several hundred dollars and looking for good alternatives can save you money that can better be spent on rocks. Two 75 watt flood lights (3,200 kelvin) in clip-on holders will do instead of studio lights. If you like to work in small spaces, two OttLites™ give a nice balanced daylight (5,600 kelvin). These are smaller fluorescent lights that take up very little space.

The camera is mounted on a tripod, and a remote control is used to prevent camera shake. I set the camera on "aperture priority" at f16 or f22 (generally one stop away from completely stopped down on the lens). Stopping down most of the way on the lens gives me decent depth of field (more of the piece is in focus). Stopping the lens all the way down, can degrade the quality of the photograph due to diffraction effects. You may want to play around with this. I let the camera figure the exposure automatically, but I control the aperture. I could put the camera on manual and control the aperture and the shutter speed, but both cameras do a great job of handling that automatically.

The UV photos are made in a similar fashion as the daylight photos, but using the Olympus camera and a UV System SuperBright UV lamp as my light source. I place the specimen on a piece of flat black board and hold the light above and to the sides during the two to six second exposure. Again, I let the camera decide the exposure while I control the aperture.

The specimen is held in place with tacky clay, or with the aid of a popsicle stick or a small nonfluorescent rock. The tacky clay product is sold in stationery stores as "FunTac™" or a similar name. Years ago, I used to roll up masking tape to hold items in place, but once I began using the tacky clay, I stuck with it (pun intended).

Once the photos are taken, I upload them to my computer (I use a Macintosh). I open the photos in Photoshop. Daylight photos are adjusted so that they have good "levels." Levels show the range of shades of black through white in a photo. (It helps to think of all photos as shades of black and white, even when they are in color). The photo should have a nice range of black (shadows) to white (highlights).

Photoshop also allows you to further balance the color cast of your photographs in Levels. If something in the picture is supposed to be white and it looks slightly yellowish or bluish, you can remove the color cast. There are three eyedroppers in the Levels box. These eyedroppers are used to remove color casts. One eyedropper is used on black, one is used on white, and the third is used on gray. You use one of the eyedroppers, not all three. Find an area that should be black, white, or gray and choose the appropriate eyedropper. Click on your area of black, white, or gray and the color casts usually vanish. The amazing thing about removing the color cast is that it restores the other colors to their correct color and hue.

My photos are used in print, so I save them as a five-inch-wide 300 dpi (dots per inch) TIFF file. TIFF is a format that does not compress the information in a digital photograph. The 5-to-6 megapixel cameras generally give me a thirty to forty-inch wide JPEG photograph at 72 dpi. JPEG is a compressed format. When I put a photo on the Internet, I save it as a JPEG since it compresses (cuts down the digital size of) the photo. Photoshop allows you to easily control the size of the photograph. I could shoot in the newer RAW format, but, so far, I think the quality of the photos is adequate working from the "very fine" JPEG output of the camera.

When I work on the photos taken under UV light, the first thing I usually do is go to "Levels" in Photoshop and move the "white" slider to the left to brighten the highlights. I then move the "mid-tone" slider to the left to lighten the mid-tones and bring out a bit of the nonfluorescing parts of the piece and then move the "shadow" slider to the right to darken the background. This makes the background black with no detail. (Tip: never use the "brightness" control. The three sliders under Levels give you control over specific ranges of brightness and darkness that are unattainable in "brightness" alone.) I exit Levels and look at the specimen on the computer screen. If the colors are not right, I adjust colors using the "hue" or "saturation" controls. Hue changes the color to another color while saturation intensifies or reduces the same color. I work on each color individually until it is as accurate as I can get it (see the section at the beginning of the book on The Colors of the Minerals. Sometimes I decrease the saturation of the green so that the other colors seem more visible). If I am trying to get an area of esperite to its original yellow, I go to the "color picker" and the "air brush". I pick the shade of yellow that I want and then set the "blending" mode of the brush to "color" and the pressure of the brush to seven to twelve percent so that it leaves the texture of the specimen alone and only changes the color. White areas and black areas will not change under the "color" mode, so if the yellow color is blown out (that is, it turns to white) in the photograph, I change the blending mode to "normal" and adjust my brush size (again pressure at seven to twelve percent) so that I can paint within the lines and restore the color that I saw under the ultraviolet lamp.

I have been using Photoshop for over ten years and continue to read books, attend courses, and buy video learning tapes to help me make my photographs as realistic and accurate as I possibly can. Photoshop Elements is a less complex version of the larger program, and it offers the beginner or intermediate digital photographer many ways to improve their photographs. If you use either version of Photoshop, buy a video that will teach you some shortcuts.

Most of the photographs were taken for this book using a new SuperBright ultraviolet lamp. If you have an older (pre-1995) ultraviolet lamp, the colors you see with the same minerals may not be as bright and may have a slightly bluer look. There are now several good handheld UV lamps on the market. As you build your collection of fluorescent minerals, you will probably want to invest in a current, quality ultraviolet lamp.

A Gallery of Minerals

Throughout these pages, the first photo generally shows the mineral under daylight while the subsequent photo or photos show the mineral under shortwave (SW) ultraviolet (UV) light, midwave (MW), and/or longwave (LW) UV light. One of the keys to successful collecting is to see enough variations of what you are collecting so that when you find a good collectible specimen, you don't overlook it. The book often shows several examples of a particular fluorescent mineral.

Franklin Mine, Sterling Hill Mine, and the Rest of New Jersey

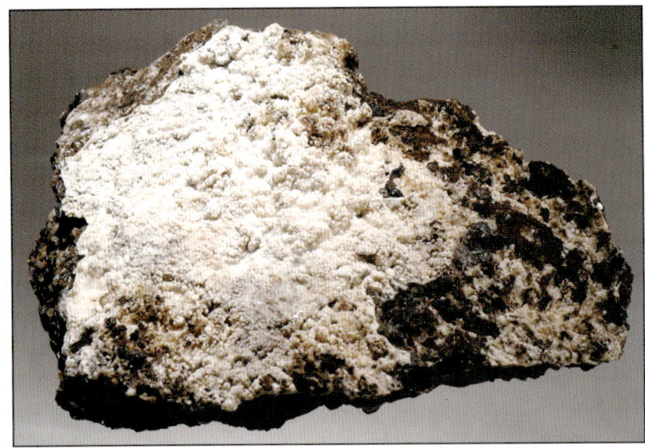

ARAGONITE. Aragonite on calcite from the Sterling Hill mine, Ogdensburg, Sussex County, from a small find in the summer of 2005. The aragonite fluoresces pale blue SW and the calcite fluoresces orange-red SW. The piece weighs 2 lb. 2.0 oz. and is 6.0 x 4.0 x 2.3 inches. Value $40-45

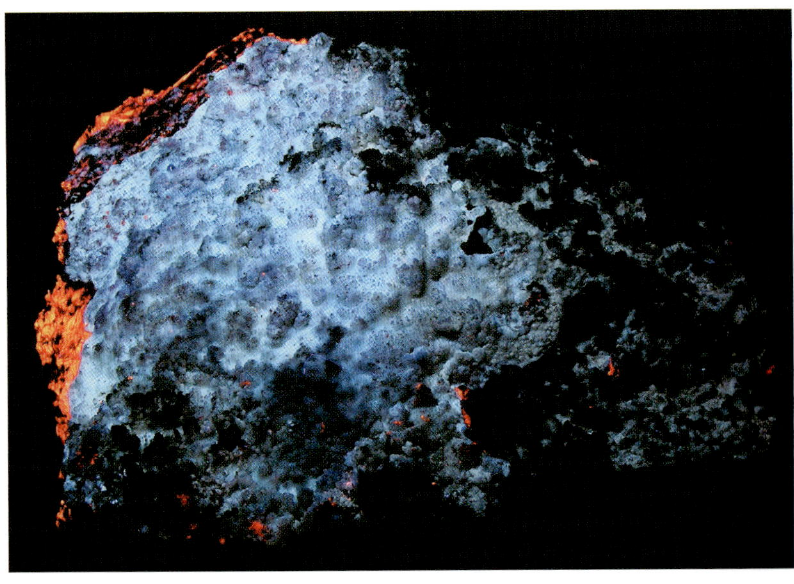

Same under SW

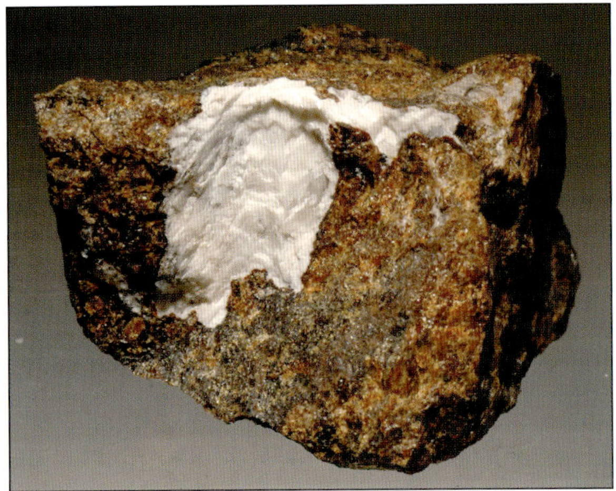

BARITE. Rare, striated or fibrous white barite in andradite found at the Mill site, Franklin, Sussex County. This barite fluoresces red SW. An EDS analysis was done in August 2004 (thanks to Dick Bostwick). The analysis found barium and sulfur in the right proportions for barium sulphate. The piece weighs 6.0 oz. and is 2.0 x 1.8 x 1.5 inches. This is the other half of the piece mistakenly identified in my first book as wollastonite.

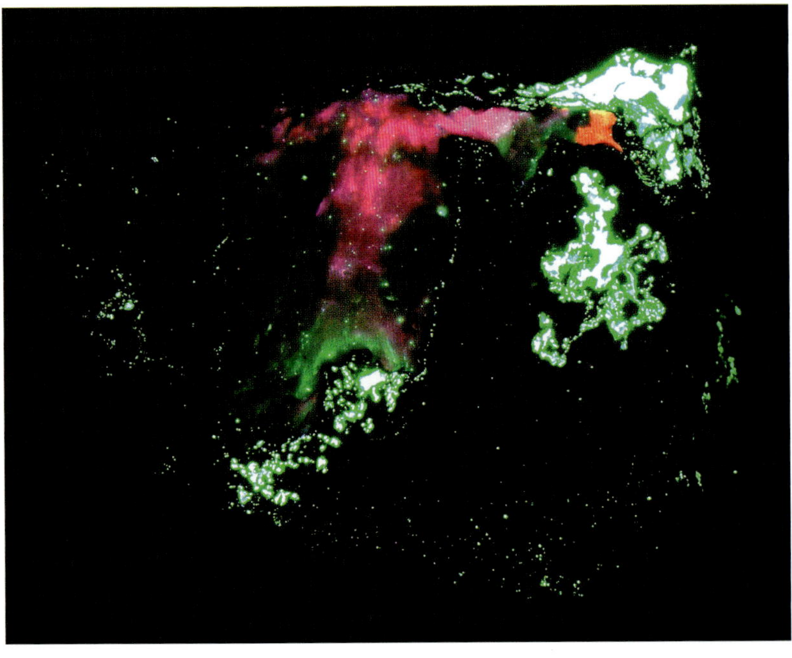

Same under SW

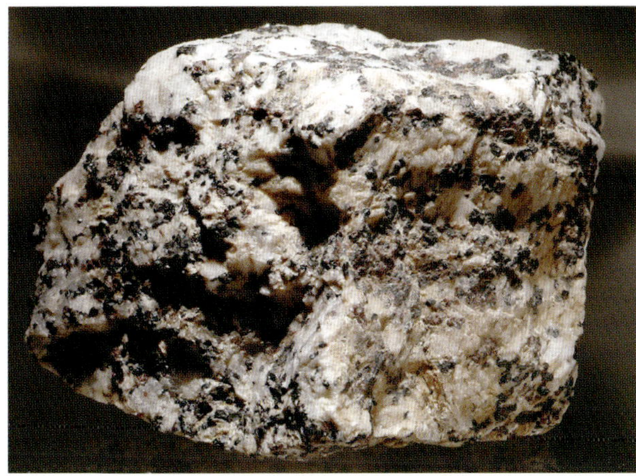

BARITE, SPHALERITE. Barite with sphalerite, calcite and willemite from the Sterling Hill mine, Ogdensburg, Sussex County. The barite fluoresces pale yellow LW, the sphalerite fluoresces orange and blue LW, the willemite fluoresces green SW, and the calcite fluoresces orange-red SW. The piece weighs 1 lb. 13.0 oz. and is 3.3 x 3.0 x 2.8 inches. Courtesy of Ed Letscher. Value $55-65

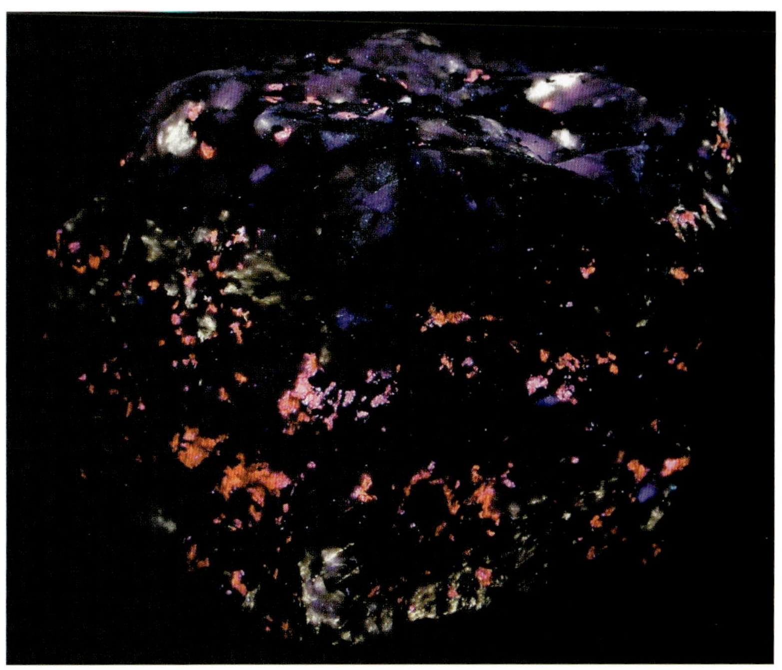

Same under LW

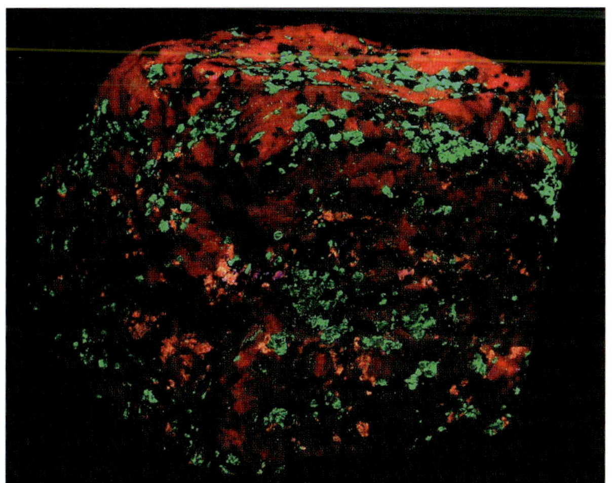

Same under SW

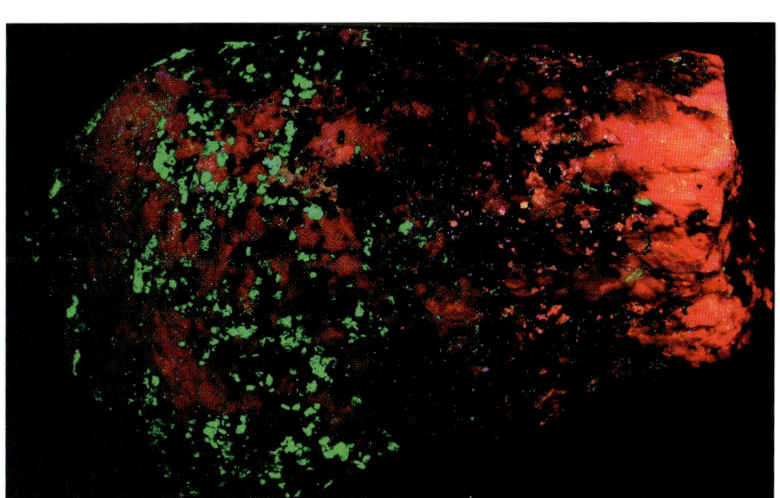

Same under SW

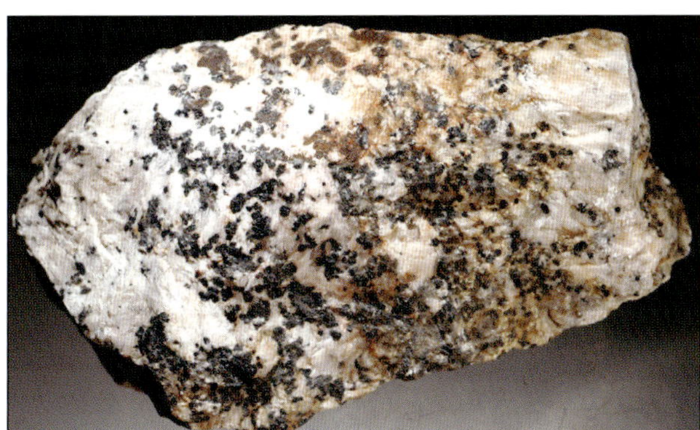

BARITE, SPHALERITE. Barite and sphalerite from the Sterling Hill mine, Ogdensburg, Sussex County. The sphalerite fluoresces orange and blue LW and the barite fluoresces pale yellow LW. The piece weighs 1 lb. 12.0 oz. and is 5.0 x 2.5 x 2.5 inches. Value $55-65

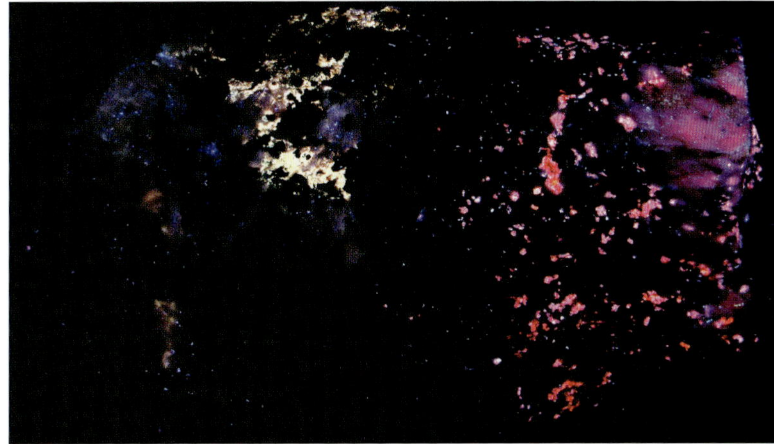

Same under LW

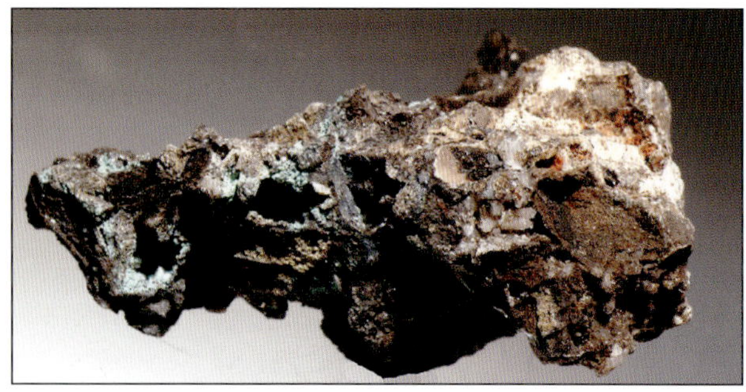

BETA-WILLEMITE. A yellow-fluorescing willemite made up of many small, sparkling crystals with calcite from the Andover mine, Andover, Sussex County. Beta-willemite fluoresces yellow SW. Note: although the name "beta-willemite" is used by collectors, the mineral is properly referred to as "yellow-fluorescing willemite". The piece weighs 5.5 oz. and is 3.5 x 2.8 x 1.5 inches. Value $45-55

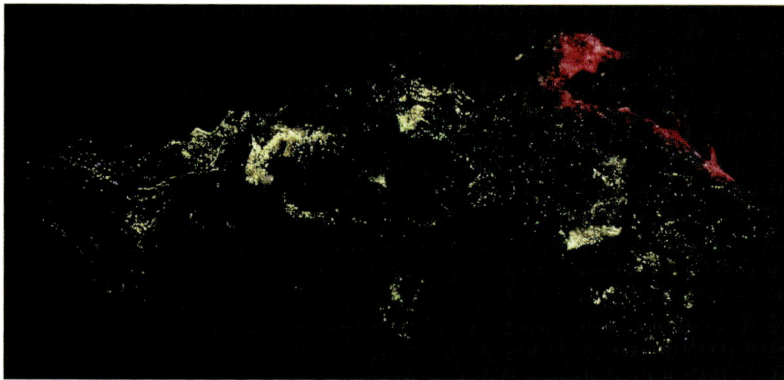

Same under SW

BUSTAMITE. Weathered bustamite from the Mill site, Franklin, Sussex County. The bustamite fluoresces pink-red SW. The piece weighs 7.0 oz. and is 2.8 x 2.5 x 1.3 inches. Value $70-75

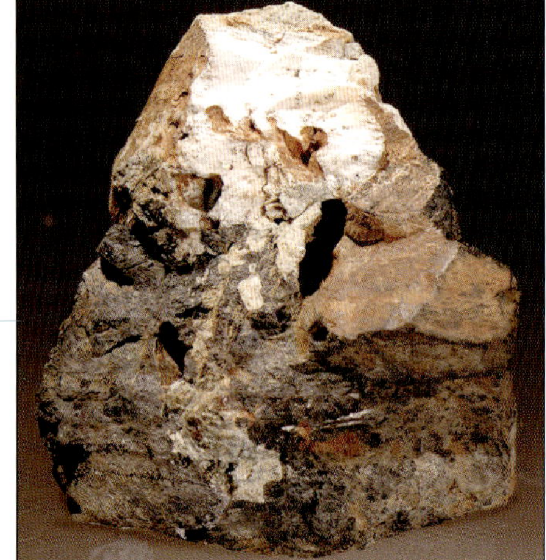

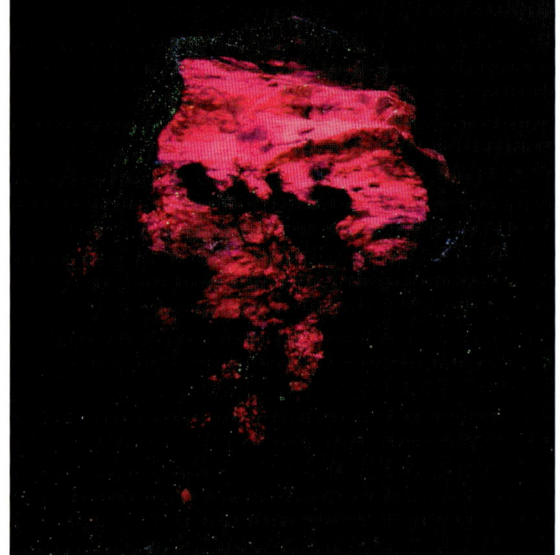

Same under SW

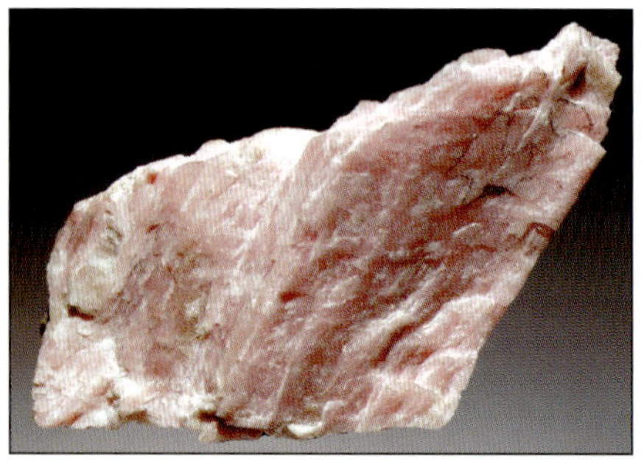

BUSTAMITE. Fluorescent bustamite with hardystonite, clinohedrite, and willemite from the Franklin mine, Franklin, Sussex County. The bustamite fluoresces red LW, the clinohedrite fluoresces orange SW, hardystonite fluoresces purple-blue SW and LW, willemite fluoresces green SW. The piece weighs 3.8 oz. and is 2.0 x 2.0 x 1.5 inches. Value $245-265

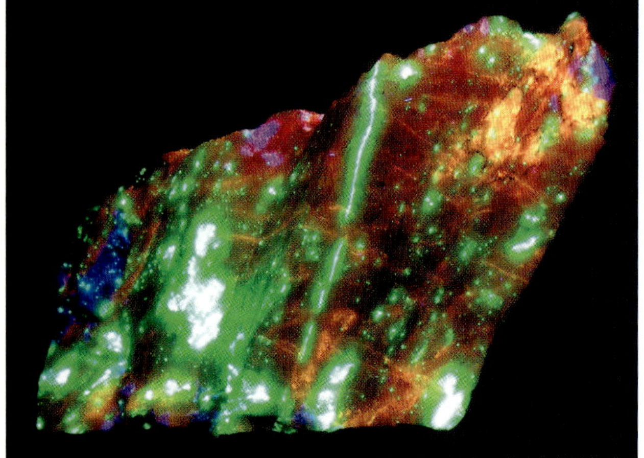

Same under SW

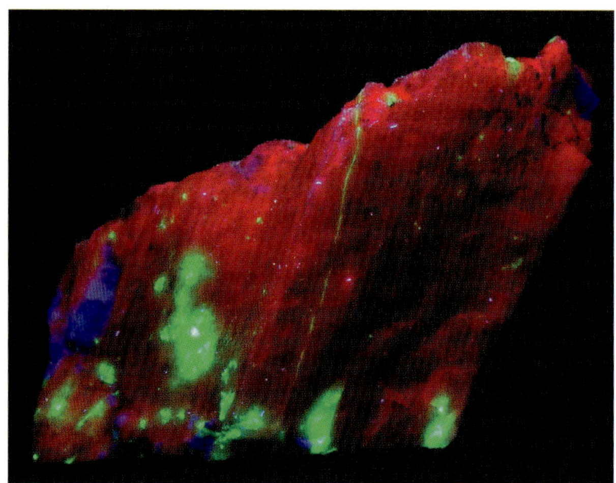

Same under LW

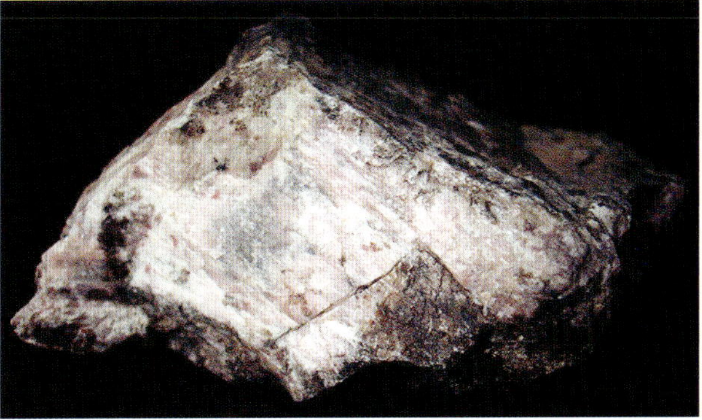

BUSTAMITE, HARDYSTONITE. Hardystonite and bustamite from the Franklin mine, Franklin, Sussex County. Hardystonite fluoresces purple-blue SW and bustamite fluoresces red SW and brighter red LW. The piece weighs 12.0 oz. and is 4.0 x 2.5 x 2.0 inches. Courtesy of Dru Wilbur.

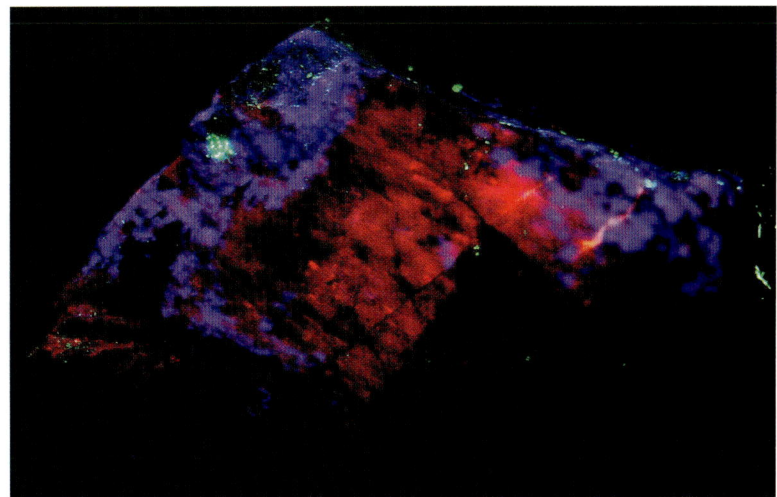

Same under SW

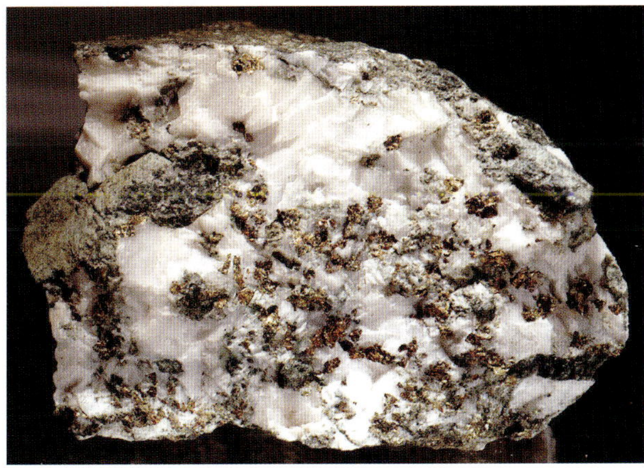

CALCITE. Calcite and chalcopyrite from the Snake Hill quarry, Secaucus, Hudson County. The calcite fluoresces pale red SW. The piece weighs 1 lb. 5.0 oz. and is 4.0 x 2.6 x 2.0 inches. Value $30-35

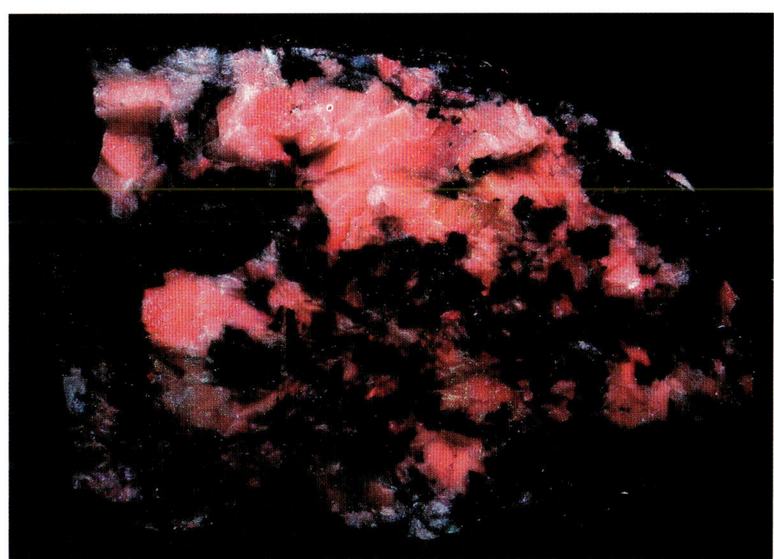

Same under SW

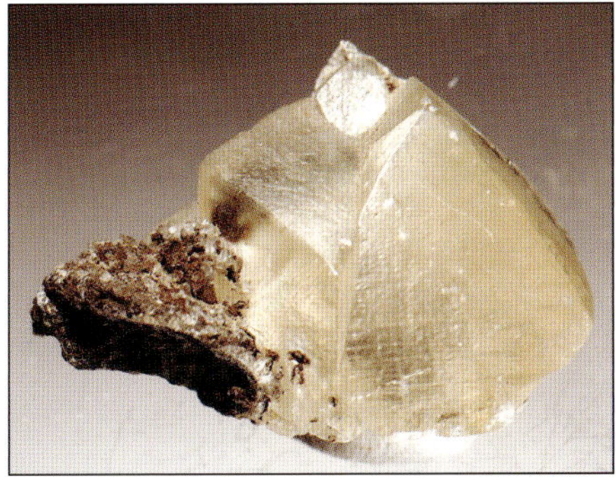

CALCITE. Calcite crystals from Prospect Park, Passaic County. The calcite fluoresces dark pink SW and tan LW. The piece weighs 1.0 oz. and is 1.5 x 1.3 x 1.0 inches. Value $15-20

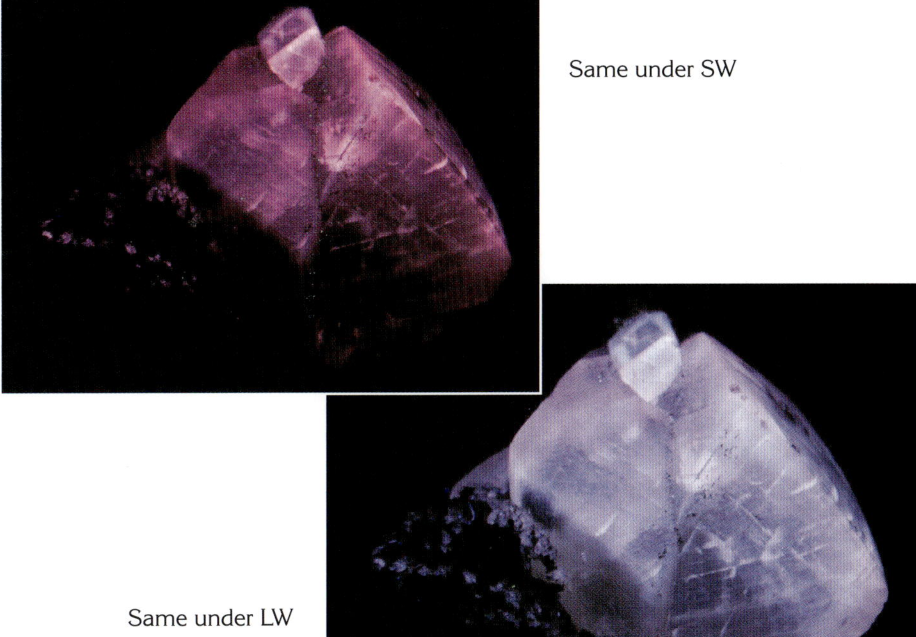

Same under SW

Same under LW

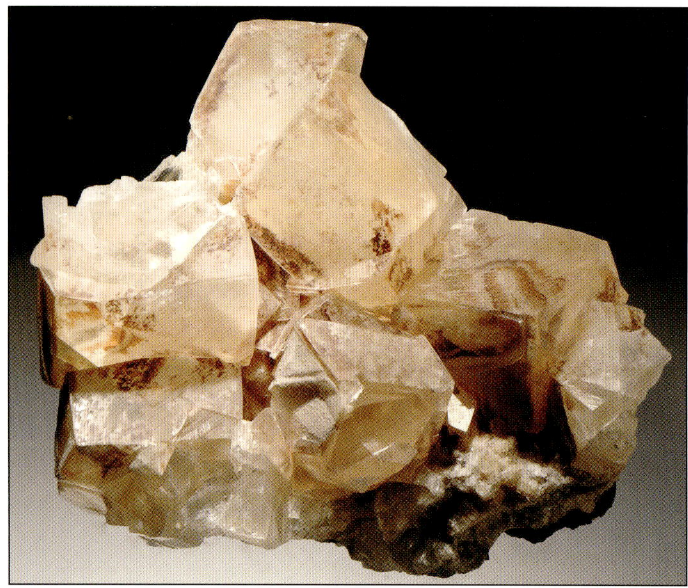

CALCITE. Calcite crystals from Fanwood, Union County. The surfaces of the crystals fluoresce peach LW and pale blue SW. The piece weighs 12.0 oz. and is 3.5 x 3.0 x 2.5 inches. Value $25-30

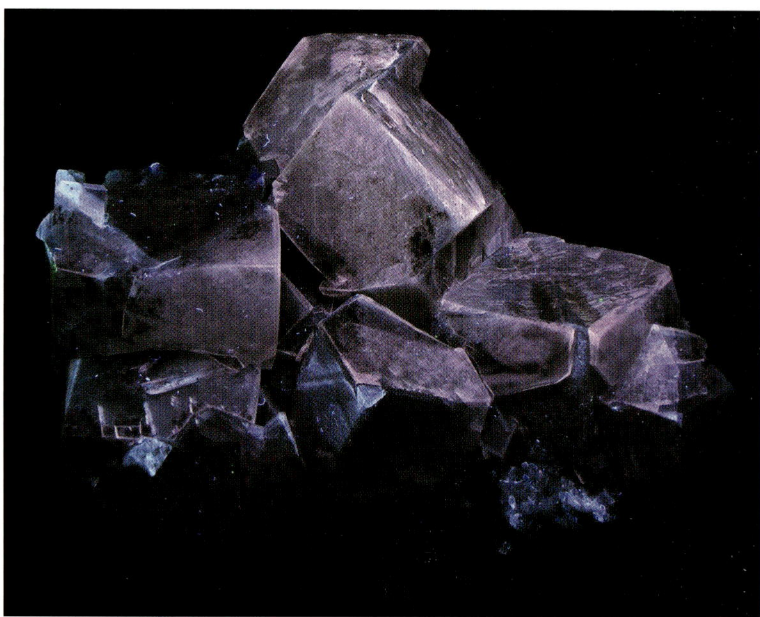

Same under LW

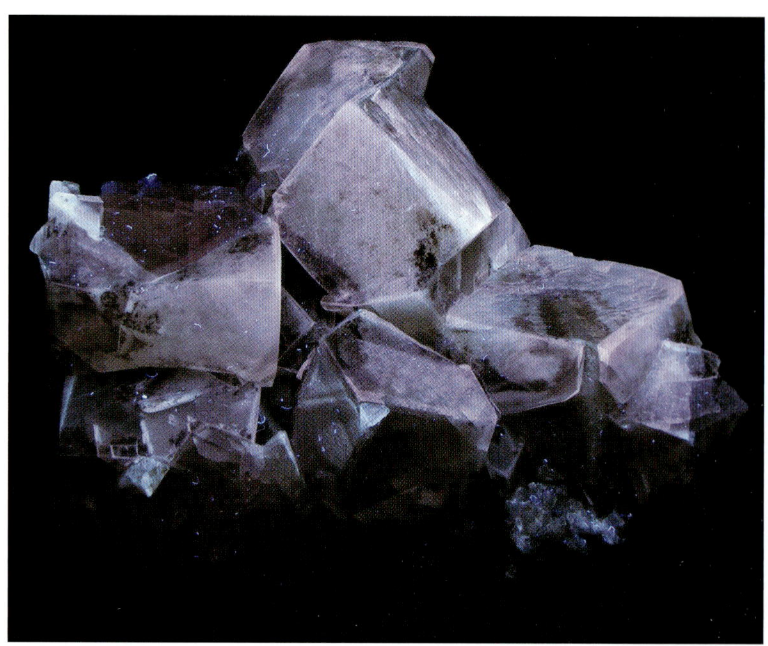

Same under SW

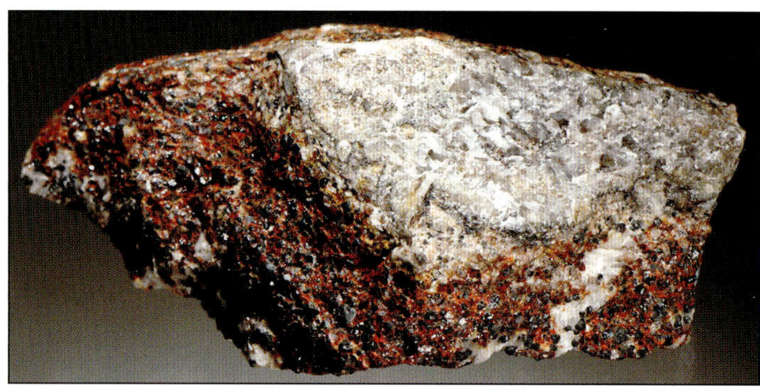

CALCITE, ZINCITE. Calcite with franklinite and zincite from the Sterling Hill mine, Ogdensburg, Sussex County. The piece contains rare blue-fluorescing calcite LW and grains of zincite that fluoresce pale yellow. The piece weighs 1 lb. 3.0 oz. and is 4.3 x 2.3 x 2.0 inches. Value $90-100

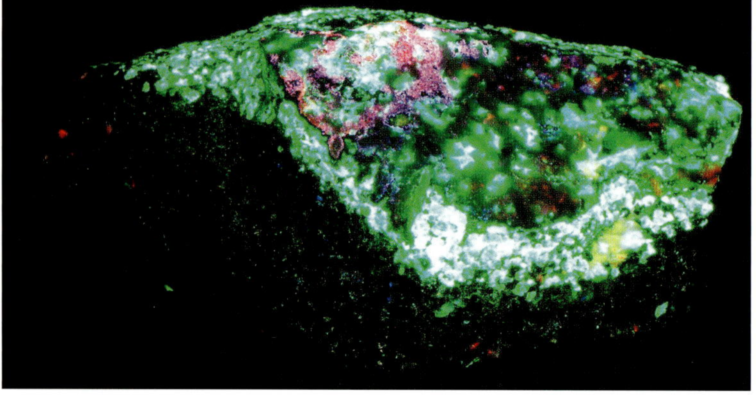

Same under SW

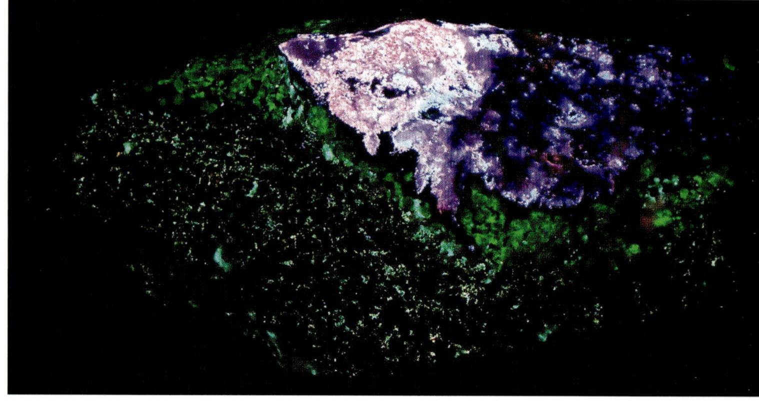

Same under LW

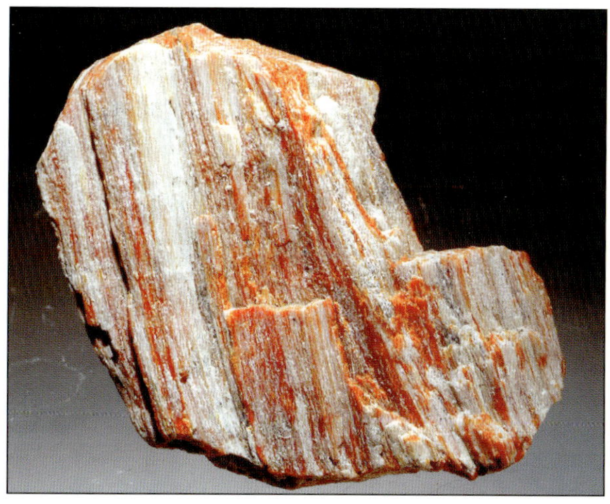

CALCITE. Calcite and willemite from the Sterling Hill mine, Ogdensburg, Sussex County. This unusual piece looks like petrified wood. The calcite fluoresces orange-red SW. The willemite fluoresces green SW and has a long-lasting phosphorescence. The piece weighs 5.8 oz. and is 3.0 x 2.5 x 0.8 inches. Value $40-55

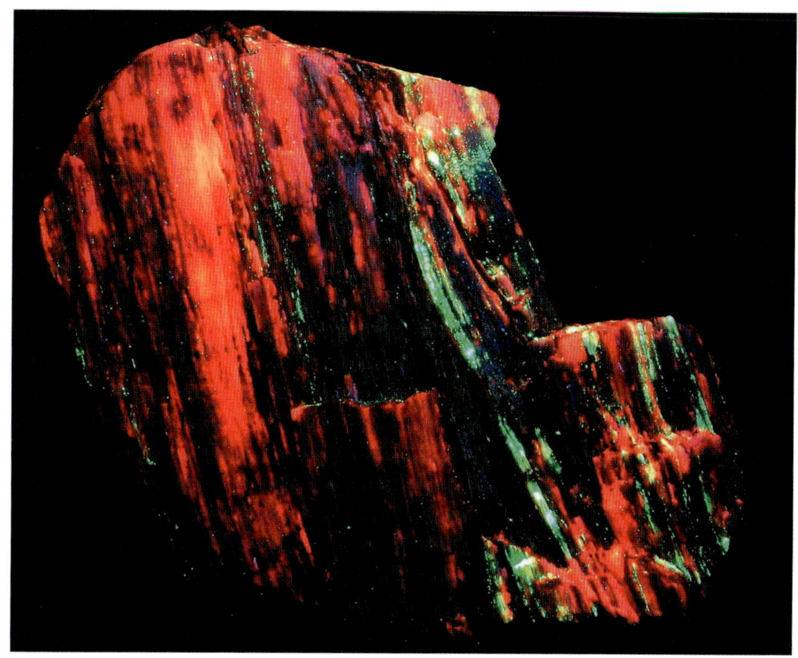

Same under SW

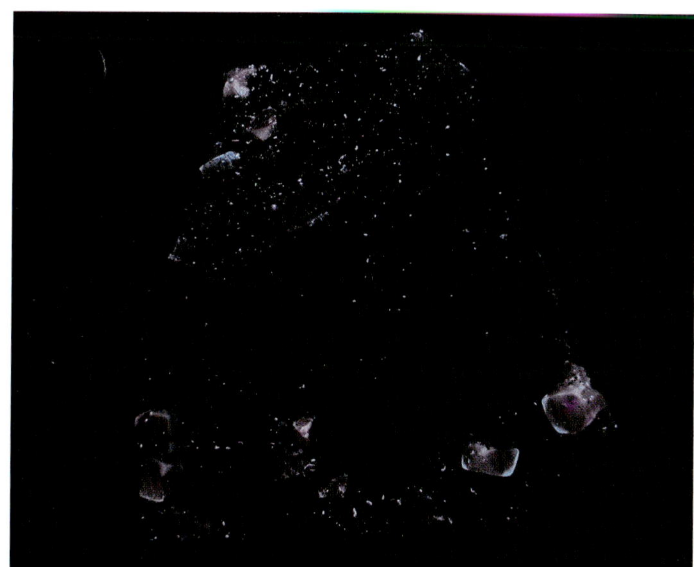

Same under SW

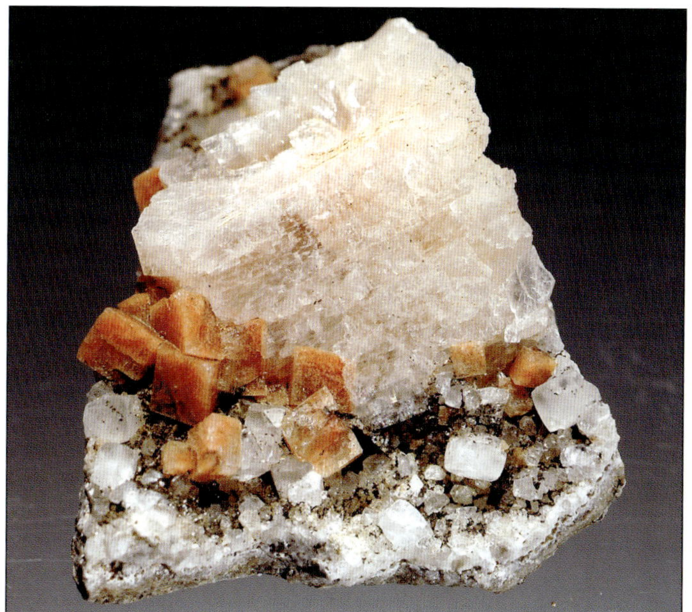

CALCITE. Calcite with tan chabazite crystals with heulandite on quartz from Upper New Street, Paterson, Passaic County. The piece was collected by Viktor Tandard in the early 1980s. The calcite fluoresces weak pink LW and weak white SW. The piece weighs 6.0 oz. and is 4.3 x 2.3 x 1.3 inches. Value $15-18

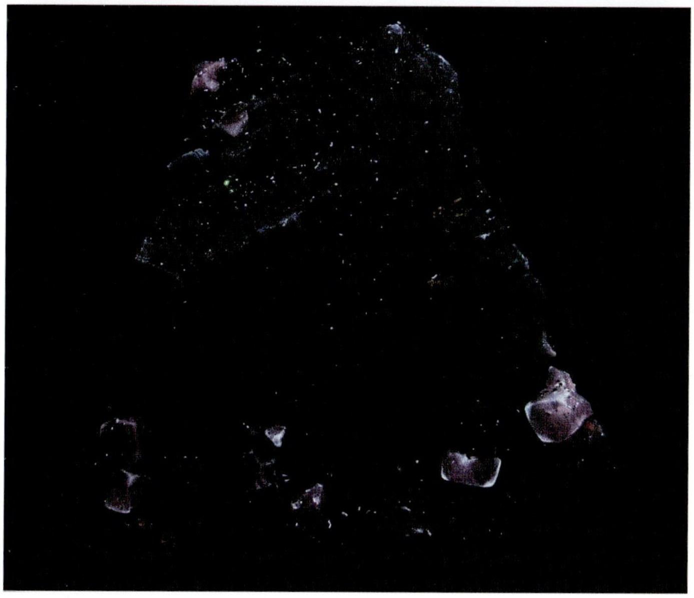

Same under LW

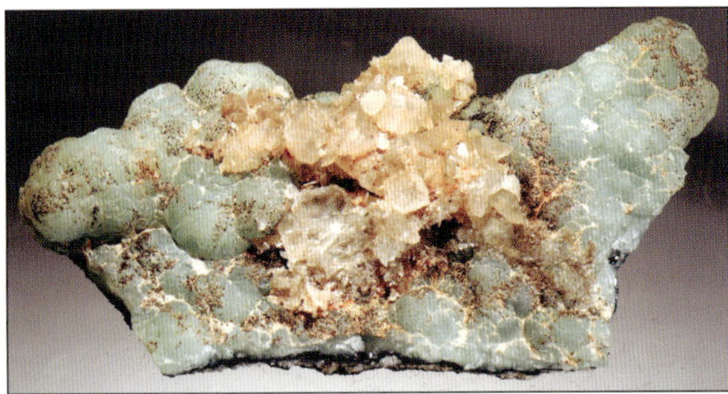

CALCITE. Calcite crystals on prehnite from Upper New Street, Paterson, Passaic County, that were collected by John Lassy in the early 1980s. Calcite fluoresces pink to tan SW and pale tan LW. The piece weighs 6.0 oz. and is 4.3 x 2.3 x 1.3 inches. Value $35-40

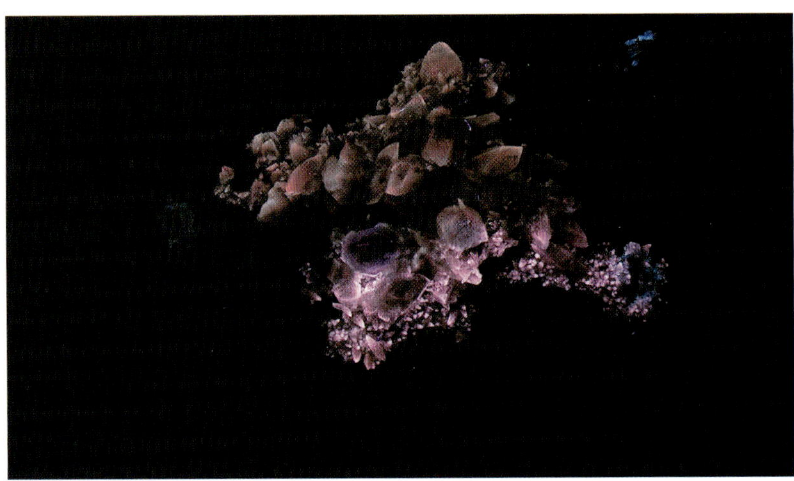

Same under SW

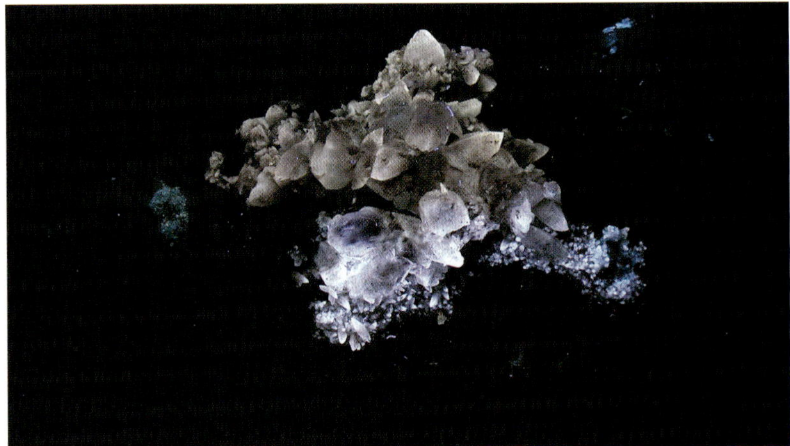

Same under LW

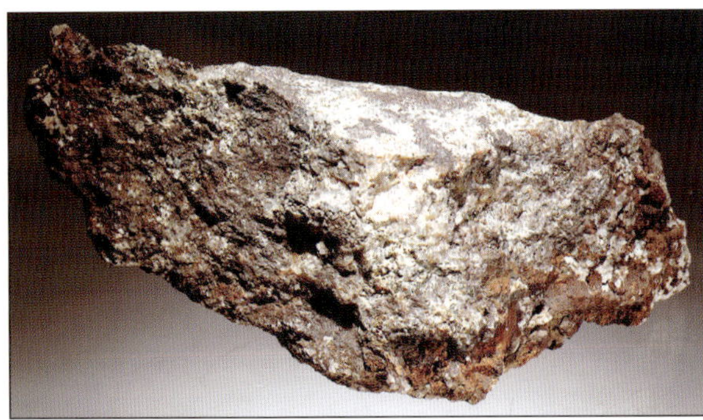

CERUSSITE. White patches of cerussite on a dark matrix with galena from the Sterling Hill mine, Ogdensburg, Sussex County. The cerussite fluoresces pale yellow LW. The piece weighs 1 lb. 12.0 oz. and is 5.0 x 2.5 x 2.5 inches. Value $45-50

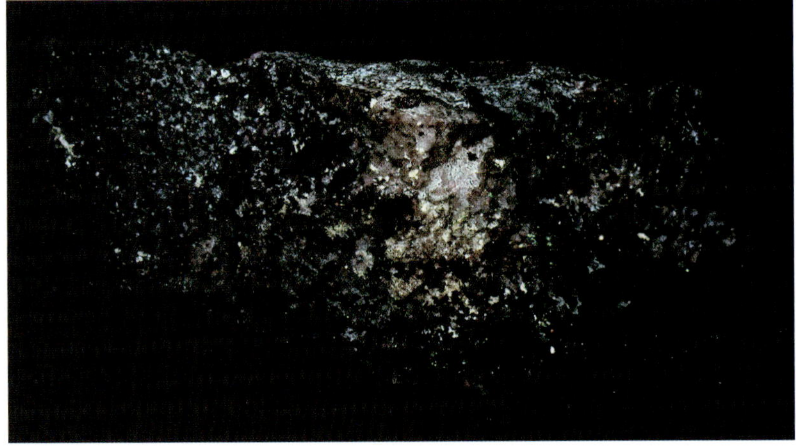

Same under LW

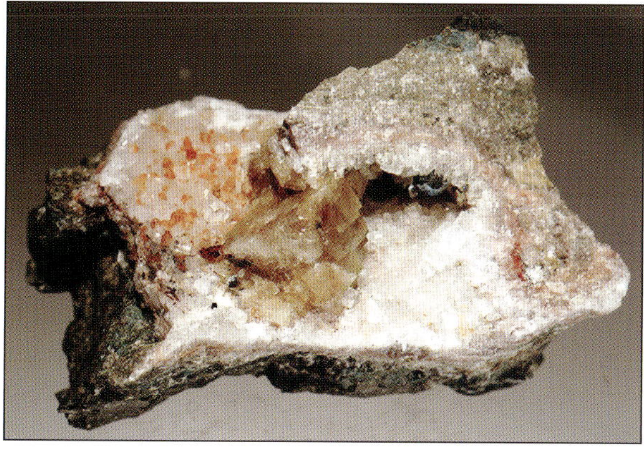

CHABAZITE. Chabazite crystals on quartz from Upper New Street, Paterson, Passaic County. This chabazite fluoresces a very bright green SW and LW. Many pieces of chabazite were examined from this location and this was the only one to fluoresce bright green. The piece weighs 3.5 oz. and is 2.5 x 1.5 x 1.5 inches. Value $50-60

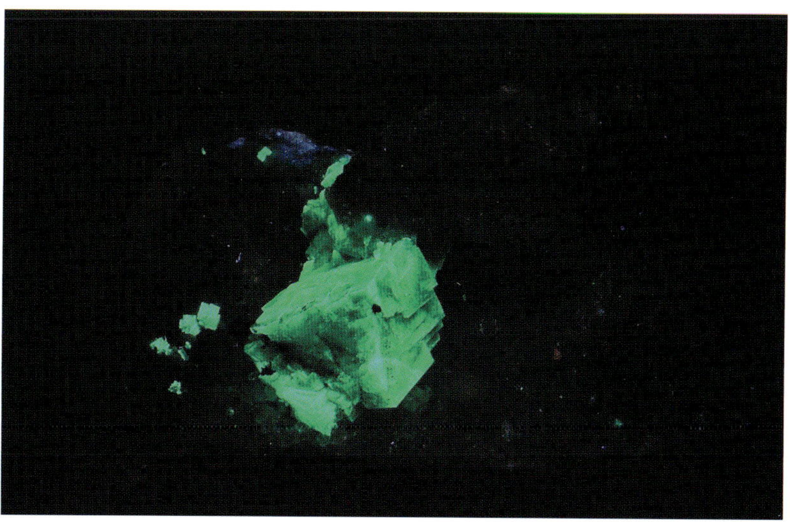

Same under SW

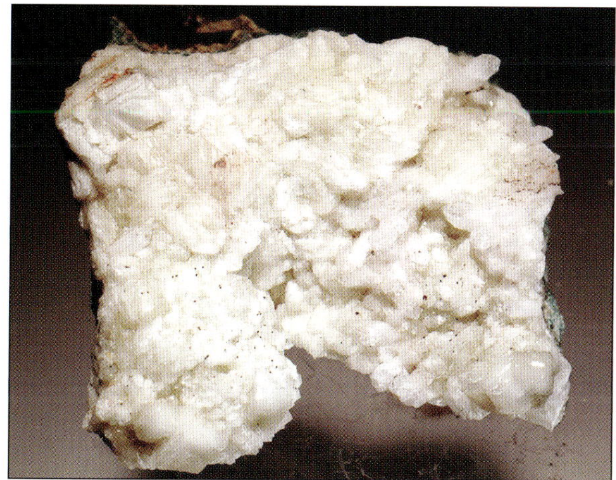

CHABAZITE. Chabazite crystals on datolite crystals from Upper New Street, Paterson, Passaic County. This chabazite fluoresces pink SW and white LW. There is an unidentified mineral fluorescing pale green SW. The piece weighs 6.3 oz. and is 2.5 x 2.5 x 1.5 inches. Value $20-25

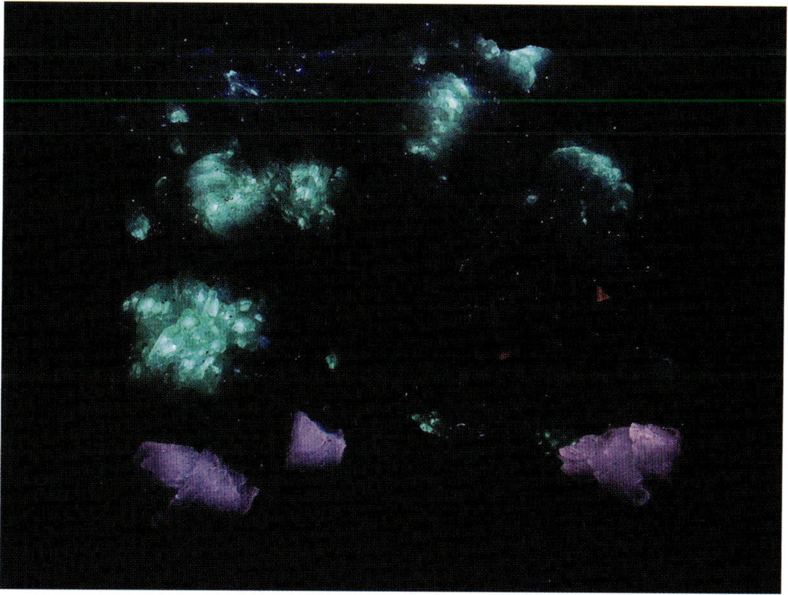

Same under SW

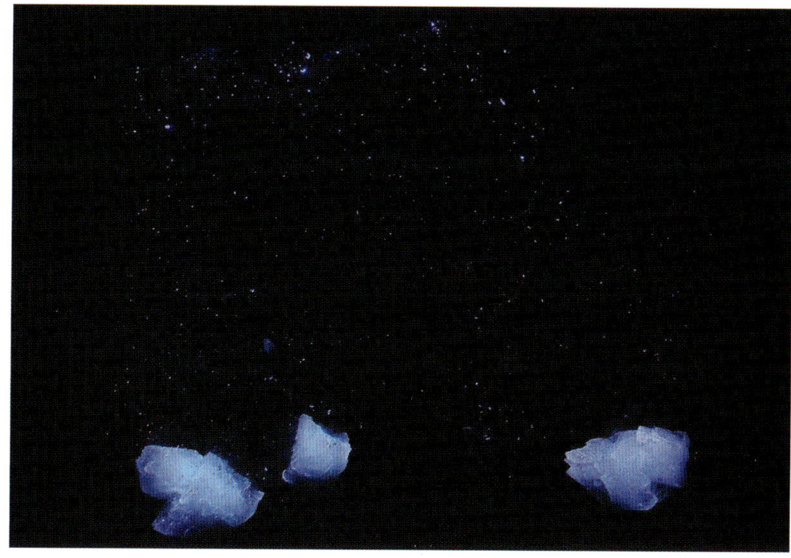

Same under LW

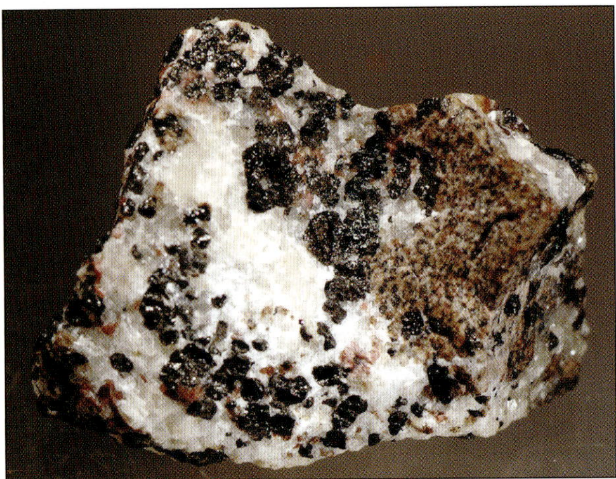

CUSPIDINE. Cuspidine is a rare Franklin mineral. This piece is from a 2005 find on the Buckwheat dump, Franklin, Sussex County. Cuspidine fluoresces orange SW and rose-peach MW. The piece weighs 4.0 oz. and is 2.0 x 1.8 x 1.4 inches. Value $295-325

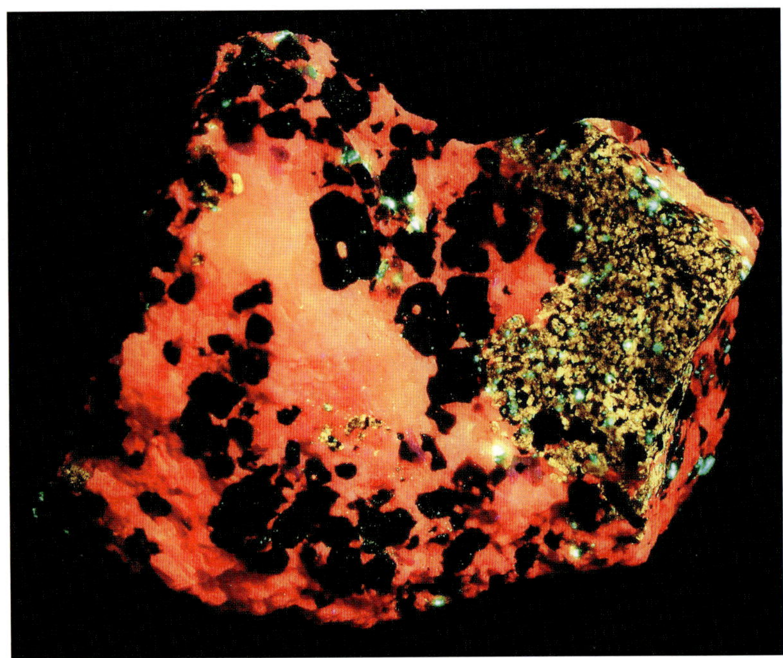

Same under SW

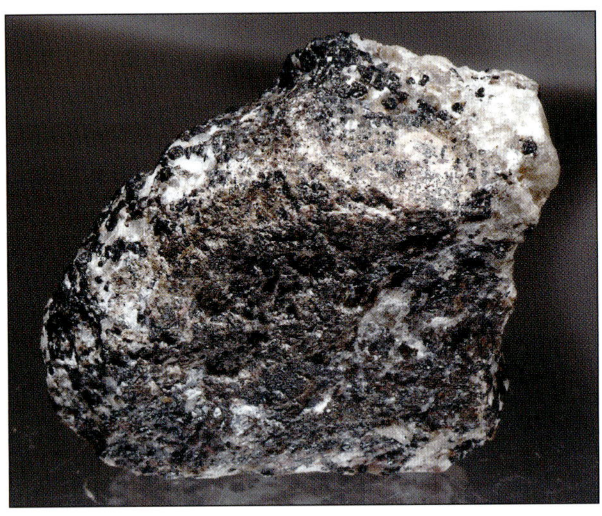

CUSPIDINE. Cuspidine from a 2005 find on the Buckwheat dump, Franklin, Sussex County. Cuspidine fluoresces orange SW and rose-peach MW. The piece weighs 9.0 oz. and is 2.5 x 2.3 x 1.8 inches. Value $450-500

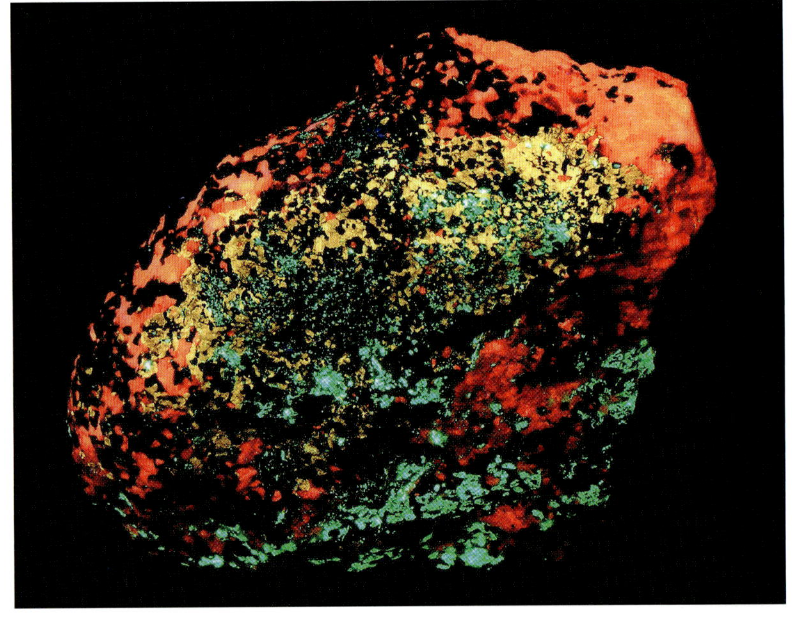

Same under SW.

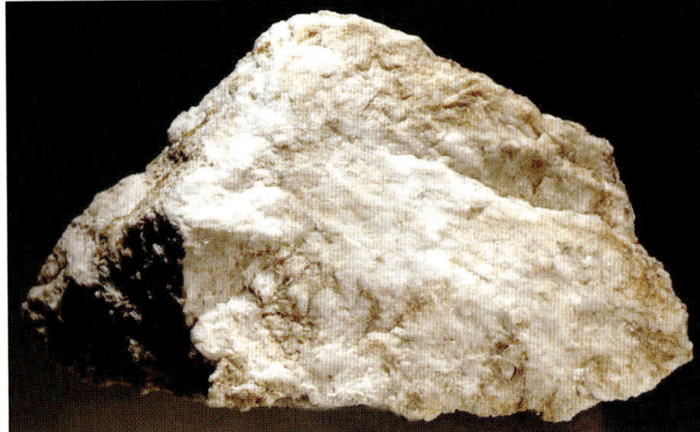

DIOPSIDE. A rich specimen of diopside from near Lake Valhalla, Montville, Morris County. The diopside fluoresces bright pale blue SW. The piece weighs 1 lb. 1.0 oz. and is 4.5 x 2.8 x 2.0 inches. Value $30-35

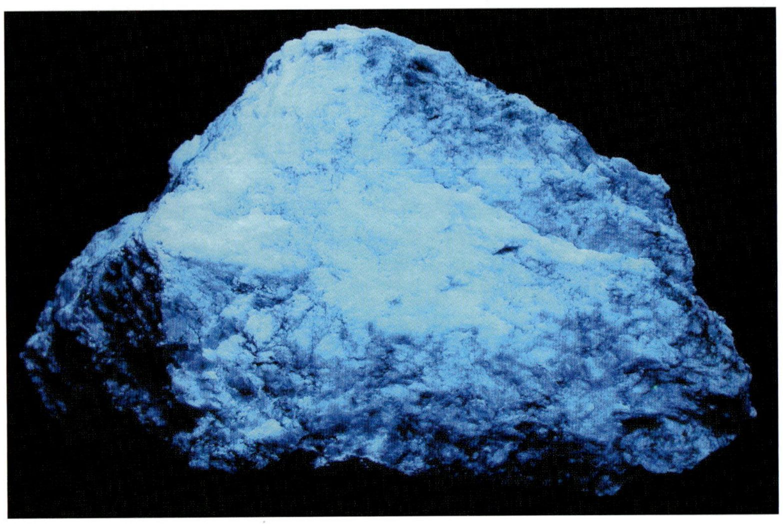

Same under SW

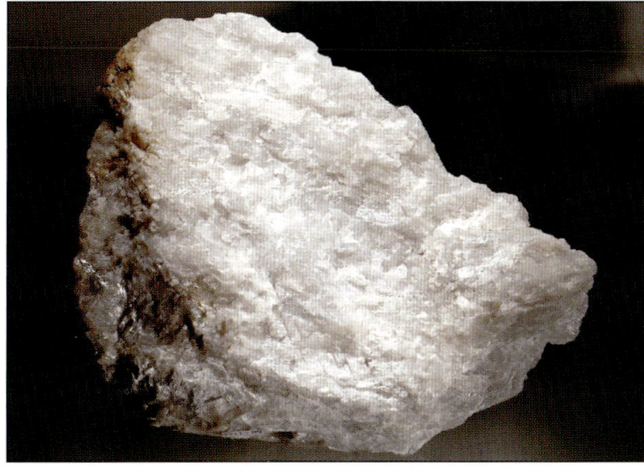

DIOPSIDE. Another example of diopside from near Lake Valhalla, Montville, Morris County. The diopside fluoresces bright pale blue SW. The piece weighs 1 lb. 11.0 oz. and is 6.0 x 4.8 x 1.4 inches. Value $30-35

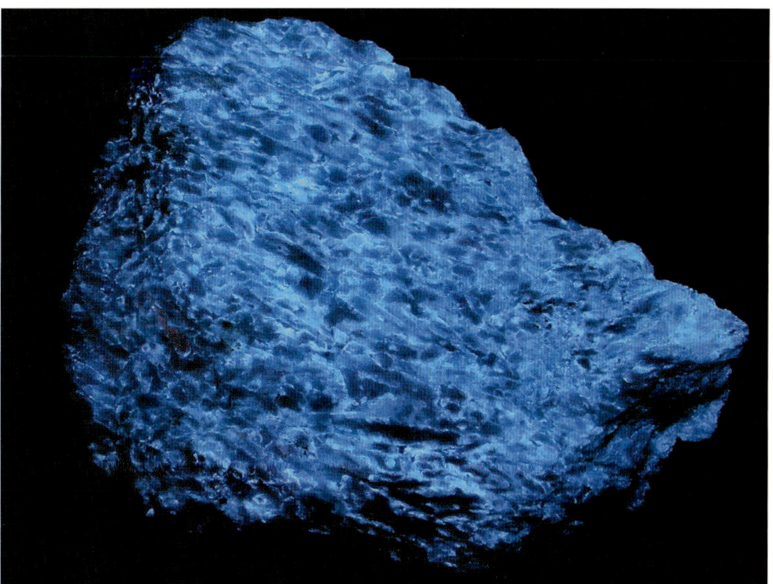

Same under SW

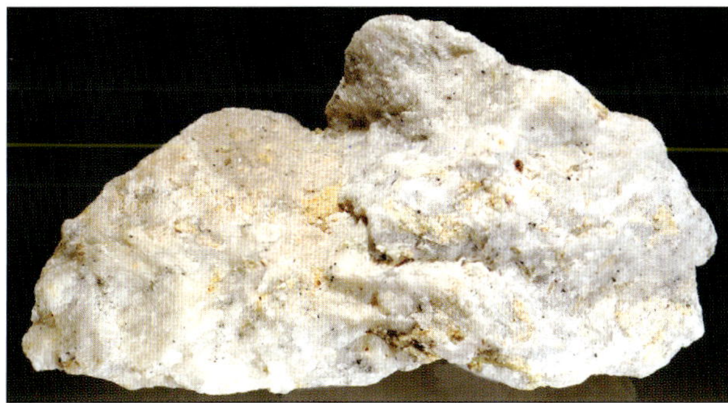

DIOPSIDE, NORBERGITE. Calcite with diopside and norbergite from the Farber quarry, Franklin, Sussex County. The diopside surrounds the norbergite. Norbergite fluoresces yellow SW and diopside fluoresces pale blue SW. The piece weighs 11.0 oz. and is 4.5 x 2.8 x 2.3 inches. Value $35-45

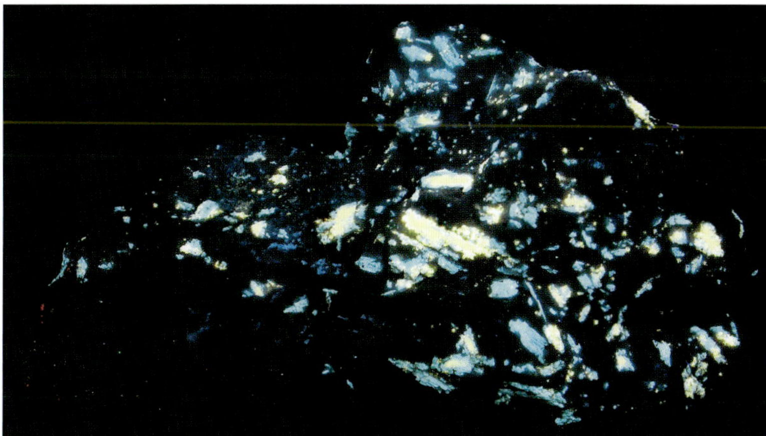

Same under SW

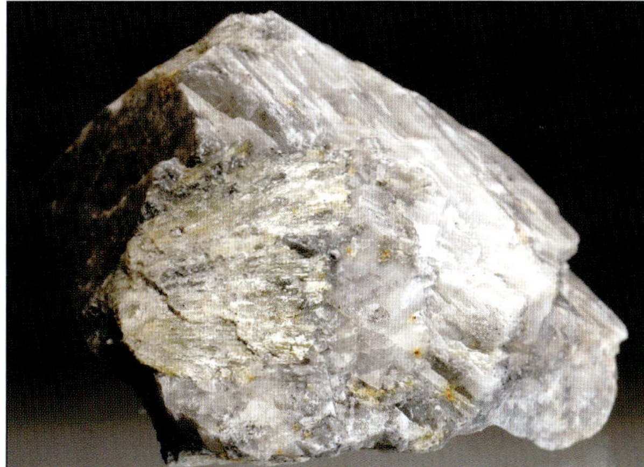

DIOPSIDE, TREMOLITE. Calcite with tremolite and diopside from the Farber quarry, Franklin, Sussex County. Tremolite fluoresces pale blue SW and diopside fluoresces bright pale blue SW. The piece weighs 12.0 oz. and is 3.5 x 3.0 x 2.0 inches. Value $30-40

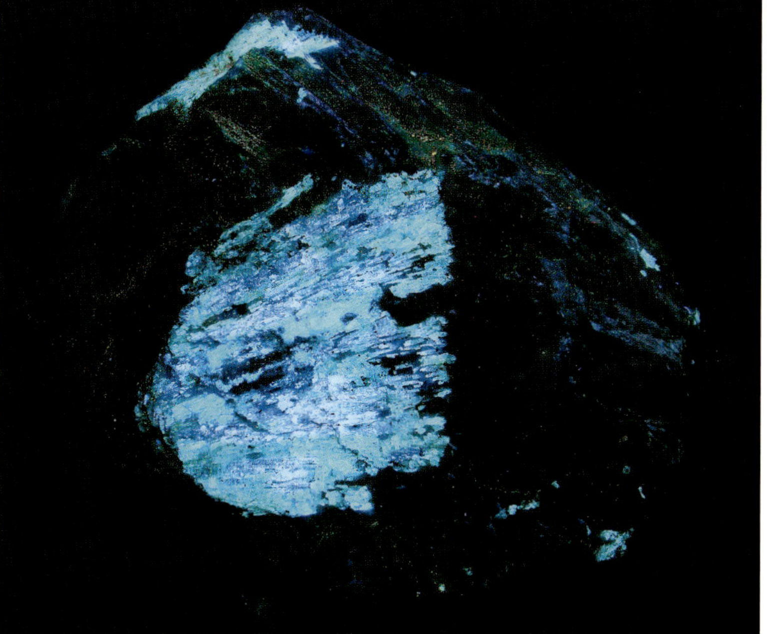

Same under SW

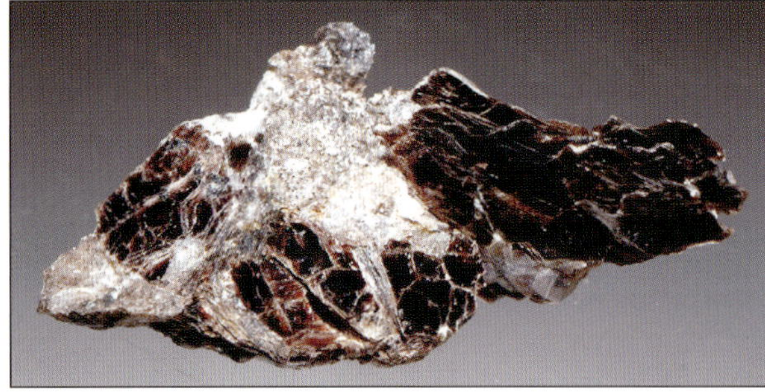

ESPERITE. Esperite, hardystonite, calcite, hendricksite, and willemite from the Franklin mine, Franklin, Sussex County. It is unusual to find esperite and hardystonite between the blades of the hendricksite. The esperite fluoresces yellow SW, the hardystonite fluoresces purple-blue SW, the calcite fluoresces orange-red SW, and the willemite fluoresces green SW. This is a small piece weighing about 1.0 oz. Value $18-20.

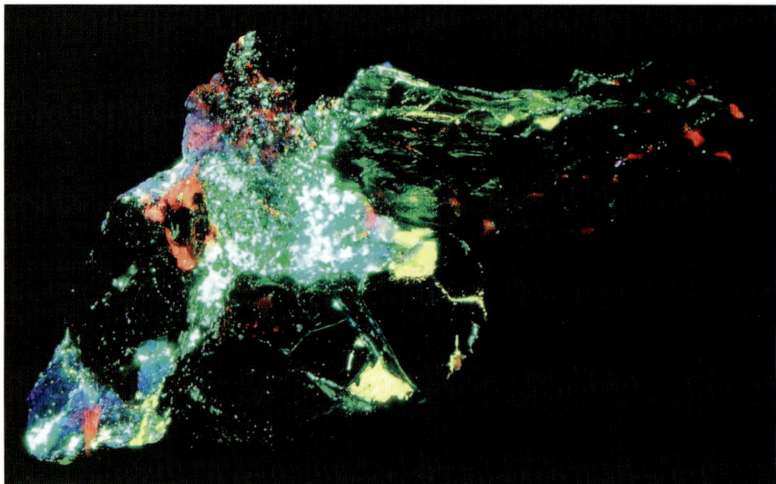

Same under SW

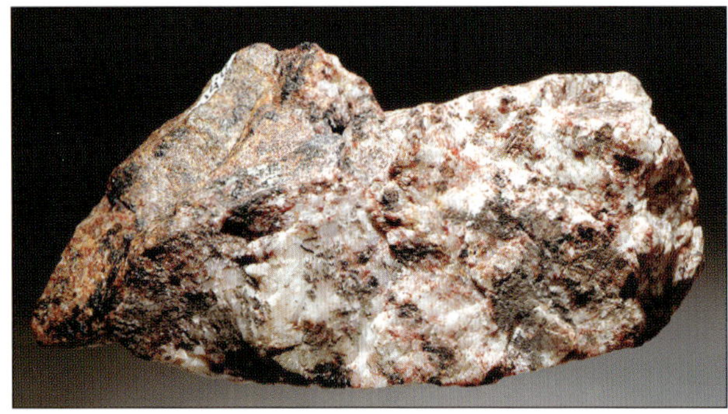

ESPERITE. Esperite with willemite veins from the Franklin mine, Franklin, Sussex County. The esperite fluoresces yellow SW and the willemite fluoresces green SW. The piece weighs 14.0 oz. and is 4.5 x 2.3 x 1.0 inches. Courtesy of Gar VanTassel.

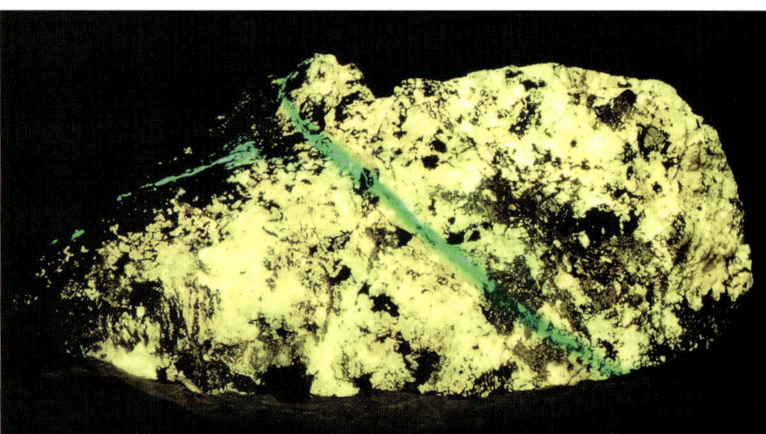

Same under SW

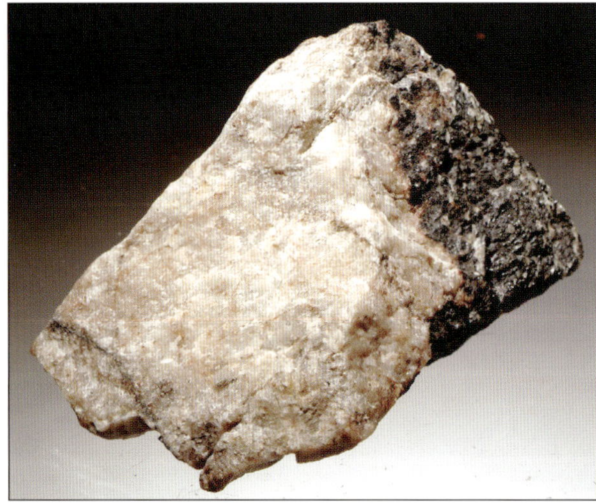

ESPERITE. Esperite with willemite from the Franklin mine, Franklin, Sussex County. The esperite fluoresces yellow SW and the willemite fluoresces green SW. The piece weighs 10.0 oz. and is 3.3 x 2.5 x 2.0 inches. Courtesy of Gar VanTassel.

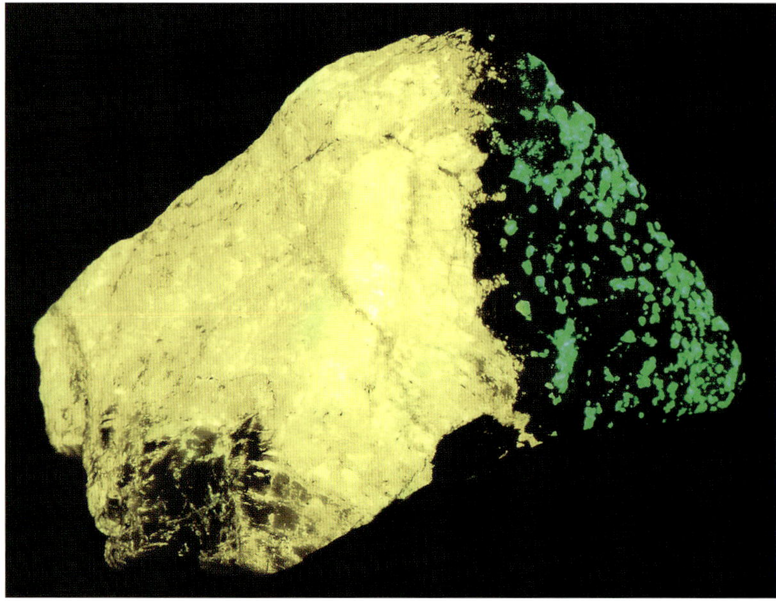

Same under SW

ESPERITE. An exceptional specimen with esperite, hardystonite, clinohedrite, willemite, and a touch of calcite from the Franklin mine, Franklin. The esperite fluoresces yellow SW, the hardystonite fluoresces purple-blue SW, the clinohedrite fluoresces bright orange SW, the willemite fluoresces green SW, and the calcite fluoresces orange-red SW. The piece weighs 1 lb. 4.0 oz. and is 4.3 x 3.3 x 1.5 inches. Courtesy of Gar VanTassel.

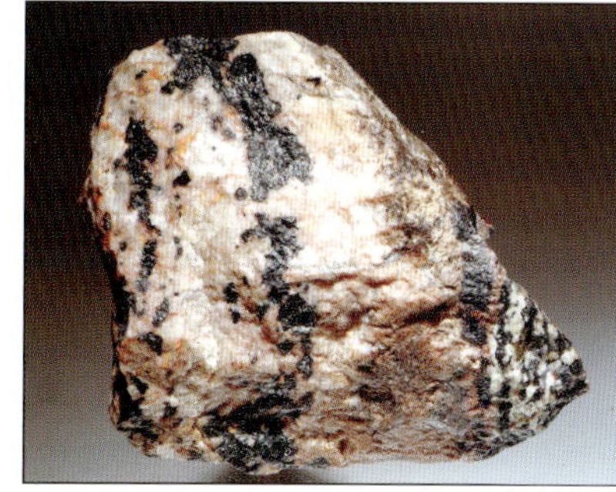

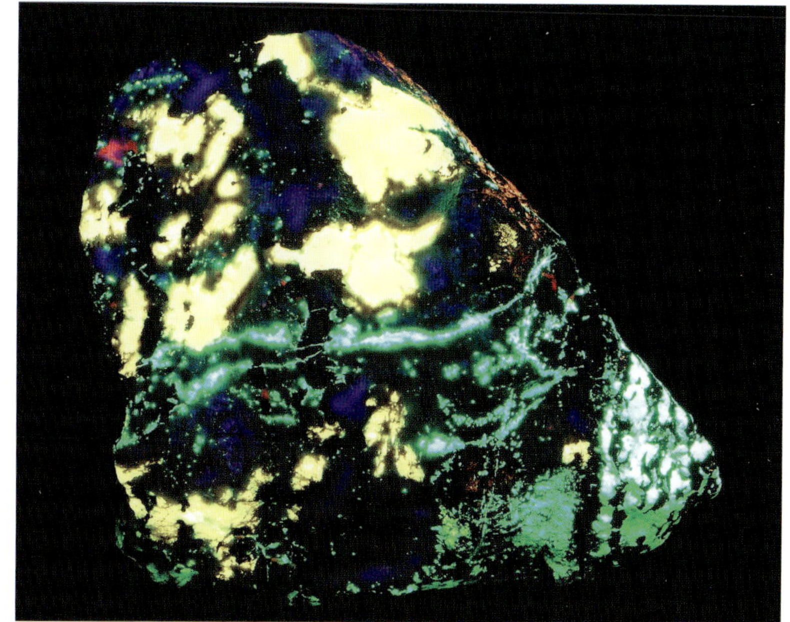

Same under SW

FELDSPAR. An example of feldspar with magnetite from the Edison mine, Ogdensburg, Sussex County. The feldspar fluoresces violet SW. The piece weighs 1 lb. 7.0 oz. and is 3.8 x 3.0 x 2.5 inches. Courtesy of Philip Persson. Value $15-17

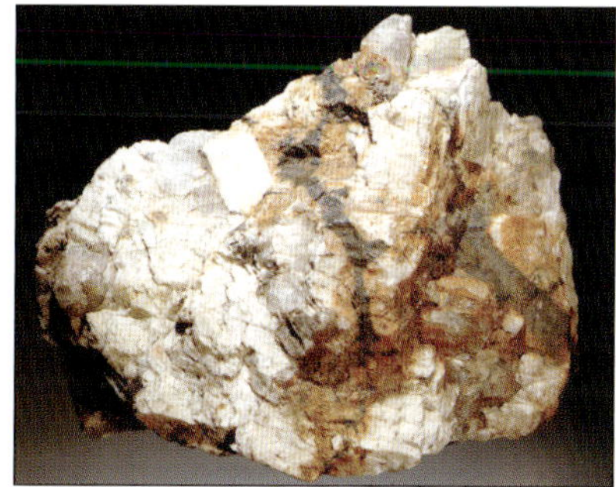

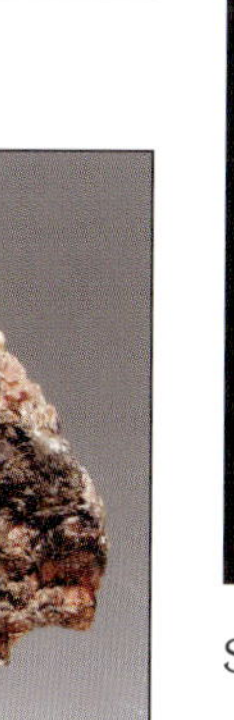

Same under SW

FELDSPAR, CALCITE. Feldspar with calcite from the Andover iron mine, Andover, Sussex County. The close-up photo shows the calcite crystals. The feldspar fluoresces velvety red SW and the calcite fluoresces a green-blue SW and pale blue LW. The piece weighs 1.0 oz. and is 2.1 x 1.8 x 0.4 inches. Value $10-15

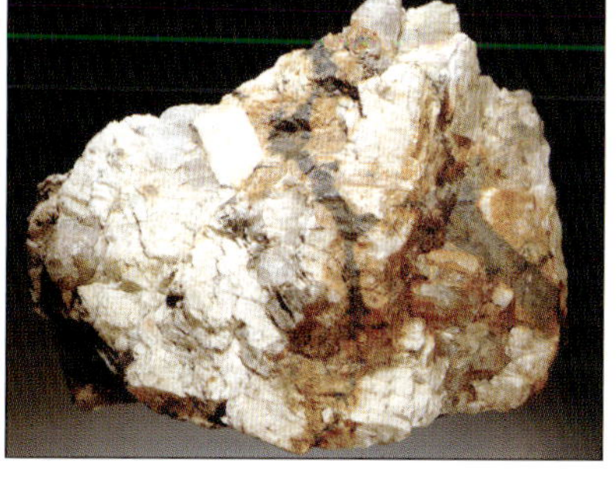

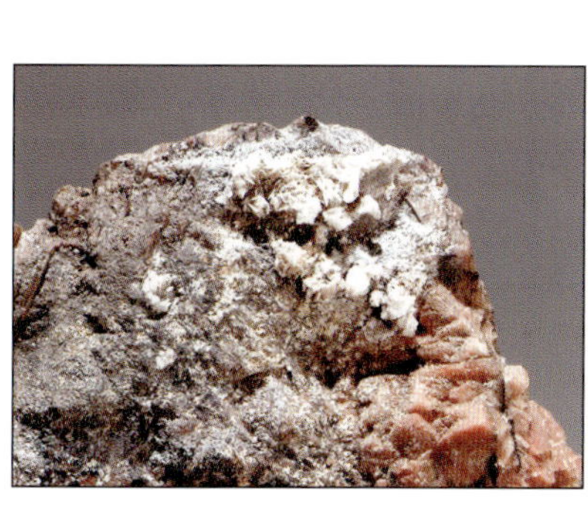

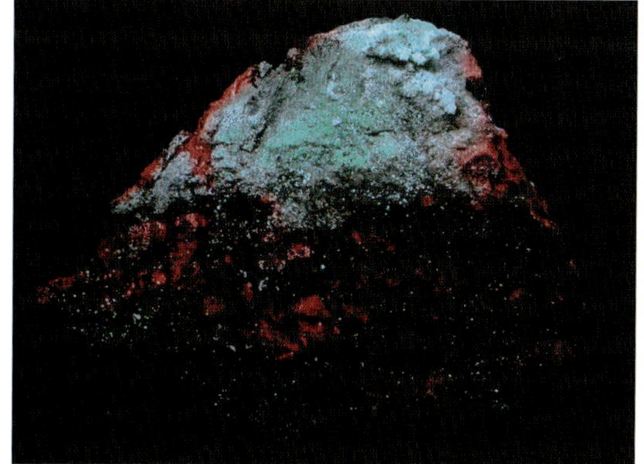

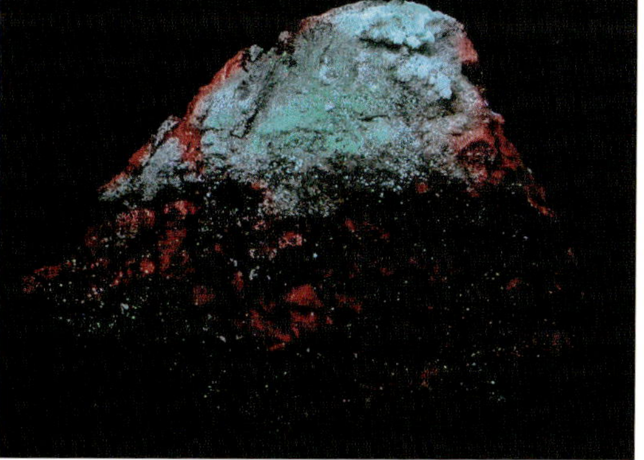

Same under SW

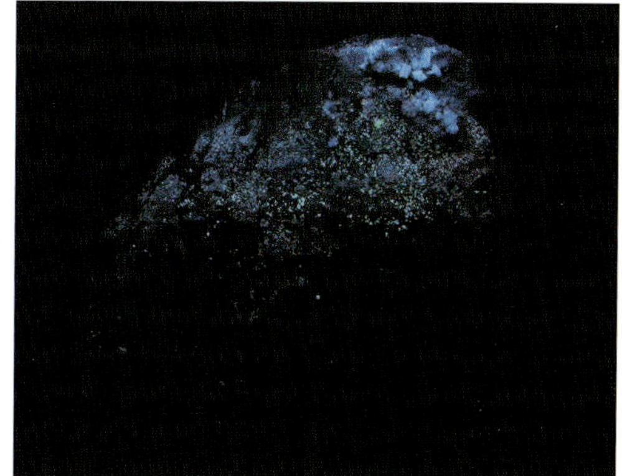

Same under LW

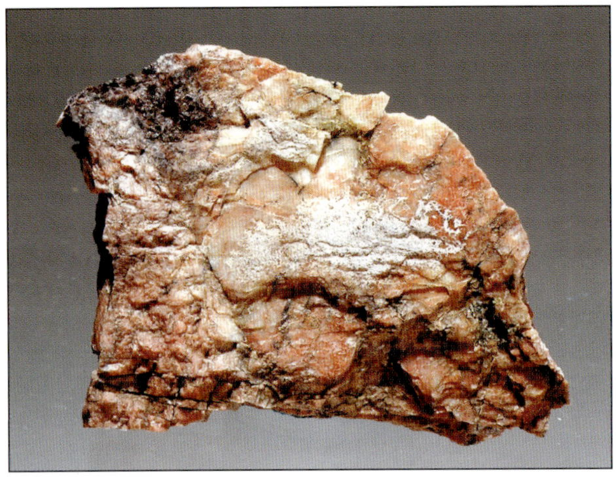

FELDSPAR. Feldspar with calcite from the Andover iron mine, Andover, Sussex County. The feldspar fluoresces velvety red SW and the calcite fluoresces green SW. The piece weighs 1.0 oz. and is 2.0 x 1.3 x 0.4 inches. Value $10-15

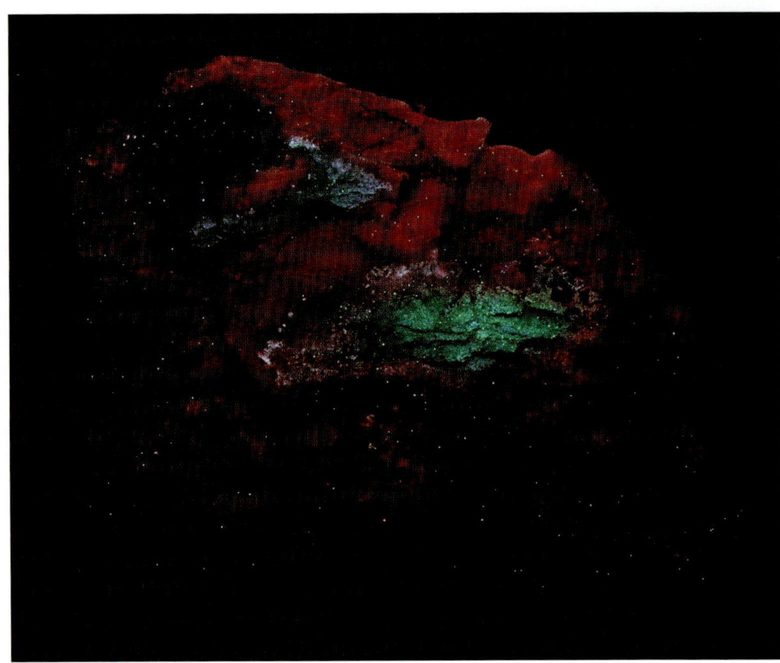

Same under SW

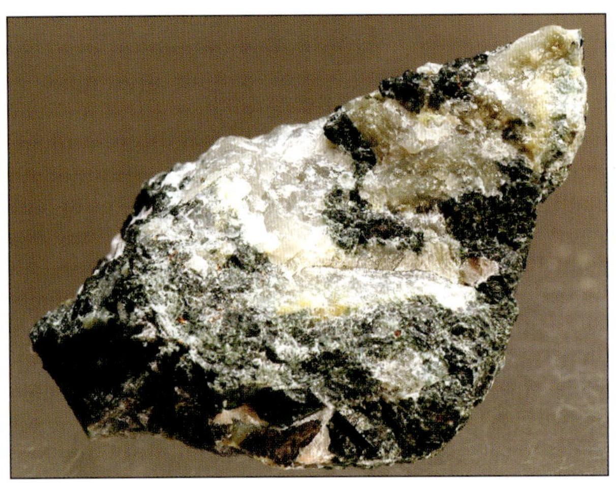

FELDSPAR. Feldspar with calcite from Hamburg quarry, Hamburg, Sussex County, that was found in summer 2005. The feldspar fluoresces velvety red SW and the calcite fluoresces orange-red SW. The piece weighs 2.5 oz. and is 2.3 x 1.6 x 1.6 inches. Value $15-18

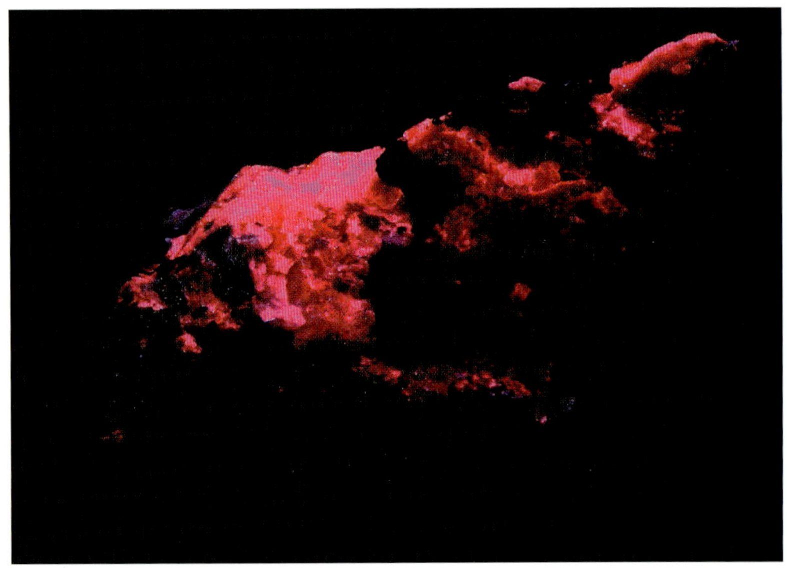

Same under SW

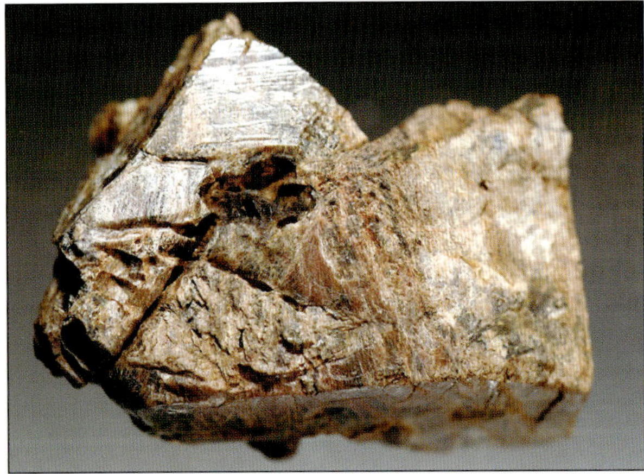

FLUORAPATITE. A weathered fluorapatite crystal from the saddle leading into the Noble pit, Sterling Hill mine, Ogdensburg, Sussex County. It fluoresces peach-orange SW. The piece weighs 4.0 oz. and is 2.0 x 1.5 x 1.5 inches Value $20-25

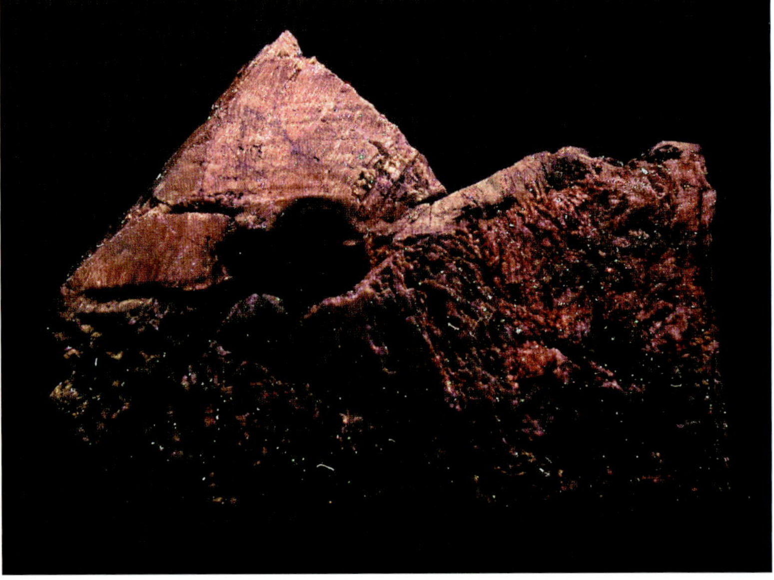

Same under SW

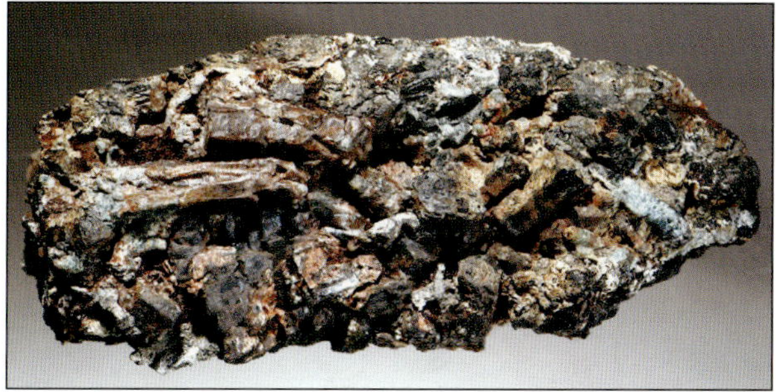

FLUORAPATITE. A group of weathered fluorapatite crystals from the saddle leading into the Noble pit, Sterling Hill mine, Ogdensburg, Sussex County. The fluorapatie fluoresces peach-orange SW. The piece weighs 9.0 oz. and is 4.0 x 1.8 x 1.5 inches. Value $20-25

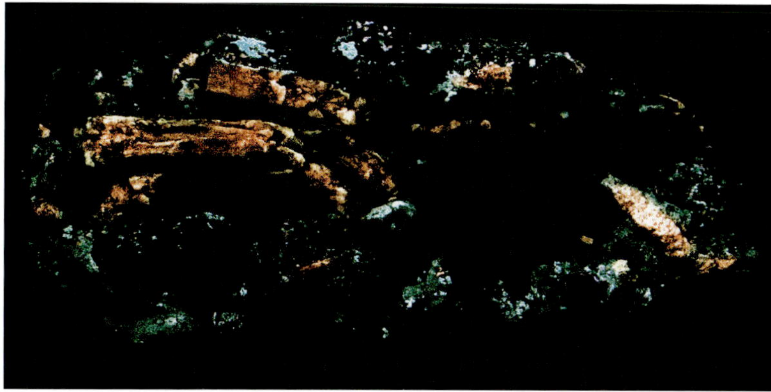

Same under SW

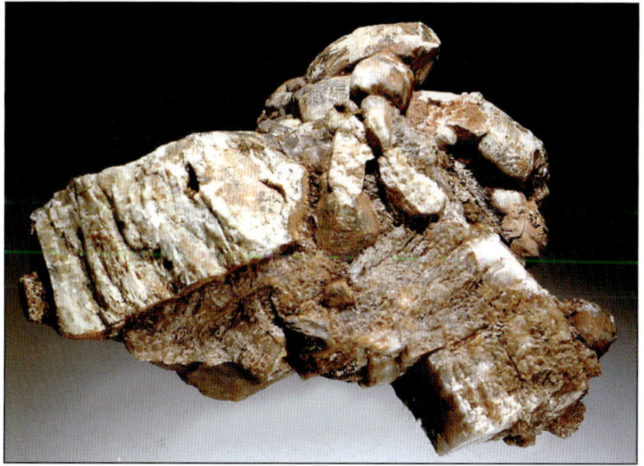

FLUORAPATITE. A large hexagonal crystal of fluorapatite and smaller crystals of fluorapatite with franklinite and calcite from the saddle leading into the Noble pit, Sterling Hill mine, Ogdensburg, Sussex County. You can sometimes find weathered franklinite crystals and weathered fluorapatite in this area. This fluorapatite fluoresces violet SW. The piece weighs 9.0 oz. and is 3.8 x 2.5 x 2.0 inches. Value $50-65

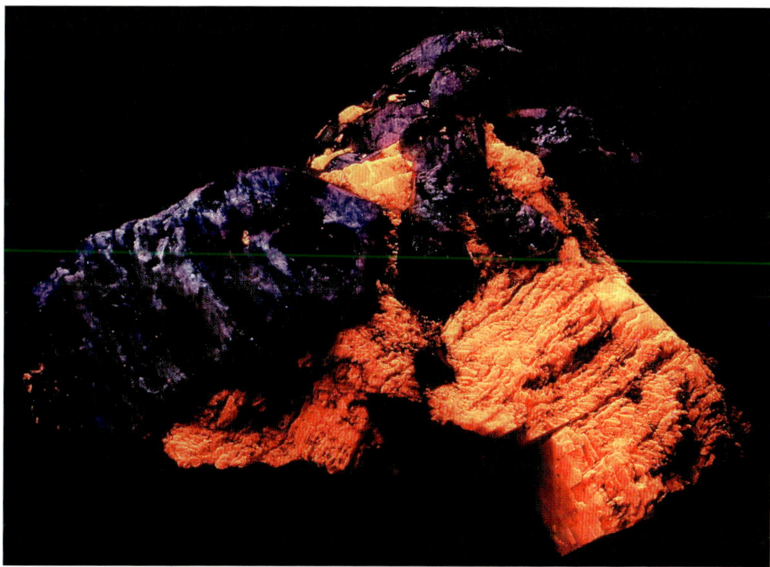

Same under SW

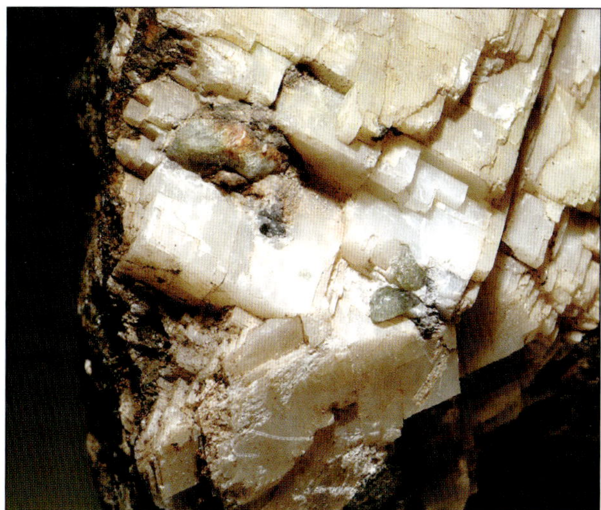

FLUORAPATITE. Small blue-green crystals of fluorapatite in calcite from the Franklin quarry, Franklin, Sussex County. This fluorapatite fluoresces pale blue SW. The piece weighs 16.0 oz. and is 3.5 x 3.0 x 3.0 inches. Value $10-15

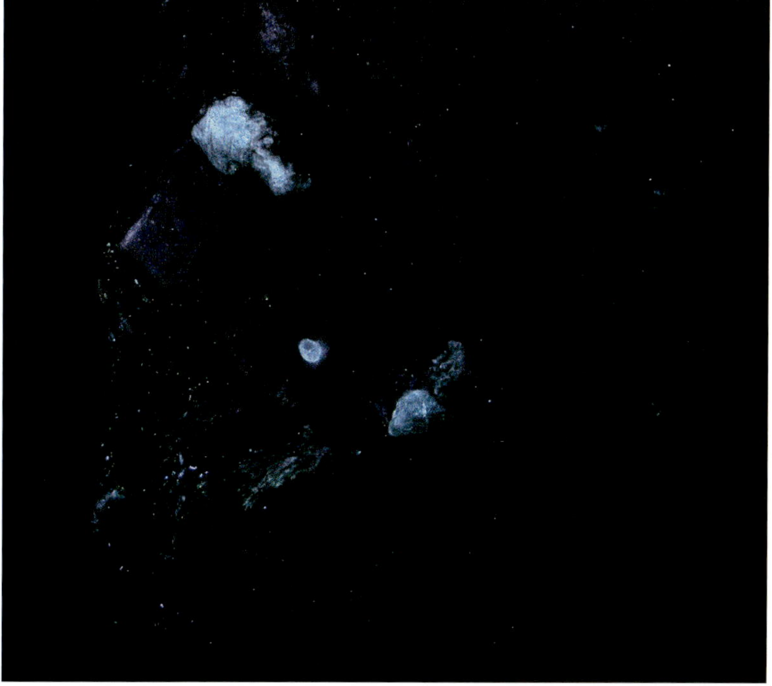

Same under SW

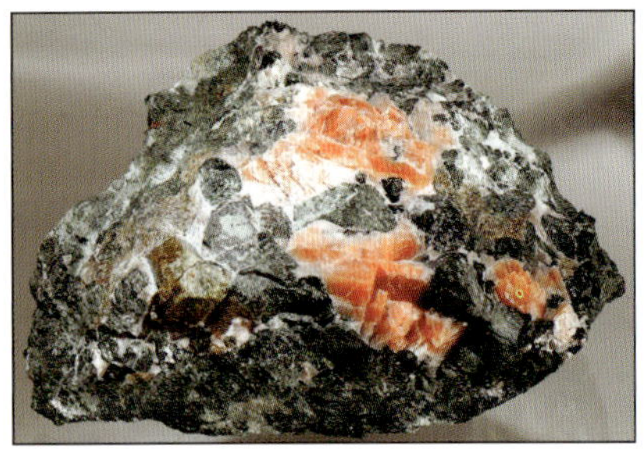

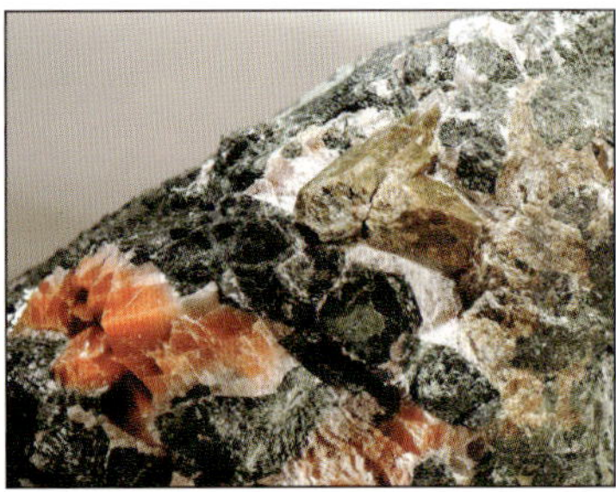

FLUORAPATITE. An attractive specimen with a group of tan fluorapatite crystals, dark green augite crystals, and salmon calcite from the Hamburg quarry, Hamburg, Sussex County. This was found in summer 2005. This fluorapatite fluoresces pale purple MW, less brightly SW. The calcite fluoresces orange-red MW. The piece weighs 8 lbs and is 7.5 x 6.0 x 3.0 inches. Value $65-70

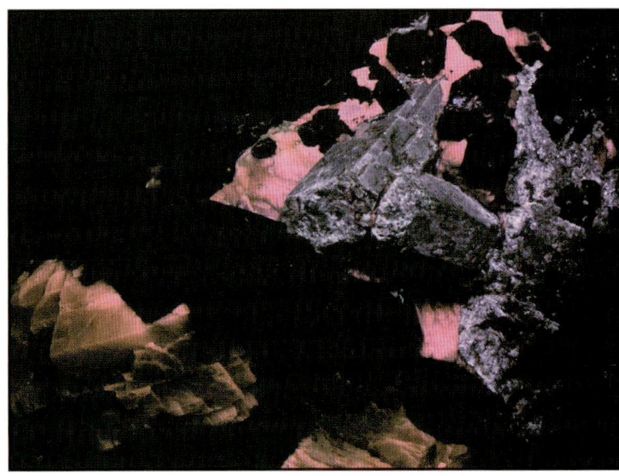

Same under MW

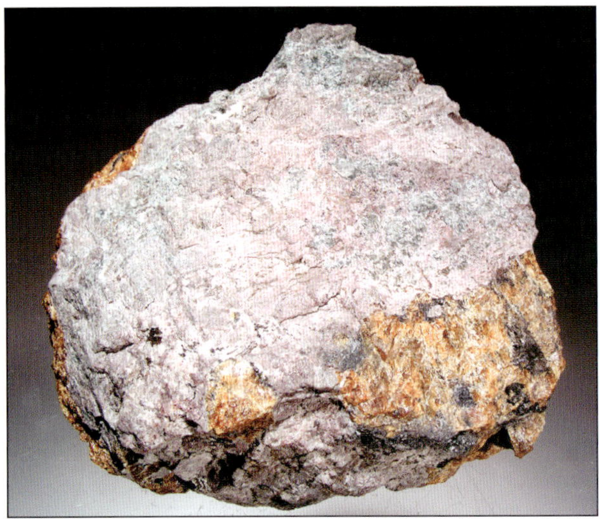

FLUORAPATITE. Massive fluorapatite from the Mill site, Franklin, Sussex County. The fluorapatite fluoresces bright orange SW. The piece weighs 4.6 oz. and is 2.5 x 2.4 x 1.0 inches. Value $30-35

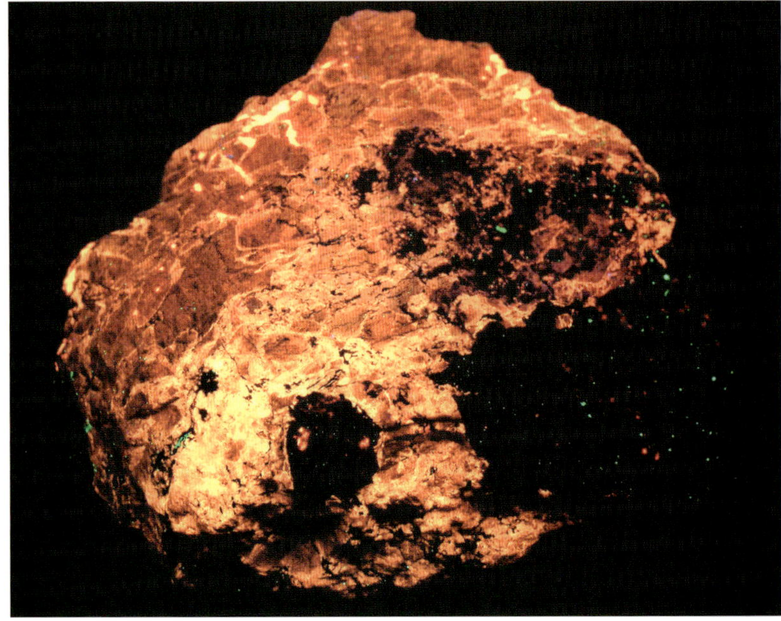

Same under SW

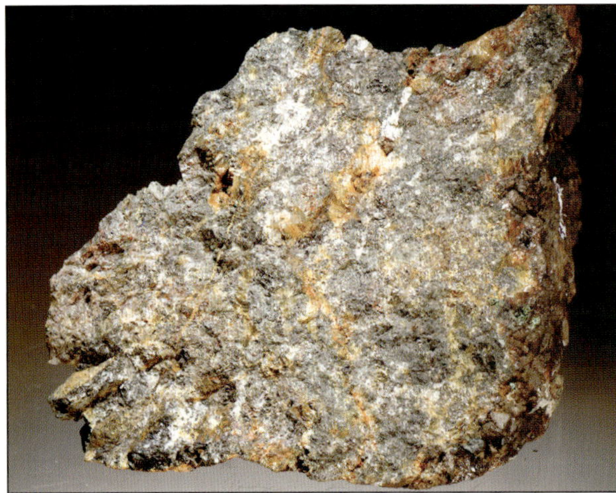

FLUORAPATITE. Fluorapatite and chlorophane (a varietal name for green-fluorescing, thermoluminescent fluorite) with magnetite from the Edison mine, Ogdensburg, Sussex County. The fluorapatite fluoresces orange-yellow SW, the chlorophane fluoresces pale blue SW and LW with a long-lasting phosphorescence. The piece weighs 9.0 oz. and is 4.0 x 3.0 x 1.3 inches. Value $25-30

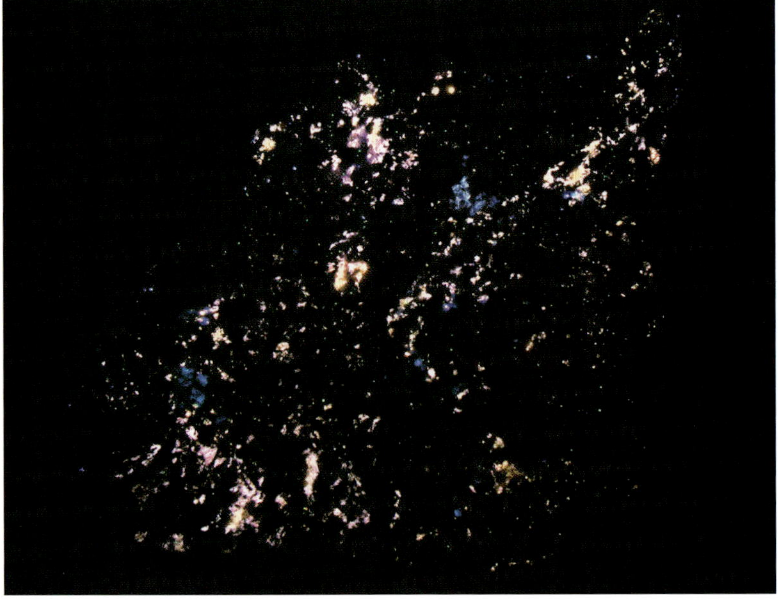

Same under SW

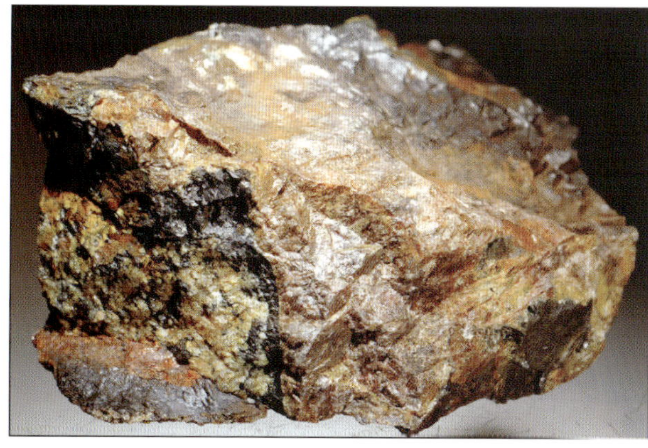

FLUORAPATITE, MICROCLINE. Fluorapatite with microcline and magnetite from the Edison mine, Ogdensburg, Sussex County. The fluorapatite fluoresces orange-yellow SW and the microcline fluoresces violet SW. The piece weighs 1 lb. 11.0 oz. and is 4.5 x 3.0 x 2.0 inches. Courtesy of Philip Persson. Value $35-40

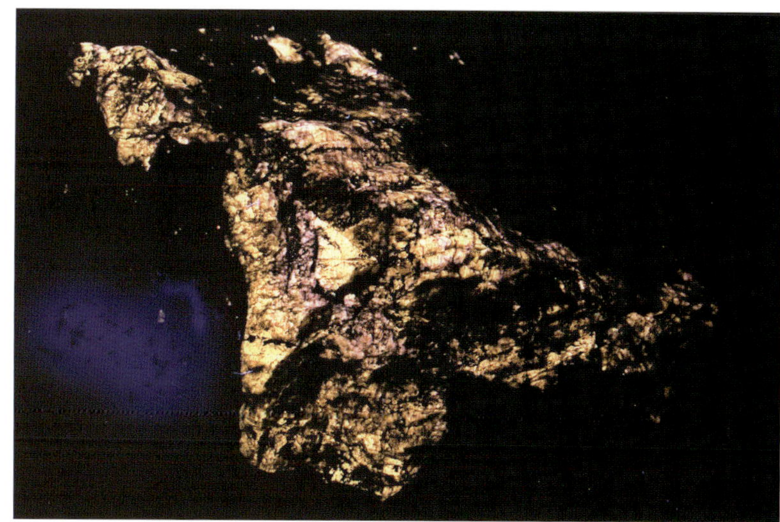

Same under SW

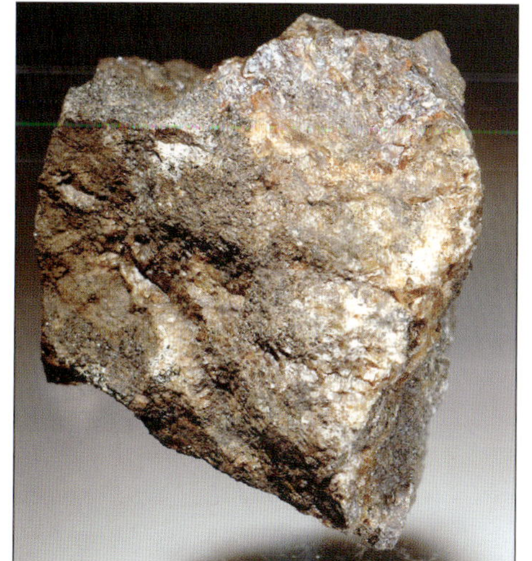

FLUORITE, FLUORAPATITE. An interesting piece with brown chlorophane (a varietal name for green-fluorescing, thermoluminescent fluorite), purple fluorite, and fluorapatite from the Taylor Road dump, Franklin, Sussex County. Chlorophane loses its phosphorescence and green-blue fluorescence on exposure to light. To preserve the green-blue fluorescence, phosphorescence, and triboluminescence, it is necessary to keep the specimens away from light. Chlorophane fluoresces green-blue SW and LW and has a long-lasting phosphorescence, fluorite fluoresces violet LW and fluorapatite fluoresces orange SW. The piece weighs 1 lb. 3.0 oz. and is 3.3 x 2.3 x 2.3 inches. Value $50-60

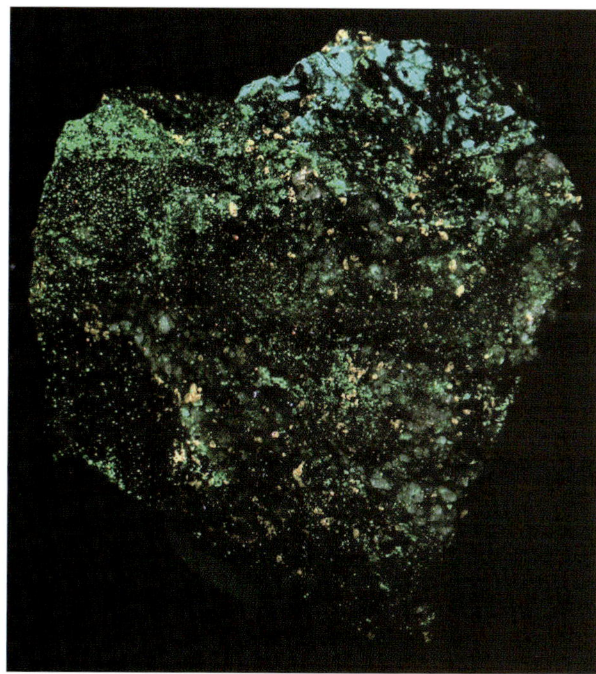

Same under SW

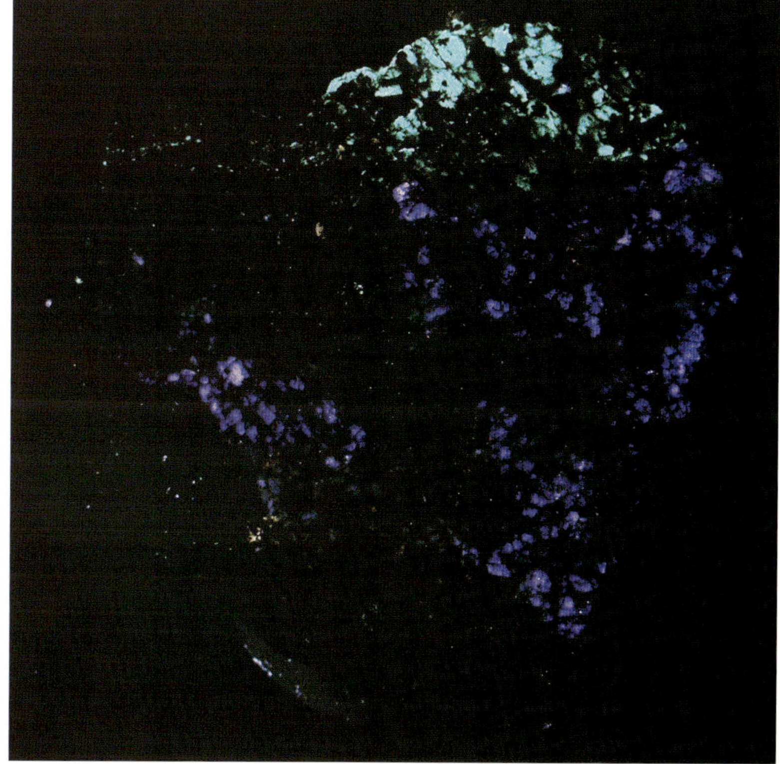

Same under LW

FLUORITE, FLUORAPATITE. Purple fluorite, brown chlorophane (a varietal name for green-fluorescing, thermoluminescent fluorite), willemite, and fluorapatite from the Taylor Road dump, Franklin, Sussex County. The chlorophane fluoresces green-blue SW and LW and has a long-lasting phosphorescence. The fluorite fluoresces violet LW, fluorapatite fluoresces orange SW and willemite fluoresces green SW. The piece weighs 1 lb. 11.0 oz. and is 4.0 x 4.0 x 2.0 inches. Value $50-60

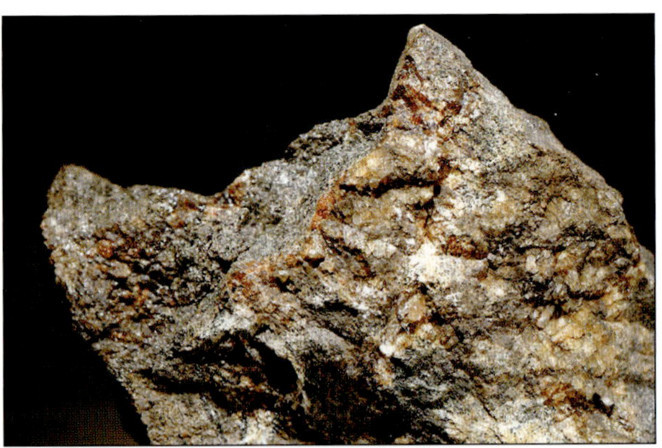

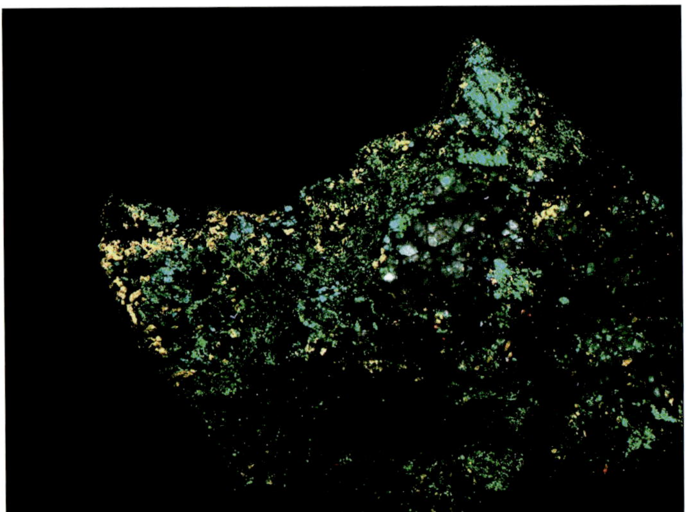

Same under SW

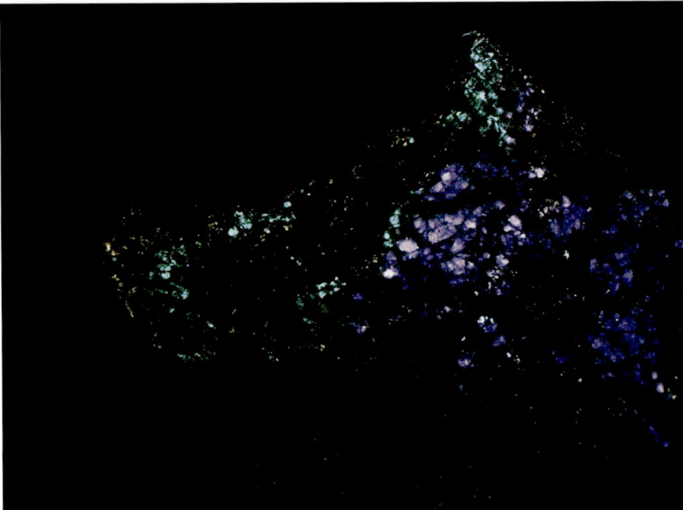

Same under LW

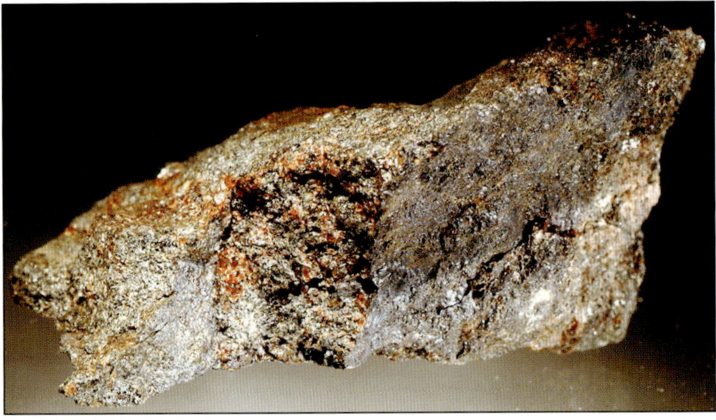

FLUORITE, FLUORAPATITE. Chlorophane (a varietal name for green-fluorescing, thermoluminescent fluorite) with fluorapatite from the Taylor Road dump, Franklin, Sussex County. Chlorophane fluoresces green-blue SW and LW and has a long-lasting phosphorescence and fluorapatite fluoresces orange SW. The piece weighs 1 lb. 11.0 oz. and is 4.0 x 4.0 x 2.0 inches. Value $50-55

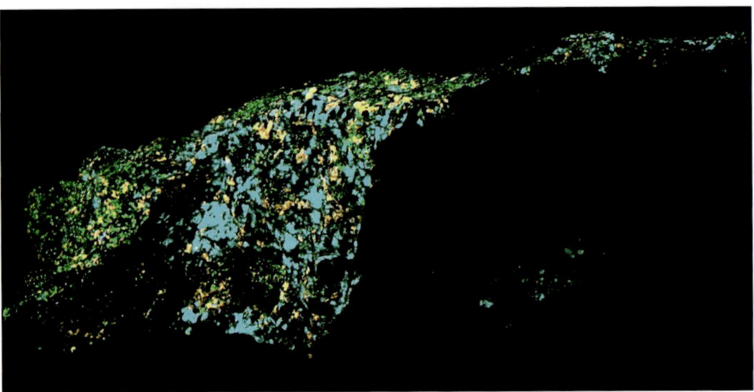

Same under SW

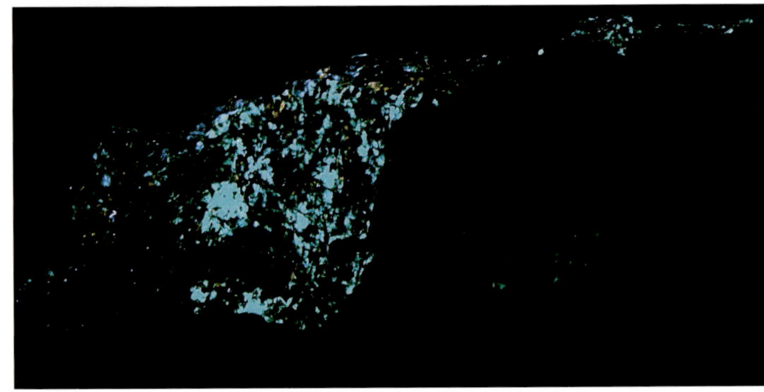

Same under LW

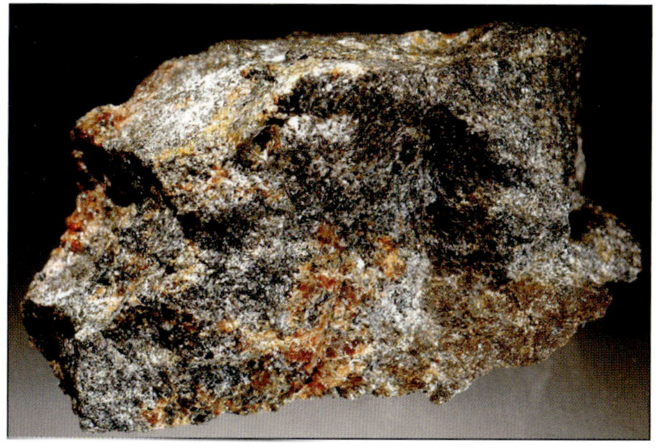

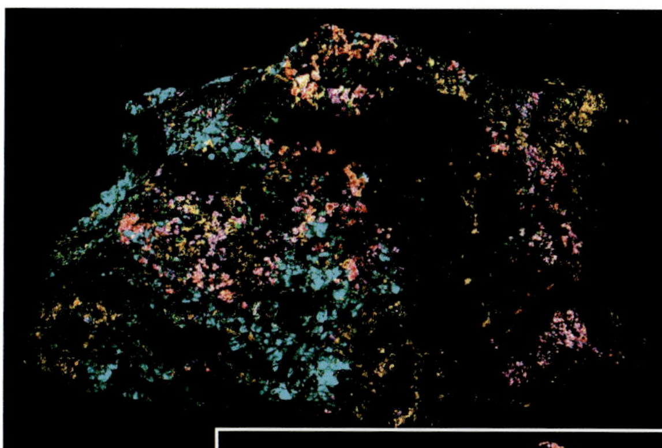

Same under SW

FLUORITE, SPHALERITE, FLUORAPATITE. Chlorophane (a varietal name for green-fluorescing, thermoluminescent fluorite) with sphalerite, willemite, and fluorapatite from the Taylor Road dump, Franklin, Sussex County. Chlorophane fluoresces green-blue SW and LW and has a long-lasting phosphorescence, fluorapatite fluoresces orange SW. Sphalerite fluoresces orange and blue LW and willemite fluoresces green SW. The piece weighs 2 lb. 2.0 oz. and is 4.0 x 3.0 x 2.0 inches. Value $75-90

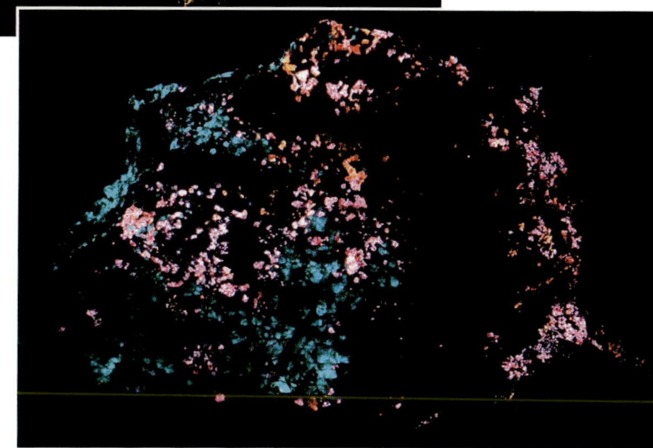

Same under LW

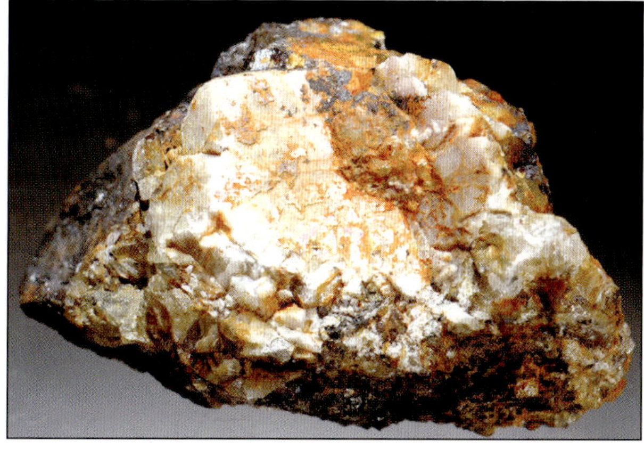

FLUORITE. Fluorite with magnetite from the Edison mine, Ogdensburg, Sussex County. The fluorite fluoresces white SW. The piece weighs 4.3 oz. and is 2.5 x 1.8 x 1.3 inches. Value $25-30

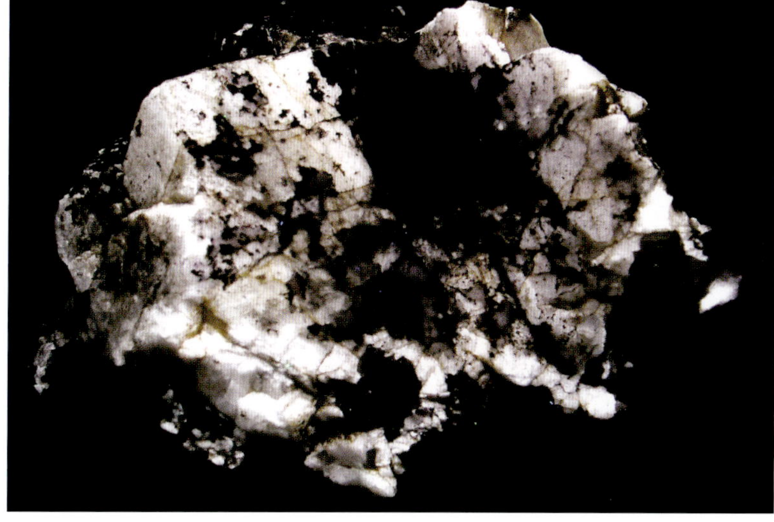

Same under SW

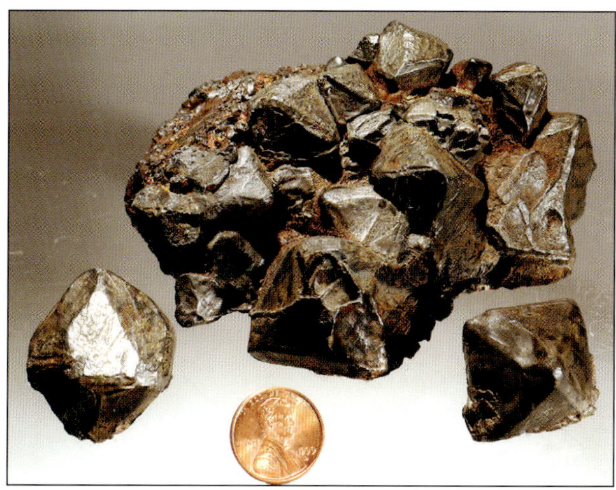

FRANKLINITE. A plate of octahedral crystals of franklinite from the saddle leading into the Noble pit, Sterling Hill mine, Ogdensburg, Sussex County. The piece weighs 1 lb. 6.0 oz. and is 4.0 x 3.0 x 2.0 inches. Value $50-55

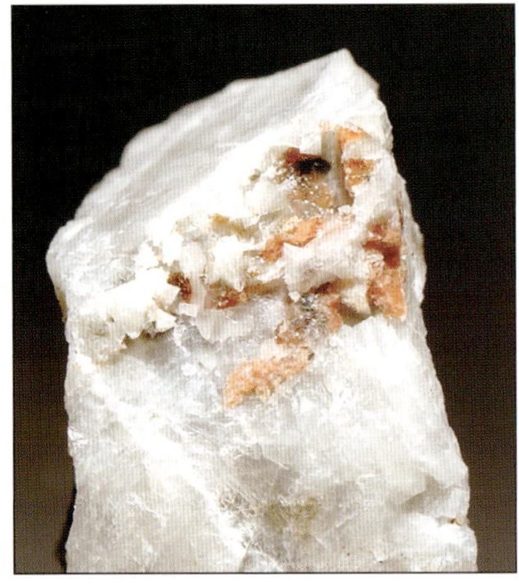

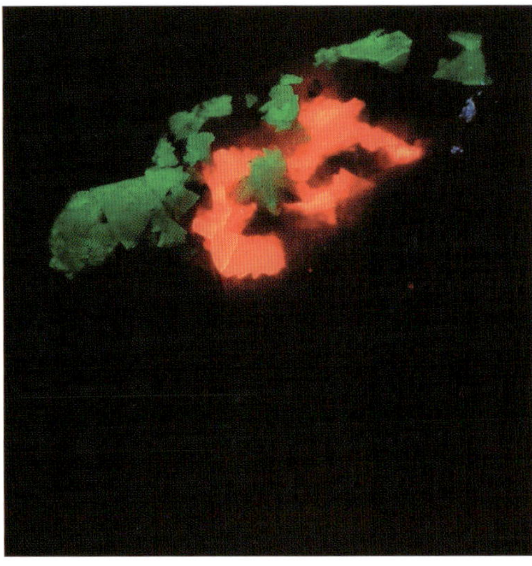

Same under SW

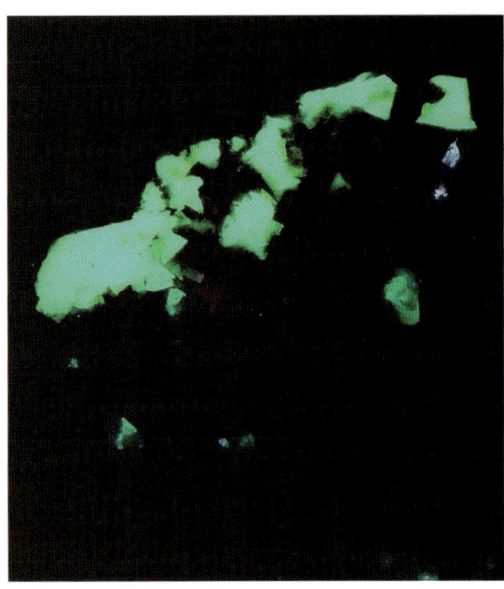

Same under LW

GENTHELVITE. Tan genthelvite crystals from the Sterling Hill mine, Ogdensburg, Sussex County. Most of the genthelvite found at Sterling Hill had no discernible crystalline shape. This small piece was etched in weak hydrochloric acid. The acid removed the surrounding calcite and revealed the tiny tetrahedral crystals of genthelvite. Genthelvite fluoresces lime-green LW and less brightly SW. The piece weighs 0.8 oz. and is 1.5 x 1.3 x 0.6 inches. Value $80-100

HARDYSTONITE, BARITE. An attractive piece with hardystonite, barite, willemite, calcite, and franklinite from the Franklin mine, Franklin, Sussex County. Hardystonite fluoresces purple-blue SW, barite fluoresces white SW, calcite fluoresces orange-red SW, and willemite fluoresces green SW. The piece weighs 8.0 oz. and is 3.0 x 2.0 x 2.0 inches. Courtesy of Dru Wilbur.

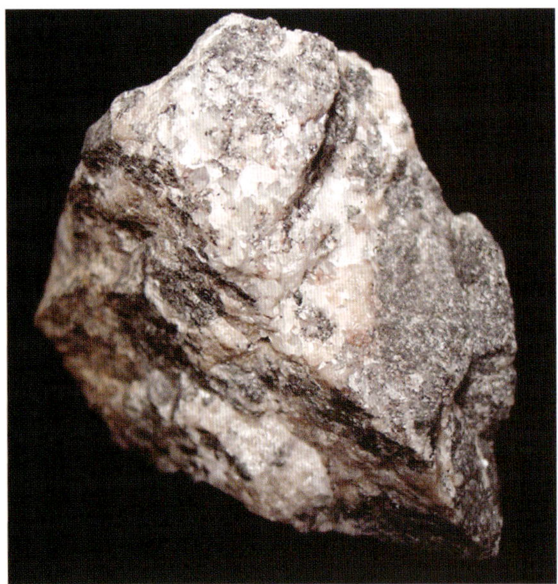

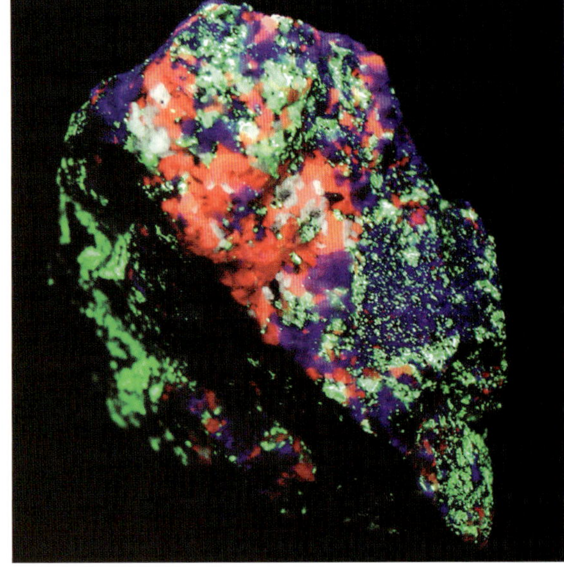

Same under SW

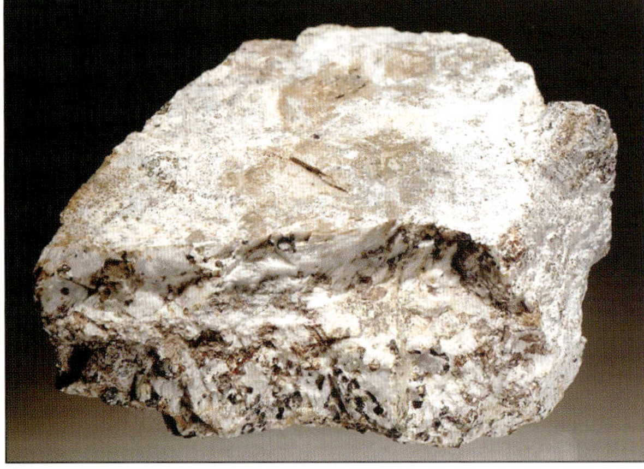

HYDROZINCITE. Hydrozincite on calcite from the May 2004 exploration of the west wall of the sand lot at the Passaic pit, Sterling Hill mine, Ogdensburg, Sussex County. Hydrozincite fluoresces bright pale blue SW and calcite fluoresces orange-red SW. The piece weighs 15.0 oz. and is 4.0 x 3.0 x 1.5 inches. Value $40-45

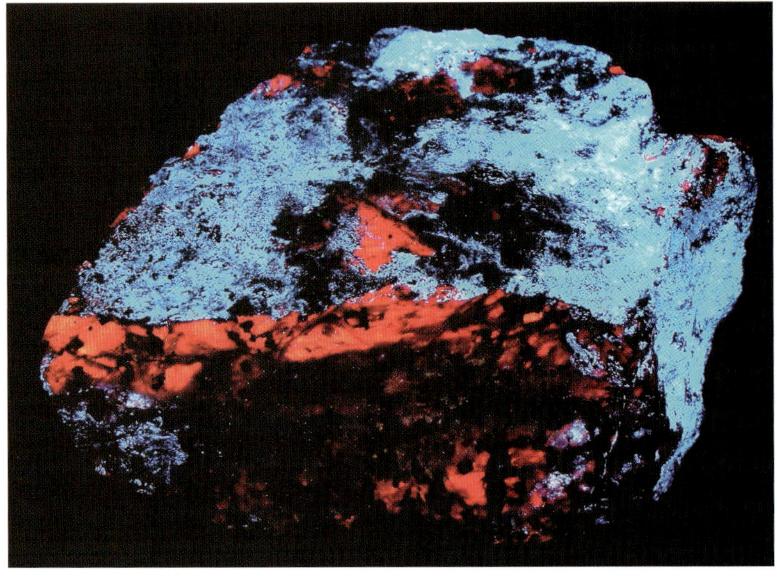

Same under SW

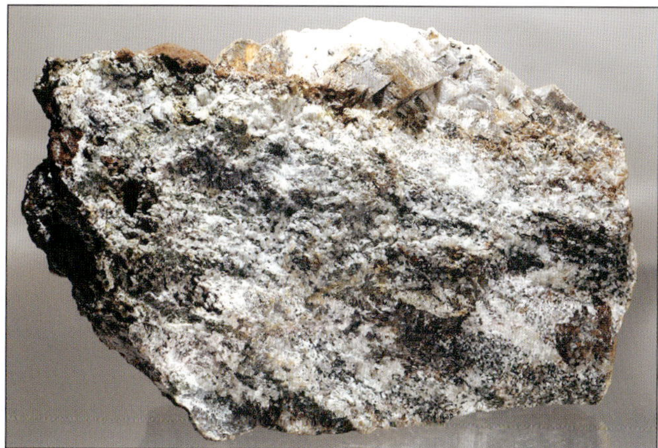
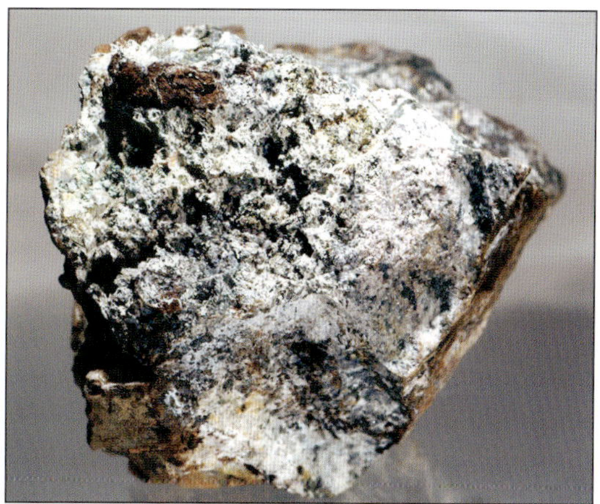

HYDROZINCITE, MALACHITE. Hydrozincite, sphalerite, calcite, malachite, galena, and cerussite from the Sterling Hill mine, Ogdensburg, Sussex County. Material was removed from the wall at the SE corner of the Passaic pit leaving boulders at the base of the wall. Exploration of these boulders during the summer of 2005 yielded large gahnite crystals, salmon-colored calcite, and very brightly fluorescing hydrozincite. One broken boulder had a vug coated in malachite and hydrozincite. There was also galena and sphalerite on the piece. The sphalerite fluoresces orange LW, the cerussite fluoresces yellow-cream LW, the calcite fluoresces orange-red SW, and the hydrozincite fluoresces bright pale blue SW. The piece weighs 2 lb. 15.0 oz. and is 5.5 x 3.5 x 3.5 inches. Value $75-85

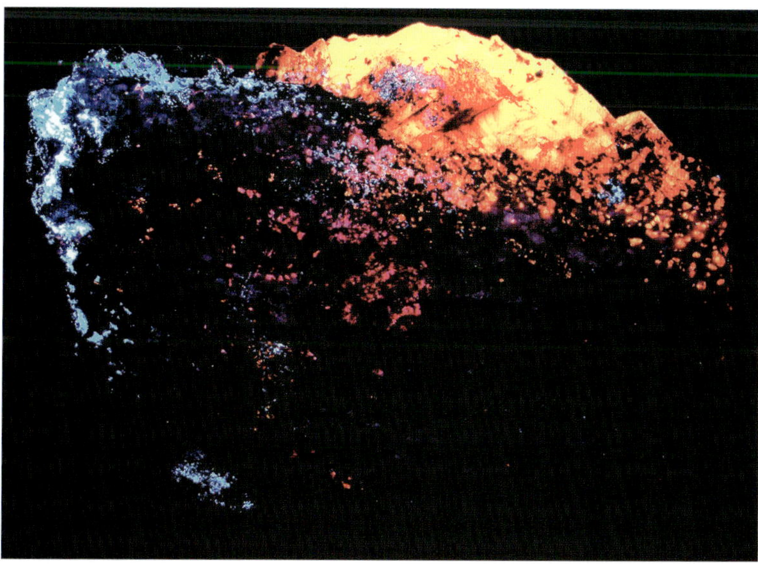
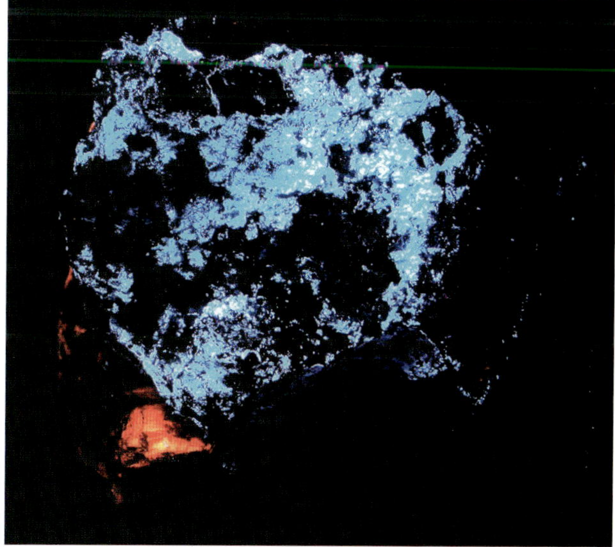

Same under SW

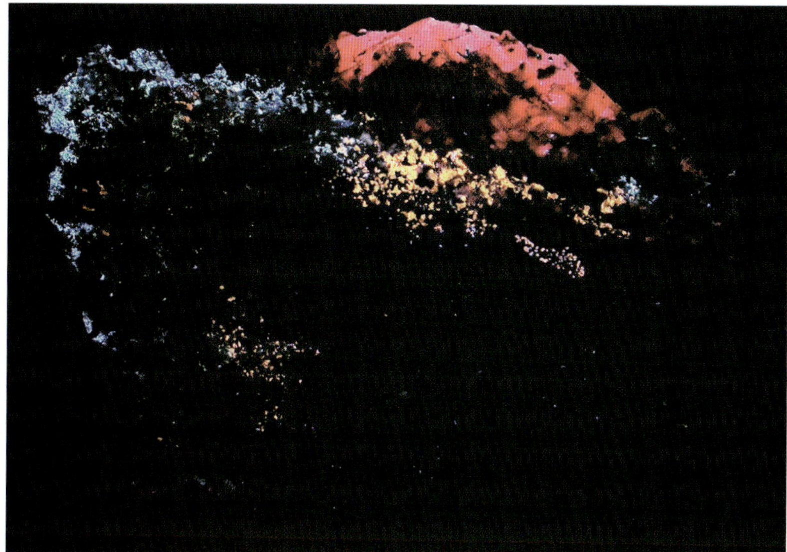
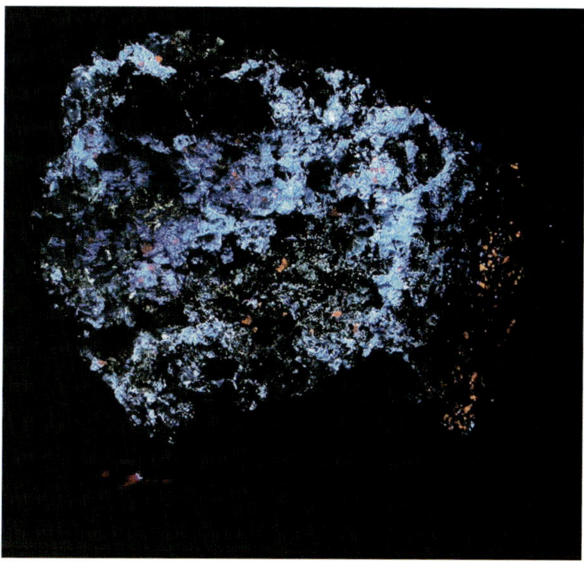

Same under LW

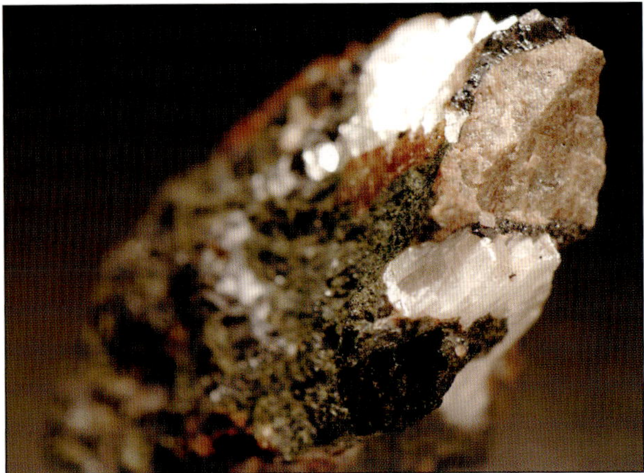

JOHNBAUMITE. Johnbaumite on calcite and andradite from Franklin, Sussex County. The johnbaumite fluoresces orange SW and the calcite fluoresces orange-red SW. The piece weighs 1.0 oz. and is 1.8 x 0.9 x 0.4 inches. Value $75-85

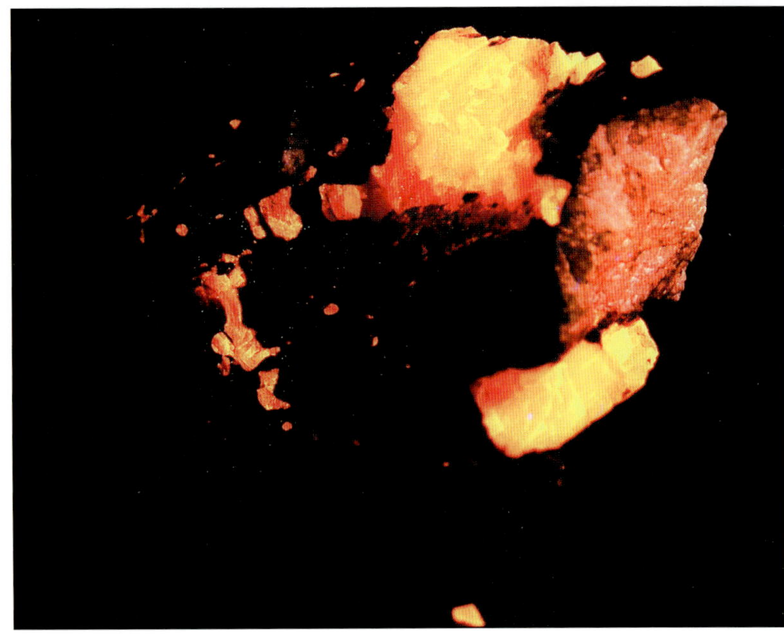

Same under SW

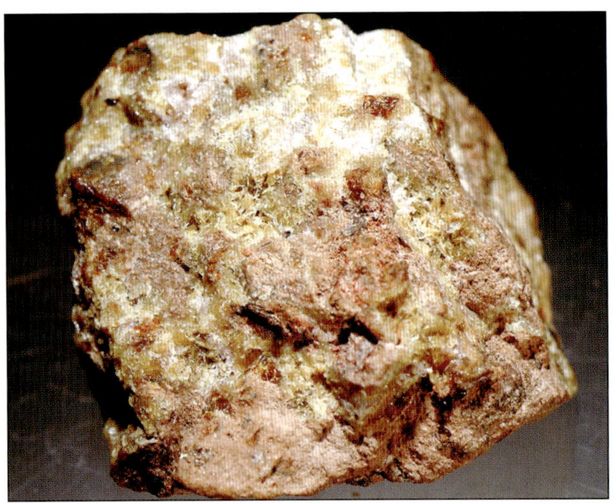

MANGANAXINITE. Manganaxinite crystals, clinohedrite, willemite, and prehnite on hancockite from Franklin, Sussex County. The clinohedrite fluoresces bright orange SW the willemite fluoresces green SW, the prehnite fluoresces peach with a bluish tint SW, and the manganaxinite crystals do not fluoresce. The piece weighs 3.0 oz. and is 2.0 x 1.8 x 1.0 inches. Value $80-100

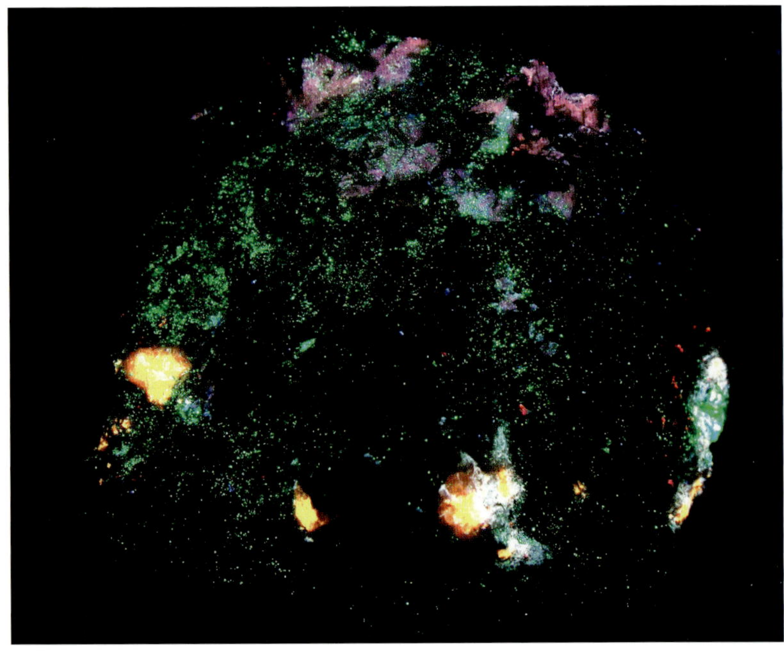

Same under SW

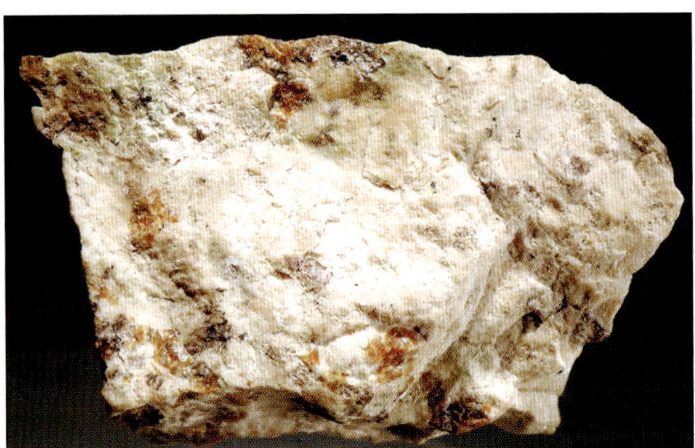

MANGANAXINITE. Manganaxinite with clinohedrite and willemite from Franklin, Sussex County. The manganaxinite fluoresces red SW, the clinohedrite fluoresces bright orange SW, and the willemite fluoresces green SW. The piece weighs 6.8 oz. and is 3.0 x 2.0 x 1.1 inches. Value $225-250

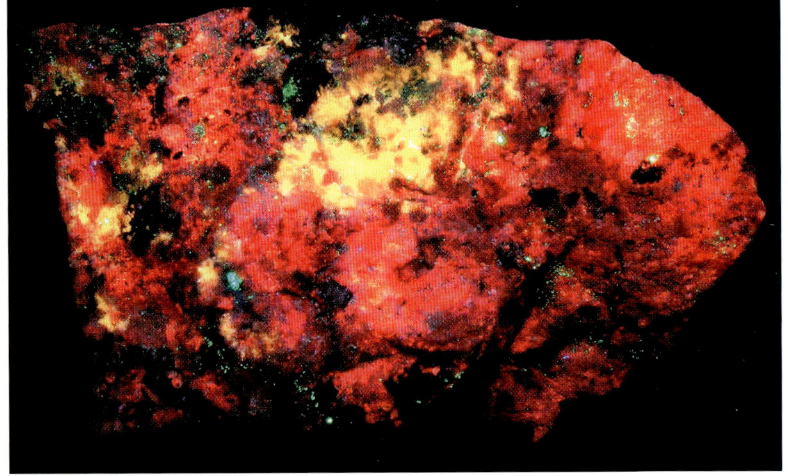

Same under SW

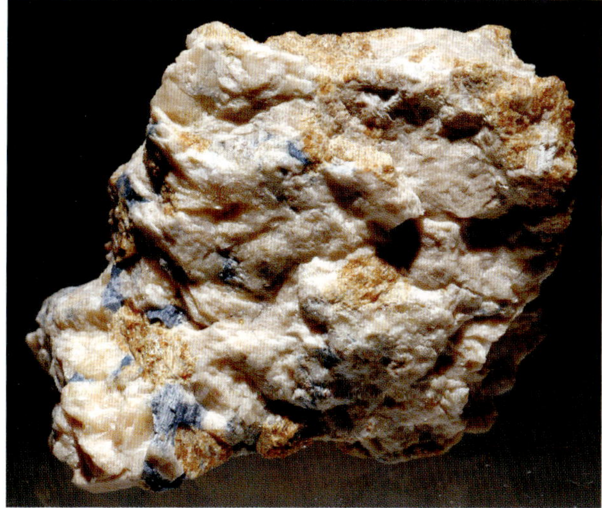

MANGANAXINITE. Manganaxinite with blue vesuvianite, var. cyprine, and willemite from the Franklin mine, Franklin, Sussex County. The manganaxinite fluoresces red SW and the willemite fluoresces green SW. The piece weighs 5.0 oz. and is 2.3 x 1.8 x 1.4 inches. Value $200-230

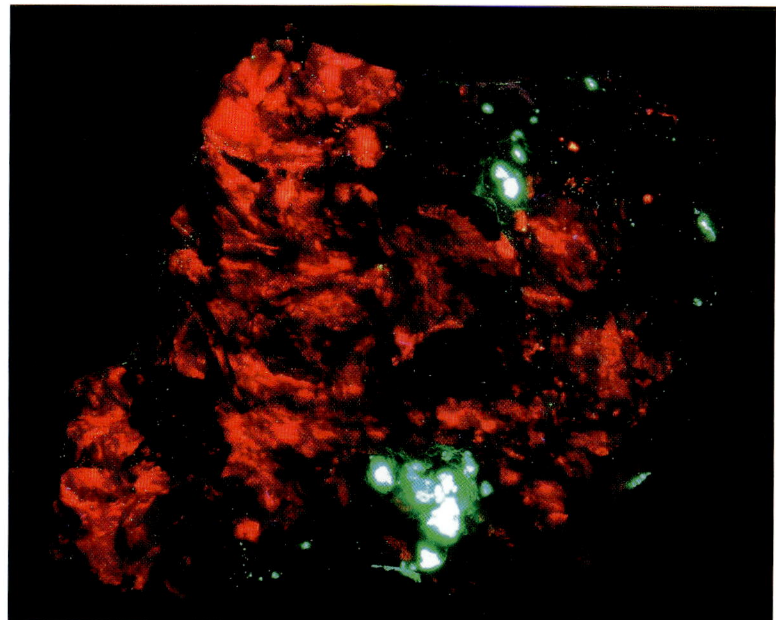

Same under SW

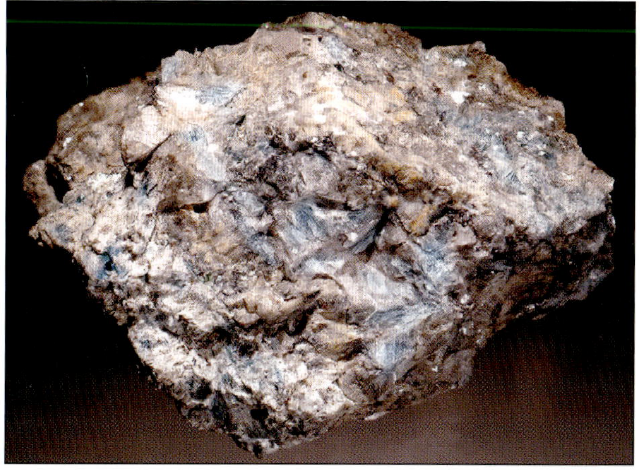

MARGARITE. Margarite is a rare fluorescent mineral from Franklin. A small group of margarite plates from the Franklin quarry, Franklin, Sussex County. Margarite is a blue-green brittle mica that was sparsely found in a 75-foot-wide area of this quarry. Margarite fluoresces white SW and LW. The piece weighs 6.0 oz. and is 2.8 x 2.3 x 1.3 inches. Value $200-225

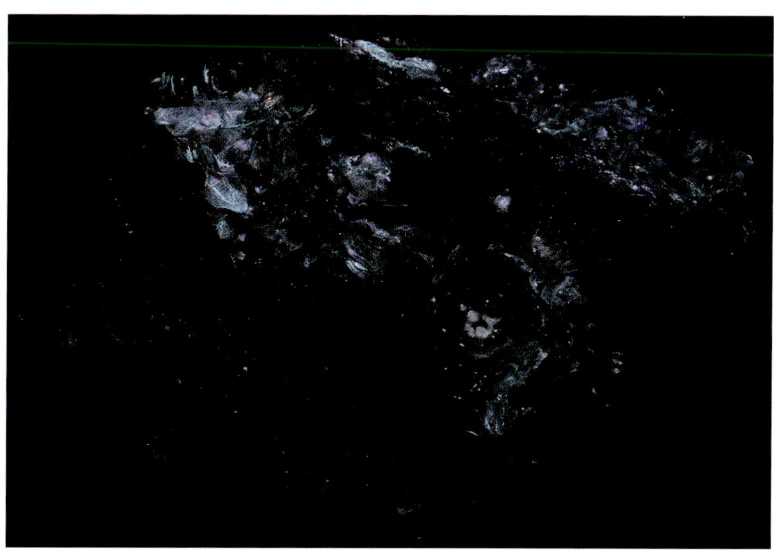

Same under SW

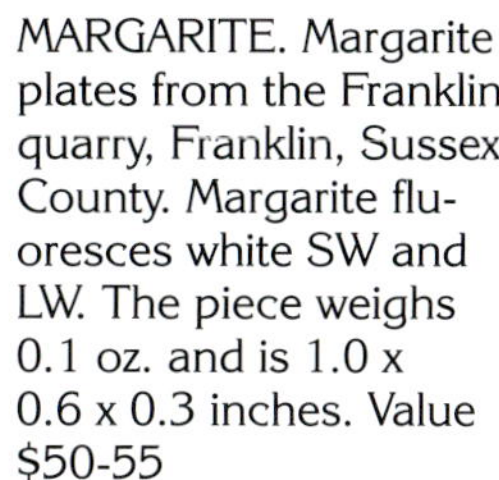

MARGARITE. Margarite plates from the Franklin quarry, Franklin, Sussex County. Margarite fluoresces white SW and LW. The piece weighs 0.1 oz. and is 1.0 x 0.6 x 0.3 inches. Value $50-55

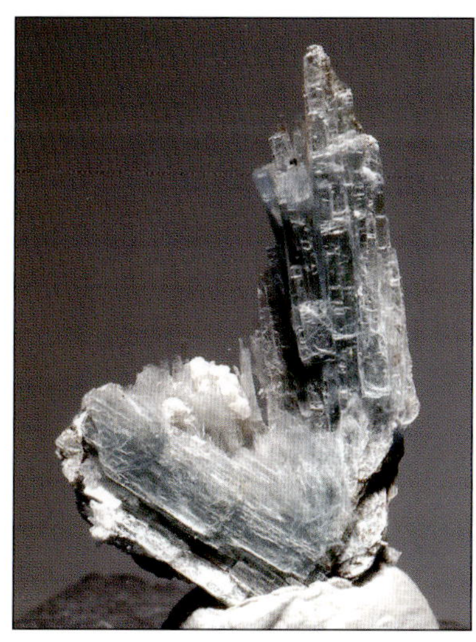

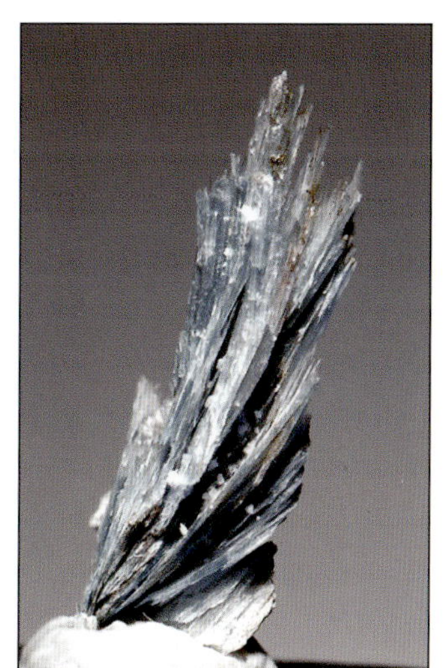

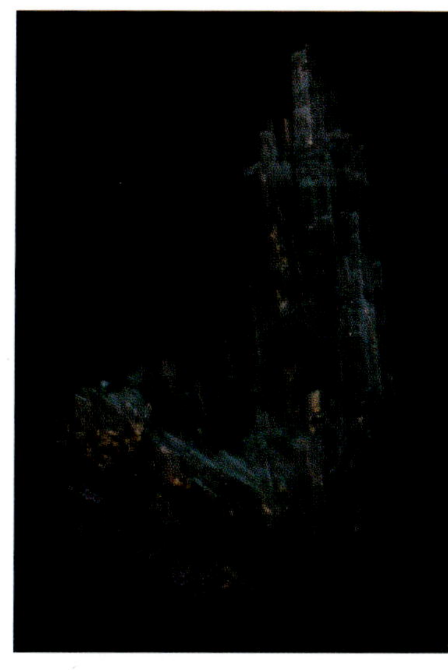

Same under SW

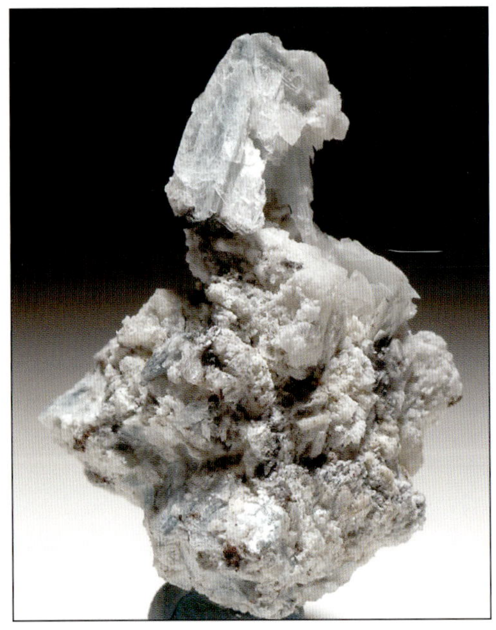

MARGARITE. Margarite plates on matrix containing phlogopite, rutile, and calcite from the Franklin quarry, Franklin, Sussex County. Margarite fluoresces white SW and LW. The piece weighs 0.2 oz. and is 1.3 x 0.8 x 0.5 inches. Value $35-40

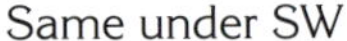

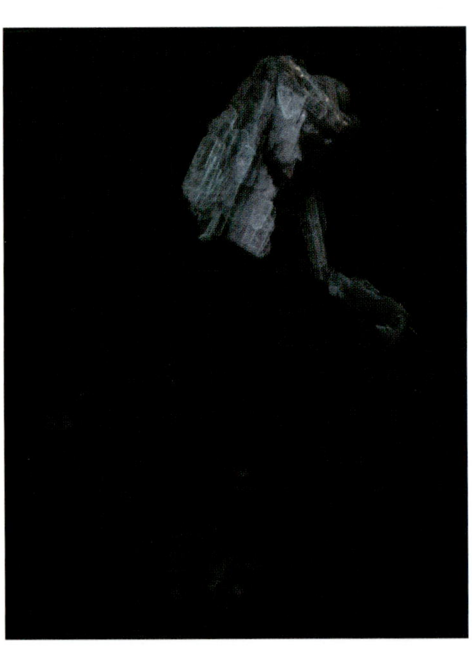

Same under SW

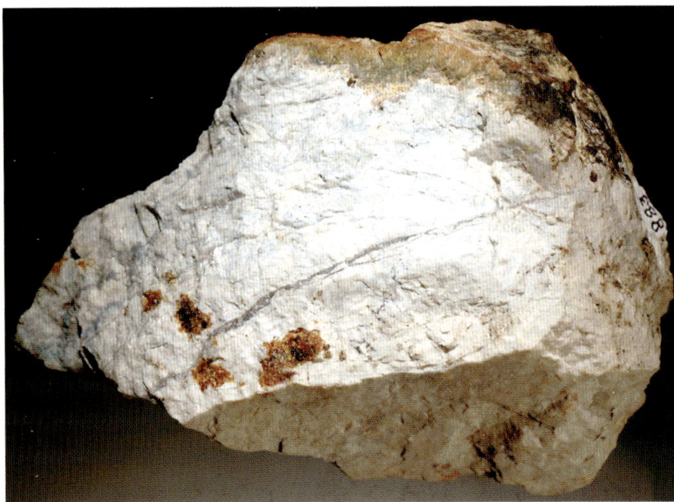

MARGAROSANITE. An exceptionally attractive specimen with margarosanite and willemite from Franklin, Sussex County. The margarosanite fluoresces bright pale blue and red SW and the willemite fluoresces green SW. The piece weighs 15.5 oz. and is 4.5 x 3.0 x 1.6 inches. Value $2,250-2,500

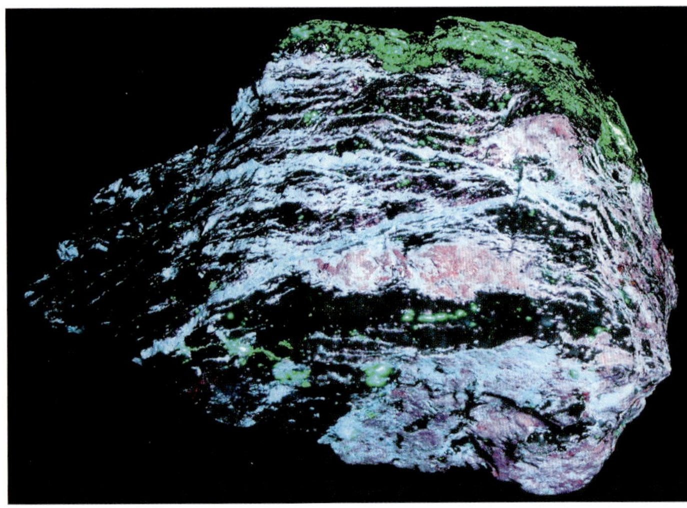

Same under SW

MARGAROSANITE. Platy margarosanite with willemite from Franklin, Sussex County. This piece is almost solid margarosanite with platy areas. The margarosanite fluoresces bright pale blue and red SW and the willemite fluoresces green SW. The piece weighs 8.5 oz. and is 2.5 x 2 x 1.8 inches. Value $3,500-4,000

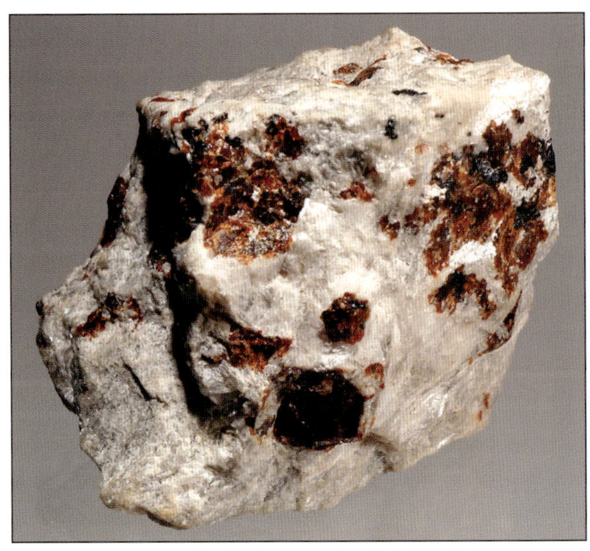

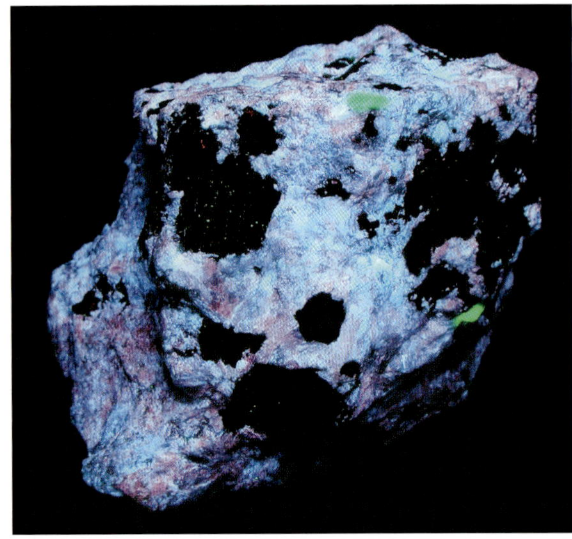

Same under SW

MARGAROSANITE. Margarosanite and pectolite with willemite from Franklin, Sussex County. The margarosanite fluoresces bright pale blue SW, the pectolite fluoresces orange SW, and the willemite fluoresces green SW. Both sides are shown. The piece weighs 2.3 oz. and is 3.0 x 1.5 x 1.0 inches. Value $300-400

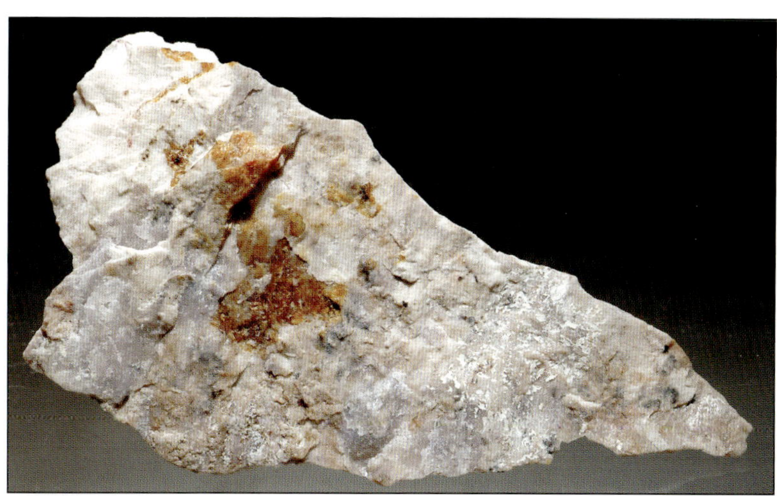

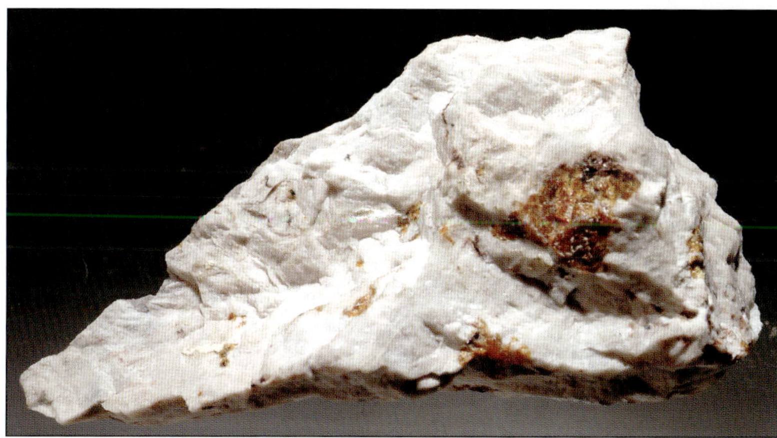

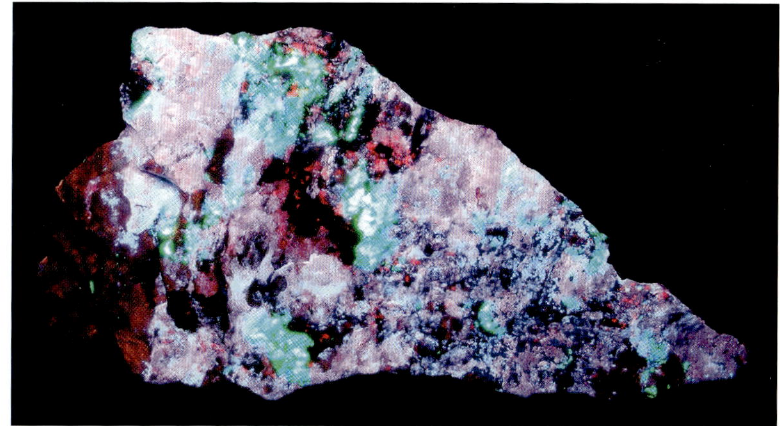

Same under SW

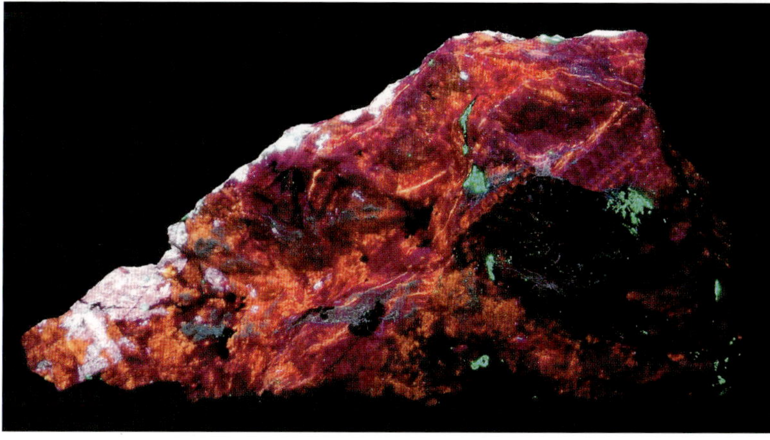

Same under SW

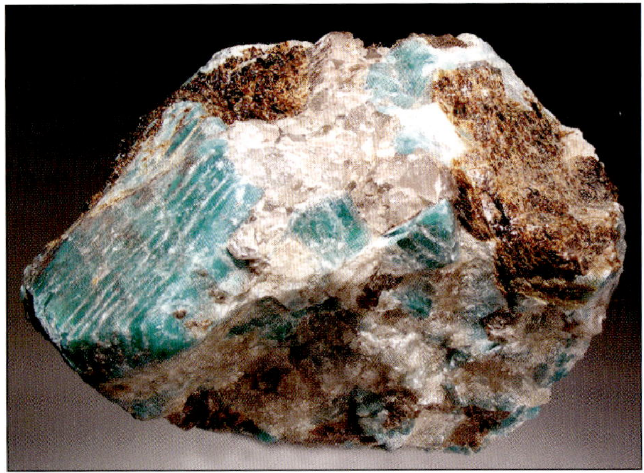

MICROCLINE. Microcline, var. amazonite, willemite, and calcite from the Mill site, Franklin, Sussex County. This microcline, var. amazonite, fluoresces pinkish red SW. The calcite fluoresces orange-red SW and the willemite fluoresces green SW. The piece weighs 3.5 oz. and is 3.5 x 1.5 x 1.0 inches. Value $20-25

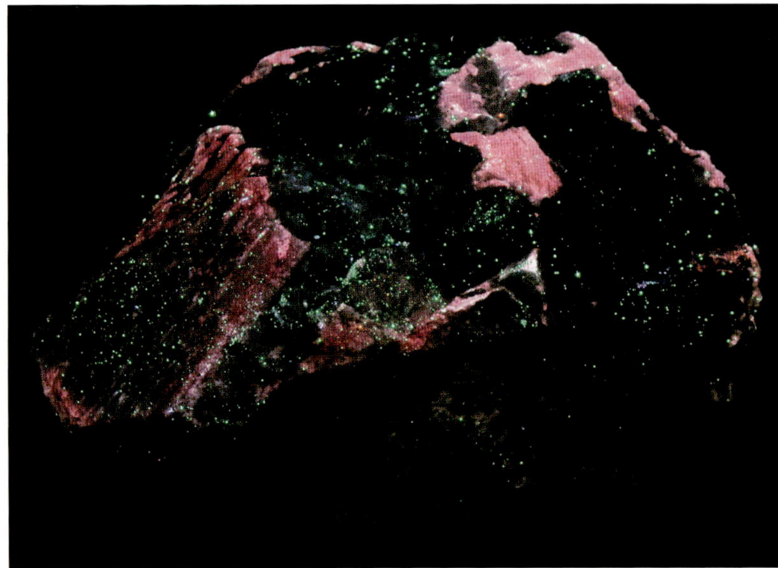

Same under SW

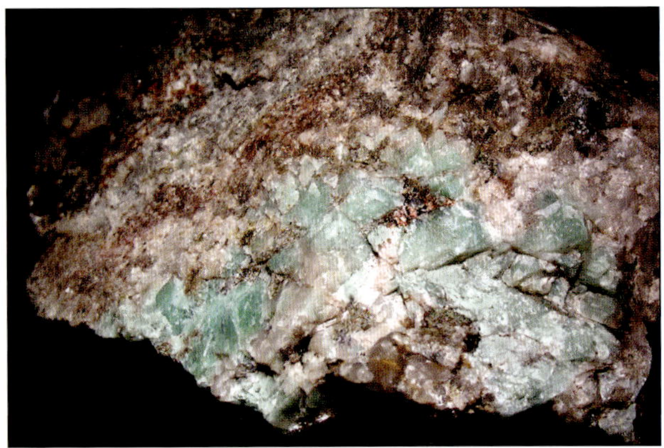

MICROCLINE, POWELLITE. A five-color piece with blue-green microcline, var. amazonite, calcite, willemite, powellite, sphalerite, and calcite that fluoresces two colors from the Mill site, Franklin, Sussex County. The powellite fluoresces white SW and pale yellow LW. The sphalerite fluoresces blue and orange LW, the microcline, var. amazonite, fluoresces pale blue SW, the willemite fluoresces green SW, and the calcite fluoresces orange-red SW and purple-blue LW. The piece weighs 2 lb. 1.0 oz. and is 5.5 x 3.0 x 2.5 inches. Value $60-75

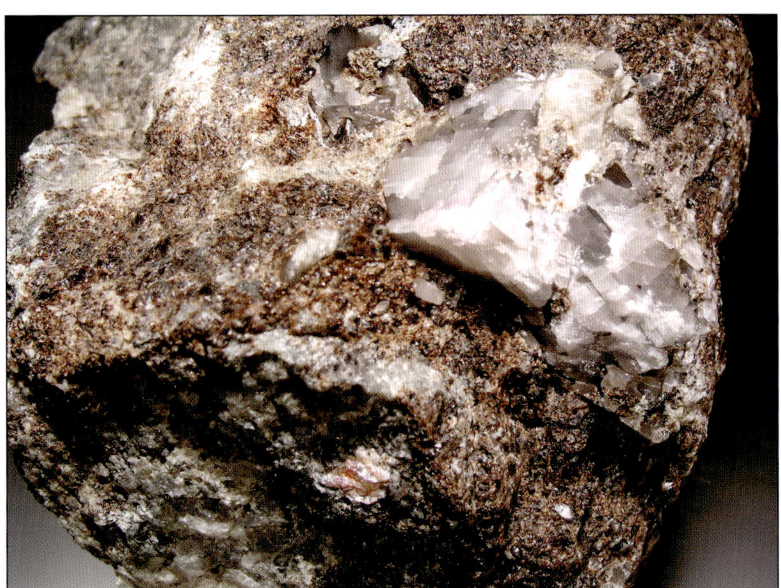

Close-up view of the calcite on the reverse side of the specimen

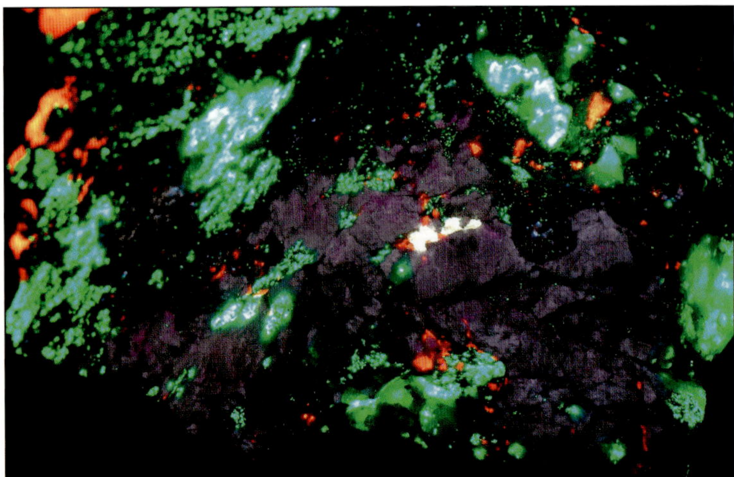

A side of the specimen showing pale blue-fluorescing microcline, var. amazonite, and powellite under SW.

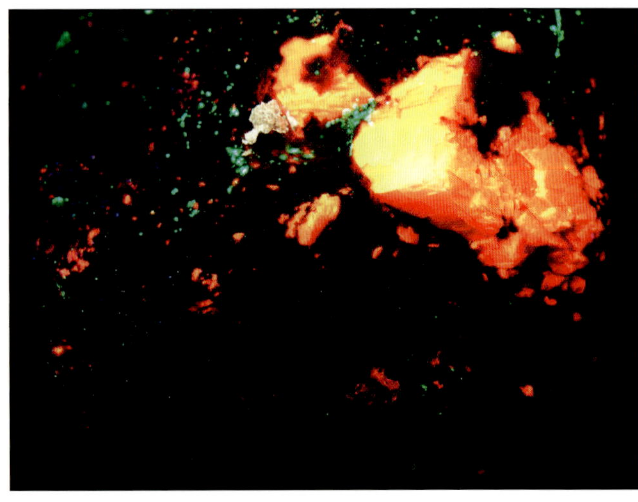

Microcline, var. amazonite, and powellite under SW

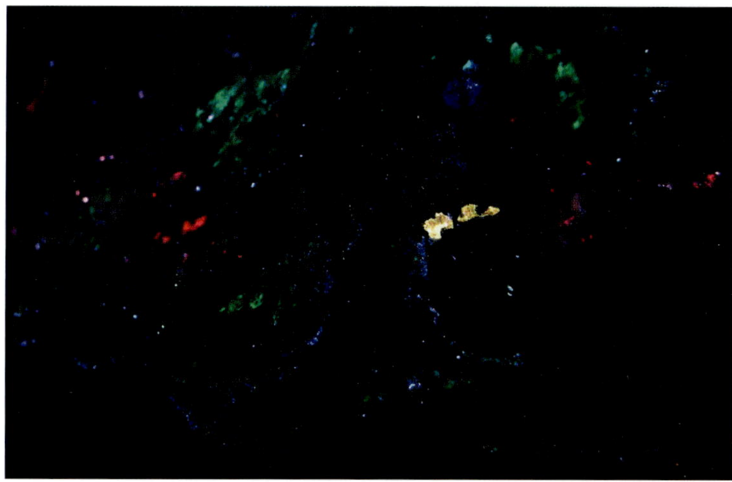

The side of the specimen showing pale blue-fluorescing microcline, var. amazonite, and powellite under LW.

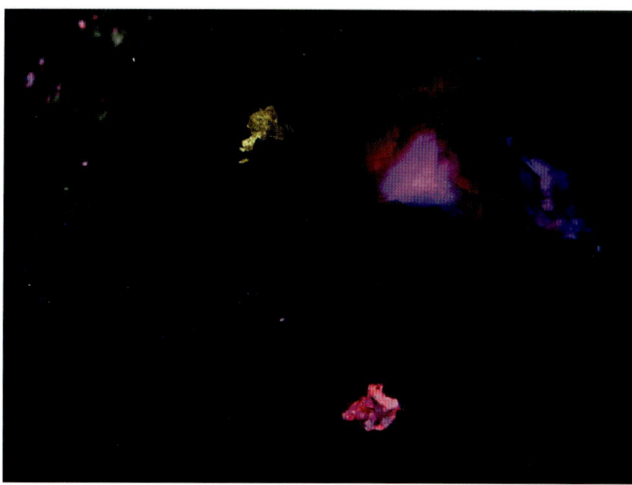

Blue-fluorescing calcite and an area of sphalerite at the bottom of the photo under LW.

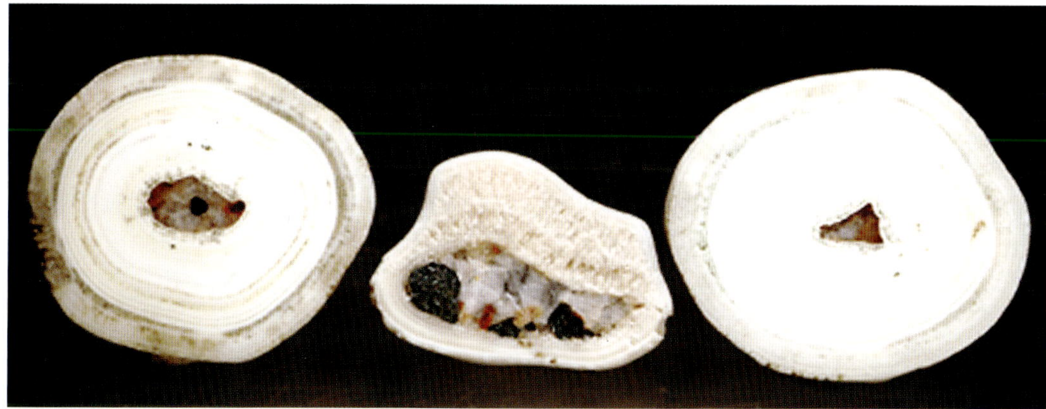

MINE PEARL. Mine pearls found by Bob Winters about 1990. Calcium carbonate accreted in layers around a fragment of ore that landed in a churning pool of calcium-rich water. These came from the 900 foot level of the Sterling Hill mine, Ogdensburg, Sussex County. The mine pearls were sliced in half and polished. They are 0.8 to 1.0 inches wide. Value $50-55 each

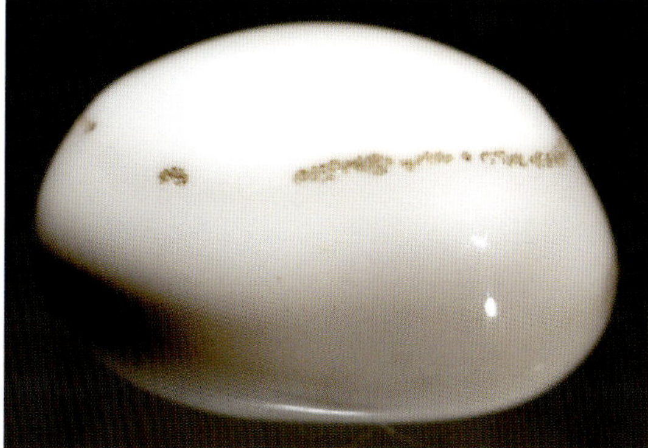

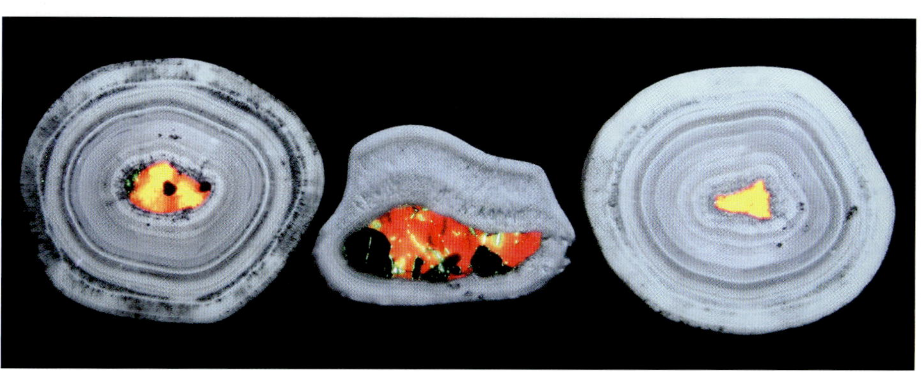

Same under SW

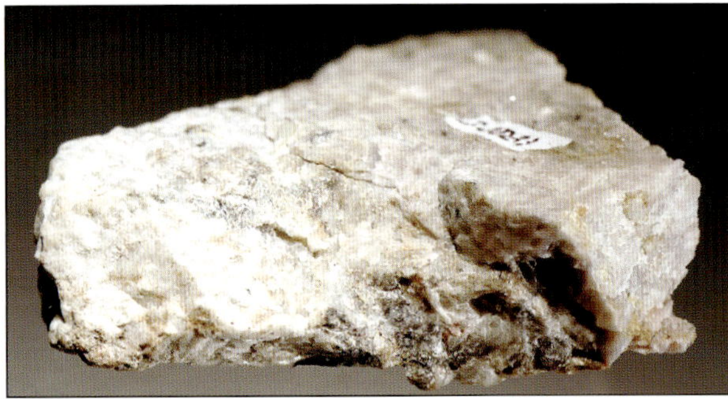

MINEHILLITE. A rare specimen with minehillite, willemite, hyalophane, and wollastonite from the Franklin minehillite assemblage, Franklin, Sussex County. The minehillite fluoresces pale blue SW and MW. The wollastonite fluoresces orange SW, the willemite fluoresces green SW, and the hyalophane fluoresces velvety red SW. The piece weighs 3.5 oz. and is 2.9 x 2.3 x .8.0 inches. Value $400-450

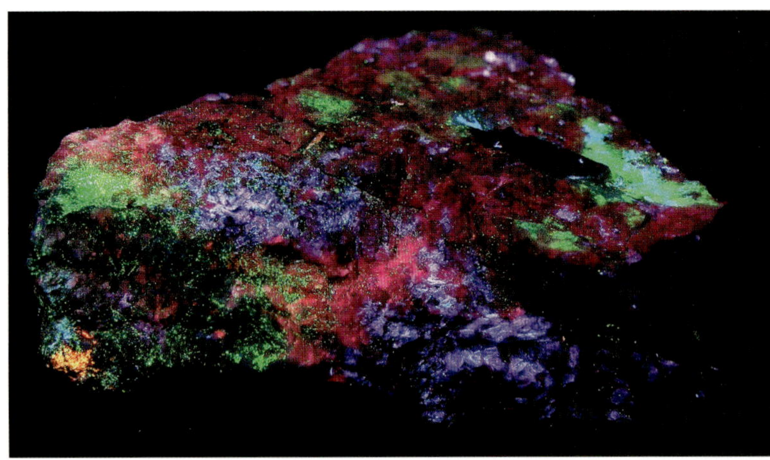

Same under SW

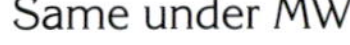

Same under MW

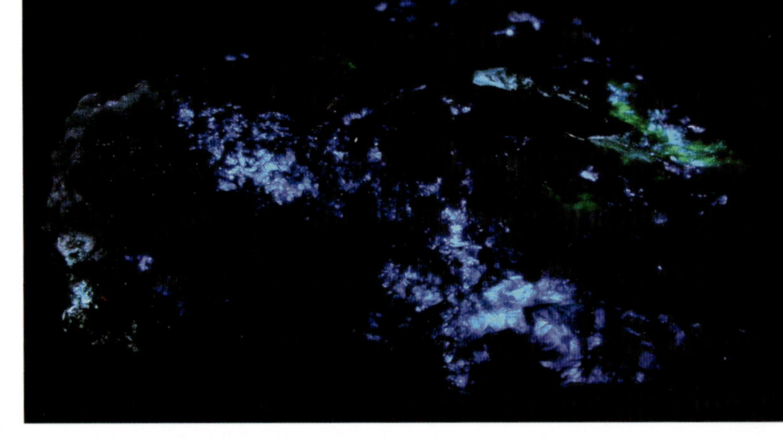

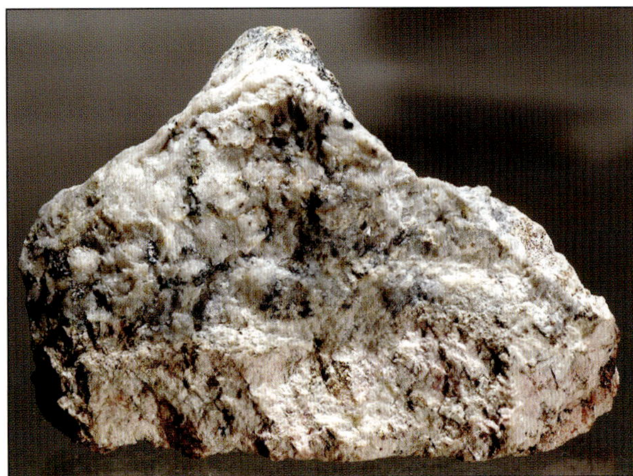

MINEHILLITE, MARGAROSANITE, WOLLASTONITE. Fibrous wollastonite, margarosanite, minehillite, and willemite from the Franklin minehillite assemblage, Sussex County. The wollastonite fluoresces yellow-orange SW, the margarosanite fluoresces bright pale blue SW and red MW, the minehillite fluoresces pale blue SW and MW, and the willemite fluoresces green SW. The piece weighs 6.8 oz. and is 3.4 x 2.5 x 1.3 inches. Value $650-700

Same under SW

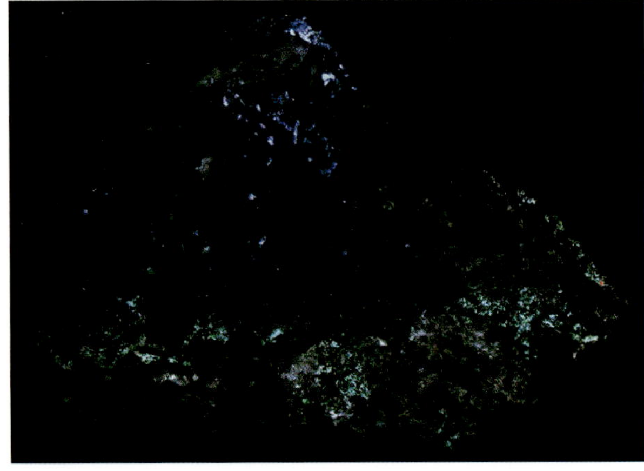

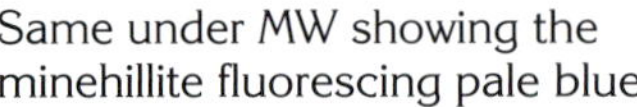

Same under MW showing the minehillite fluorescing pale blue.

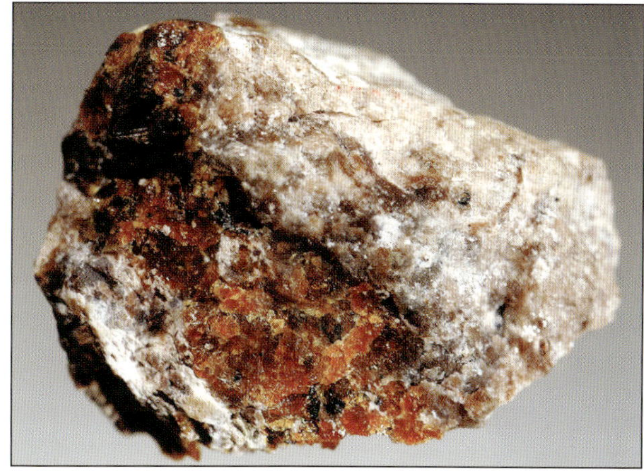

NASONITE. Nasonite, a rare Franklin mineral, with manganaxinite from Franklin, Sussex County. Nasonite fluoresces mustard-yellow SW and the manganaxinite fluoresces red SW. The piece weighs 0.5 oz. and is 1.1 x 0.9 x 0.6 inches. Value $130-145

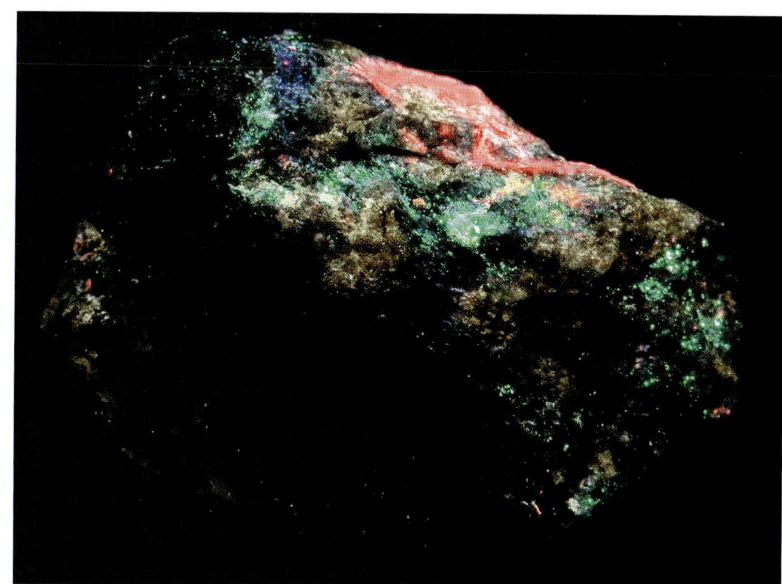

Same under SW

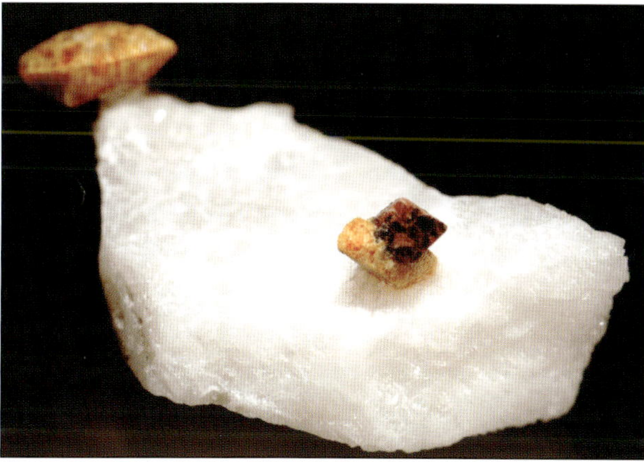

NORBERGITE, SPINEL. Norbergite crystals etched out of the surrounding marble from the Limecrest quarry, Sparta, Sussex County. One crystal has an octahedral spinel crystal attached. The norbergite fluoresces bright yellow SW. The piece weighs 0.5 oz. and is 1.8 x 0.8 x 0.8 inches. Value $20-25

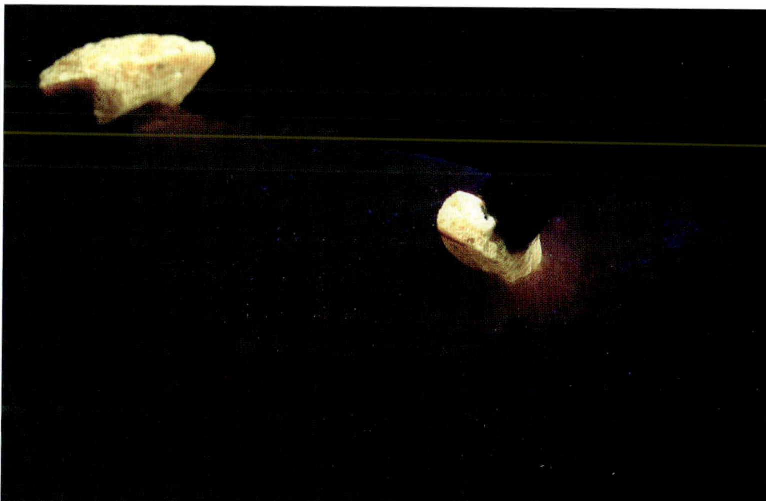

Same under SW

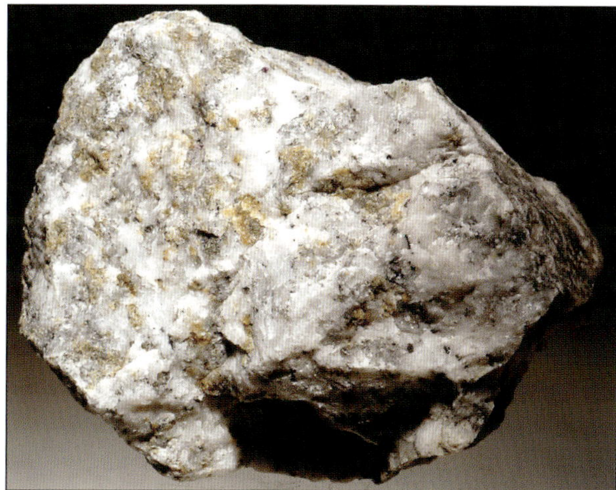

NORBERGITE. Norbergite in marble with diopside from the Franklin quarry, Franklin, Sussex County. The norbergite fluoresces yellow SW and the diopside fluoresces pale blue SW. The piece weighs 1 lb. 4.0 oz. and is 4.3 x 3.0 x 2.3 inches. Value $30-35

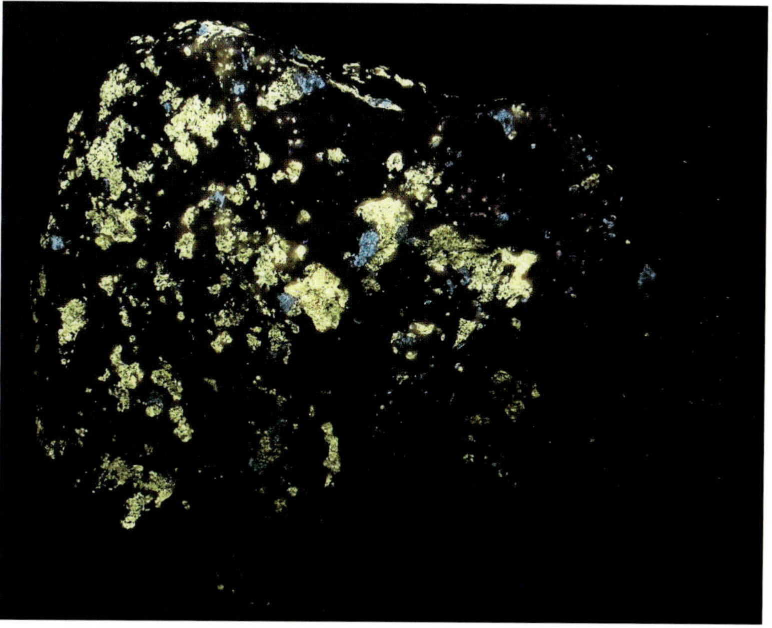

Same under SW

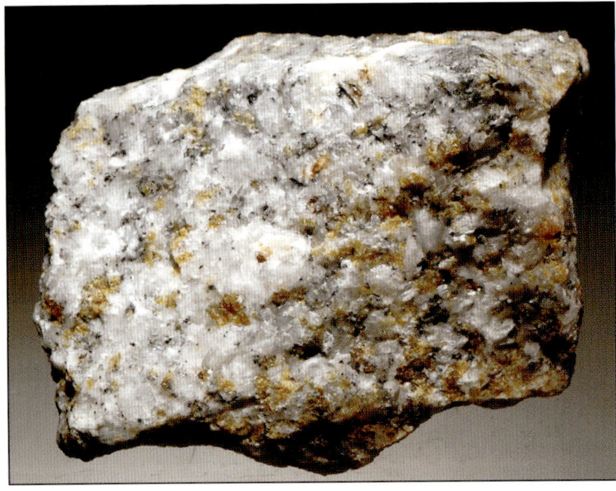

NORBERGITE. Norbergite in marble from the Franklin quarry, Franklin, Sussex County. The norbergite fluoresces yellow SW. The piece weighs 1 lb. 2.0 oz. and is 3.8 x 3 x 2.5 inches. Value $20-25

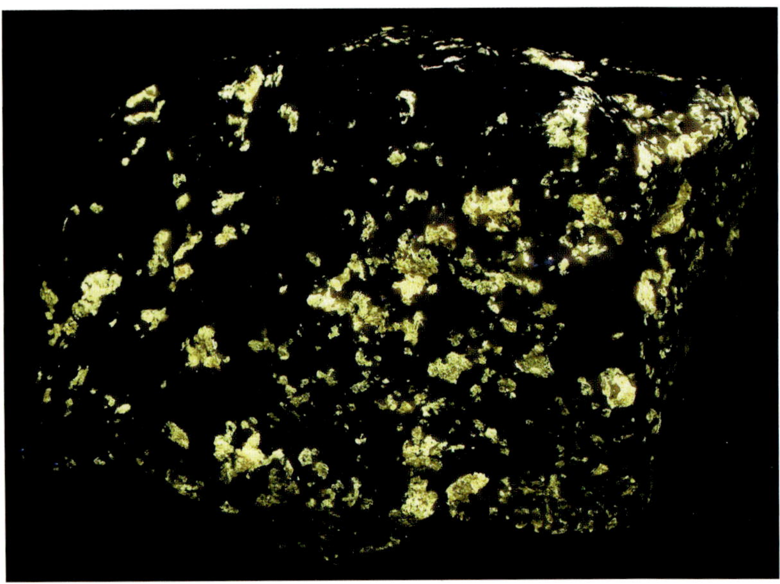

Same under SW

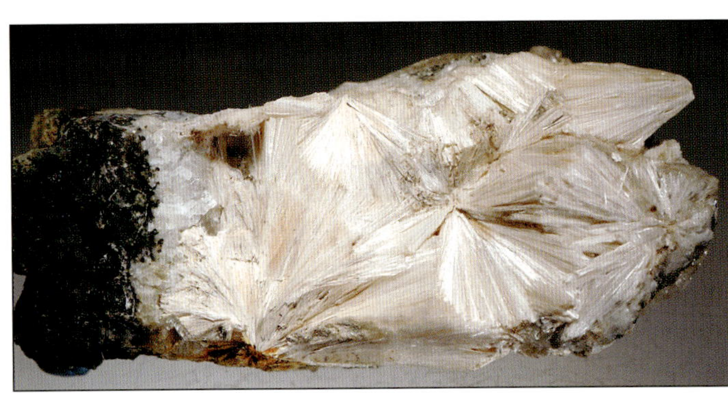

PECTOLITE. Pectolite from Lower New Street, Paterson, Passaic County. The pectolite fluoresces tan and dark pink LW and less brightly SW. The piece weighs 4.8 oz. and is 3.3 x 2.0 x 1.5 inches. Value $35-40

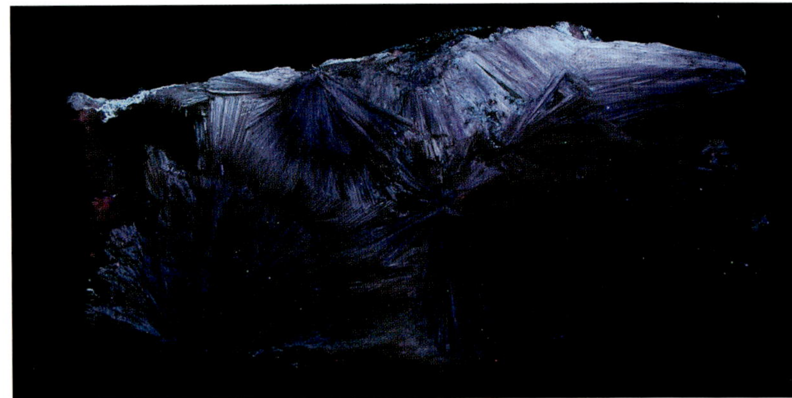

Same under SW

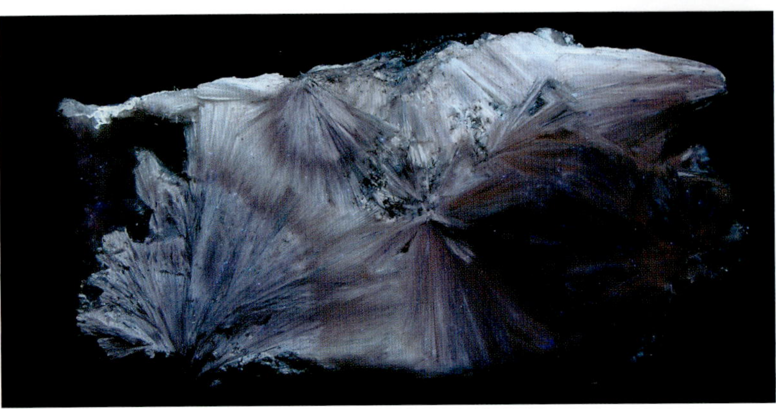

Same under LW

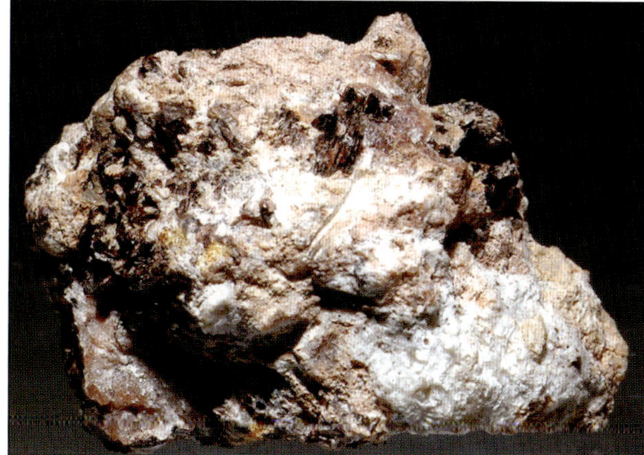

PECTOLITE. Pectolite, roeblingite, willemite, and fibrous xonotlite from Franklin, Sussex County. The pectolite fluoresces orange SW, the roeblingite fluoresces red SW, the xonotlite fluoresces violet SW, and the willemite fluoresces green SW. The piece weighs 0.8 oz. and is 1.3 x 1.0 x 1.0 inches. Value $100-125

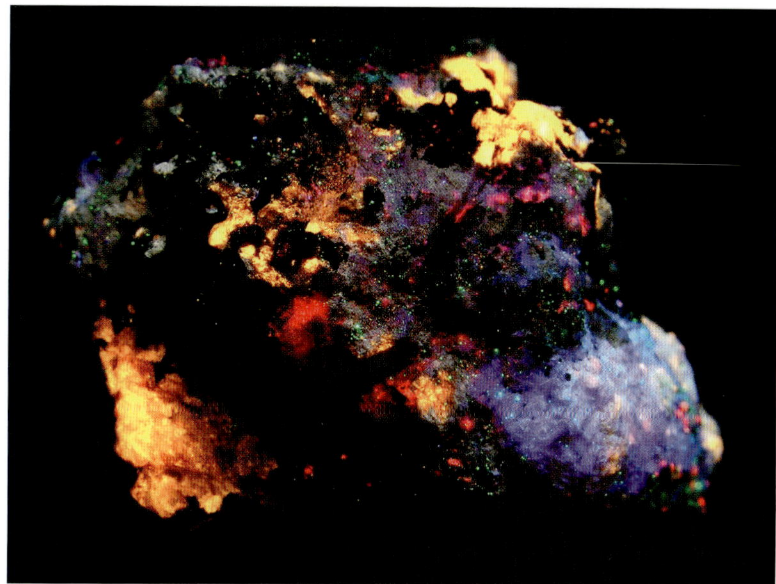

Same under SW

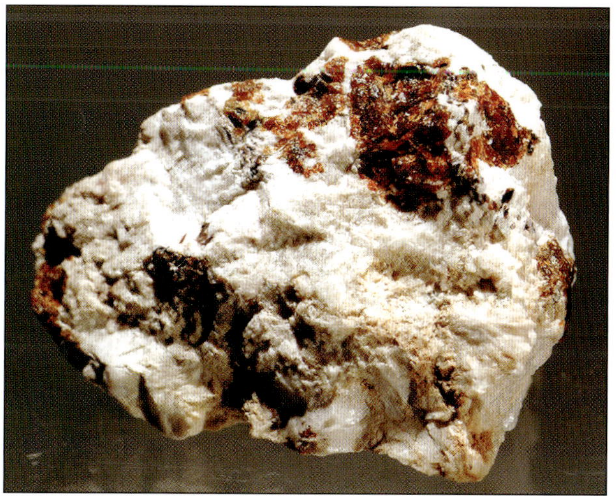

PECTOLITE, PREHNITE. Pectolite, prehnite, margarosanite, and willemite from Franklin, Sussex County. The pectolite fluoresces orange SW, the prehnite fluoresces pale violet SW, the margarosanite fluoresces bright pale blue SW, and willemite fluoresces green SW. This is a very attractive fluorescent specimen. The piece weighs 4.5 oz. and is 2.4 x 1.8 x 1.8 inches. Value $750-800

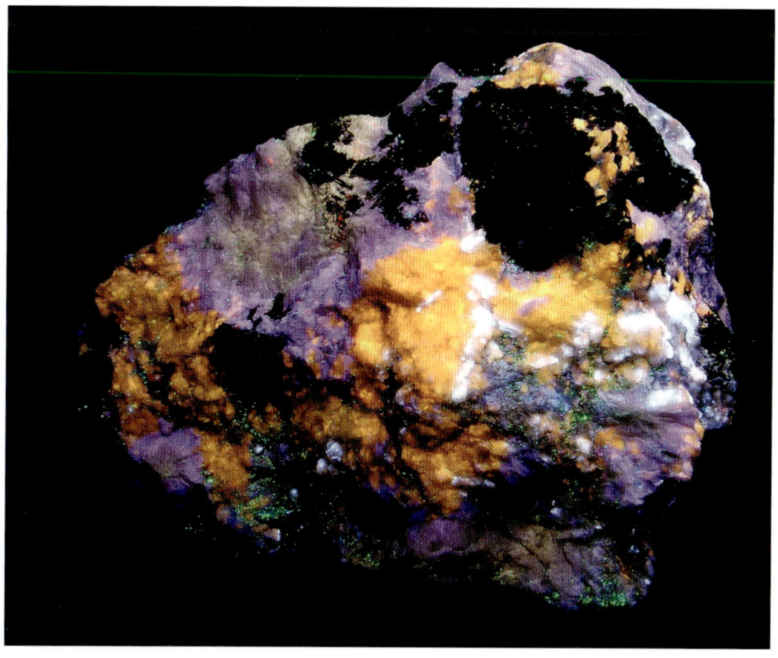

Same under SW

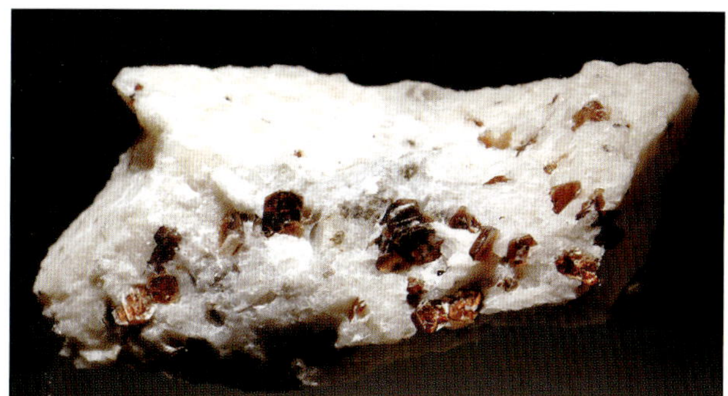

PHLOGOPITE. Orange plates of phlogopite in calcite from the Franklin quarry, Franklin, Sussex County. The phlogopite fluoresces yellow SW. The piece weighs 7.0 oz. and is 3.5 x 2.3 x 1.3 inches. Value $10-15

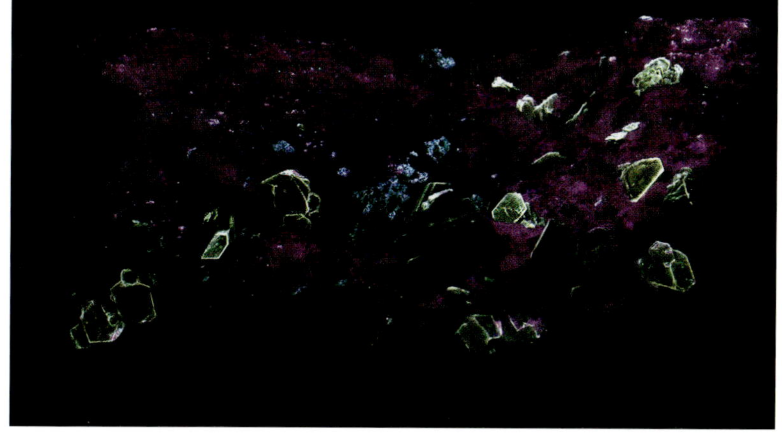

Same under SW

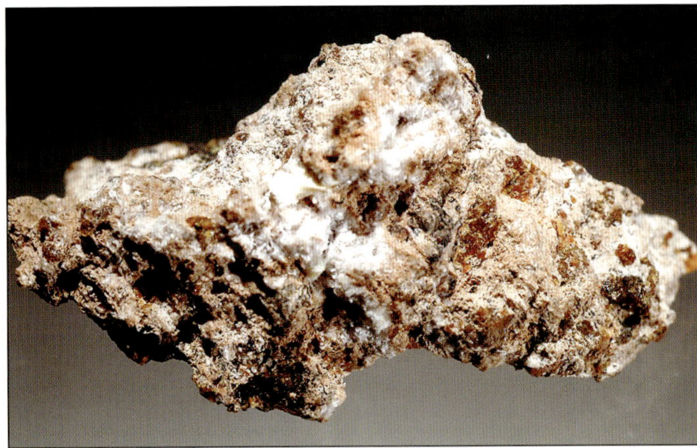

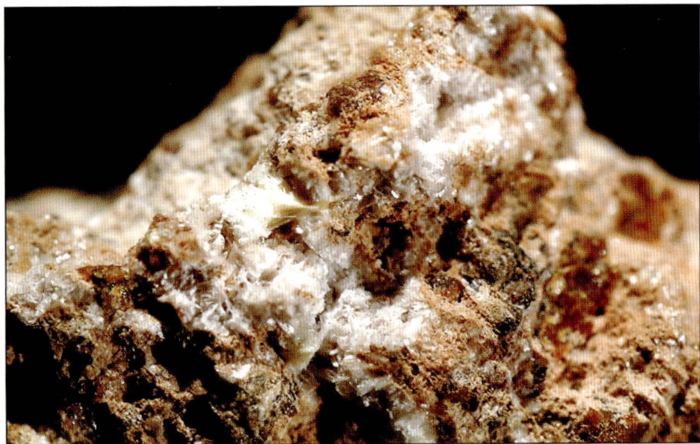

PREHNITE, XONOTLITE. Prehnite crystals, fibrous xonotlite, and clinohedrite on altered hancockite from the Franklin mine, Franklin, Sussex County. Crystals of prehnite and fibrous xonotlite are rare. The prehnite fluoresces peach SW, the xonotlite fluoresces violet SW and the clinohedrite fluoresces bright orange SW. The piece weighs 0.8 oz. and is 2.0 x 1.0 x 1.0 inches. Value $200-225

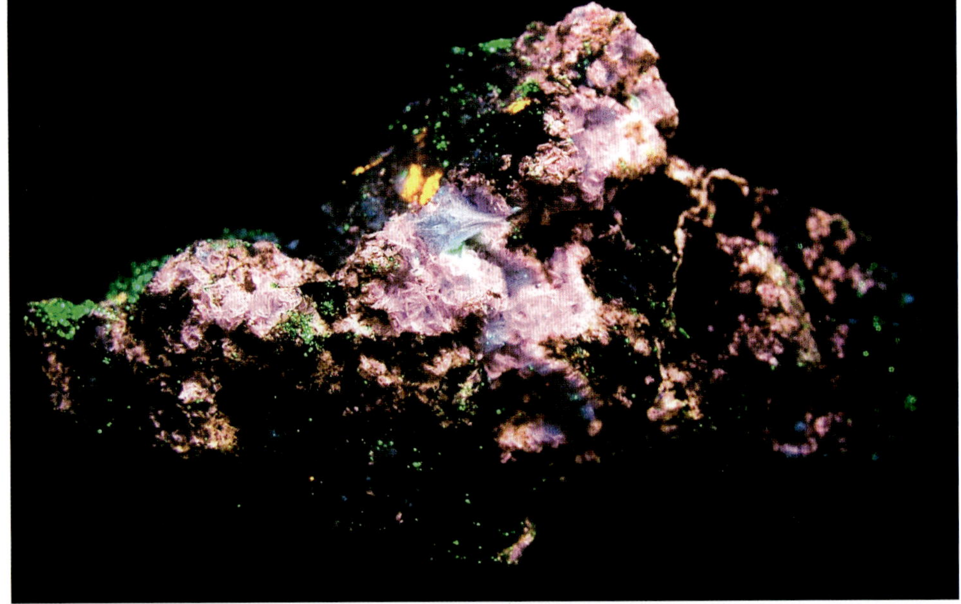

Same under SW

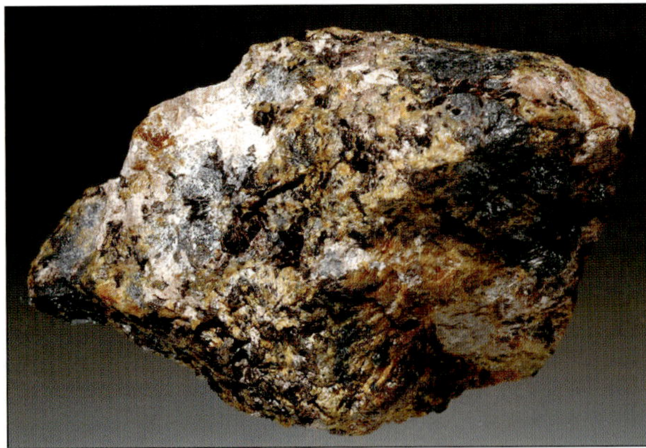

PREHNITE, XONOTLITE. Prehnite with xonotlite, willemite, hendricksite, barite, and clinohedrite from the Franklin mine, Franklin, Sussex County. The prehnite fluoresces peach SW, the clinohedrite fluoresces orange SW, the xonotlite fluoresces violet SW, the barite fluoresces white SW, and the willemite fluoresces green SW. The piece weighs 9.3 oz. and is 2.8 x 2.0 x 1.5 inches. Value $230-250

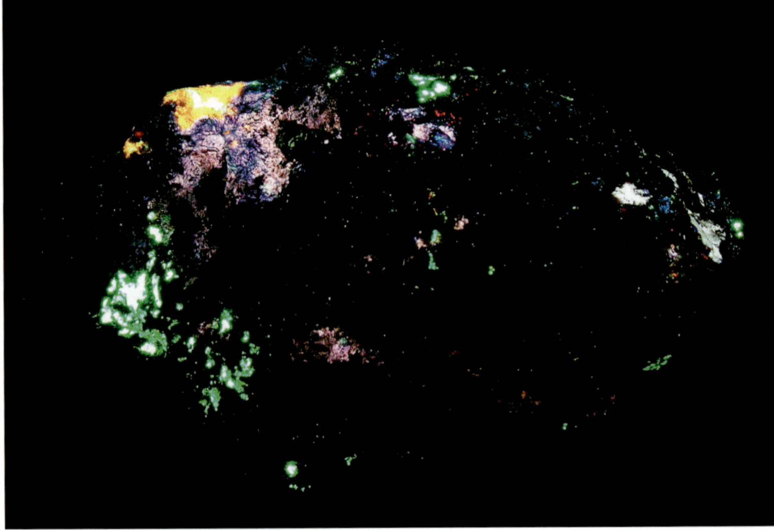

Same under SW

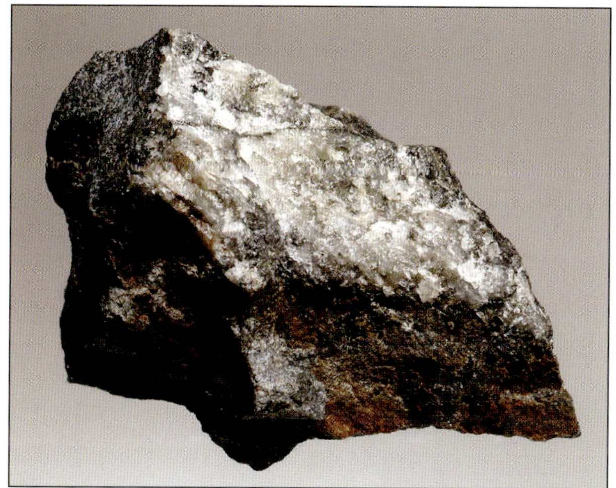

QUARTZ. Quartz with sphalerite on a dark matrix from the Taylor Road dump, Franklin, Sussex County. It is very uncommon to find fluorescing quartz in Franklin. The quartz fluoresces pale yellow SW and the sphalerite fluoresces blue and orange SW and LW. The piece weighs 8.0 oz. and is 3.0 x 2.0 x 2.0 inches. Value $80-100

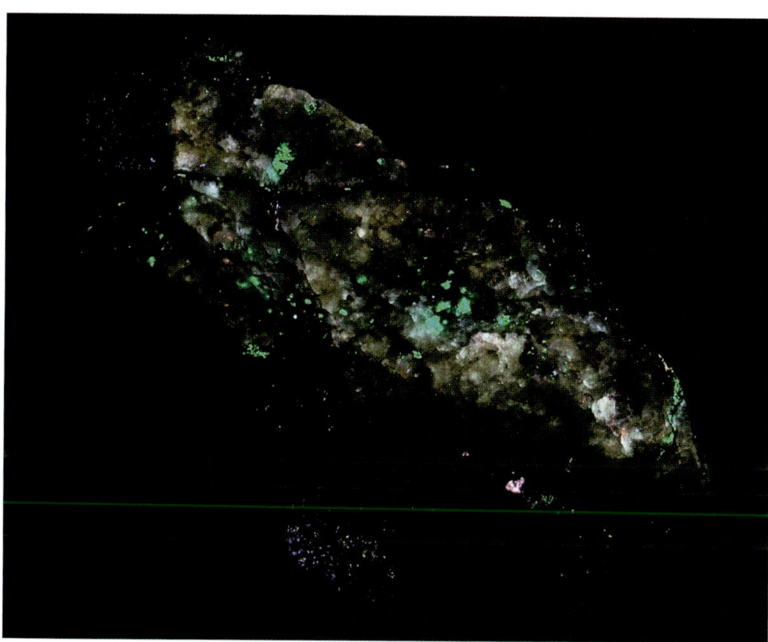

Same under SW

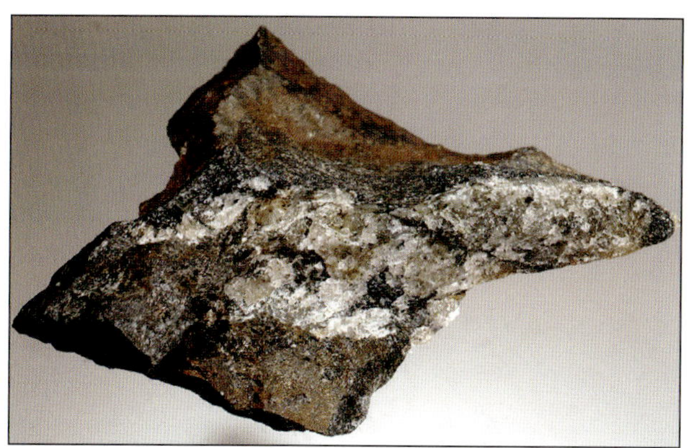

QUARTZ. Quartz with sphalerite on a dark matrix from the Taylor Road dump, Franklin, Sussex County. It is very uncommon to find fluorescing quartz in Franklin. The quartz fluoresces pale yellow SW and the sphalerite fluoresces blue and orange SW and LW. The piece weighs 8.0 oz. and is 4.0 x 3.0 x 1.9 inches. Value $80-100

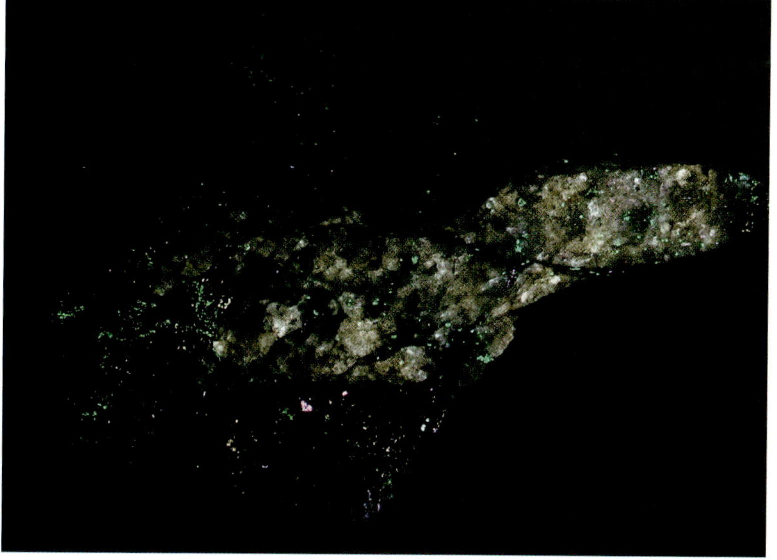

Same under SW

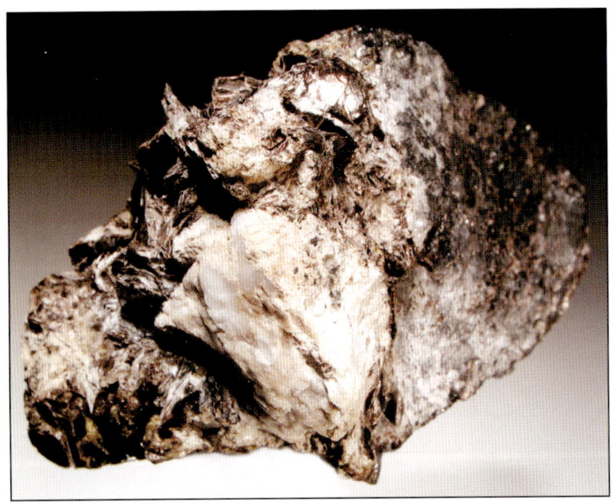

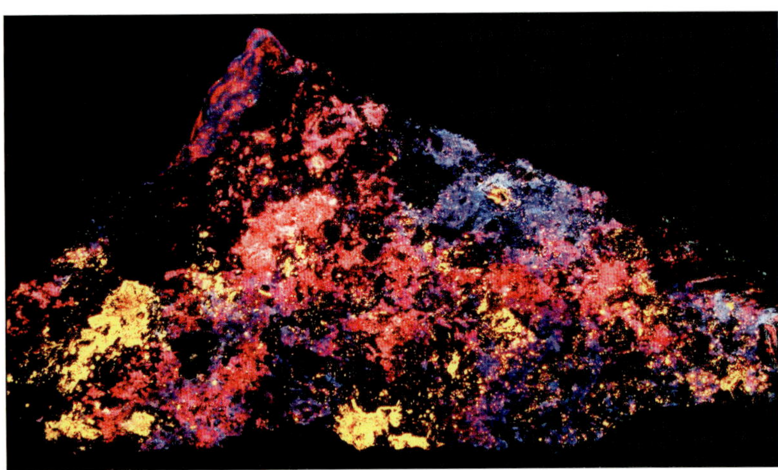

Same under SW, showing the bottom of the specimen

ROEBLINGITE, XONOTLITE. Roeblingite, xonotlite, and clinohedrite on altered hancockite with hendricksite from Franklin, Sussex County. The roeblingite appears in two forms, a china-like nodule and a plaster-like coating. The roeblingite fluoresces red SW, the xonotlite fluoresces violet SW, and the clinohedrite fluoresces orange SW. It weighs 9.0 oz. and is 3.5 x 2.0 x 1.5 inches. Value $2,250-2,500

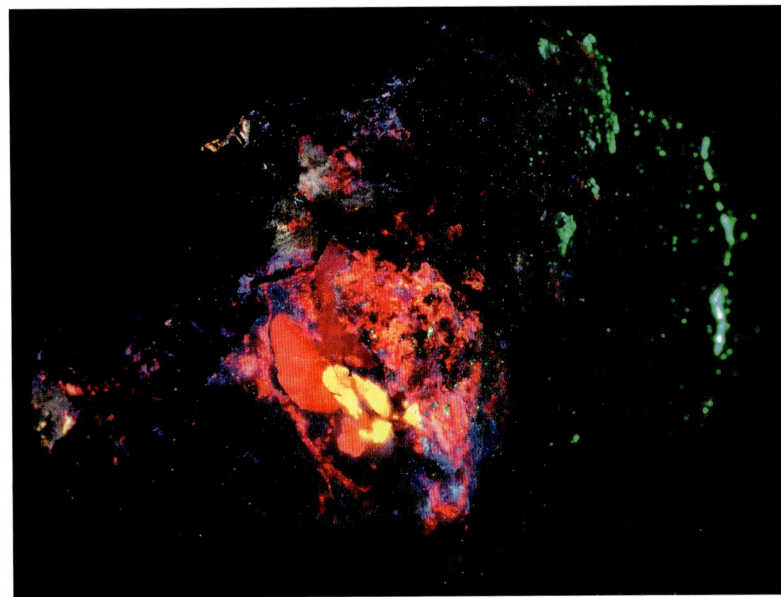

Same under SW, showing the nodule side

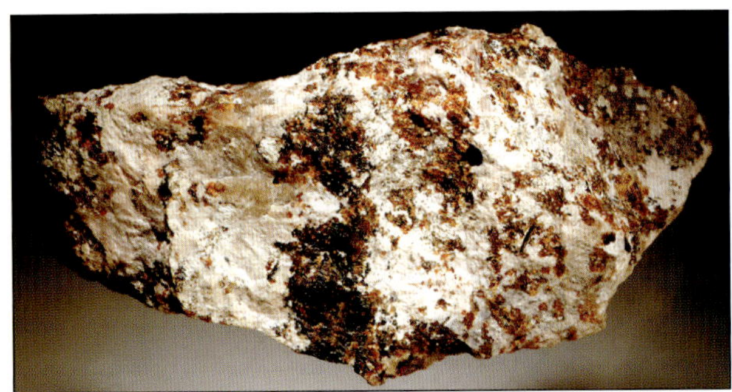

ROEBLINGITE. Roeblingite, prehnite, pectolite, xonotlite, and hyalophane from Franklin, Sussex County. This piece is exceptional because of its size and the grouping of rare Franklin minerals. The roeblingite fluoresces red SW, the pectolite fluoresces orange SW, the xonotlite fluoresces violet SW, the prehnite fluoresces pale lavender SW, and the hyalophane fluoresces velvety red SW. The piece weighs 1 lb. 1.0 oz. and is 5.0 x 2.5 x 2.0 inches. Value $3,000-3,500

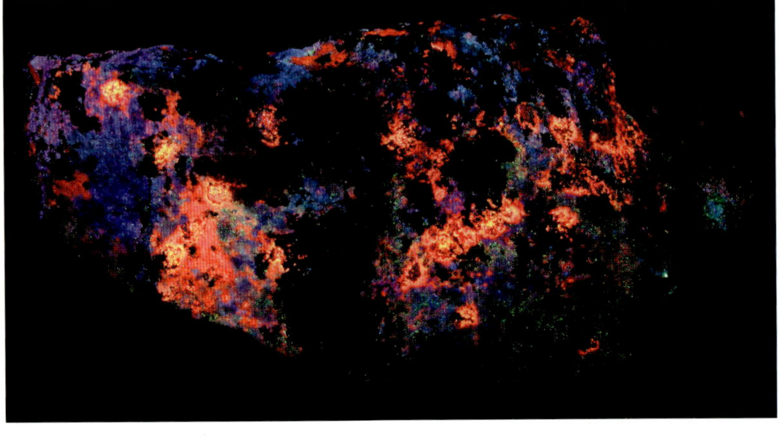

Same under SW

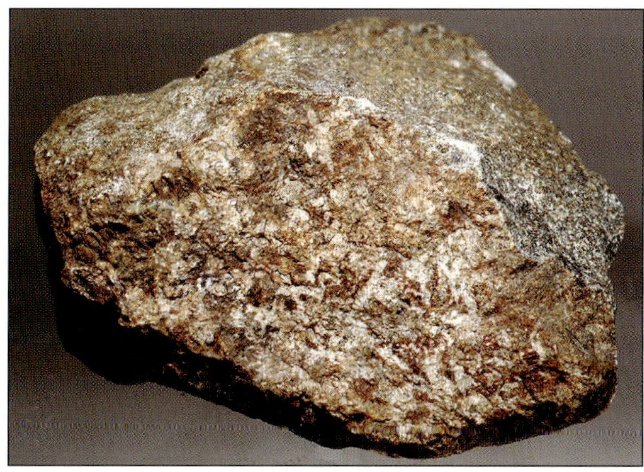

SCHEELITE. Fluorapatite and scheelite from the Taylor Road dump, Franklin, Sussex County. The fluorapatite fluoresces yellow-orange SW and the scheelite fluoresces pale yellow SW. The piece weighs 1 lb. 1.0 oz. and is 3.5 x 3.0 x 2.0 inches. Value $50-60

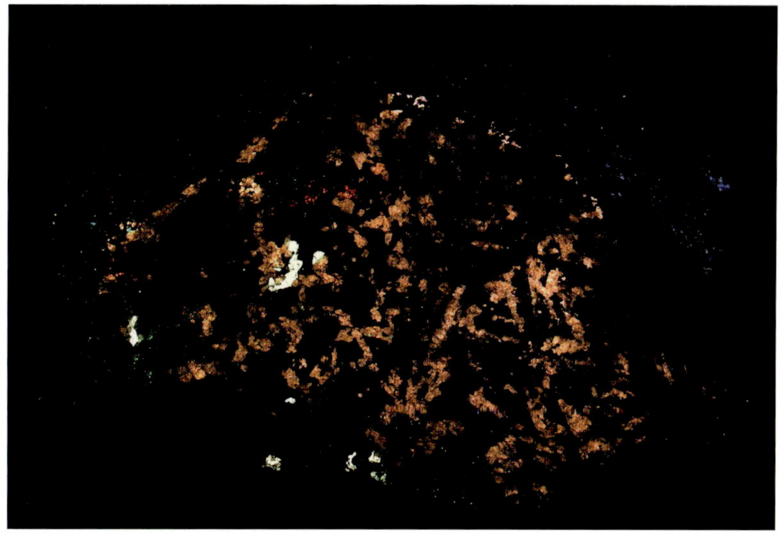

Same under SW

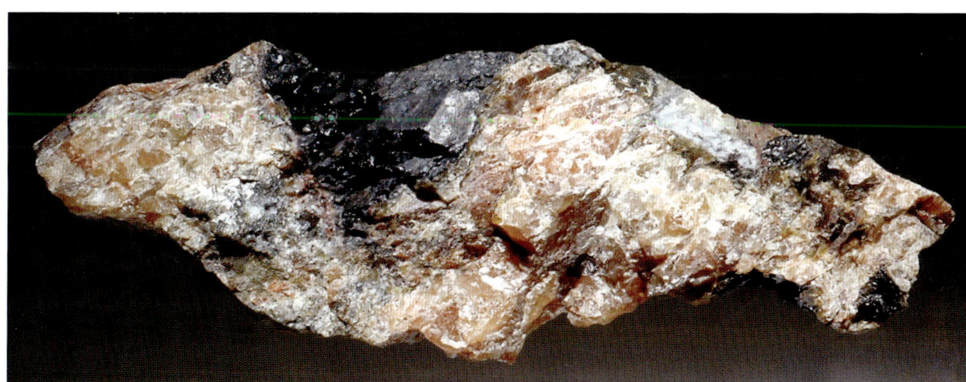

SPHALERITE. Glistening sphalerite with willemite, bustamite, and franklinite from the Franklin mine, Franklin, Sussex County. The sphalerite fluoresces blue and orange LW (which looks pink) and SW. The sphalerite also has long-lasting phosphorescence. The willemite fluoresces green SW. The piece weighs 1.8 oz. and is 2.8 x 1.3 x 1.0 inches. Value $50-55

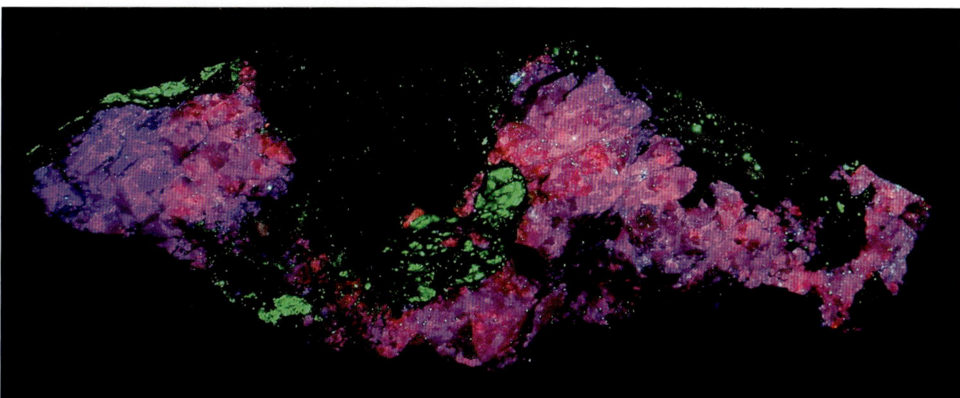

Same under SW

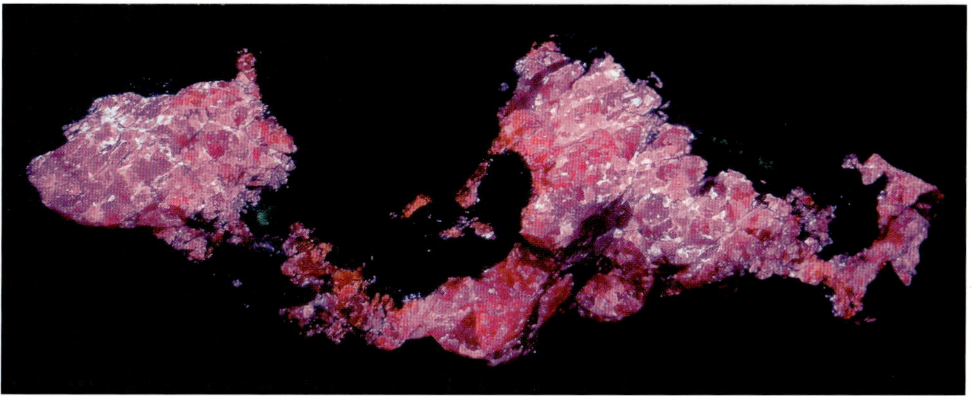

Same under LW

SPHALERITE. Sphalerite, hydrozincite, and calcite from the Franklin mine, Franklin, Sussex County. The sphalerite fluoresces orange and blue LW and SW, the calcite fluoresces orange-red SW, and the hydrozincite fluoresces bright pale blue SW. The piece weighs 1 lb. 1.0 oz. and is 4.0 x 3.3 x 2.0 inches. Value $125-140

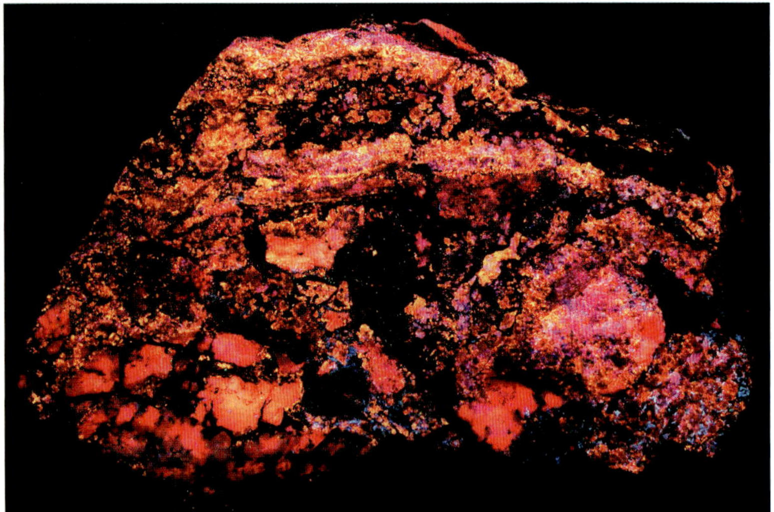

Same under SW

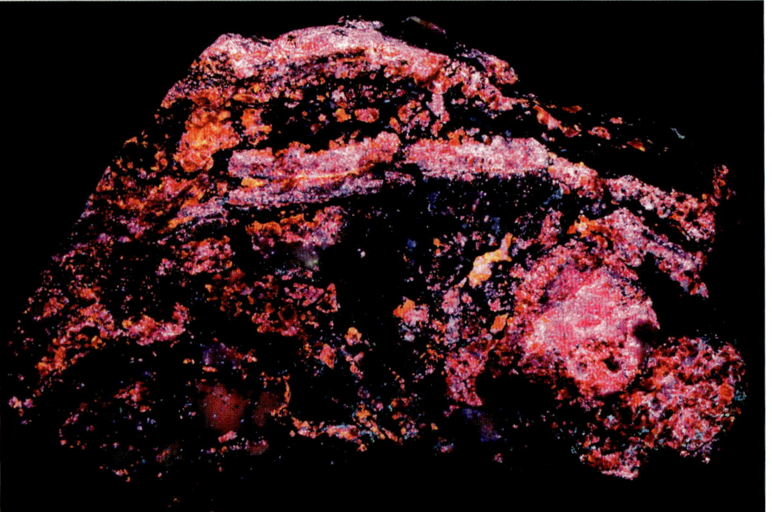

Same under LW

SPHALERITE. Sphalerite with willemite veins from the Sterling mine, Ogdensburg, Sussex County. The sphalerite fluoresces orange LW and the willemite fluoresces green SW. The piece weighs 1 lb. 4.0 oz. and is 5.0 x 2.5 x 2.0 inches. Value $75-85

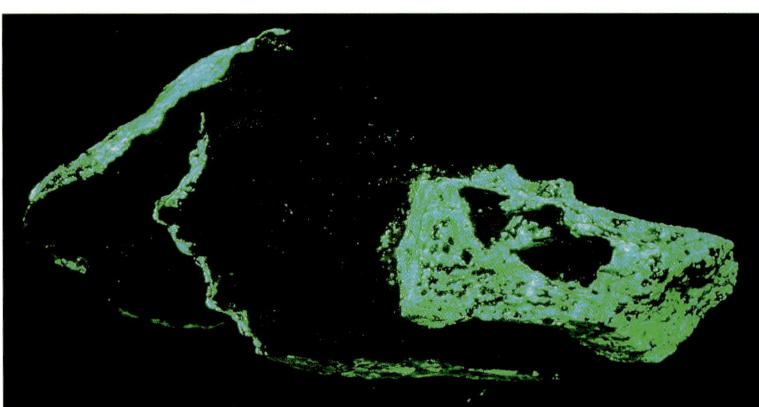

Same under SW

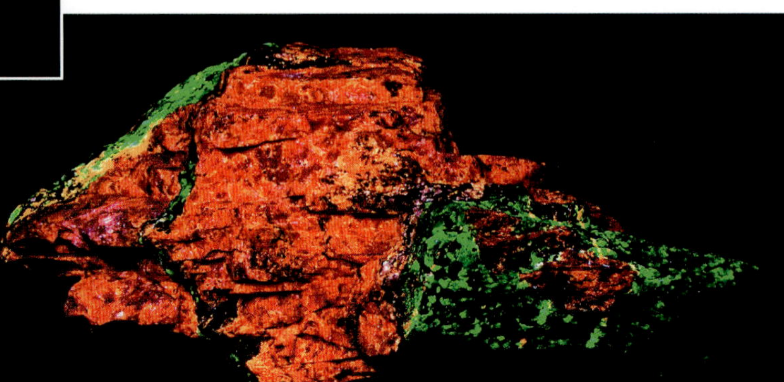

Same under LW

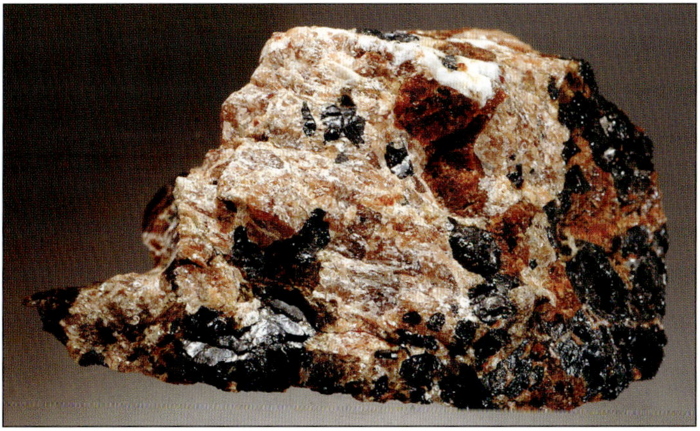

SPHALERITE. Golden sphalerite and willemite from the Franklin mine, Franklin, Sussex County. Golden sphalerite is a local term for sphalerite which fluoresces distinctive orange-yellow with blue highlights. The sphalerite fluoresces pink and blue SW and orange-yellow with blue highlights LW. The willemite fluoresces green SW. The piece weighs 1 lb. 6.0 oz. and is 4.5 x 3.0 x 2.5 inches. Value $265-285

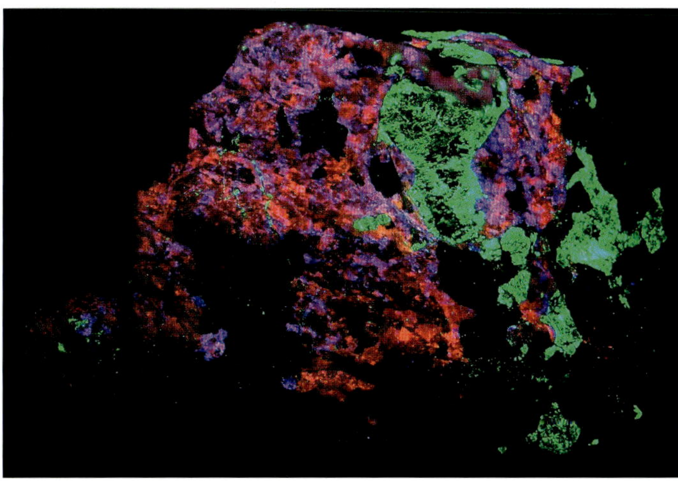

Same under SW

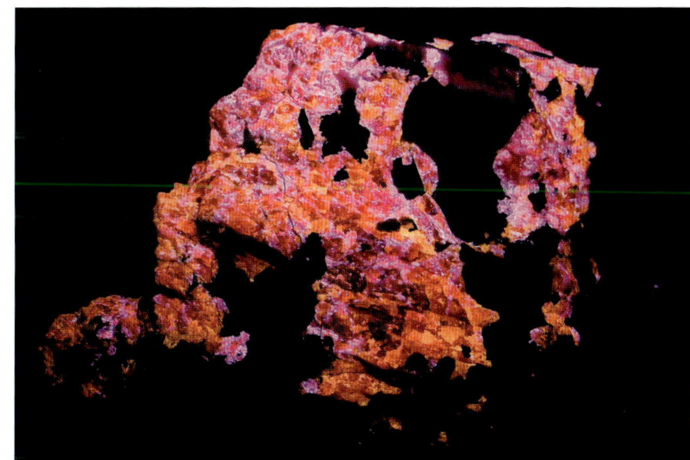

Same under LW

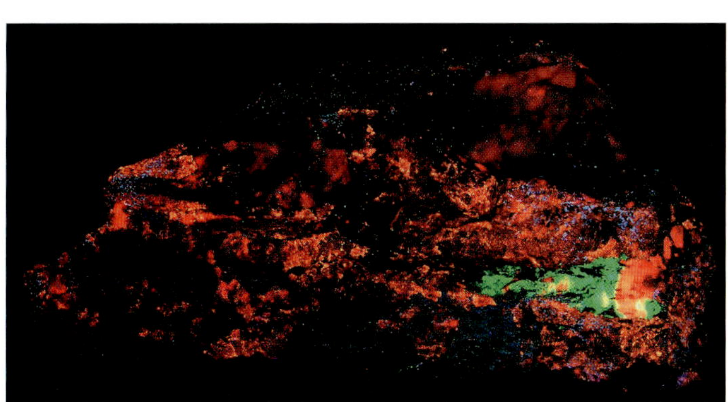

Same under SW

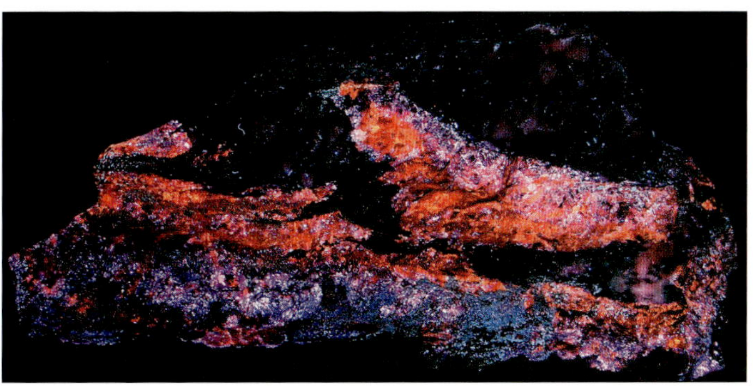

Same under LW

SPHALERITE. Sphalerite with willemite and calcite from the Franklin mine, Franklin, Sussex County. Sphalerite fluoresces orange and blue LW, calcite fluoresces orange-red SW, and willemite fluoresces green SW. The piece weighs 1.5 oz. and is 2.4 x 1.8 x 1.3 inches. Value $35-40

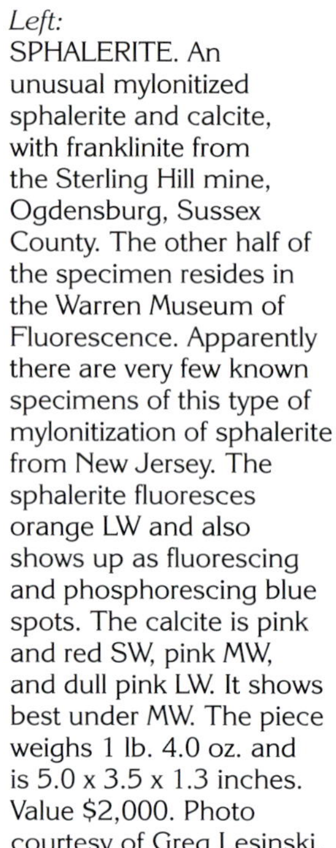

Left:
SPHALERITE. An unusual mylonitized sphalerite and calcite, with franklinite from the Sterling Hill mine, Ogdensburg, Sussex County. The other half of the specimen resides in the Warren Museum of Fluorescence. Apparently there are very few known specimens of this type of mylonitization of sphalerite from New Jersey. The sphalerite fluoresces orange LW and also shows up as fluorescing and phosphorescing blue spots. The calcite is pink and red SW, pink MW, and dull pink LW. It shows best under MW. The piece weighs 1 lb. 4.0 oz. and is 5.0 x 3.5 x 1.3 inches. Value $2,000. Photo courtesy of Greg Lesinski.

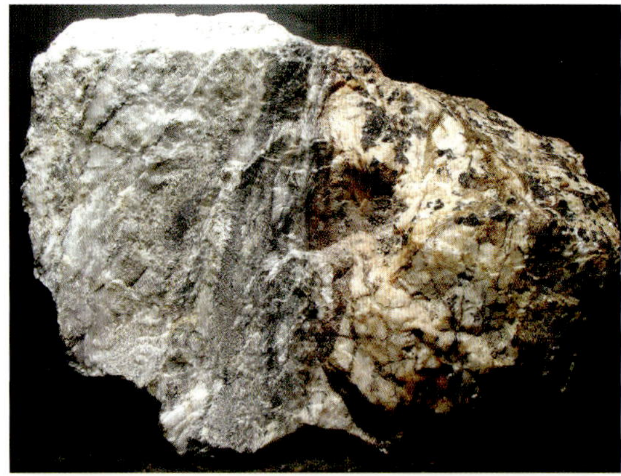

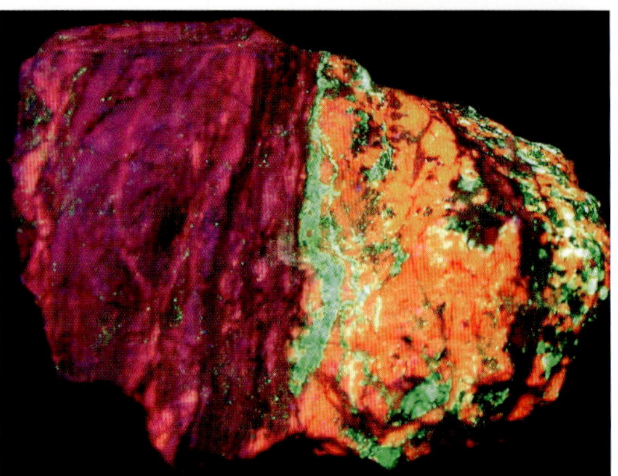

Same under SW

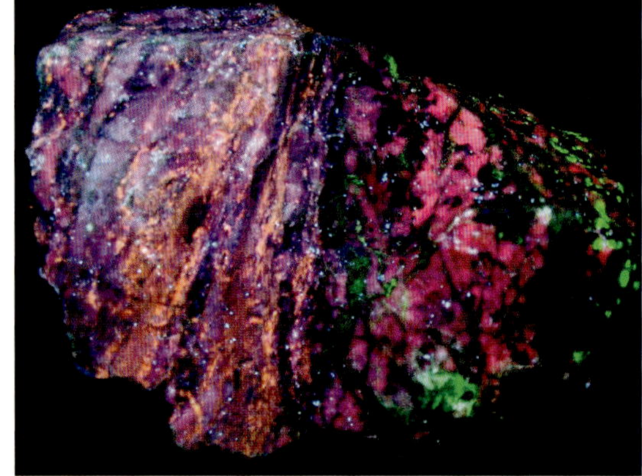

Same under LW

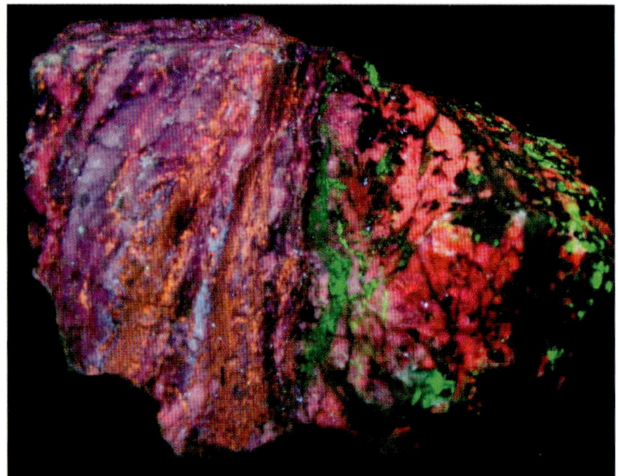

Same under MW

Right:
SPHALERITE. Another unusual mylonitized sphalerite, calcite, and willemite from the Sterling Hill mine, Ogdensburg, Sussex County. This specimen shows mylonitization of sphalerite on the left side and the contact area on the right. The sphalerite fluoresces orange and blue LW. The calcite fluoresces pink and red SW, pink MW, and dull pink LW. The piece weighs about 18 lb. and is about the size of a football. Photo courtesy of Greg Lesinski.

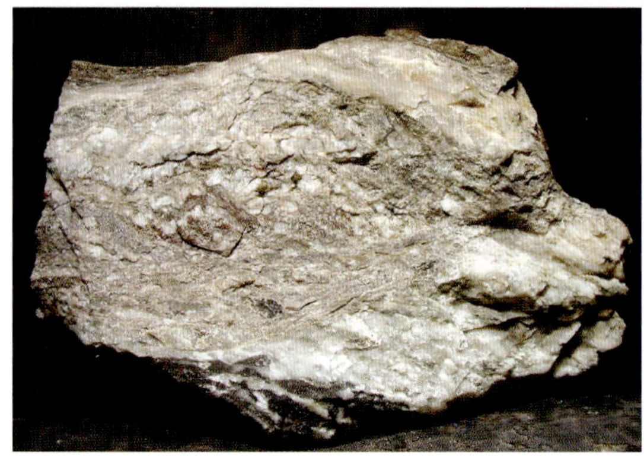

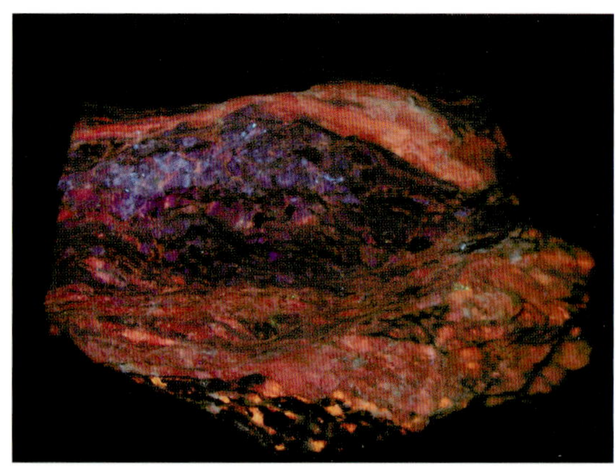

Same under SW

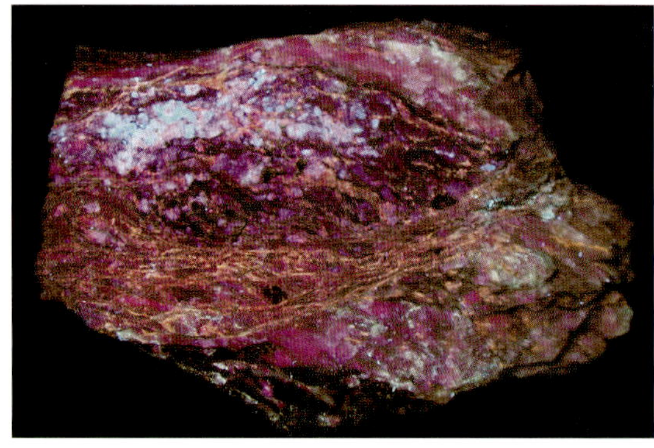

Same under LW

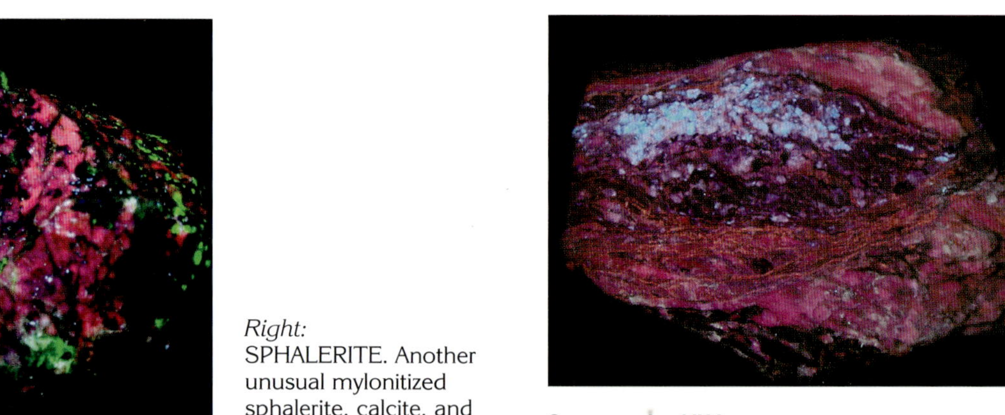

Same under MW

SPHALERITE. A fine-grained reddish-brown sphalerite, locally called "mahogany sphalerite" from the Sterling Hill mine, Ogdensburg, Sussex County. It fluoresces orange SW and LW. It weighs 1 lb. 15.0 oz. and is 4.0 x 3.8 x 2.5 inches. Value $40-50

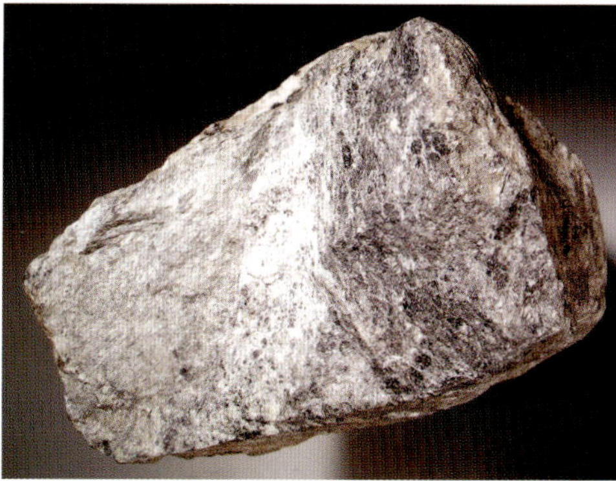

Same under SW

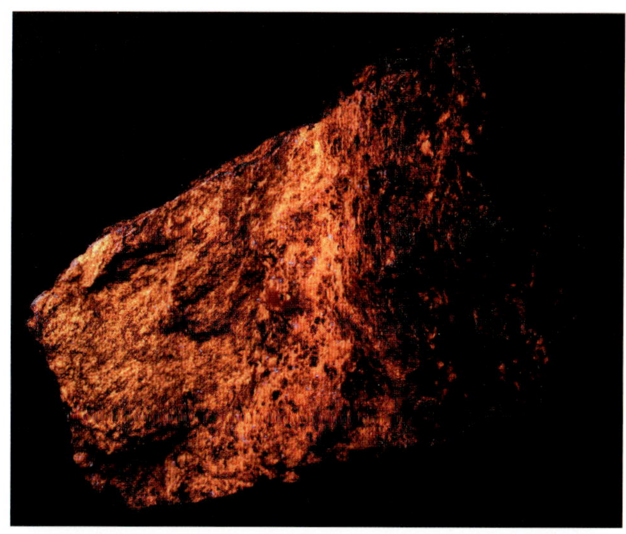

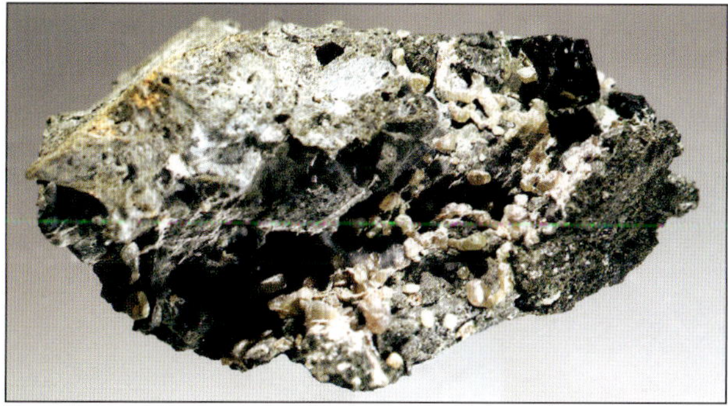

SPHALERITE. Gahnite crystals and sphalerite on a dark matrix from a 2005 find in the Passaic pit, Sterling Hill mine, Ogdensburg, Sussex County. The calcite covering the sphalerite was etched with weak hydrochloric acid to remove it and show the gahnite crystals. The sphalerite fluoresces orange LW. Gahnite is a zinc spinel. The piece weighs 4.5 oz. and is 2.5 x 2.3 x 1.1 inches. Value $25-30

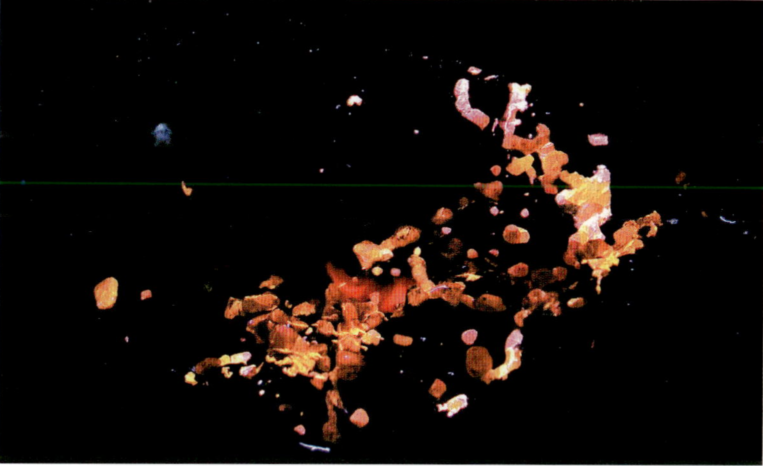

Same under LW

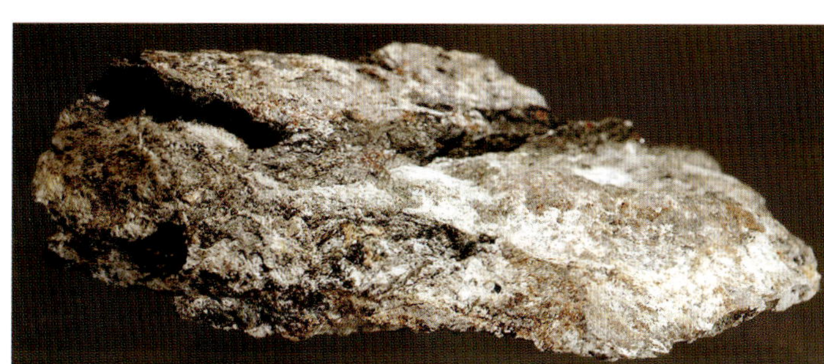

SPHALERITE, HYDROZINCITE. Sphalerite and hydrozincite from a 2005 find in the Passaic pit of the Sterling Hill mine, Ogdensburg, Sussex County. The hydrozincite is a product of the weathering of sphalerite. Sphalerite fluoresces orange LW and hydrozincite fluoresces bright pale blue SW. The piece weighs 10.3 oz. and is 4.3 x 2.5 x 1.3 inches. Value $45-55

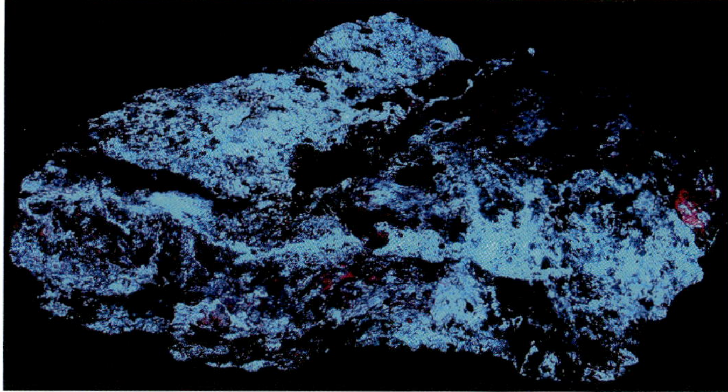

Same under SW

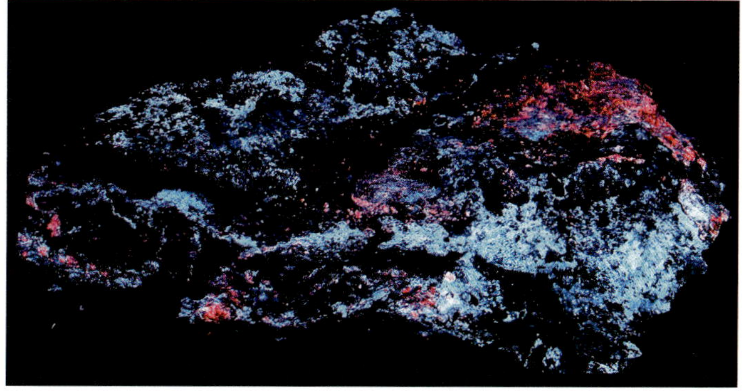

Same under LW

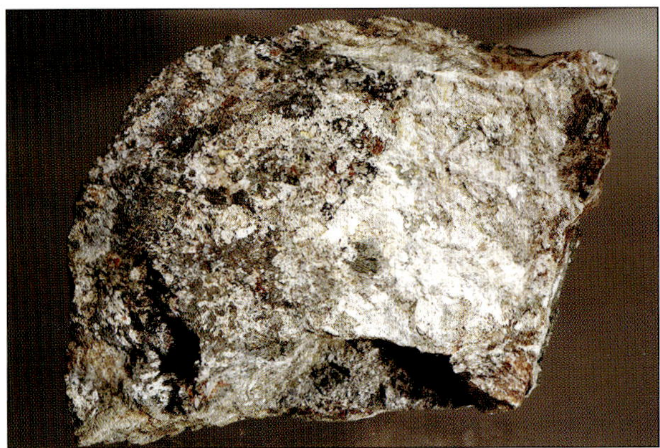

SPHALERITE, HYDROZINCITE. Sphalerite and hydrozincite from a 2005 find in the Passaic pit, Sterling Hill mine, Ogdensburg, Sussex County. Sphalerite fluoresces orange LW and hydrozincite fluoresces bright pale blue SW. The piece weighs 1 lb. 7.0 oz. and is 5.0 x 3.5 x 1.5 inches. Value $55-65

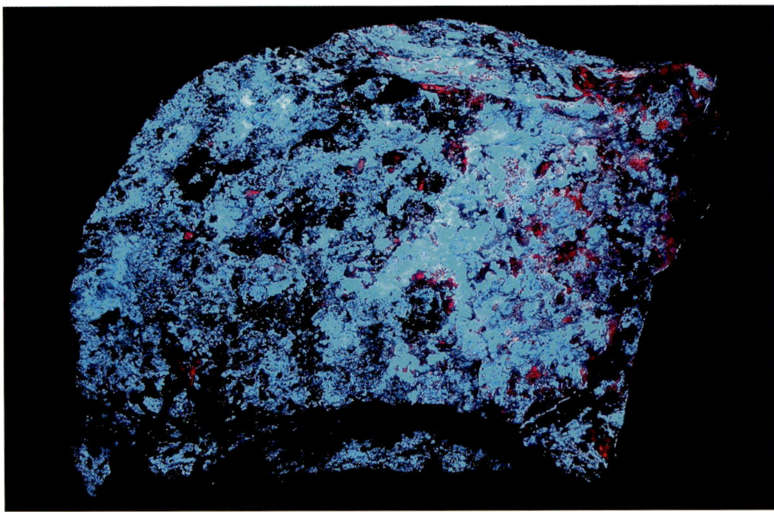

Same under SW

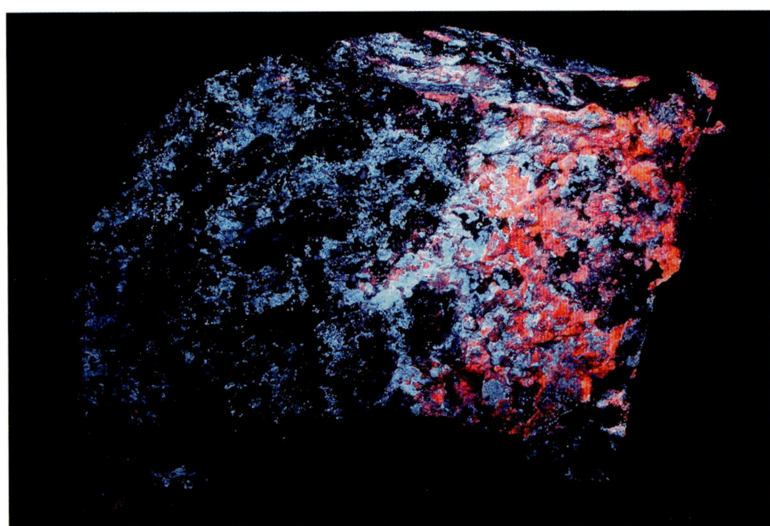

Same under LW

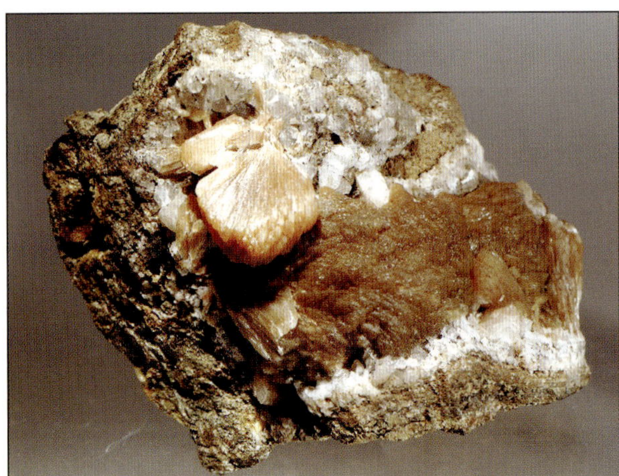

STILBITE, CALCITE. Stilbite crystals on calcite from Upper New Street, Paterson, Passaic County. Stilbite fluoresces tan LW and SW and calcite fluoresces a weak pink SW. The piece weighs 14.0 oz. and is 4.0 x 3.0 x 2.3 inches. Value $25-30

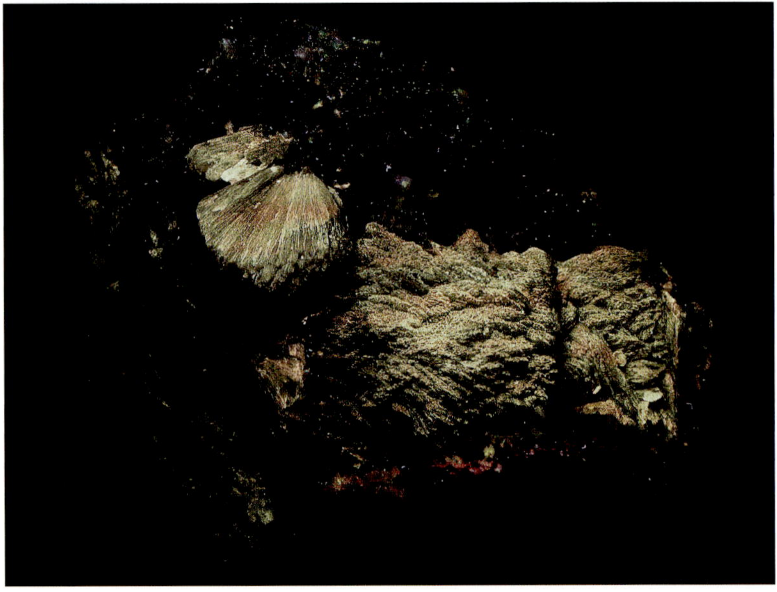

Same under SW

TREMOLITE. Tremolite from the Franklin quarry, Franklin, Sussex County. The tremolite fluoresces pale blue SW. The piece weighs 5.3 oz. and is 2.5 x 2.0 x 1.3 inches. Value $20-30

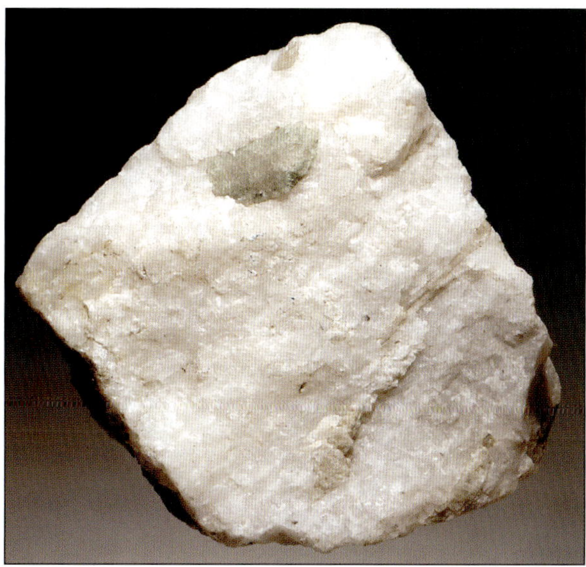

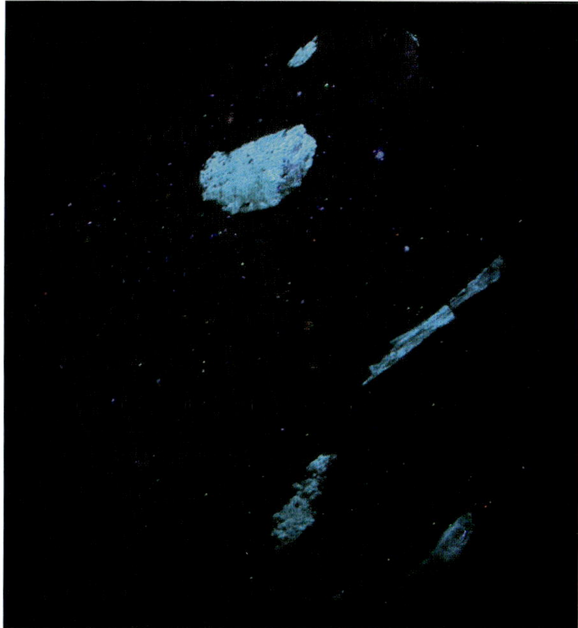

Same under SW

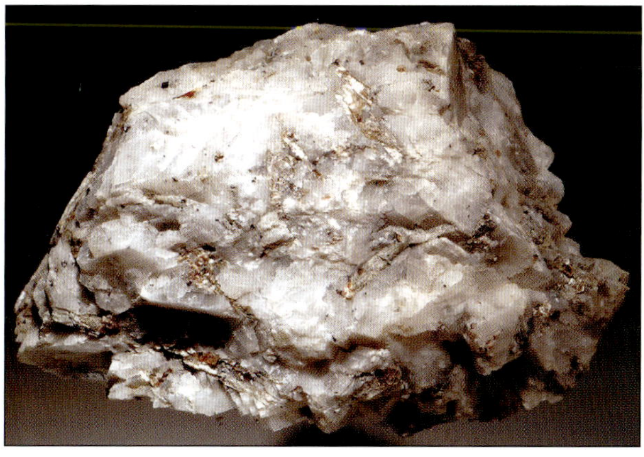

TREMOLITE. Tremolite in calcite with phlogopite from the Franklin quarry, Franklin, Sussex County. The tremolite crystals have been partially replaced by fine-grained aggregates of phlogopite. The tremolite fluoresces pale blue SW and the phlogopite fluoresces yellow SW. The piece weighs 16.0 oz. and is 4.3 x 3.3 x 2.0 inches. Value $40-45

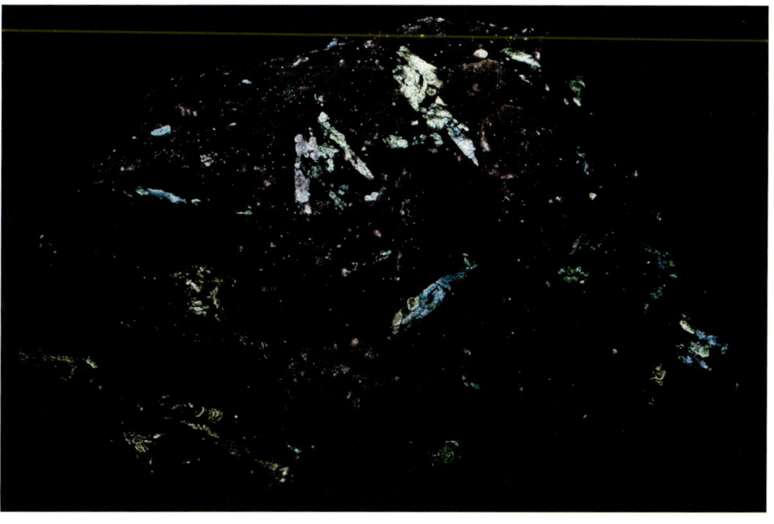

Same under SW

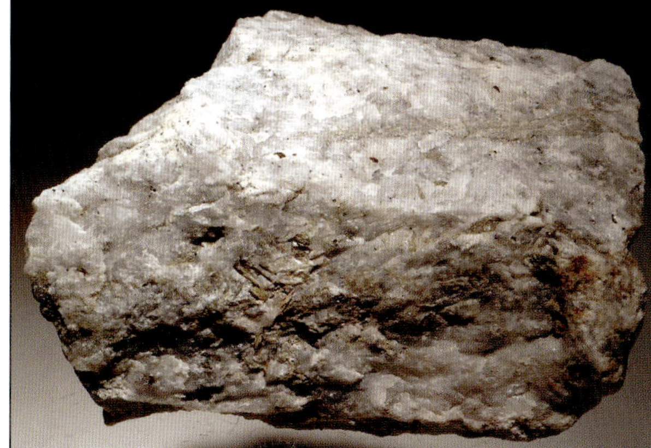

TREMOLITE. Tremolite in calcite with phlogopite from the Franklin quarry, Franklin, Sussex County. The tremolite fluoresces pale blue and the phlogopite fluoresces yellow SW. The piece weighs 1 lb. 8.0 oz. and is 4.0 x 3.3 x 2.3 inches. Value $30-35

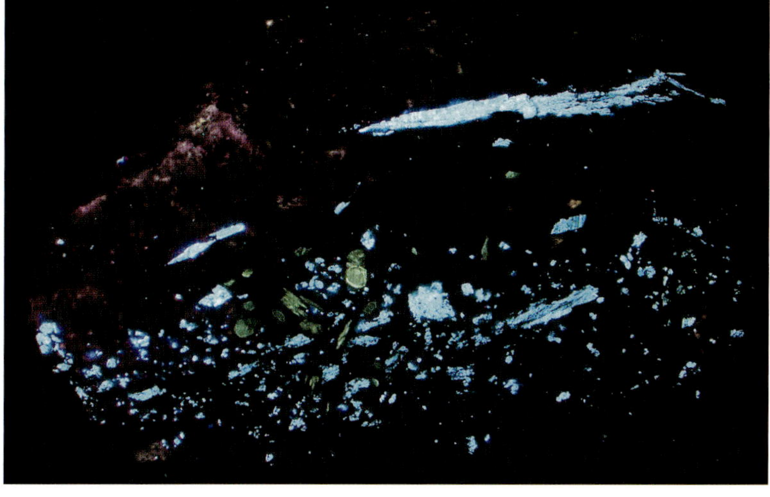

Same under SW

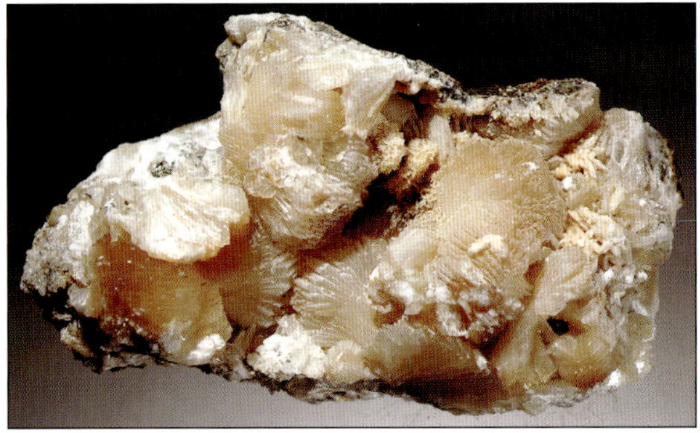

THOMSONITE. Thomsonite crystals on stilbite from Upper New Street, Paterson, Passaic County. Thomsonite fluoresces white LW and SW and has a long-lasting phosphorescence. The piece weighs 9.0 oz. and is 4.0 x 2.5 x 2.0 inches. Value $20-25

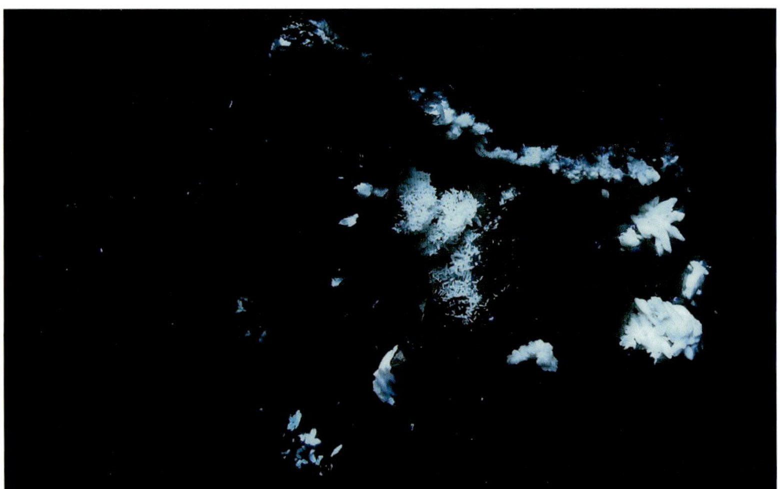

Same under LW

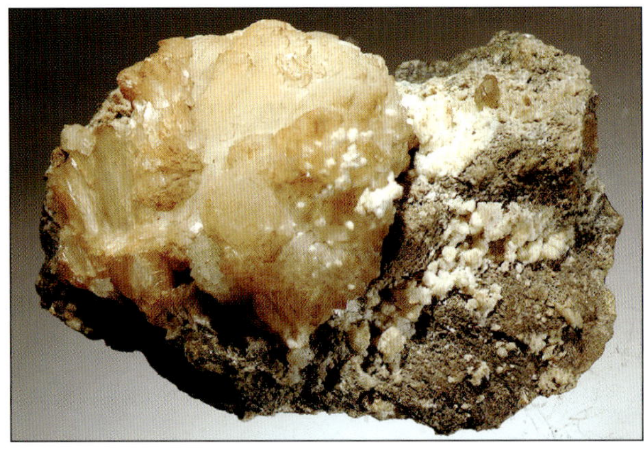

THOMSONITE. Thomsonite crystals on heulandite from Lower New Street, Paterson, Passaic County. Thomsonite fluoresces pale blue LW and SW and has a long-lasting phosphorescence. The piece weighs 10.5 oz. and is 3.5 x 2.5 x 2.3 inches. Value $22-28

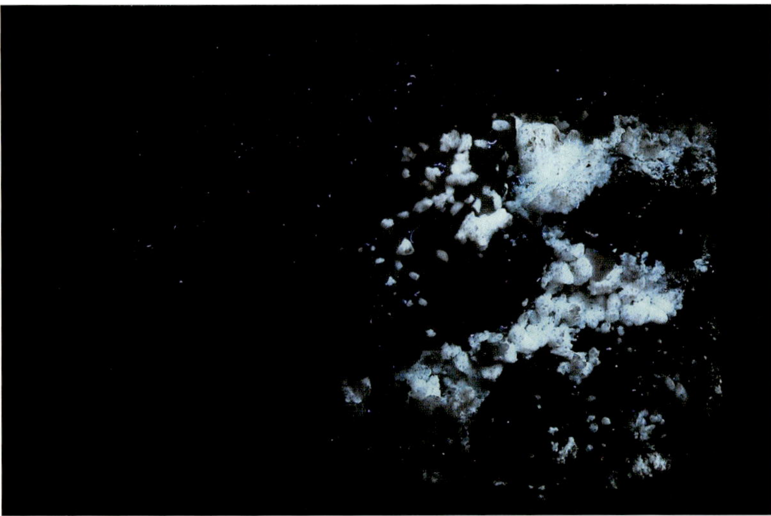

Same under SW

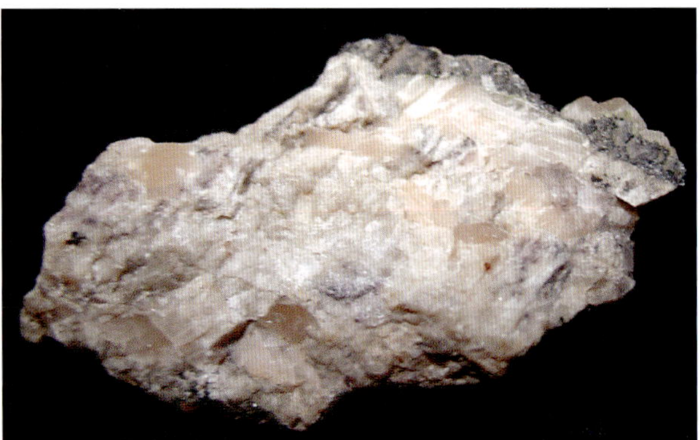

TURNEAUREITE. Turneaureite and calcite from the Franklin mine, Franklin, Sussex County. Turneaureite is a very rare Franklin mineral and fluoresces orange SW. The calcite fluoresces orange-red SW. The piece weighs 4.0 oz. and is 3.0 x 1.5 x 1.0 inches. Courtesy of Dru Wilbur.

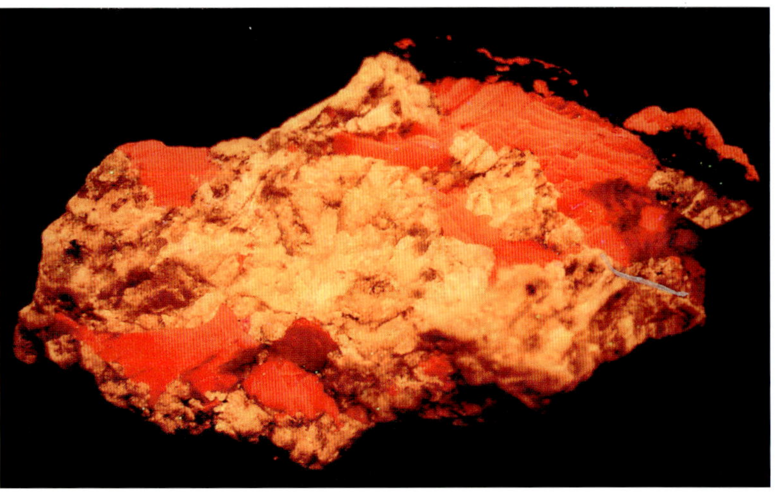

Same under SW

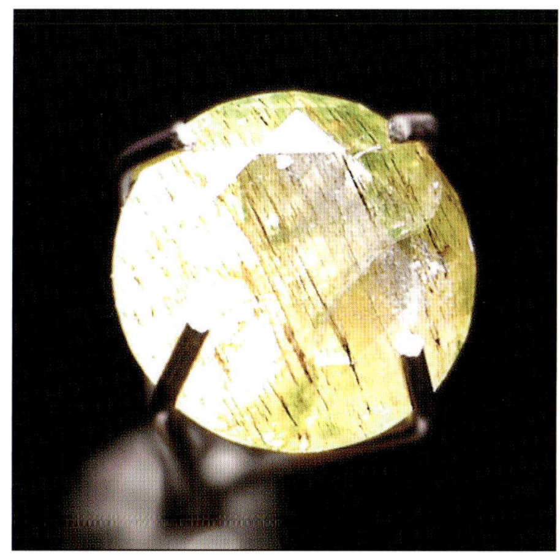

WILLEMITE. A green willemite crystal from Franklin, Sussex County faceted by Kurt Hennig. Willemite fluoresces green SW. It weighs 0.8 ct. and is 5.0 mm wide. Value $135-145

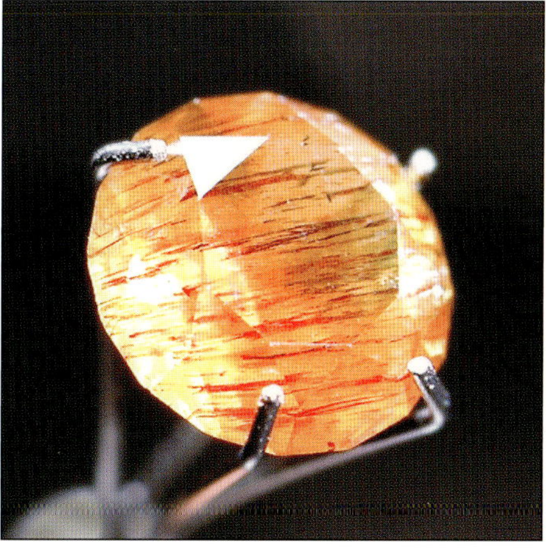

WILLEMITE. A honey-colored willemite crystal from Franklin, Sussex County faceted by Kurt Hennig. Willemite fluoresces green SW. It weighs 1.2 ct. and is 6.5 mm wide. Value $250-275

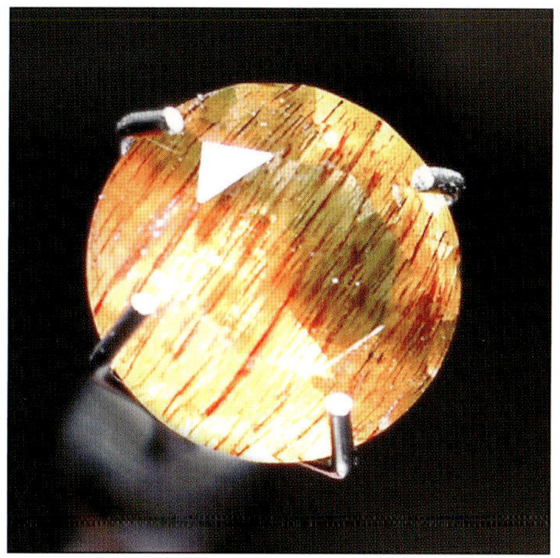

WILLEMITE. A honey-colored willemite crystal from Franklin, Sussex County faceted by Kurt Hennig. Willemite fluoresces green SW. It weighs 1.0 ct. and is 6.0 mm wide. Value $160-185

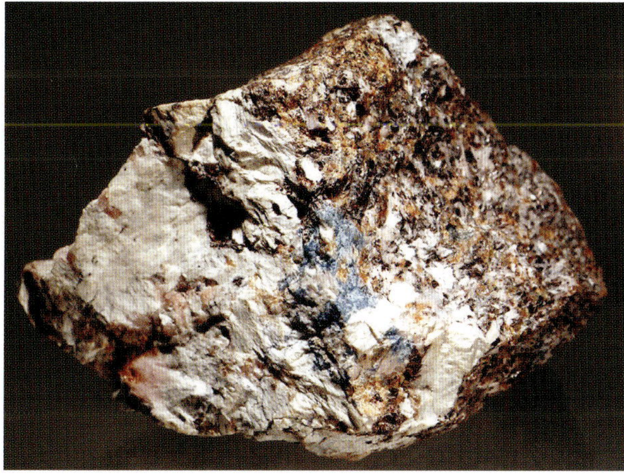

WILLEMITE, MARGAROSANITE. Blue willemite, margarosanite, bustamite, pectolite, and prehnite make up this superb specimen from Franklin, Sussex County. The willemite fluoresces green SW, the margarosanite fluoresces bright pale blue and red SW, the bustamite fluoresces red SW and LW, the pectolite fluoresces orange SW, and the prehnite fluoresces pale lavender SW. The piece weighs 1 lb. 5.0 oz. and is 4.0 x 2.8 x 2.5 inches. Value $700-750

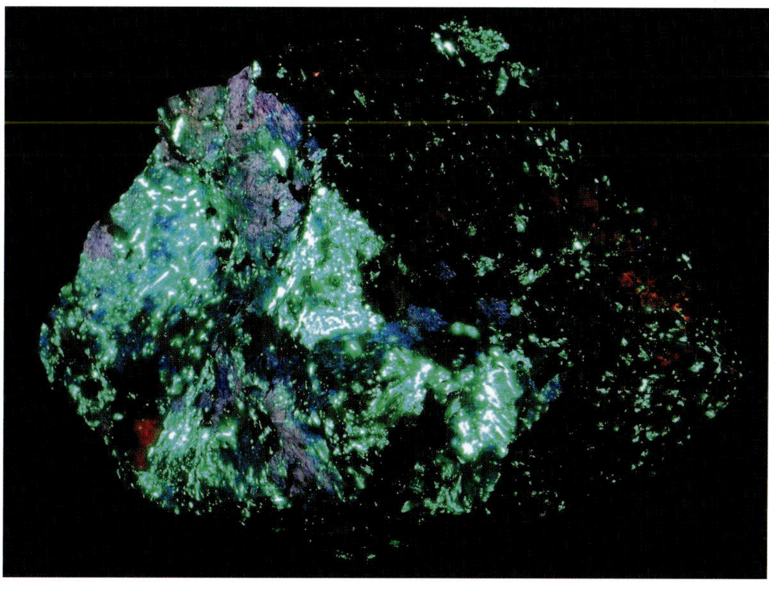

Same under SW

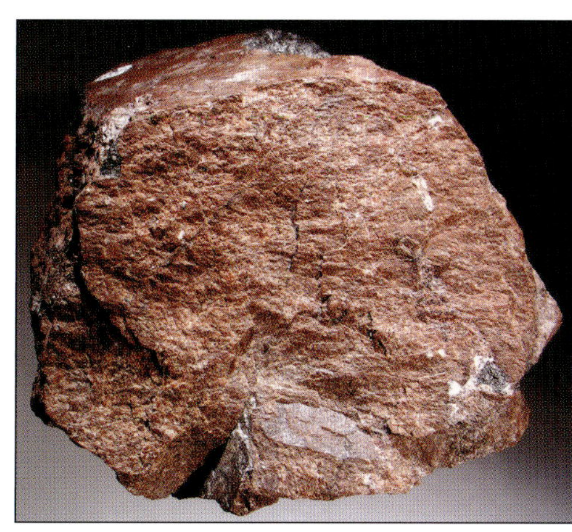

WILLEMITE. An interesting specimen of dark brown willemite with a fluorescent criss-cross pattern from the Franklin mine, Franklin, Sussex County. The willemite fluoresces green SW. The piece weighs 1 lb. 15.0 oz. and is 3.5 x 3.0 x 2.0 inches. Value $90-100

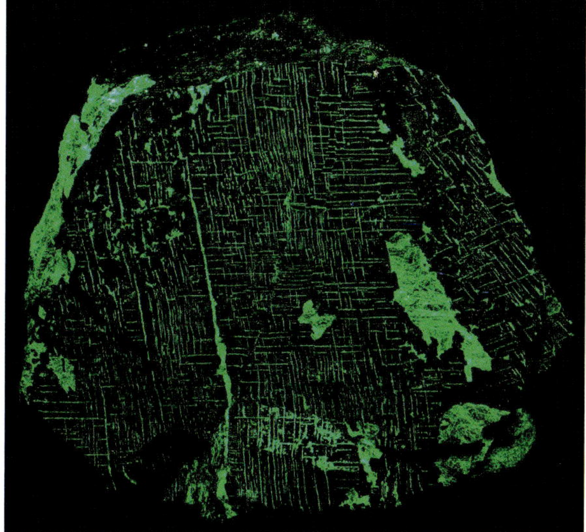

Same under SW

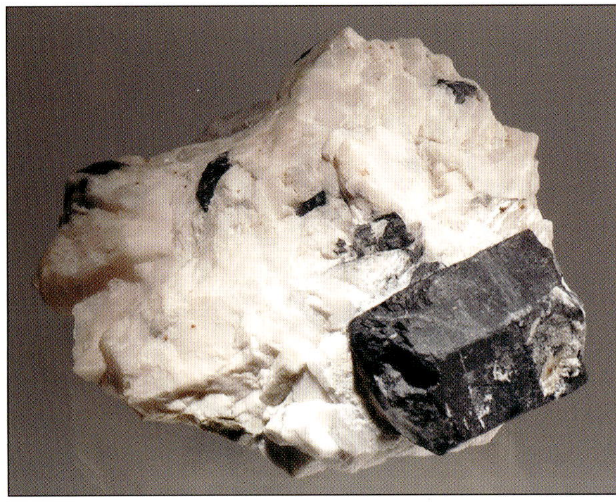

WILLEMITE, CALCITE. Large, black willemite crystals in calcite from the Sterling Hill mine, Ogdensburg, Sussex County. The willemite fluoresces green SW and the calcite fluoresces orange-red SW. The piece weighs 1 lb. 5.0 oz. and is 6.0 x 4.5 x 2.0 inches. Value $100-125

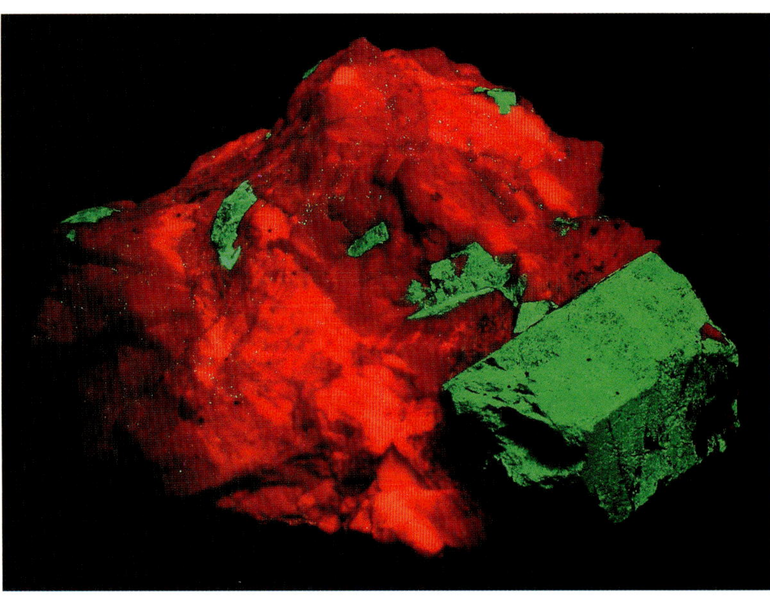

Same under SW

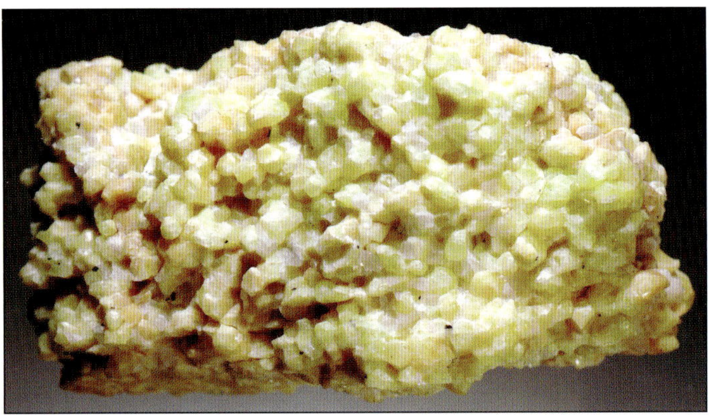

WILLEMITE. Bright lime-green willemite crystals from the Sterling Hill mine, Ogdensburg, Sussex County. The willemite fluoresces green SW. The piece weighs 7.0 oz. and is 2.5 x 1.8 x 1.3 inches. Value $45-50

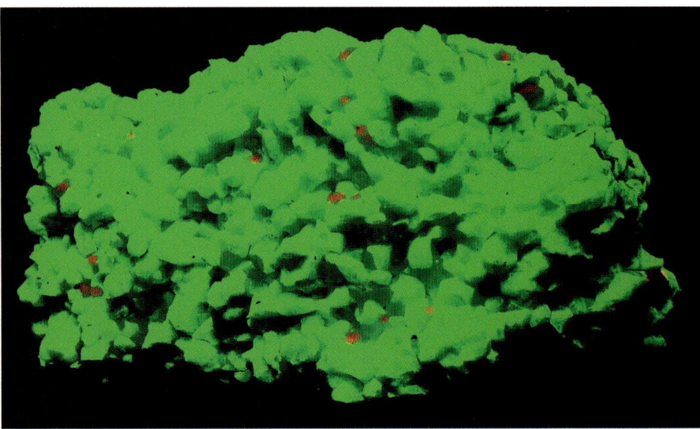

Same under SW

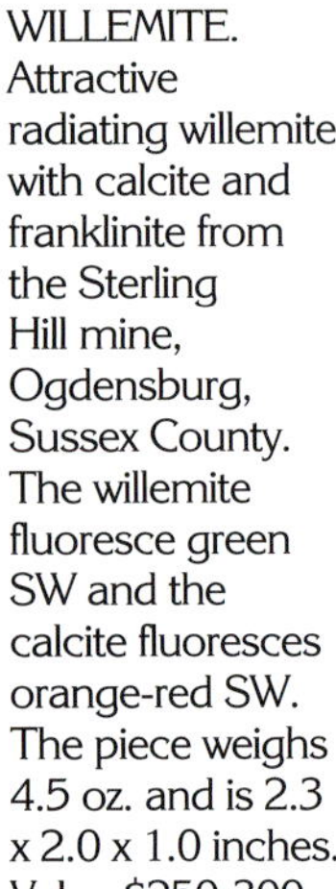

WILLEMITE. Attractive radiating willemite with calcite and franklinite from the Sterling Hill mine, Ogdensburg, Sussex County. The willemite fluoresce green SW and the calcite fluoresces orange-red SW. The piece weighs 4.5 oz. and is 2.3 x 2.0 x 1.0 inches. Value $250-300

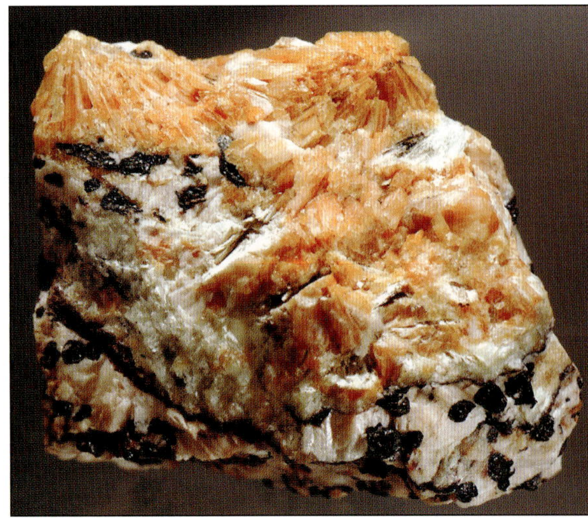

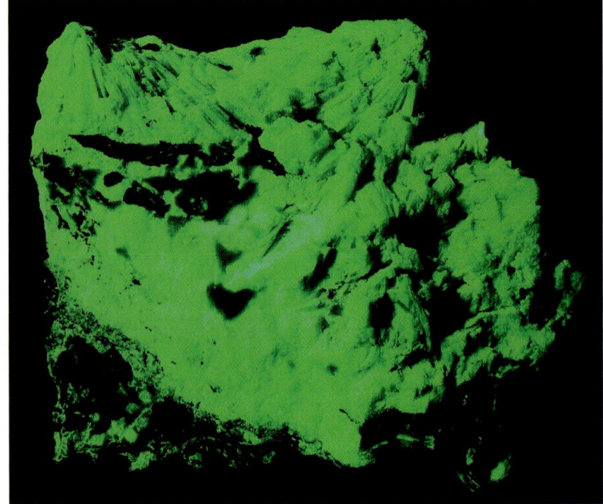

Same under SW

WILLEMITE. An interesting, radiating willemite from the Sterling Hill mine, Ogdensburg, Sussex County. It has a long-lasting phosphorescence like Franklin radiating willemite. The willemite fluoresces green SW. The piece weighs 4.3 oz. and is 3.0 x 1.8 x 0.9 inches. Value $50-60

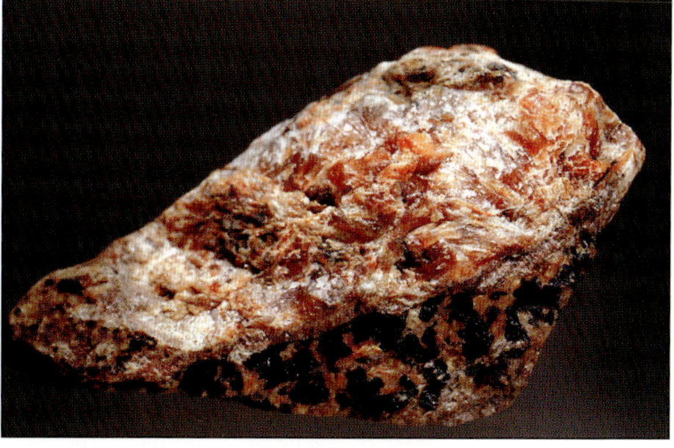

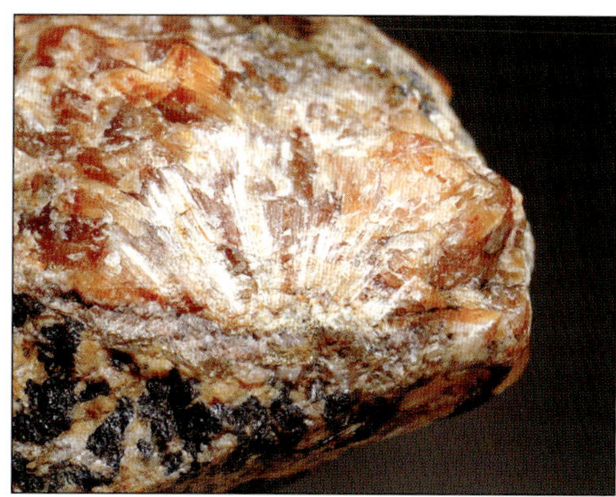

Same under SW

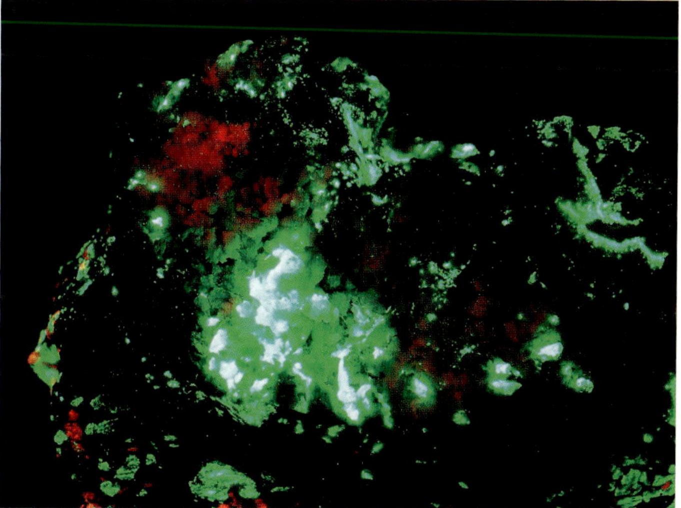

Same under SW

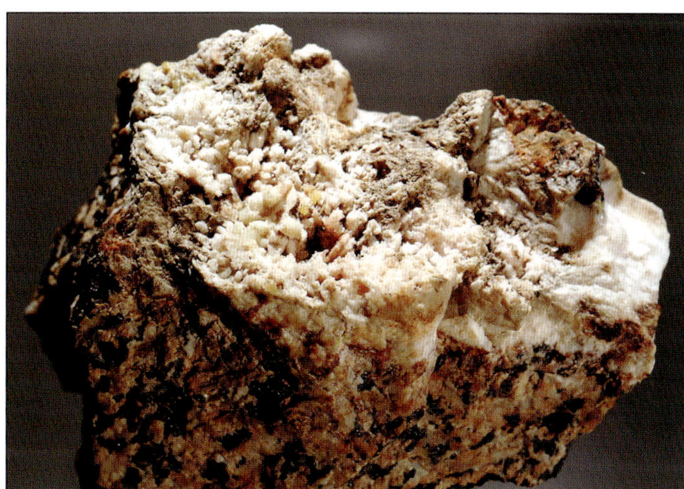

WILLEMITE, SPHALERITE. Willemite and sphalerite on calcite from the Sterling Hill mine, Ogdensburg, Sussex County. The willemite fluoresces green SW and LW and has a long-lasting phosphorescence and the sphalerite fluoresces orange LW. The piece weighs 1 lb. 1.0 oz. and is 3.8 x 2.5 x 1.8 inches. Value $40-50

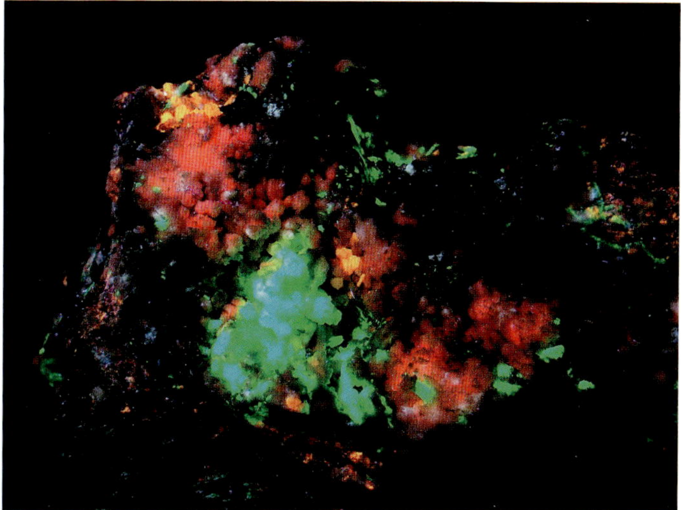

Same under LW

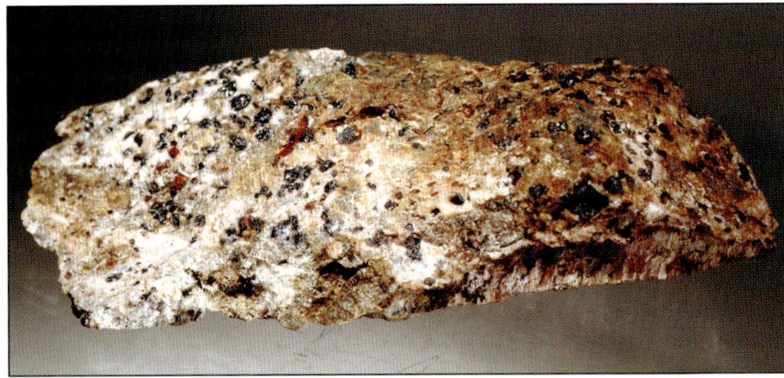

WILLEMITE. A specimen with acicular willemite crystals with sphalerite and calcite from the north ore wall of the Trotter mine, Franklin, Sussex County. The willemite crystals fluoresce green SW and have a long-lasting phosphorescence. The sphalerite fluoresces orange LW and the calcite fluoresces orange-red SW. The piece weighs 15.0 oz. and is 5.0 x 3.0 x 1.5 inches. Value $60-65

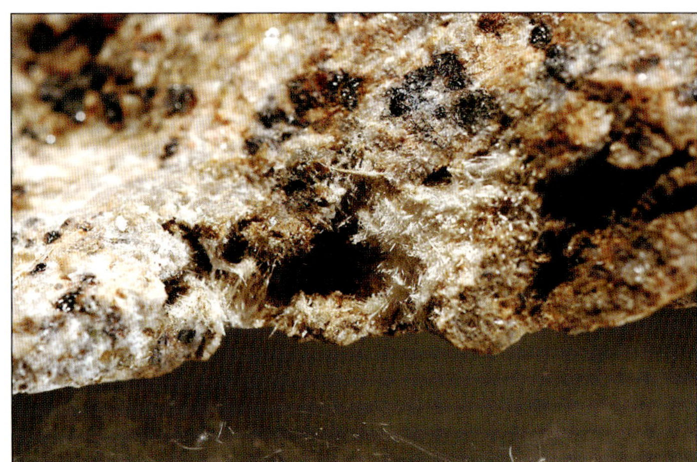

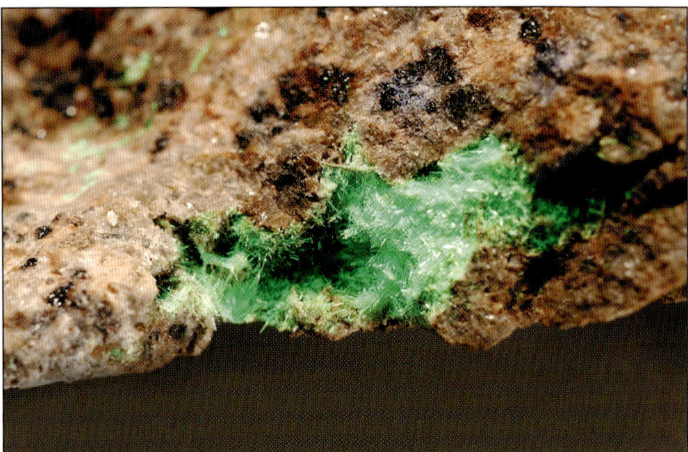

Same under SW

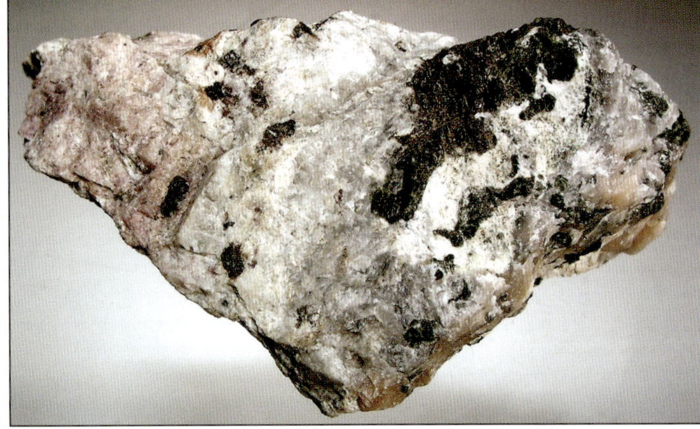

WOLLASTONITE. Fibrous wollastonite, margarosanite, and calcite from the minehillite assemblage, Franklin, Sussex County. This was found during the last few years of the mine's operation. Margarosanite fluoresces bright pale blue SW, wollastonite fluoresces orange SW, and calcite fluoresces orange-red SW. The piece weighs 2.0 oz. and is 2.8 x 1.5 x 0.8 inches. Value $350-400

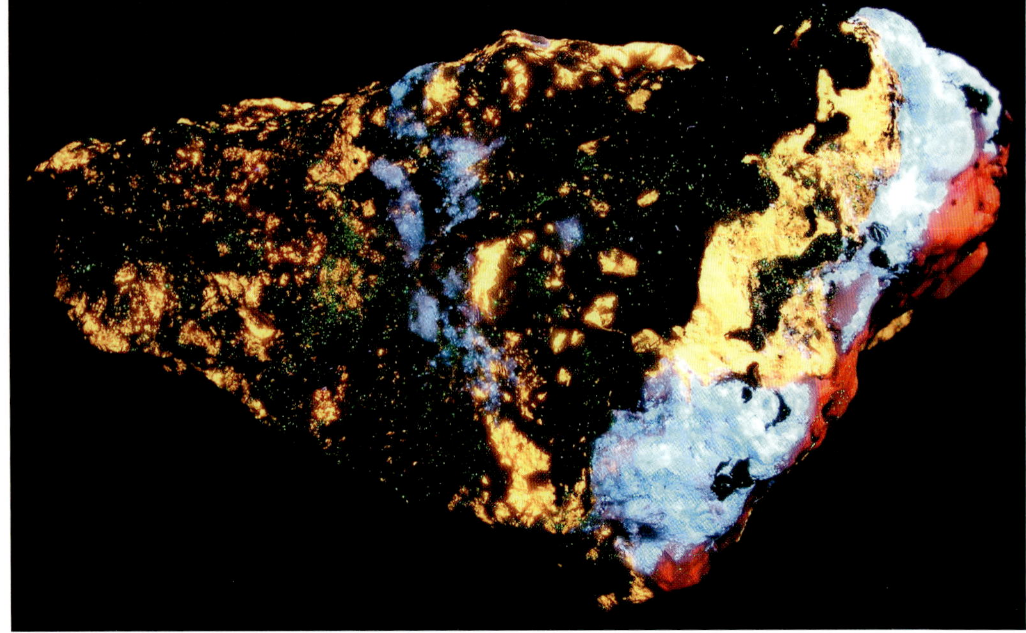

Same under SW

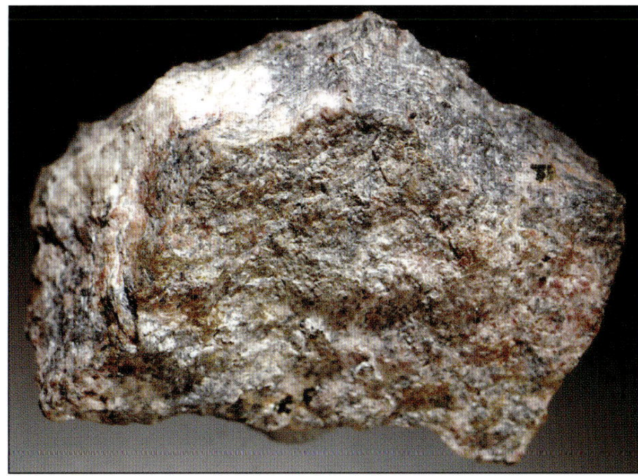

WOLLASTONITE, MARGAROSANITE. An excellent example of margarosanite and wollastonite from the minehillite assemblage, Franklin, Sussex County. The margarosanite fluoresces bright pale blue and red SW and the wollastonite fluoresces orange SW. The piece weighs 12.8 oz. and is 3.0 x 2.0 x 2.0 inches. Value $2,250-2,500

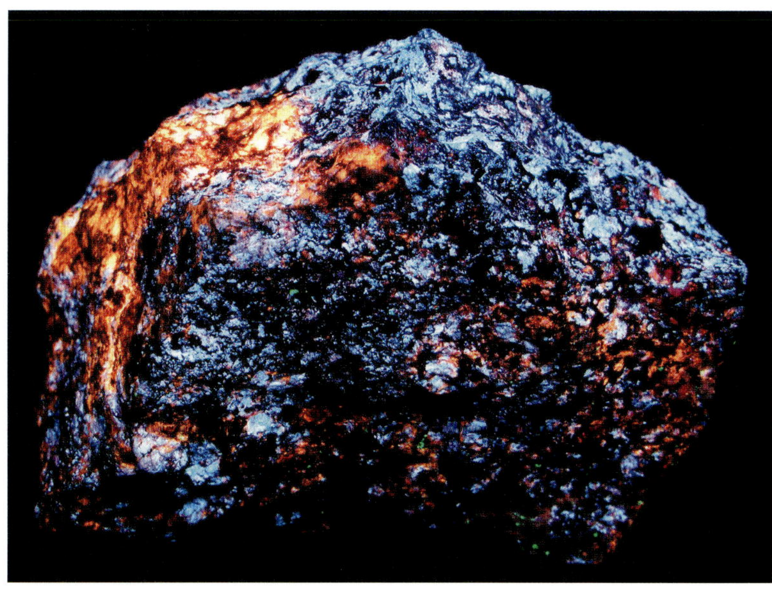

Same under SW

WOLLASTONITE. Third find wollastonite with barite on calcite from the Franklin mine, Franklin, Sussex County. "Third find" is a term that collectors use to describe wollastonite in calcite with barite. "Third find" is based upon when and where the wollastonite was found. (There is an excellent article by Dick Bostwick in the FOMS "Picking Table" journal, vol. 45, #2, Fall, 2004, that illustrates and differentiates the finds of wollastonite in the Franklin and Sterling Hill mines.) Wollastonite fluoresces orange-yellow SW and barite fluoresces white SW. The piece weighs 8.5 oz. and is 3.5 x 2.8 x 1.3 inches. Value $135-150

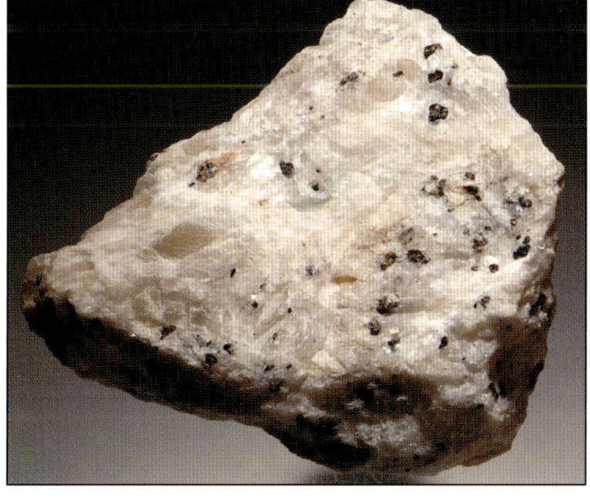

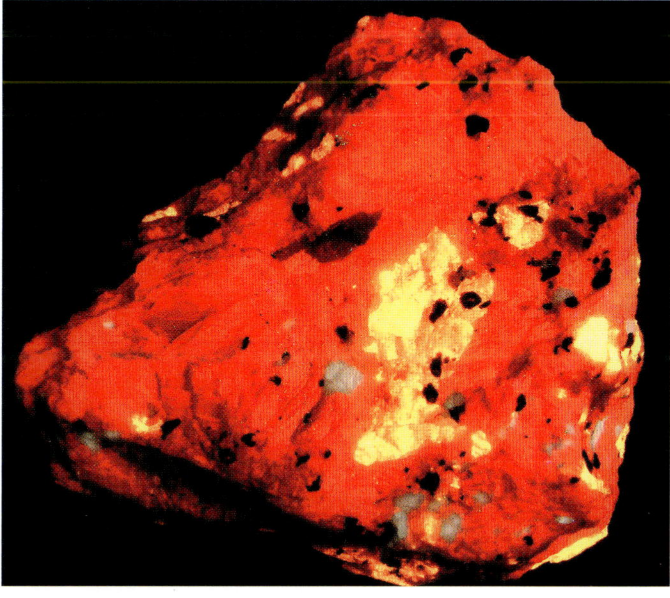

Same under SW

WOLLASTONITE. Wollastonite with calcite from the saddle leading to the Noble pit, Sterling Hill mine, Ogdensburg, Sussex County. Usually the wollastonite in this area does not have other fluorescent minerals with it. The wollastonite fluoresces yellow SW and the calcite fluoresces orange-red SW. The piece weighs 8.0 oz. and is 3.3 x 2.8 x 1.9 inches. Value $45-55

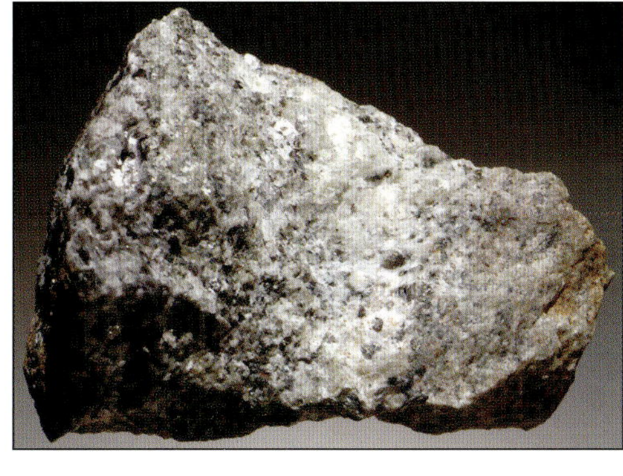

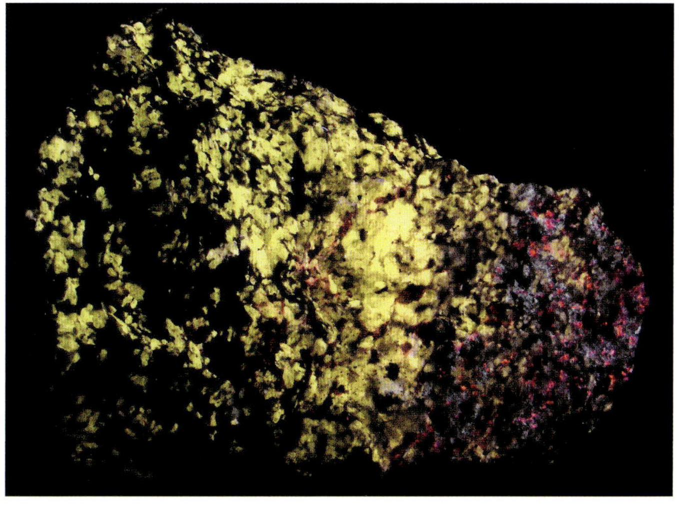

Same under SW

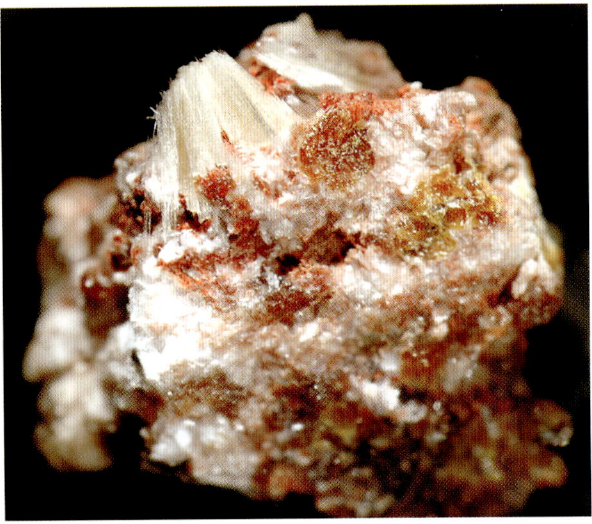

XONOTLITE. Fibrous xonotlite on prehnite and andradite from the Franklin mine, Franklin, Sussex County. The xonotlite fluoresces violet SW and the prehnite fluoresces peach SW. The piece weighs 0.3 oz. and is 0.5 x 0.6 x 0.8 inches. Value $55-65

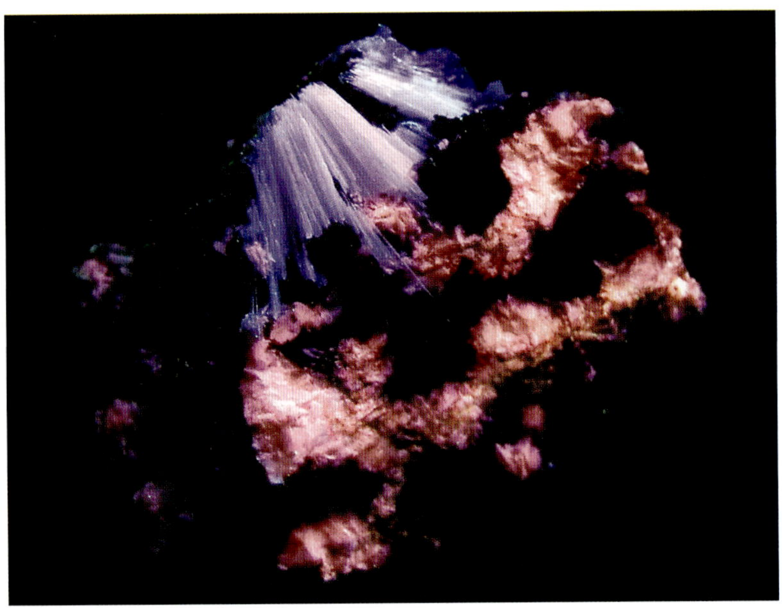

Same under SW

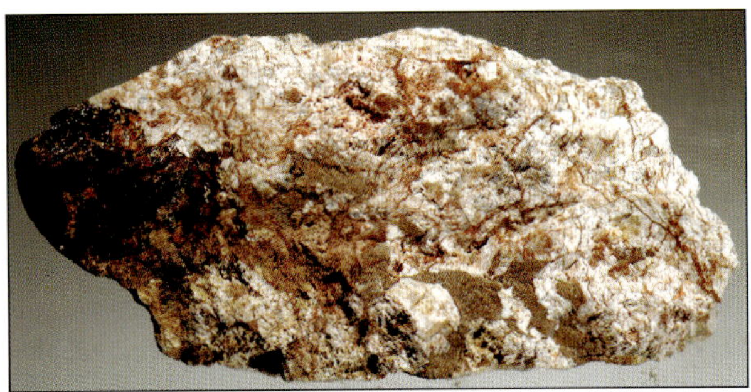

XONOTLITE. A nice example of xonotlite with andradite from the Franklin mine, Franklin, Sussex County. The piece is almost solid, massive xonotlite. The xonotlite fluoresces violet SW. The piece weighs 1.1 oz. and is 2.3 x 1.0 x 0.6 inches. Value $60-65

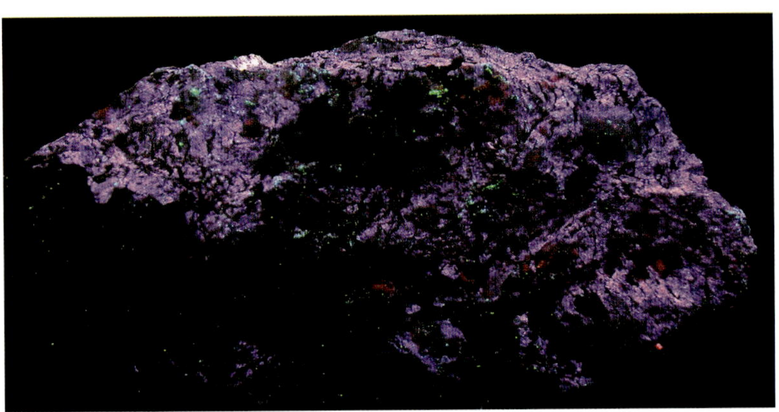

Same under SW

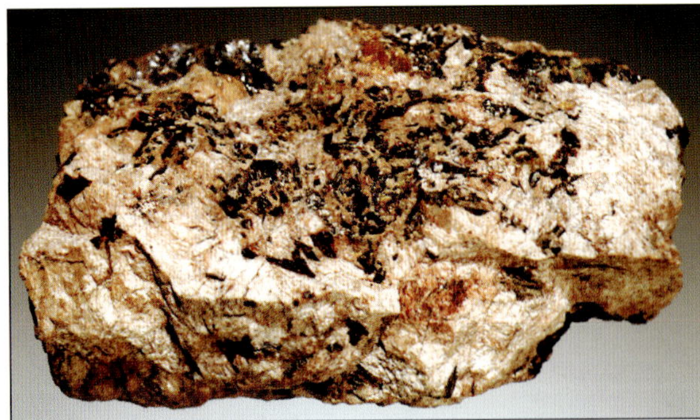

XONOTLITE, THOMSONITE. Xonotlite with thomsonite, willemite, and calcite from the Franklin mine, Franklin, Sussex County. The xonotlite fluoresces violet SW, the thomsonite fluoresces pale yellow SW, the willemite fluoresces green SW, and the calcite fluoresces orange-red SW. The piece weighs 2.3 oz. and is 2.5 x 1.5 x 0.8 inches. Courtesy of Kurt Hennig. Value $90-100

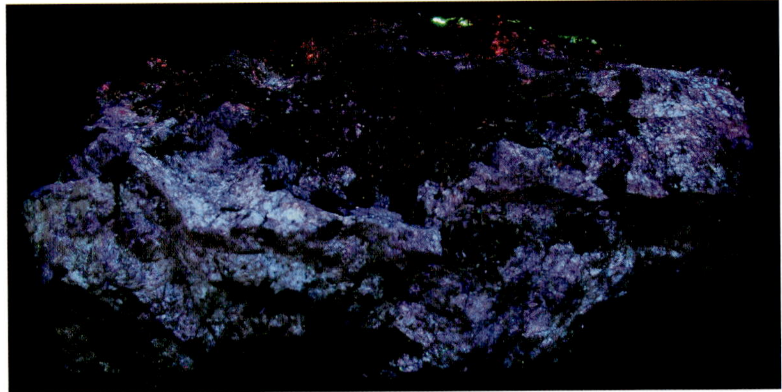

Same under SW

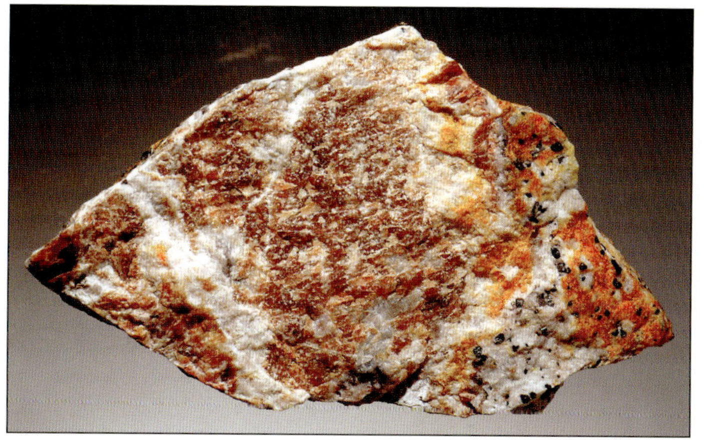

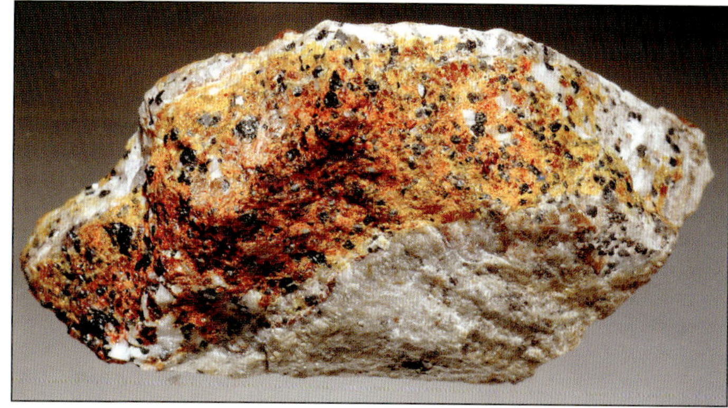

ZINCITE, SPHALERITE. An attractive piece with sphalerite and fluorescing zincite from the 340 foot level of the Sterling Hill mine, Ogdensburg, Sussex County. The sphalerite fluoresces bright orange and blue SW and LW, the zincite fluoresces pale yellow LW. The piece weighs 5.5 oz. and is 3.0 x 1.8 x 1.0 inches. Value $80-90

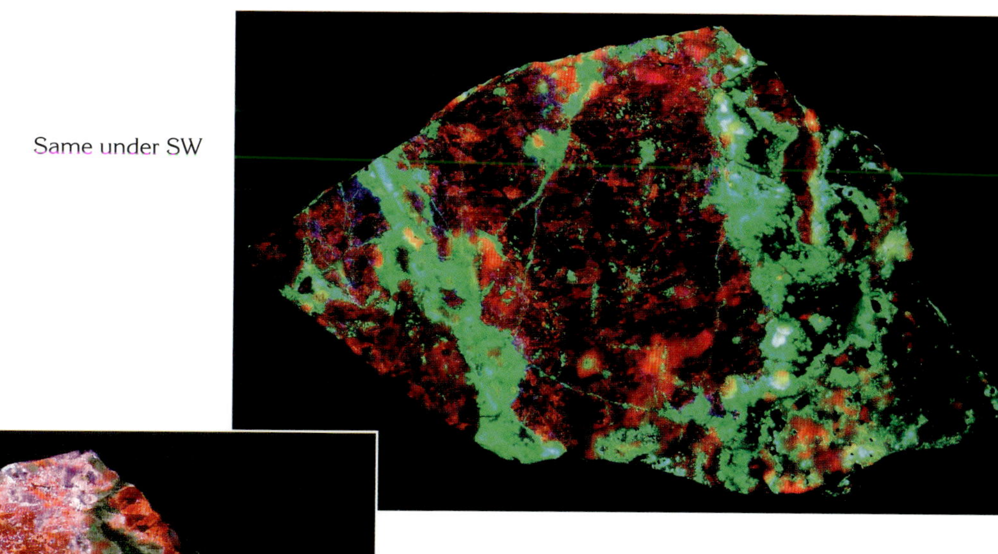

Same under SW

Same under LW and showing the fluorescing zincite.

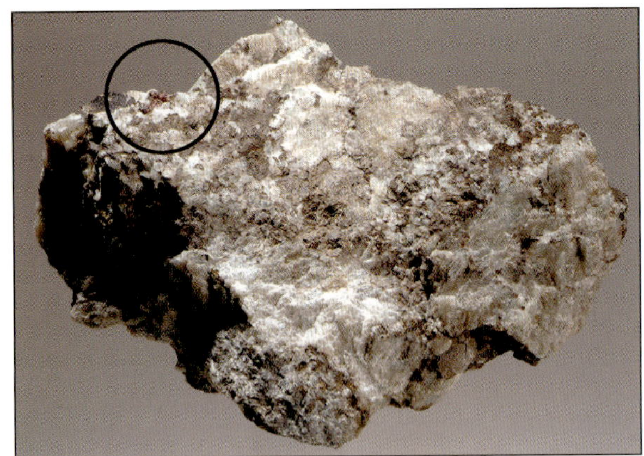

ZIRCON. Tiny zircon crystals from the Mill site, Franklin, Sussex County. Fluorescent zircons are rare from Franklin. The zircons fluoresce yellow SW. The piece weighs 12.0 oz. and is 3.5 x 2.8 x 2.5 inches. Value $60-65

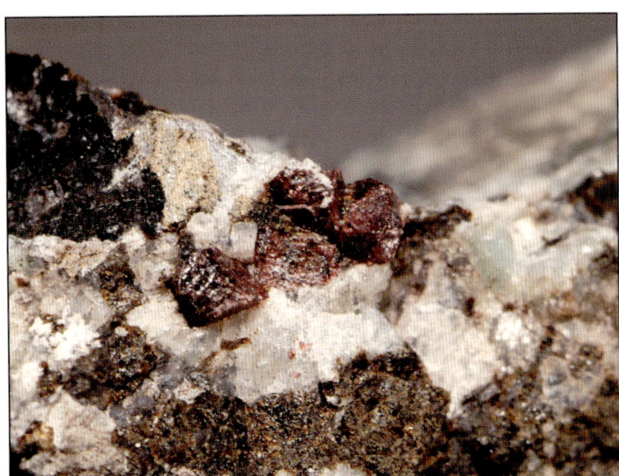

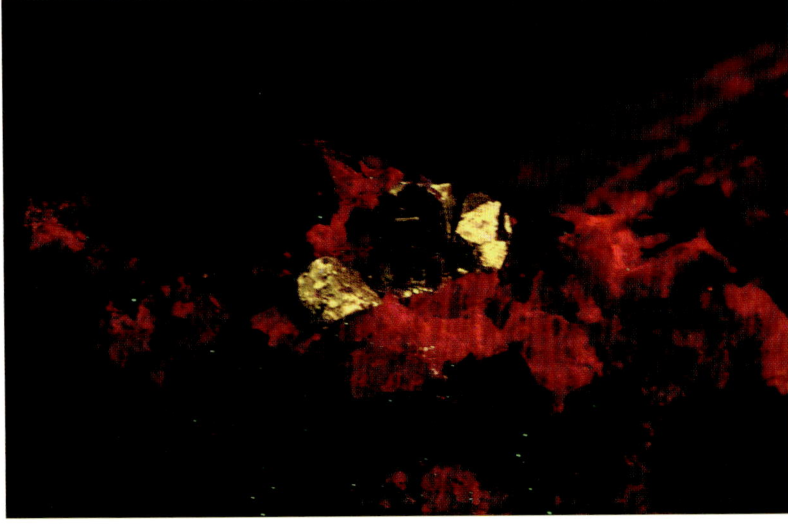

Same under SW

Eastern & Midwestern U.S.

Connecticut

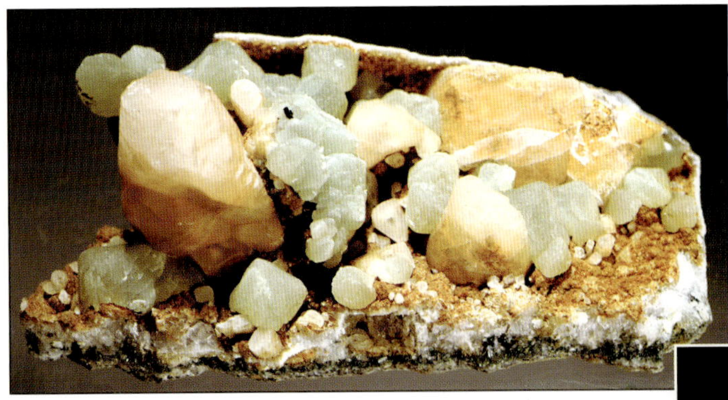

CALCITE. Calcite crystals on green prehnite crystals from the Roncari quarry, East Granby, Hartford County. The calcite fluoresces orange-red SW. The piece weighs 2.5 oz. and is 2.5 x 1.5 x 1.1 inches. Value $20-25

Same under SW

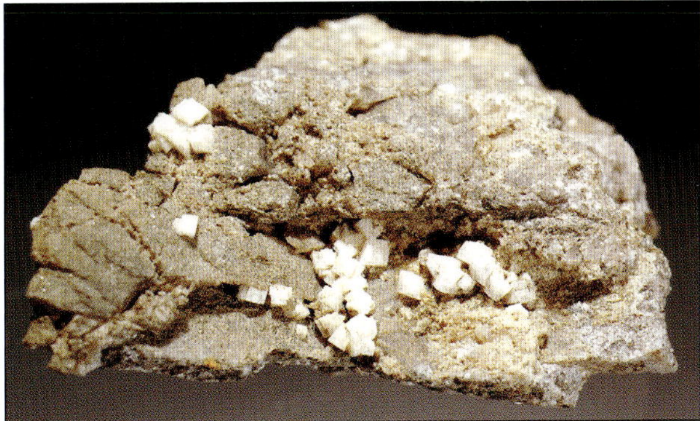

CHABAZITE. Small pseudocubic crystals of chabazite on matrix from Harwinton, Litchfield County. The chabazite fluoresces green SW. The piece weighs 1.0 oz. and is 2.0 x 1.5 x 0.8 inches. Value $25-30

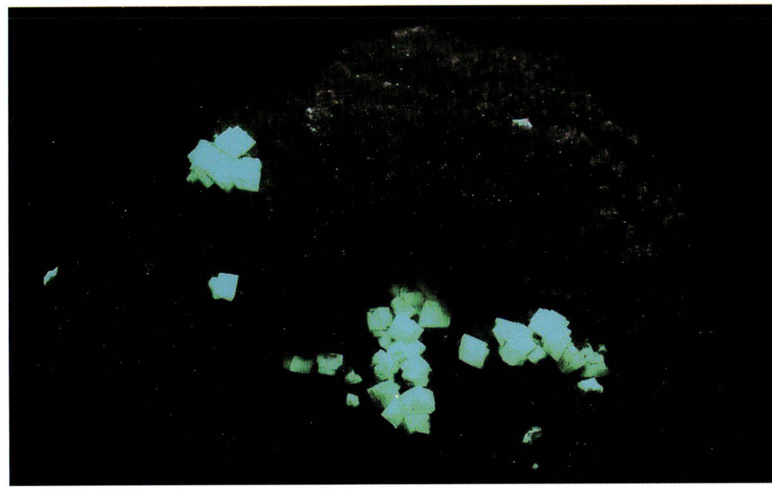

Same under SW

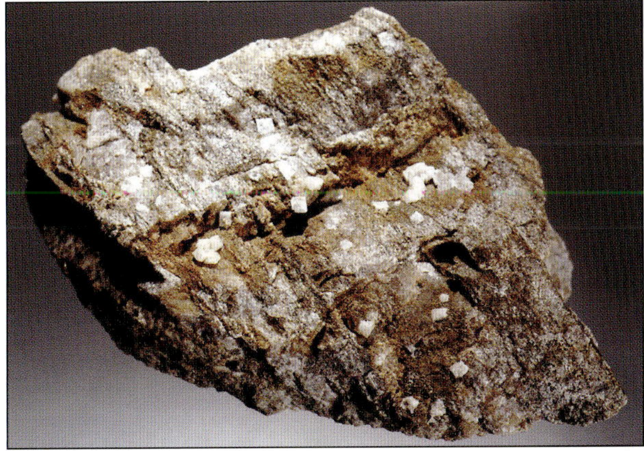

CHABAZITE. Small pseudocubic crystals of Chabazite on matrix from Harwinton, Litchfield County. The Chabazite fluoresces green SW. The piece weighs 1 lb. 2.0 oz. and is 5.0 x 2.8 x 3.0 inches. Value $35-40

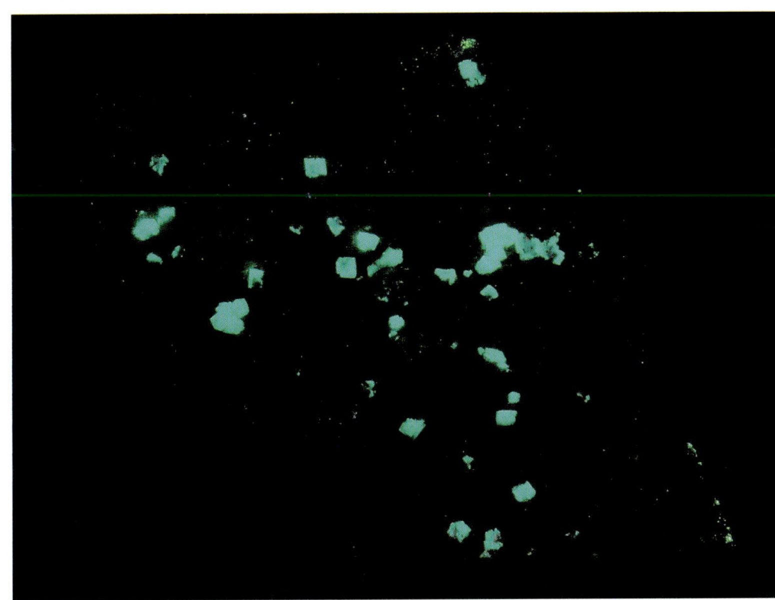

Same under SW

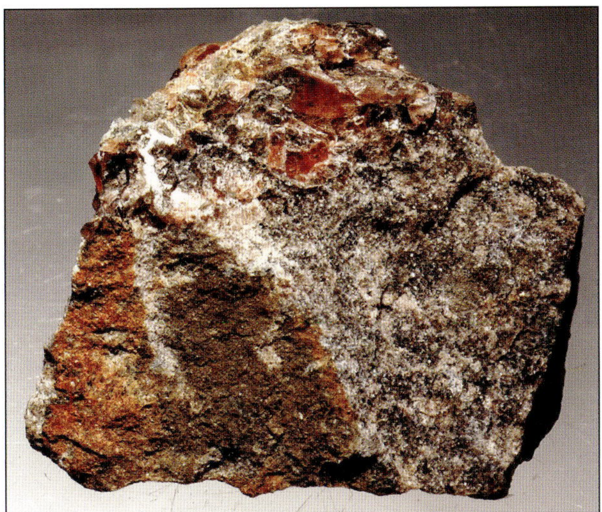

FLUORITE. Chlorophane (a varietal name for green-fluorescing, thermoluminescent fluorite) and fluorapatite from Trumbull, Fairfield County. This fluorite fluoresces green-blue under SW and LW and has a long-lasting phosphorescence and the fluorapatite fluoresces yellow LW and SW. This fluorite is sensitive to light, heat, and UV exposure and will fluoresce if heated. It should be kept in the dark (some people wrap it in aluminum foil) as it loses its ability to fluoresce after repeated exposure to light or UV. The piece weighs 1.5 oz. and is 1.5 x 1.5 x 0.9 inches. Value $6-8

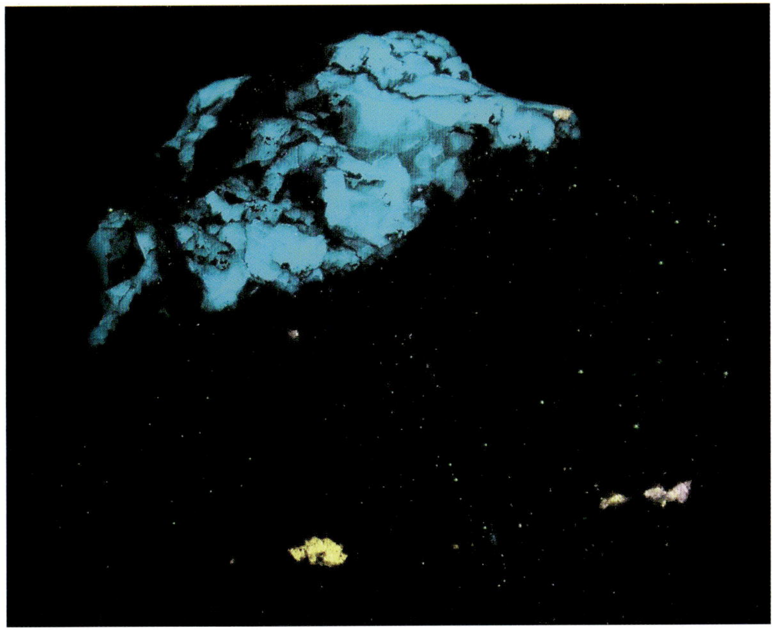

Same under SW

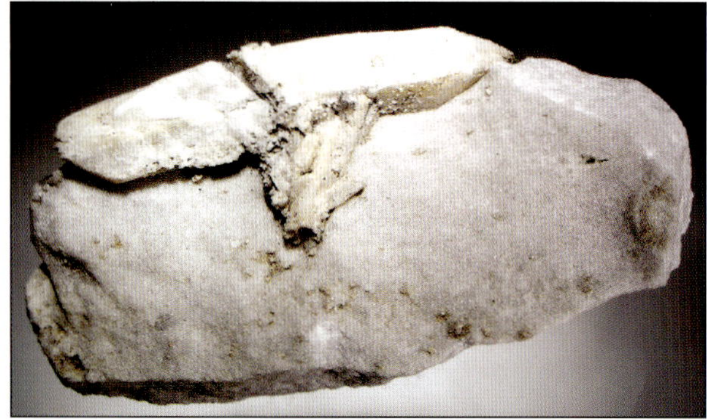

TREMOLITE. Tremolite pseudomorph after diopside crystal from the Conklin quarry, Falls Village, Litchfield County. The crystal fluoresces white under SW. The piece weighs 6.5 oz. and is 3.5 x 1.8 x 1.5 inches. Value $25-30

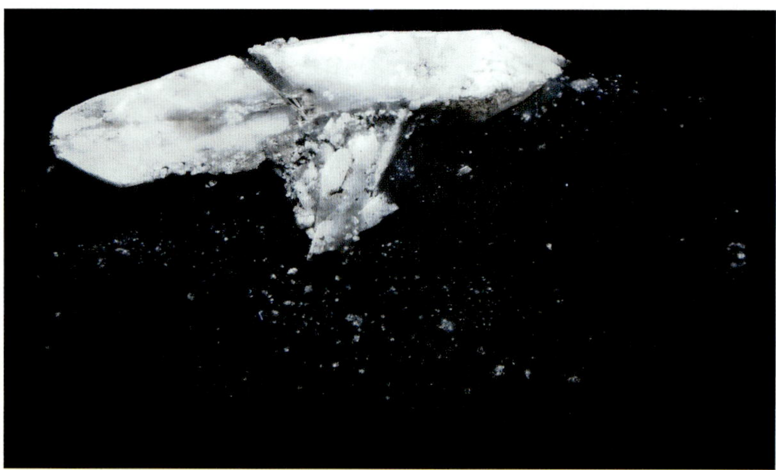

Same under SW

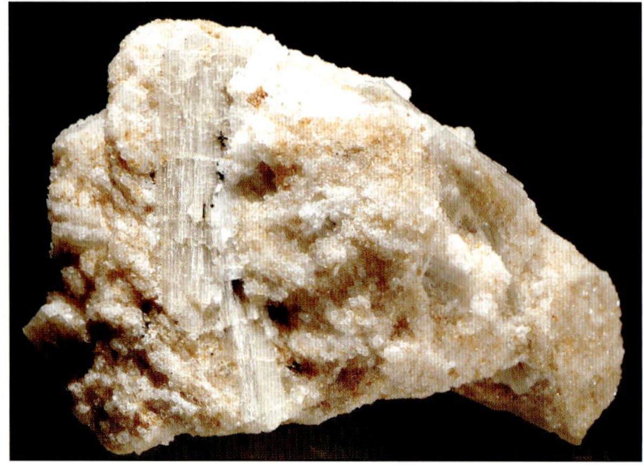

TREMOLITE. Tremolite crystals on marble with tiny chondrodite crystals from the Pfizer quarry, Canaan, Litchfield County. The tremolite fluoresces bright pale yellow and the chondrodite fluoresces yellow SW. The piece weighs 11.3 oz. and is 4.0 x 2.8 x 2.0 inches. Value $45-50

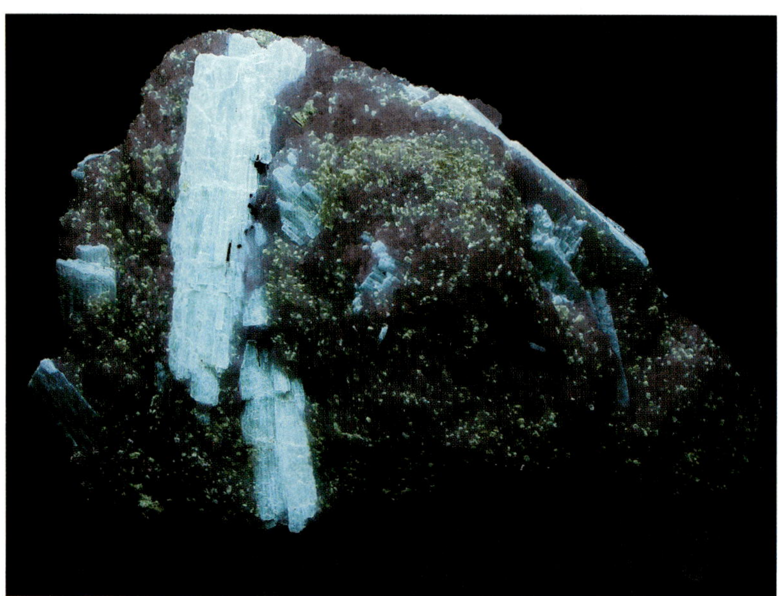

Same under SW

Florida

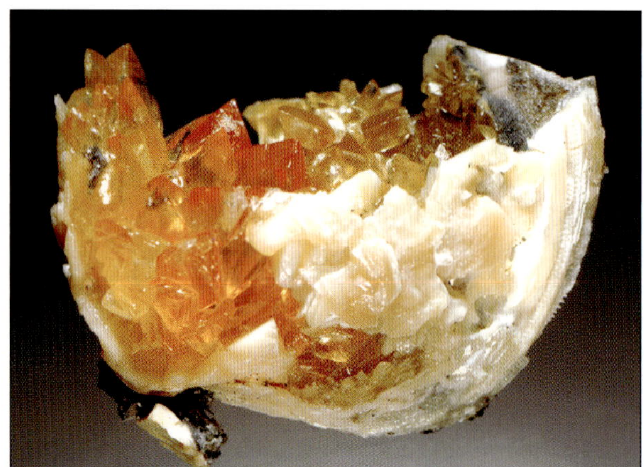

CALCITE. Dogtooth-shaped calcite inside a fossilized clamshell (Mercenaria Permanga) from Ruck's pit, Fort Drum, Okeechobee County. Time period is Pliocene- Pleistocene. The calcite crystals fluoresce bright pale yellow SW and LW and the clamshell fluoresces white SW. The piece weighs 2.3 oz. and is 2.5 x 1.5 x 1.3 inches. Value $35-40

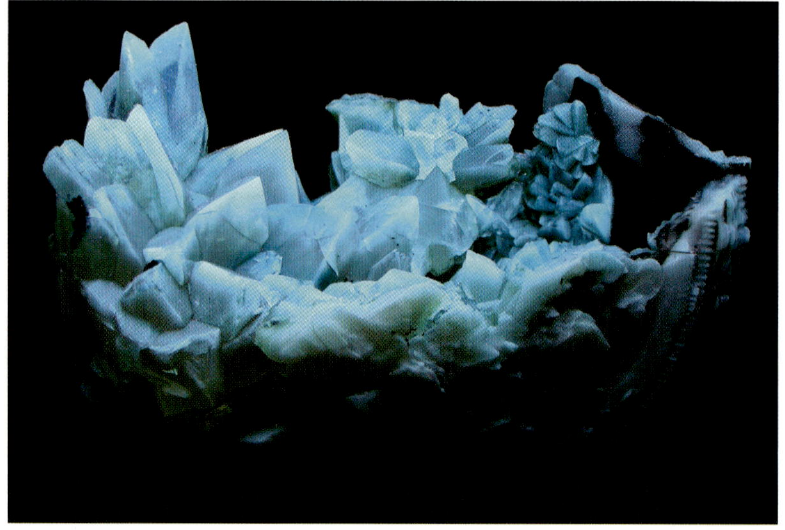

Same under SW

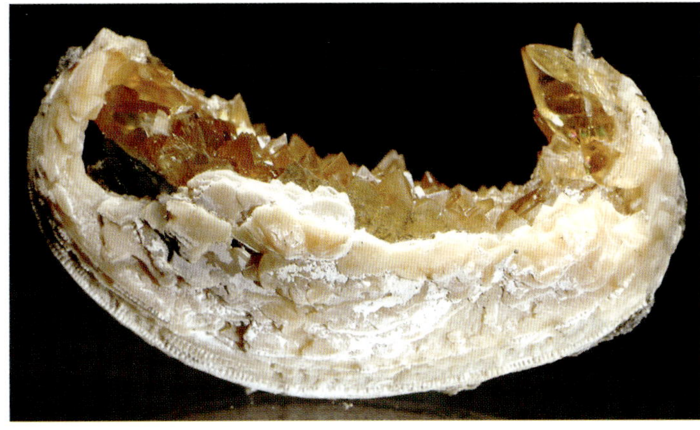

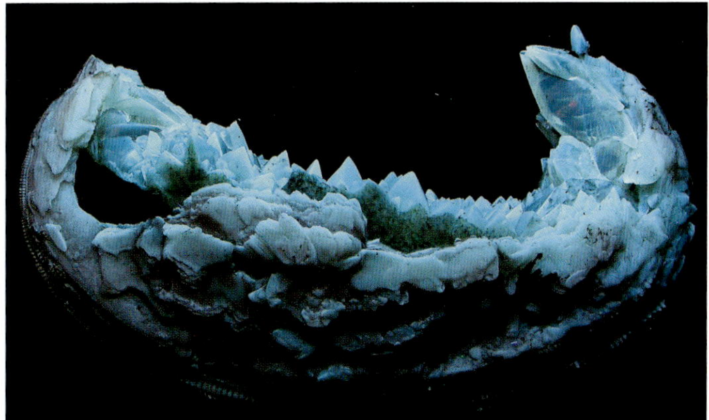

CALCITE. Dogtooth-shaped calcite inside a fossilized clamshell (Mercenaria Permanga) from Ruck's pit, Fort Drum, Okeechobee County. Time period is Pliocene- Pleistocene. The calcite crystals fluoresce bright pale yellow SW and LW and the clamshell fluoresces white SW. The piece weighs 6.0 oz. and is 4.0 x 2.0 x 1.6 inches. Value $40-45

Same under SW

Illinois

WITHERITE. White masses of crystalline witherite (a member of the aragonite group) with fluorite crystals from Cave-in-Rock, Hardin County. Witherite fluoresces pale blue SW, and cream LW and the fluorite fluoresces violet LW. The piece weighs 3 lb. 1.0 oz. and is 5.8 x 5.0 x 2.8 inches. Value $50-55

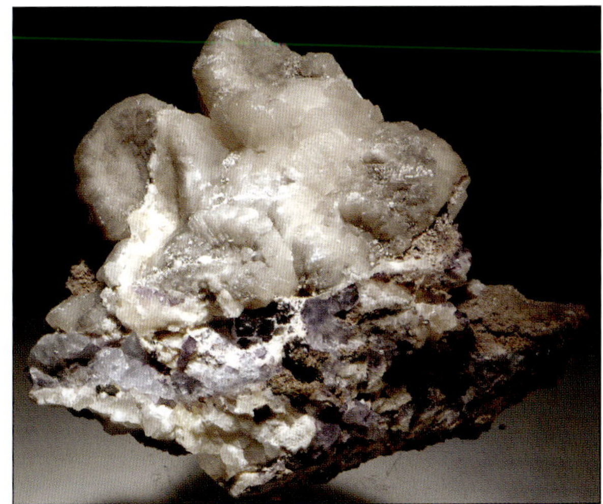

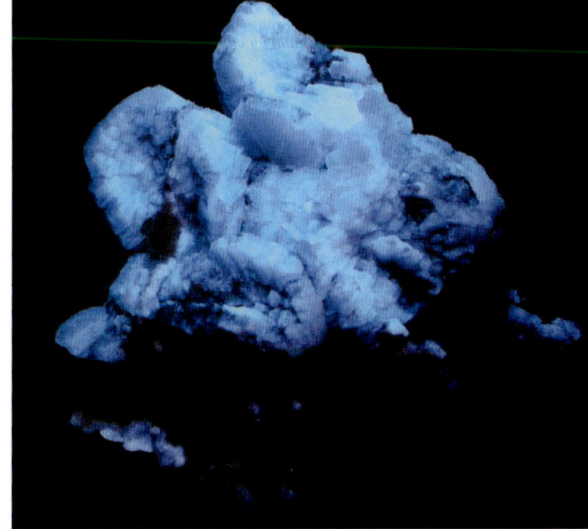

Same under SW

Maine

HYDROXYL-HERDERITE. Hydroxyl-herderite crystals with quartz crystals from the Bennett quarry, Buckfield, Oxford County. The hydroxyl-herderite fluoresces pale yellow SW. The piece weighs 1.8 oz. and is 2.8 x 2.3 x 1.0 inches. Value $25-30

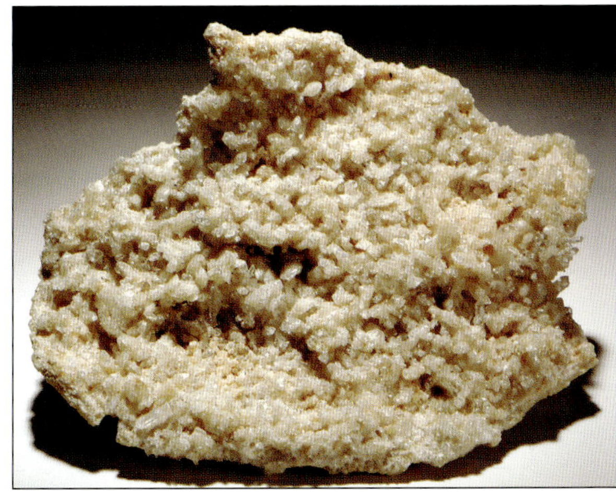

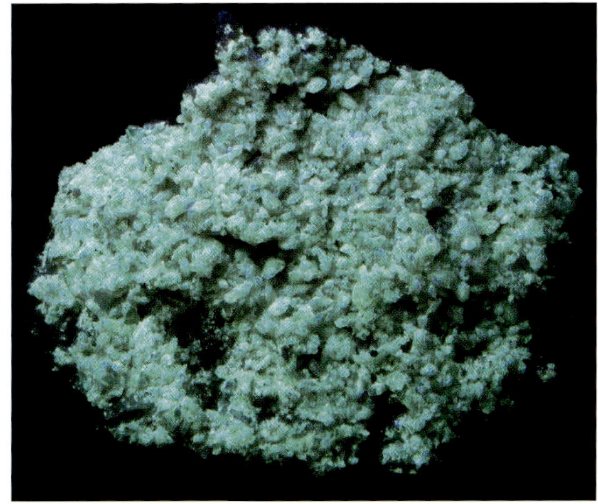

Same under SW

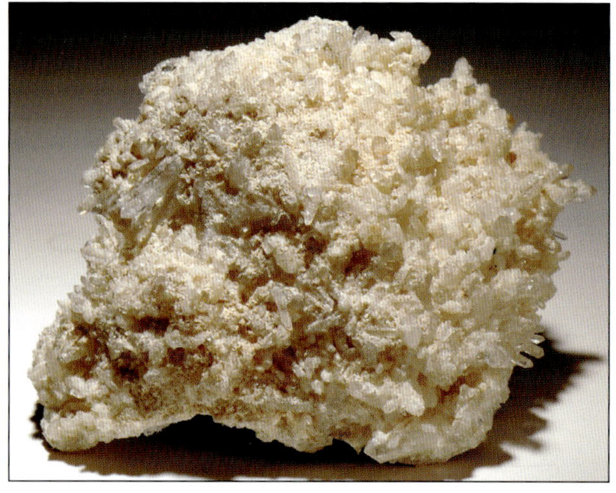

HYDROXYL-HERDERITE. Hydroxyl-herderite crystals with quartz crystals from the Bennett quarry, Buckfield, Oxford County. The hydroxyl-herderite fluoresces pale yellow SW. The piece weighs 1.8 oz. and is 2.3 x 1.8 x 1.3 inches. Value $25-30

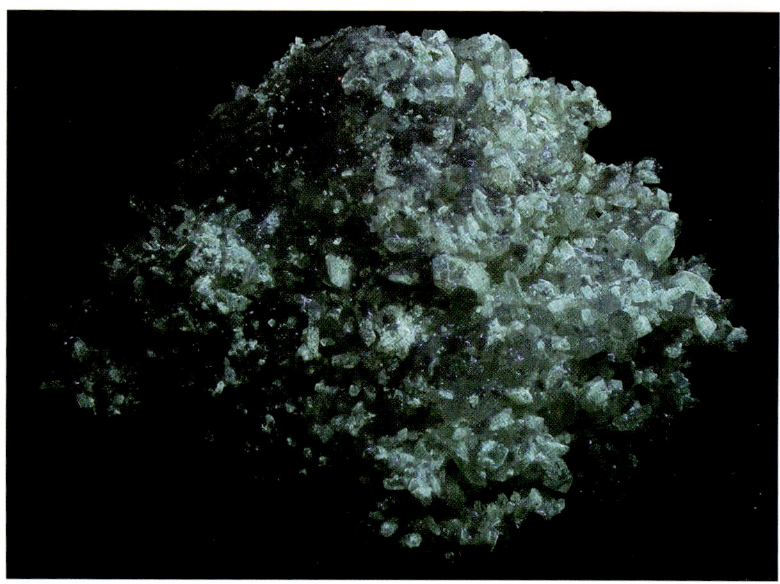

Same under SW

Massachusetts

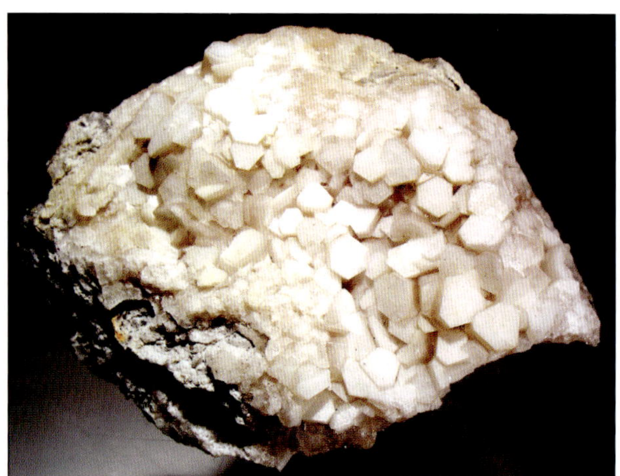

CALCITE. Nailhead-shaped calcite on calcite from the Northfield quarry, Northfield, Franklin County. The nailhead crystals fluoresce pink. The calcite under them fluoresces pink-orange SW. The piece weighs 11.3 oz. and is 3.5 x 3.3 x 2.0 inches. Value $30-35

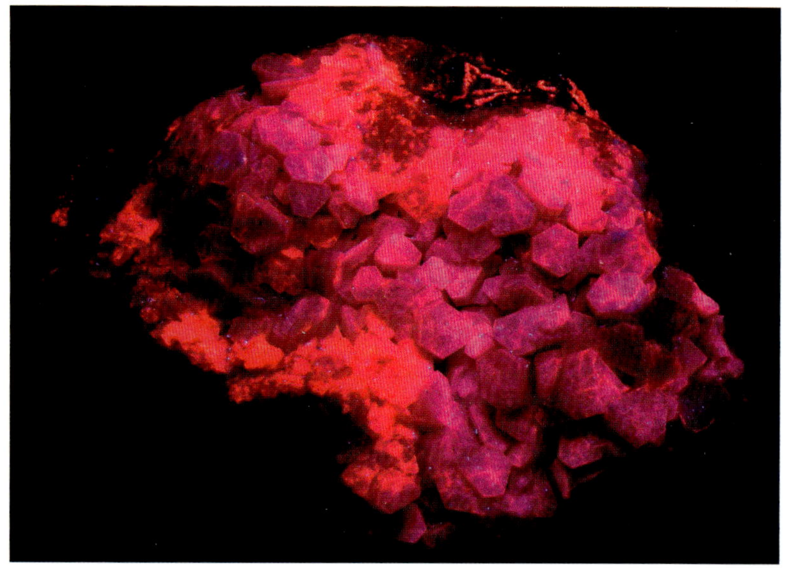

Same under SW

New Hampshire

ALBITE. A spear of smoky quartz on microcline and albite from North Conway, Carroll County. The albite fluoresces velvety red SW while a coating on the quartz fluoresces a weak white SW. The piece weighs 5.3 oz. and is 2.5 x 2.5 x 2.3 inches. Value $45-50

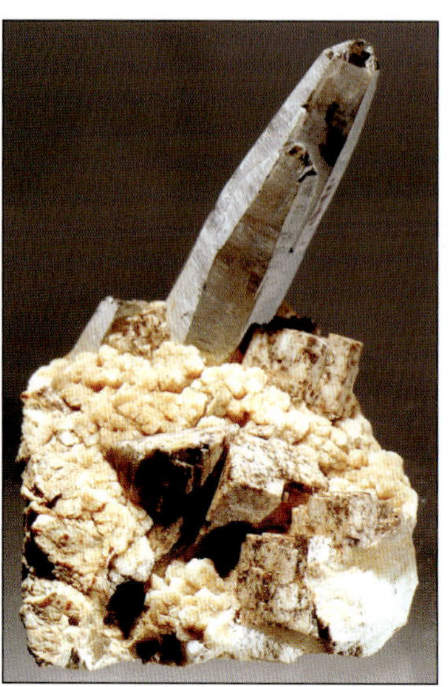

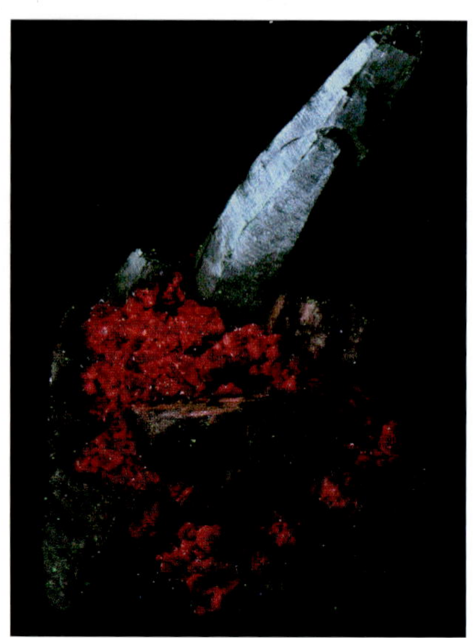

Same under SW

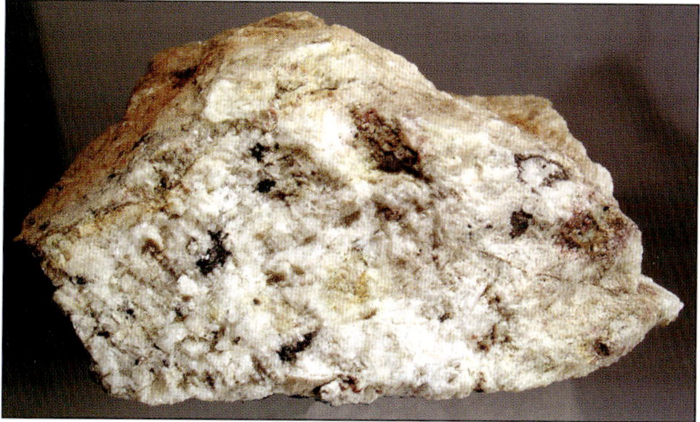

AUTUNITE. Autunite from the Palermo mine, North Groton, Grafton County. Autunite fluoresces green SW and LW and is radioactive. The piece weighs 9.0 oz. and is 3.0 x 2.8 x 1.5 inches. Value $30-35

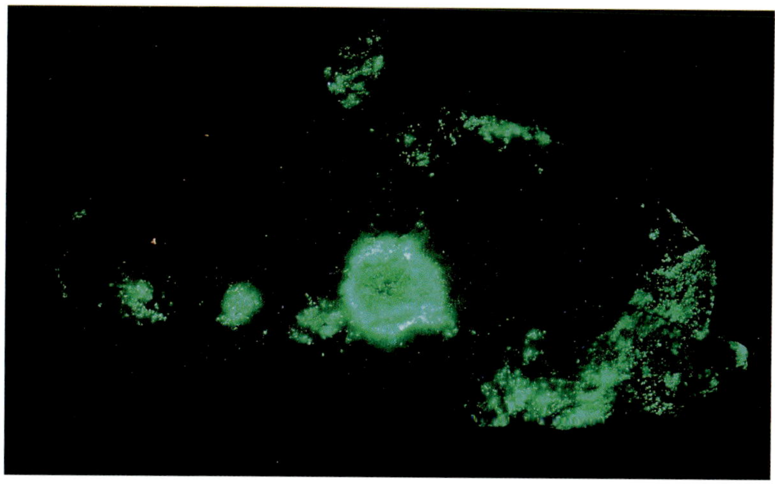

Same under SW

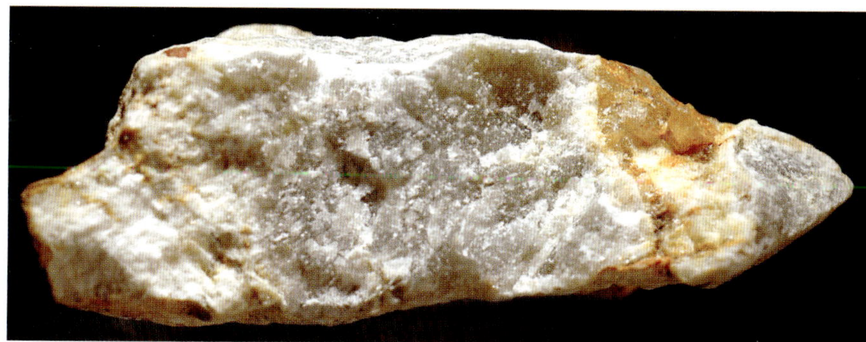

EUCRYPTITE. Eucryptite with hyalite surrounding it from the Parker Mountain mine, Center Strafford, Strafford County. Eucryptite fluoresces red SW and hyalite fluoresces green SW. The piece weighs 2.5 oz. and is 2.6 x 1.8 x 0.8 inches. Value $30-35

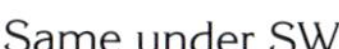

Same under SW

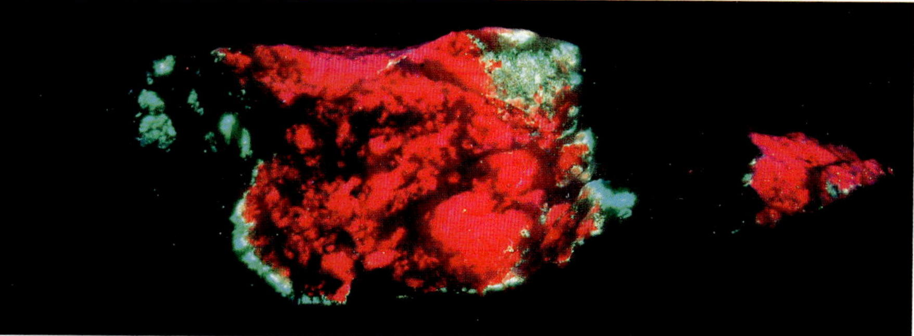

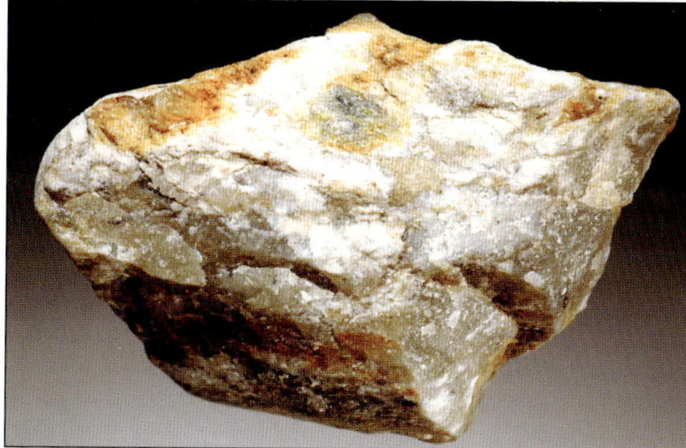

EUCRYPTITE. Eucryptite with hyalite and fluorapatite from the Parker Mountain mine, Center Strafford, Strafford County. Eucryptite fluoresces cherry-red SW, the hyalite fluoresces green SW, and the fluorapatite fluoresces orange SW. The piece weighs 1.3 oz. and is 2.3 x 1.5 x 0.8 inches. Value $30-35

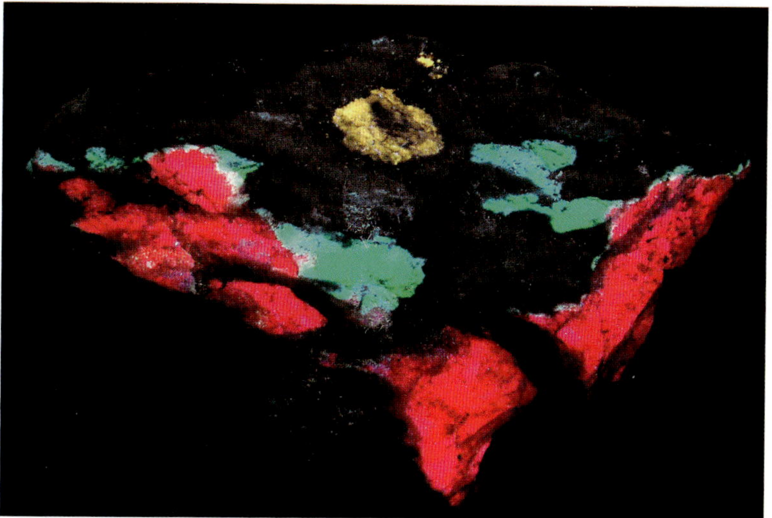

Same under SW

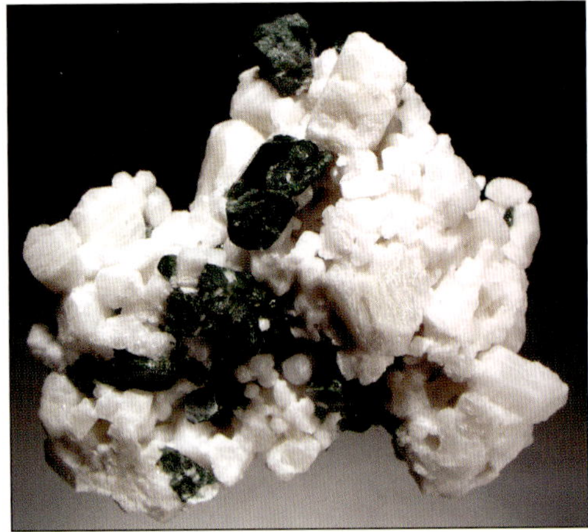

ALBITE. Albite with crystals of diopside from Pitcairn, St. Lawrence County. The albite fluoresces pink-red SW. The piece weighs 5.5 oz. and is 2.8 x 2.5 x 2.3 inches. Value $30-35

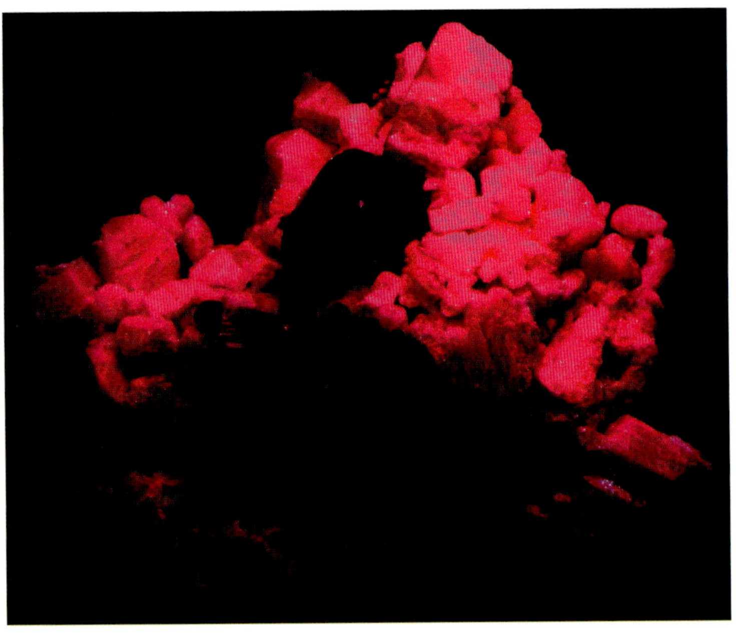

Same under SW

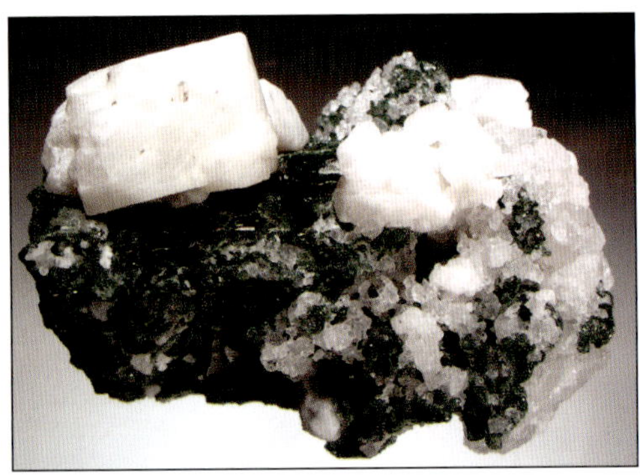

ALBITE. Albite with crystals of diopside, titanite, and quartz from Pitcairn, St. Lawrence County. The albite fluoresces pink-red SW. The piece weighs 3.0 oz. and is 2.3 x 1.5 x 1.3 inches. Value $30-35

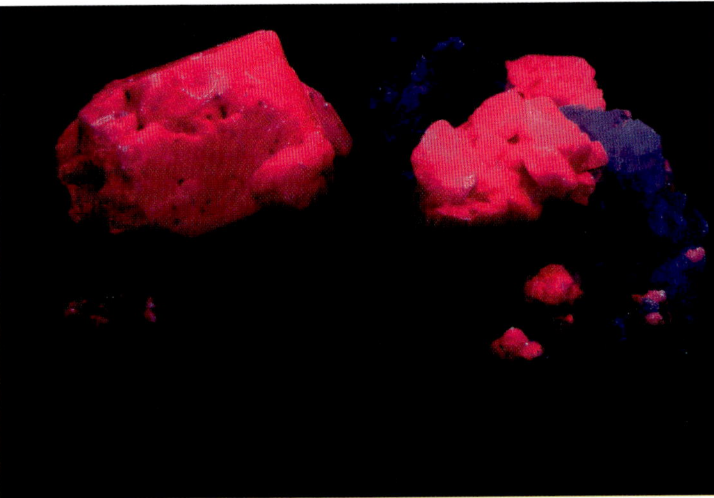

Same under SW

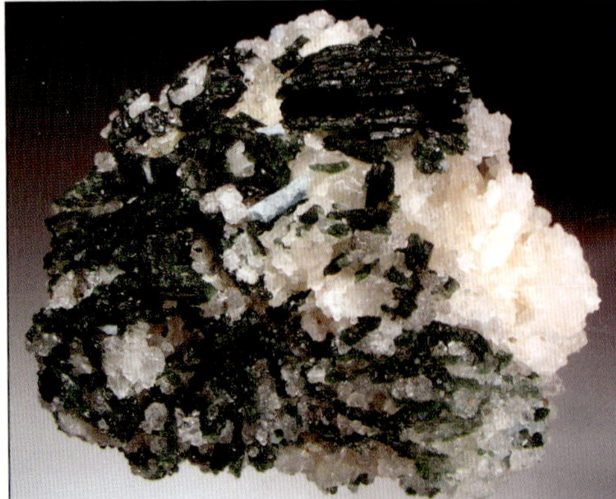

ALBITE, FLUORAPATITE. Albite with tiny pale blue crystals of fluorapatite with diopside from Pitcairn, St. Lawrence County. The albite fluoresces pink-red and the fluorapatite crystals fluoresce pale blue SW. The piece weighs 5.0 oz. and is 2.5 x 2.5 x 1.5 inches. Value $35-40

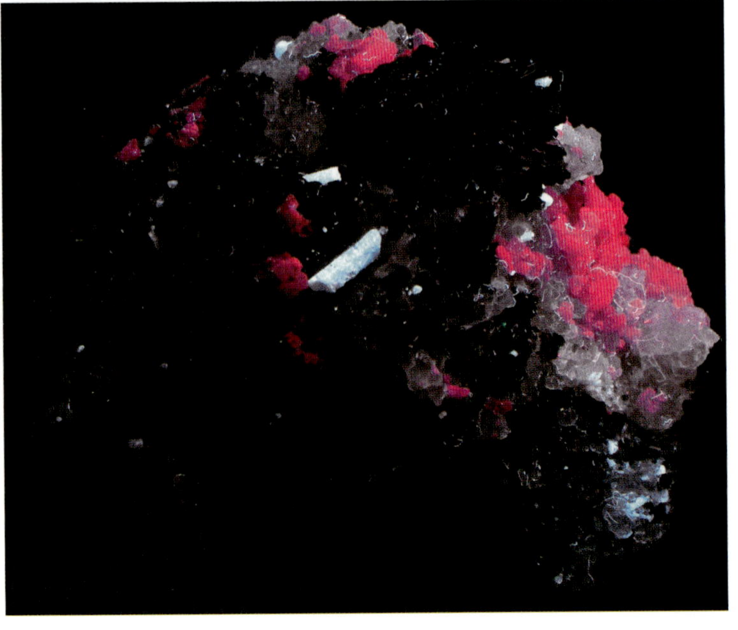

Same under SW

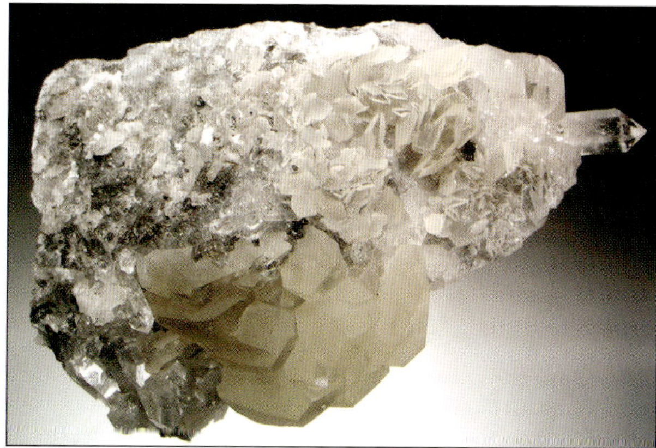

CALCITE. Nailhead-shaped calcite with quartz from the LaFarge quarry, Ravina, Albany County. The calcite fluoresces an unusual pale gray SW. The piece weighs 9.0 oz. and is 3.3 x 2.5 x 1.8 inches. Value $25-30

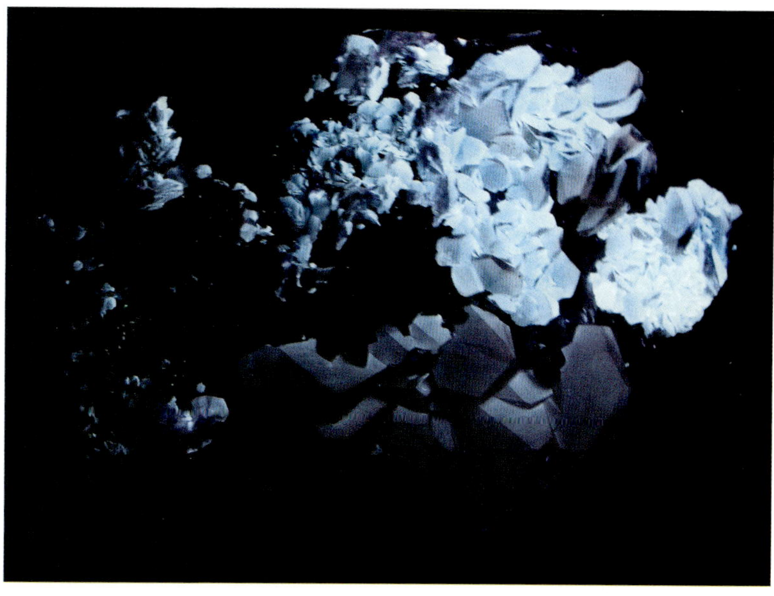

Same under SW

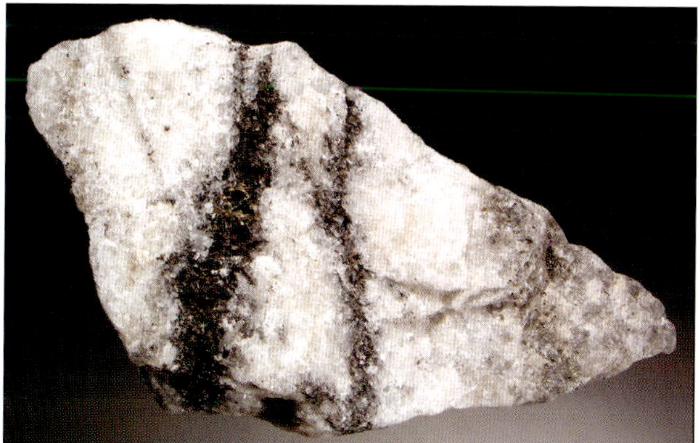

DIOPSIDE, PHLOGOPITE. Diopside and phlogopite from the Zinc Corp. of America #4 mine, Balmat, St. Lawrence County. The diopside fluoresces pale blue and the phlogopite fluoresces yellow SW. The piece weighs 5.0 oz. and is 3.5 x 2 x 1.0 inches. Value $40-45.

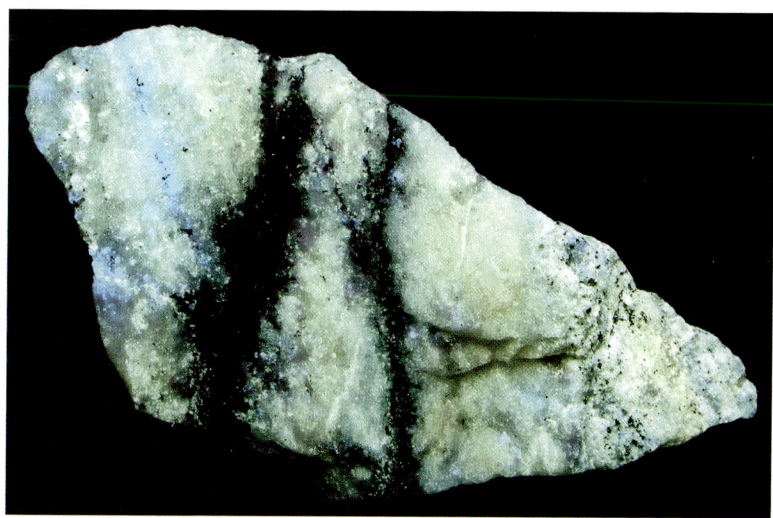

Same under SW

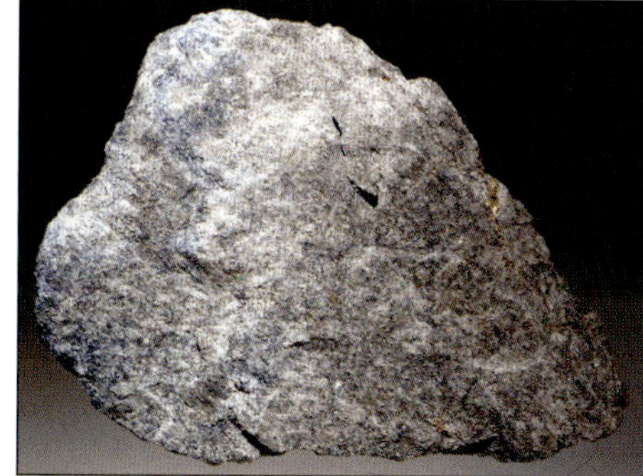

LAZURITE. Blue lazurite with dolostone (a sedimentary carbonate rock that contains a high percentage of the mineral dolomite), pyrite, and sphalerite from the 3500 foot level of the Zinc Corp. of America #4 mine, Balmat, St. Lawrence County. The lazurite fluoresces pale blue and the dolostone fluoresces pale yellow SW. The sphalerite fluoresces orange LW. The piece weighs 4.0 oz. and is 3.5 x 2.8 x 0.5 inches. Value $40-45

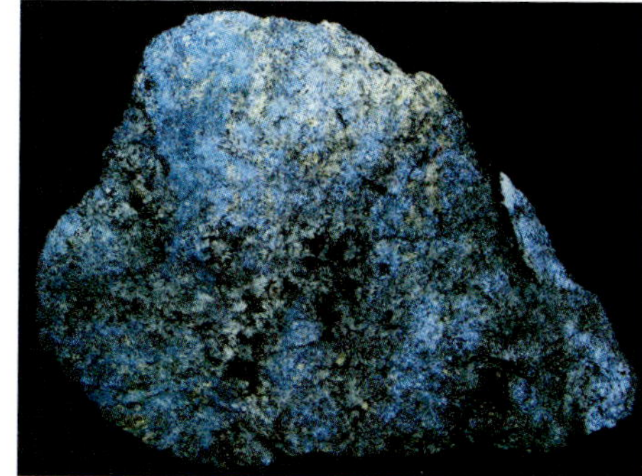

Same under SW

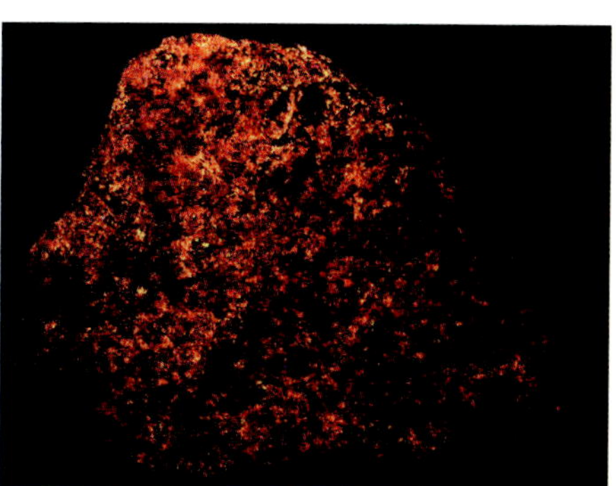

Same under LW

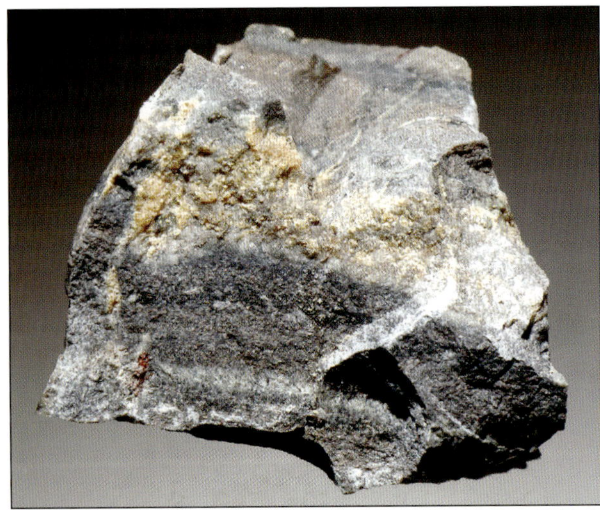

SPHALERITE. Sphalerite from the Zinc Corp. of America Pierrepont mine, Balmat, St. Lawrence County. The sphalerite fluoresces orange LW and SW and is triboluminescent and phosphorescent. The piece weighs 5.0 oz. and is 2.3 x 2.0 x 1.8 inches. Value $30-35

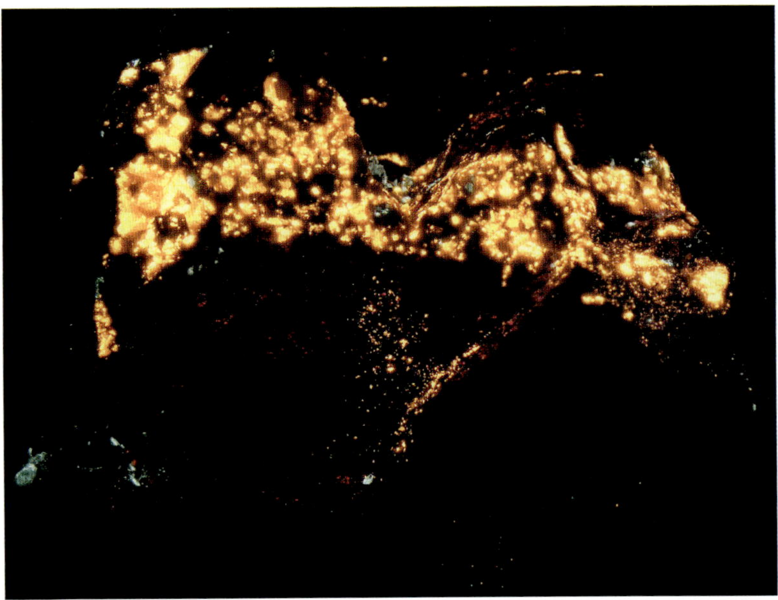

Same under SW

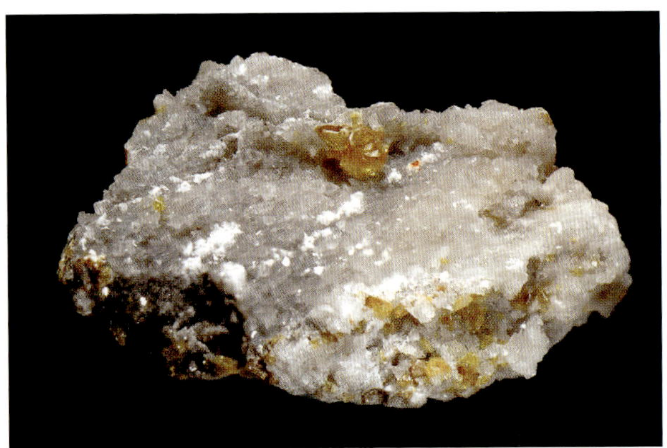

SPHALERITE. Crystals of sphalerite on calcite from the Zinc Corp. of America Hyatt mine in Talcville, St. Lawrence County. The close-up photo shows these rare, gemmy crystals of sphalerite. The sphalerite fluoresces orange LW and SW, the calcite fluoresces orange SW. The piece weighs 3.3 oz. and is 3.0 x 2.1 x 1.3 inches. Value $75-85

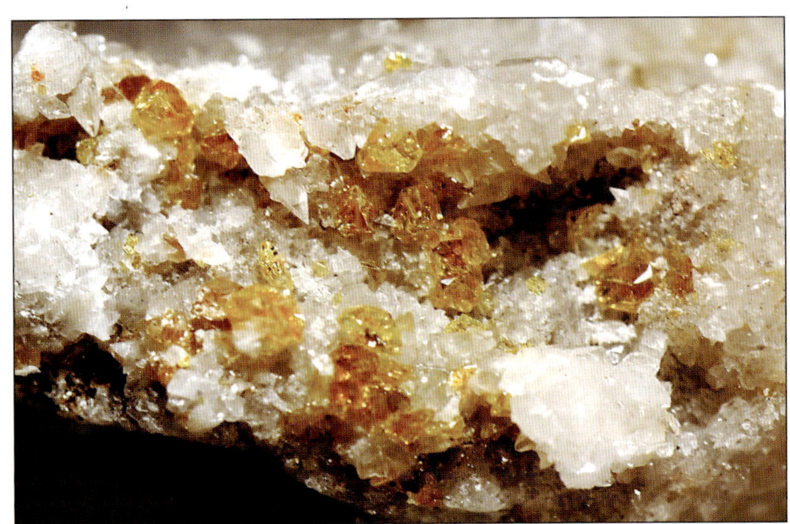

Same under LW

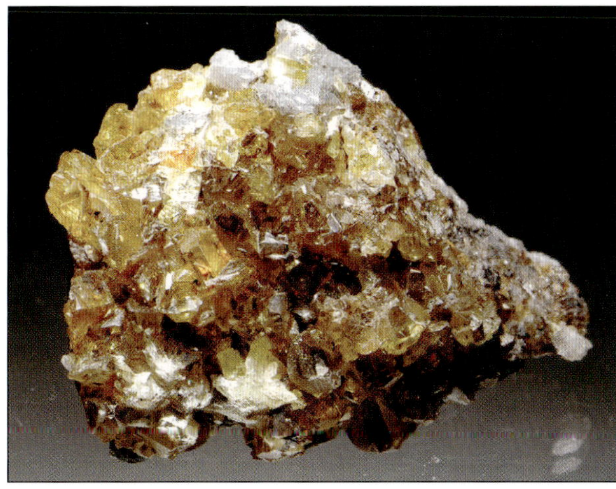

SPHALERITE. Crystals of sphalerite from the Zinc Corp. of America Hyatt mine in Talcville, St. Lawrence County. The sphalerite fluoresces orange SW and LW. The piece weighs 4.8 oz. and is 2.4 x 1.8 x 1.5 inches. Value $125-135

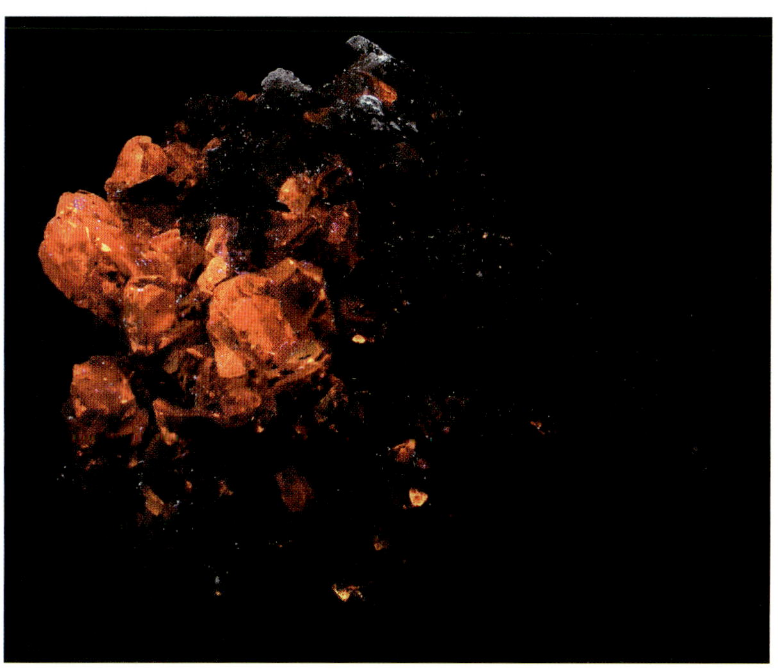

Same under SW

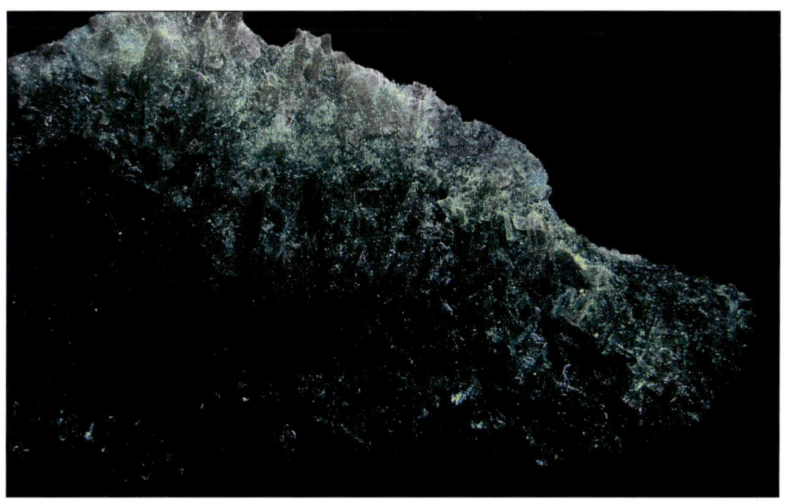

Same under SW

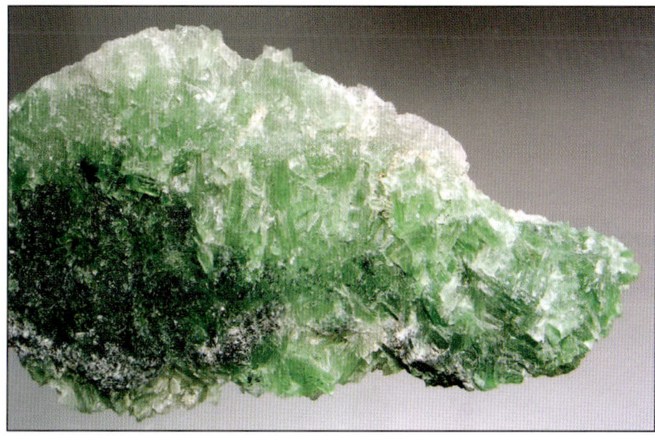

TREMOLITE. Tremolite, var. chrome-tremolite, from the Gouverneur Talc Co., Balmat, St. Lawrence County. The tremolite fluoresces pale yellow SW and LW. The piece weighs 1.0 oz. and is 3.0 x 1.5 x 0.6 inches. Value $15-20

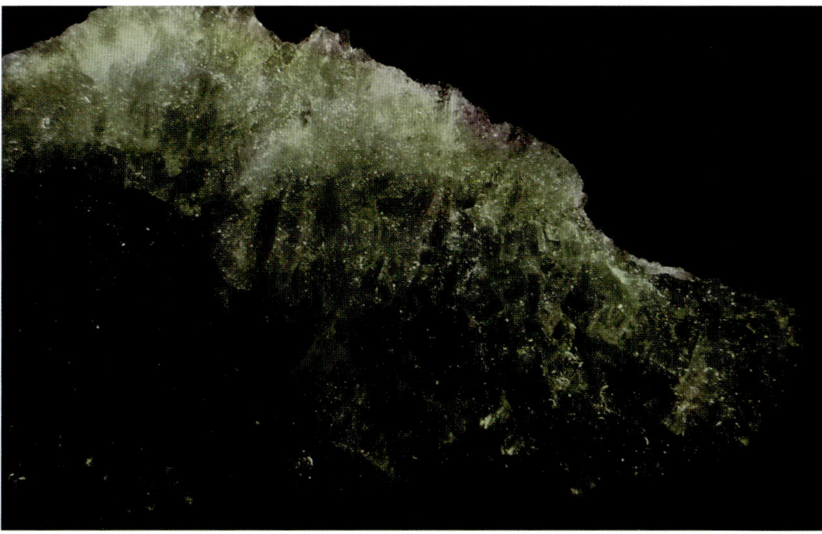

Same under LW

TURNEAUREITE. Turneaureite from the 2500 foot level, Zinc Corp. of America #4 mine, Balmat, St. Lawrence County. The turneaureite fluoresces orange SW. The piece weighs 12.0 oz. and is 3.0 x 3.0 x 1.8 inches. Value $175-200

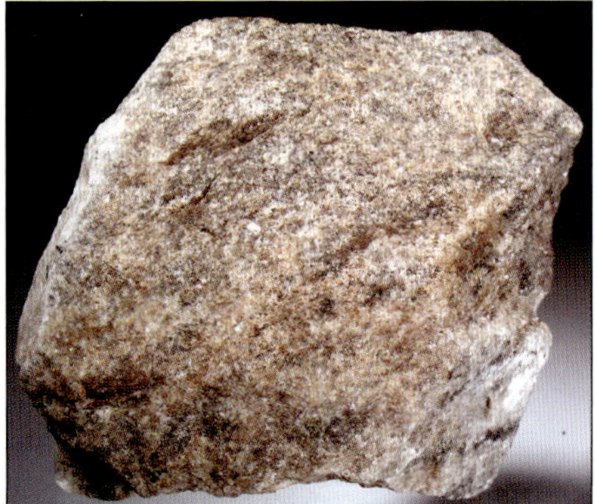

Same under SW

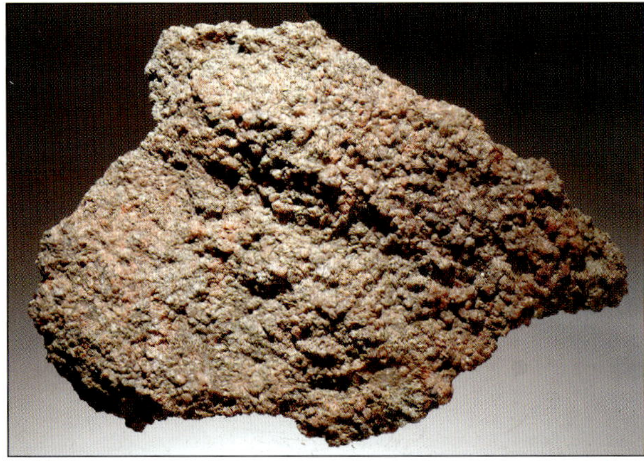

WILLEMITE. Willemite from the Zinc Corp. of America #2 mine, Balmat, St. Lawrence County. The willemite fluoresces green SW. The piece weighs 1 lb. 3.0 oz. and is 5.5 x 4.0 x 1.3 inches. Value $20-25

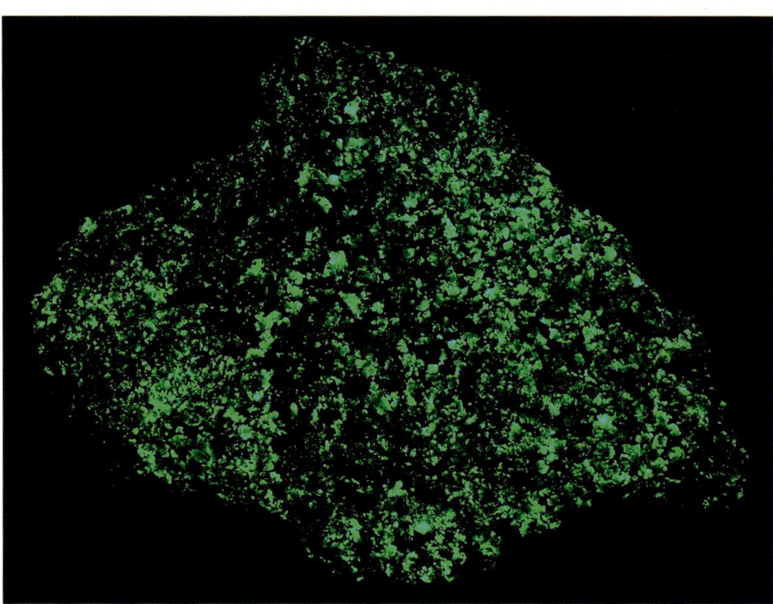

Same under SW

North Carolina

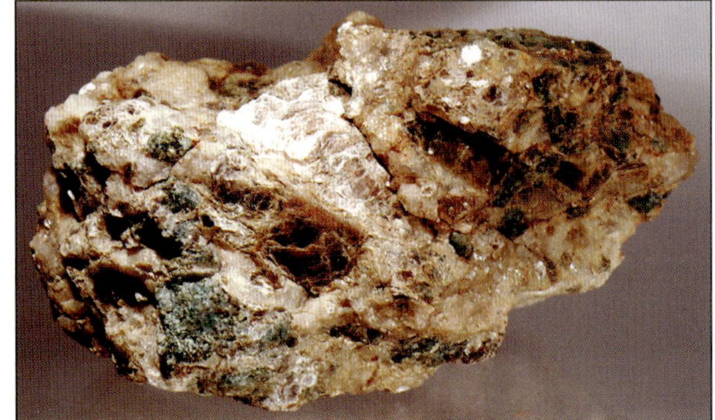

FLUORAPATITE. Fluorapatite, var. manganoan fluorapatite, with muscovite from a pegmatite vein at the Ray mica mine, near Burnsville, Yancey County. The fluorapatite on the specimen fluoresces yellow SW. The piece weighs 2 lb. 2.5 oz. and is 5.5 x 4.0 x 3.8 inches. Value $25-30

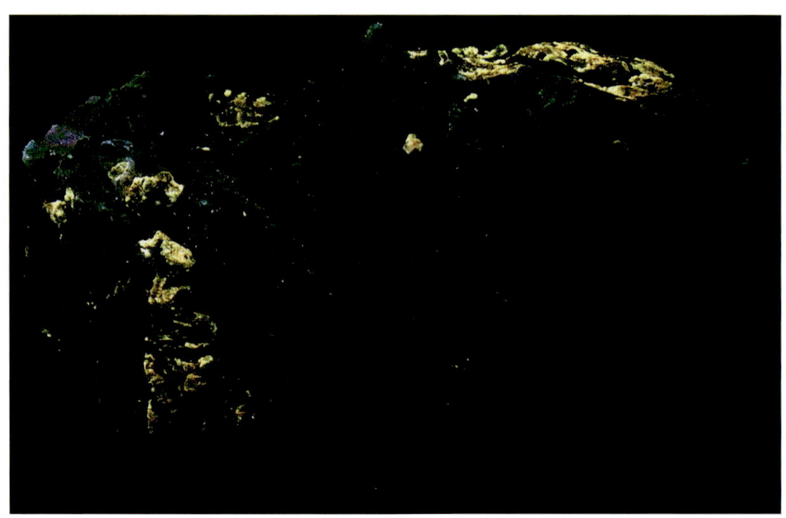

Same under SW

SPHALERITE. Massive sphalerite in a barite matrix from Hot Springs, Madison County. The sphalerite fluoresces yellow-orange LW and (less brightly) SW. The piece weighs 12.5 oz. and is 3.3 x 2.8 x 1.5 inches. Value $45-55

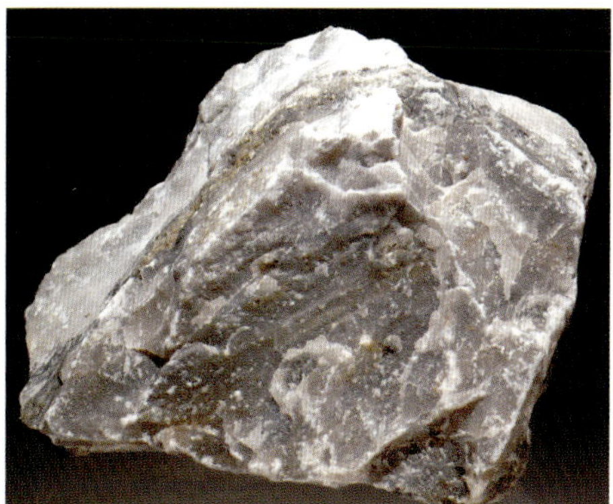

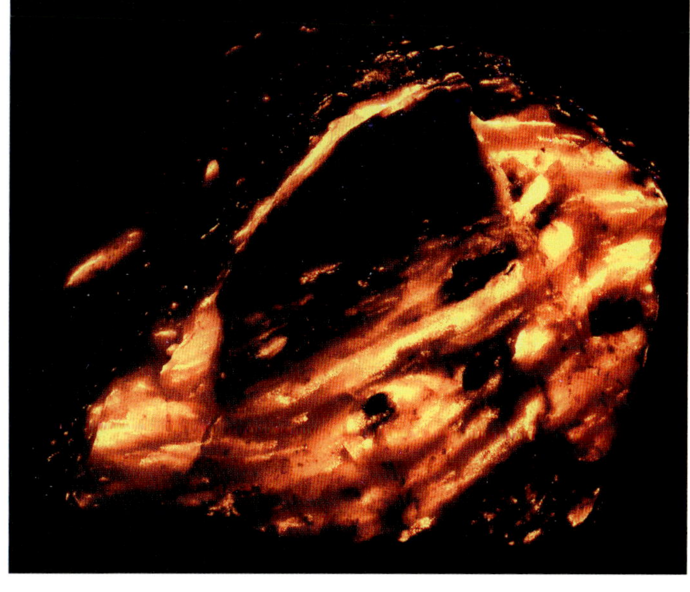

Same under LW

Ohio

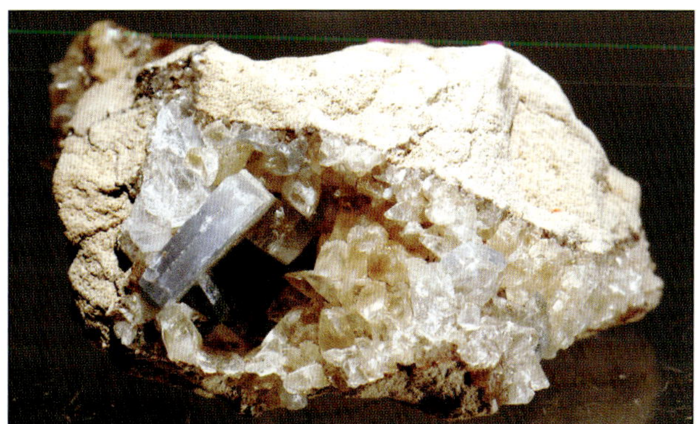

CELESTINE. Blue terminated celestine crystals from Lime City, Wood County. The celestine fluoresces mild tan SW and LW. The piece weighs 3.5 oz. and is 2.5 x 2.0 x 1.3 inches. Value $15-18

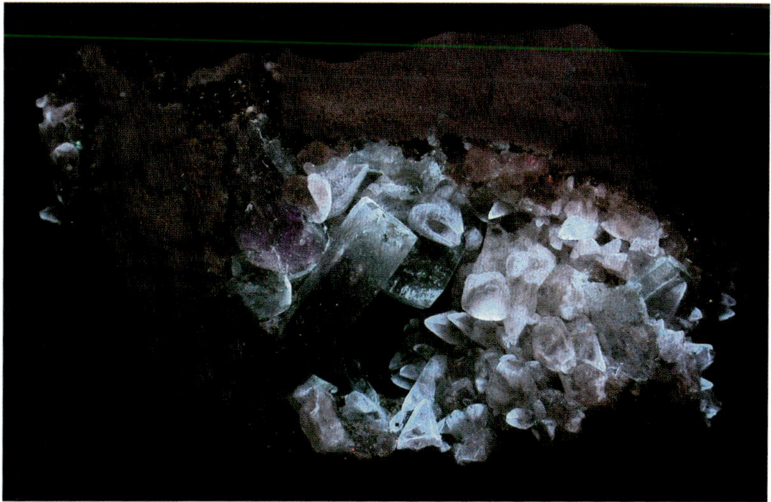

Same under SW

Pennsylvania

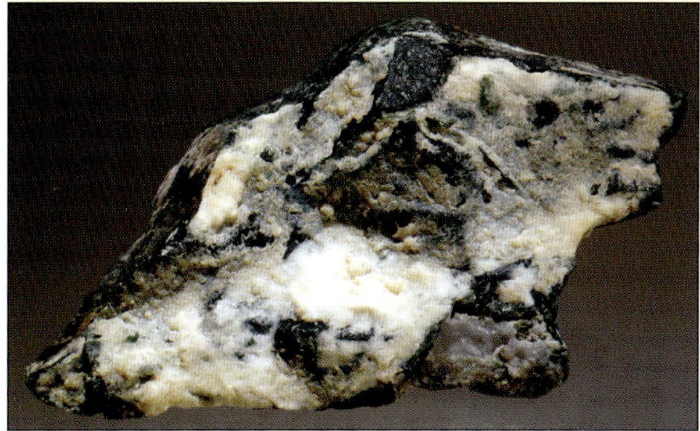

BRUCITE. Brucite on serpentine from the Cedar Hill quarry, Fulton Township, Lancaster County. The brucite fluoresces pale yellow to pale blue SW and LW. The piece weighs 3.5 oz. and is 3.3 x 1.8 x 1.3 inches. Value $15-20. Courtesy of James Poole.

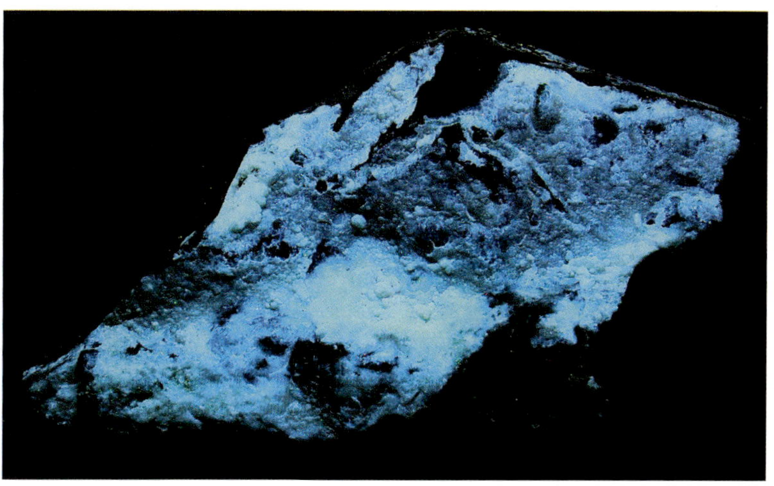

Same under SW

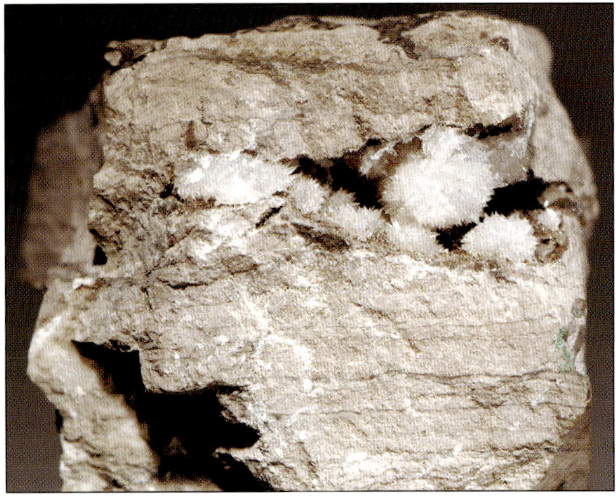

STRONTIANITE. Strontianite crystals from Winfield, Union County. This strontianite fluoresces bright pale blue LW and SW. The piece weighs 5.5 oz. and is 1.8 x 1.8 x 1.5 inches. Value $25-30

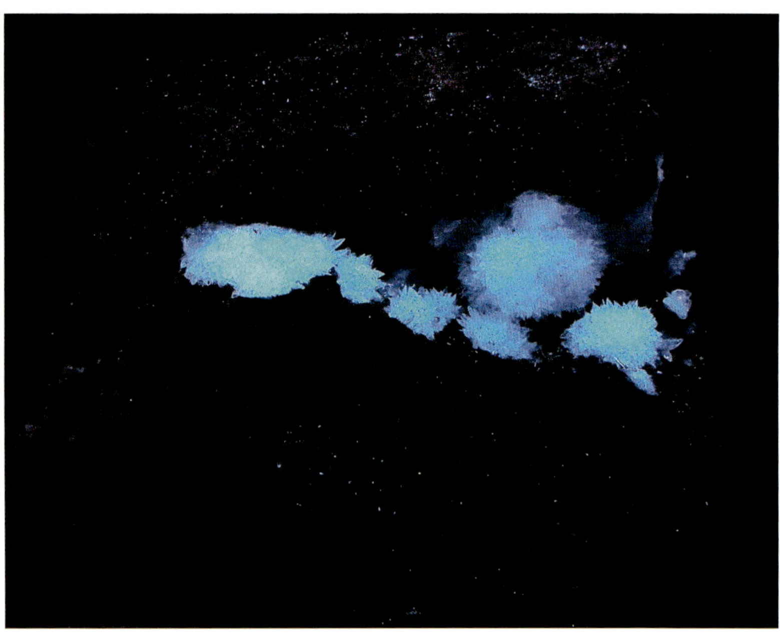

Same under SW

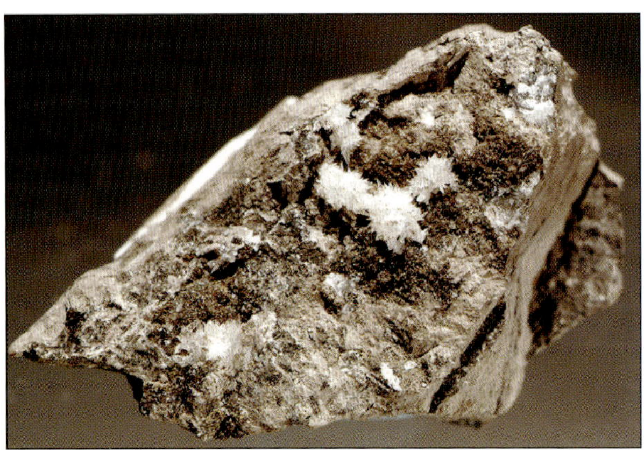

STRONTIANITE. Strontianite crystals from Winfield, Union County. This strontianite fluoresces bright pale blue LW and SW. The piece weighs 1.0 oz. and is 2.0 x 1.3 x 0.6 inches. Value $15-18

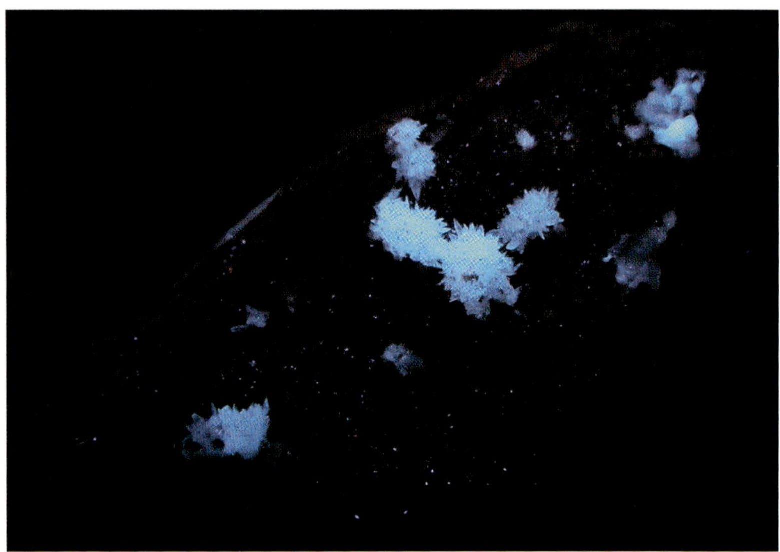

Same under SW

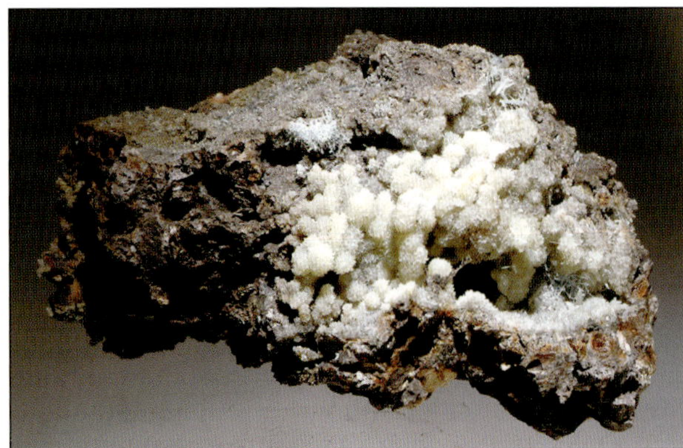

ZINCITE. A secondary fluorescing zincite and willemite from the West Bank, New Jersey Zinc Co., Palmerton, Carbon County. This is a smelter product. The zincite/willemite fluoresces pale blue SW. The piece weighs 13.3 oz. and is 4.0 x 2.5 x 2.0 inches. Value $35-40

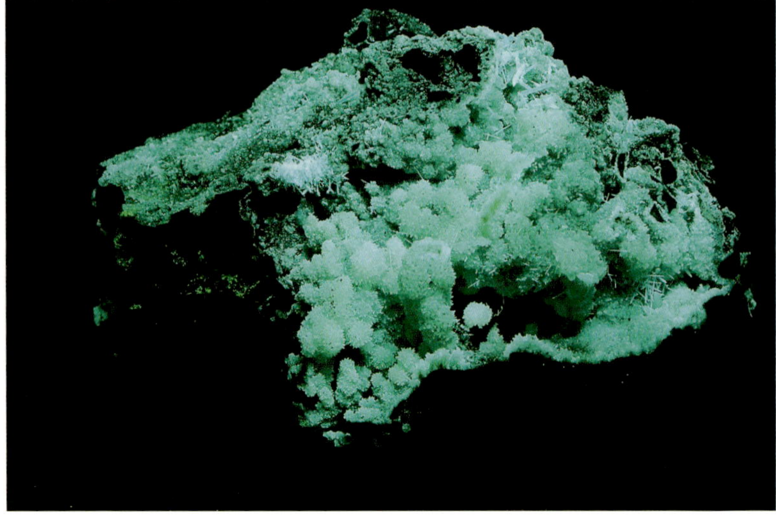

Same under SW

West Virginia

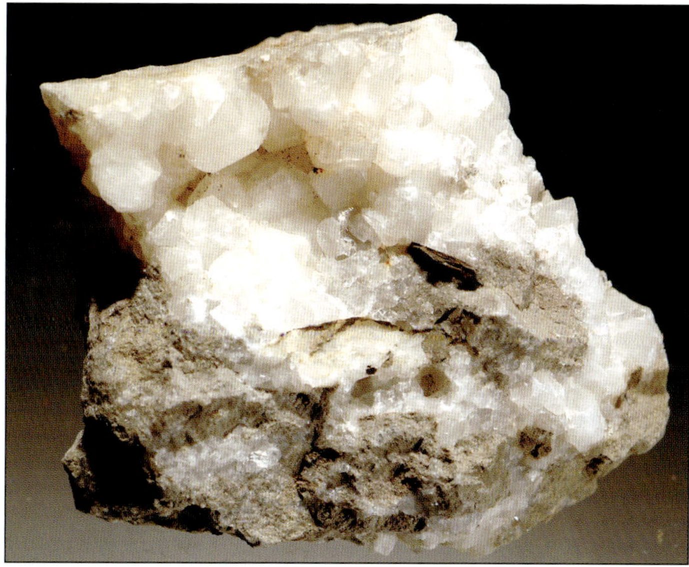

CALCITE. West Virginia is not well known for its fluorescent minerals. This is an unusual Terlingua-type calcite that fluoresces mild bluish violet under SW, mild peach under LW, and has a long-lasting phosphorescence after SW excitation. It is from near Short Gap, Mineral County. The piece weighs 16.0 oz. and is 4.5 x 2.5 x 2.5 inches. Value $14-18

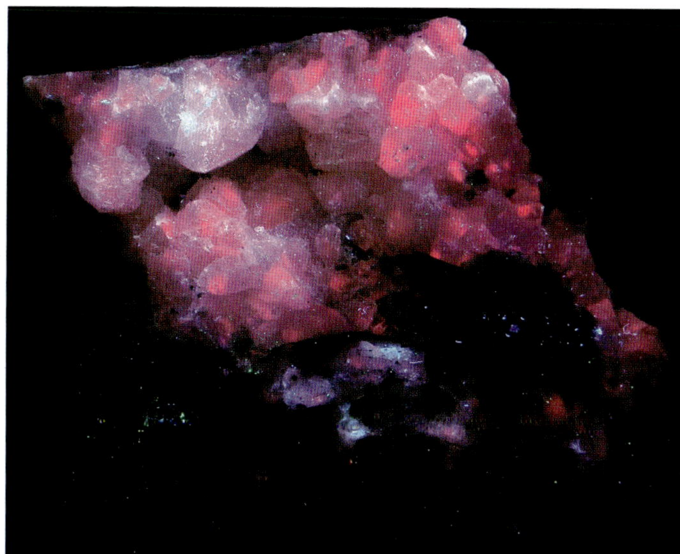

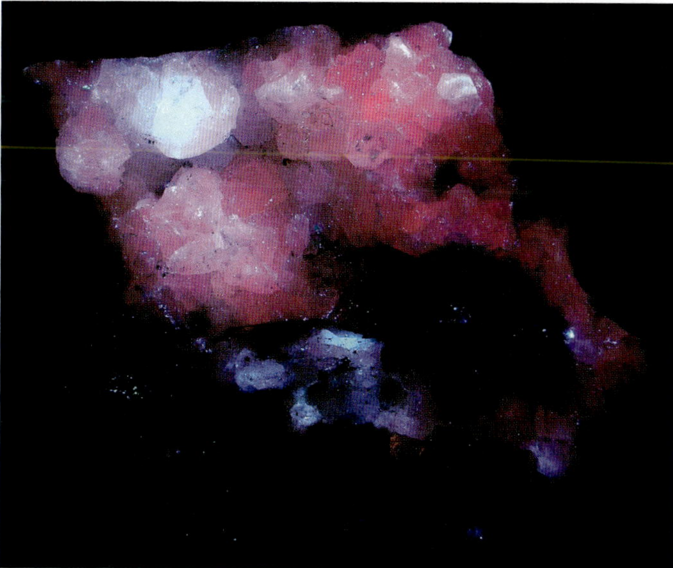

Top right:
Same under SW

Right:
Same under LW

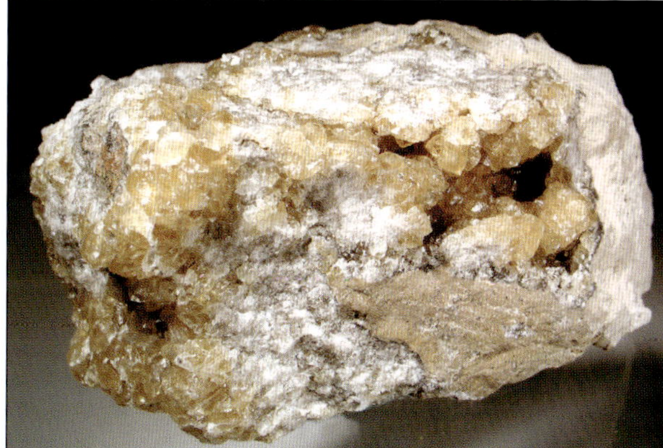

CALCITE. Calcite from Cross Lanes, Kanawha County. The calcite fluoresces yellow under LW and SW and has a long-lasting phosphorescense after SW excitation. The piece weighs 14.0 oz. and is 4.0 x 2.5 x 2.0 inches. Value $30-35

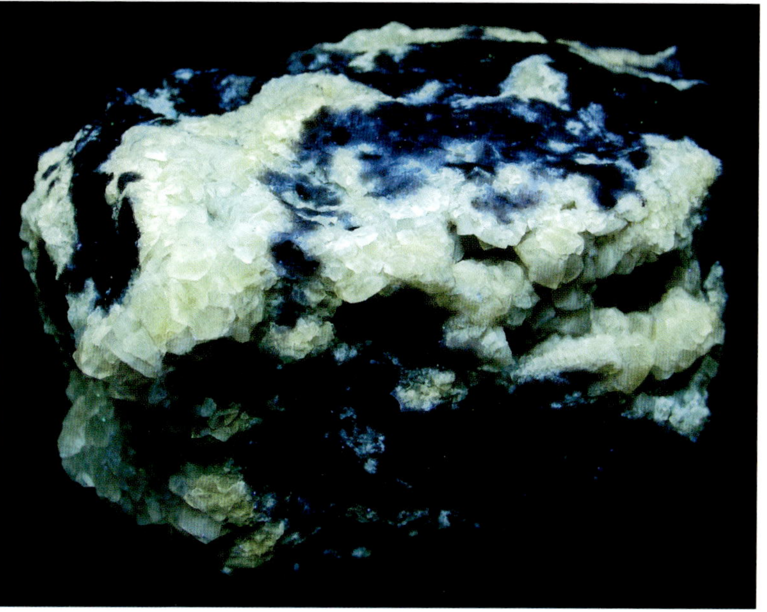

Same under SW

CALCITE. Calcite on limestone from near Romney, Hampshire County. The calcite fluoresces bright pale yellow under LW and SW and has a long-lasting phosphorescense after SW excitation The piece weighs 1.0 oz. and is 2.5 x 1.3 x 0.5 inches. Value $4-6

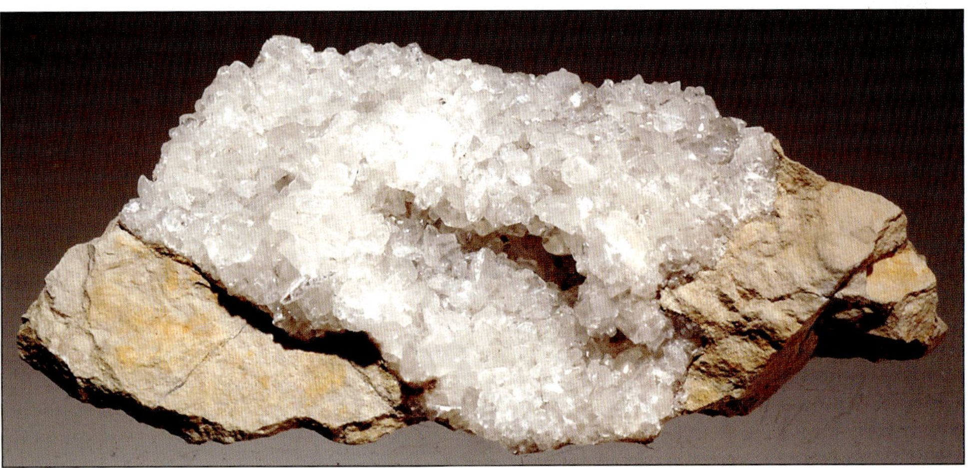

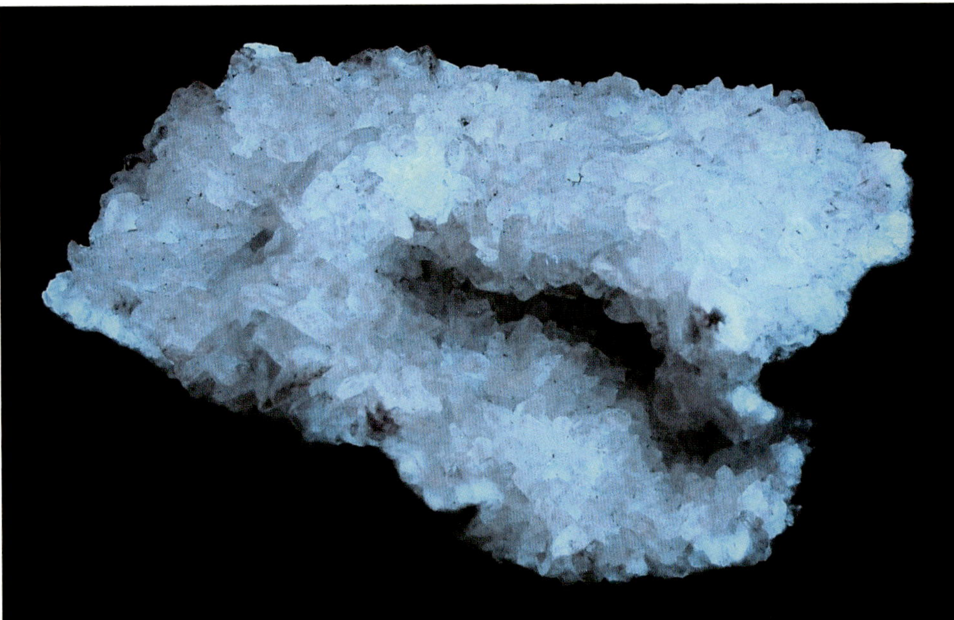

Same under SW

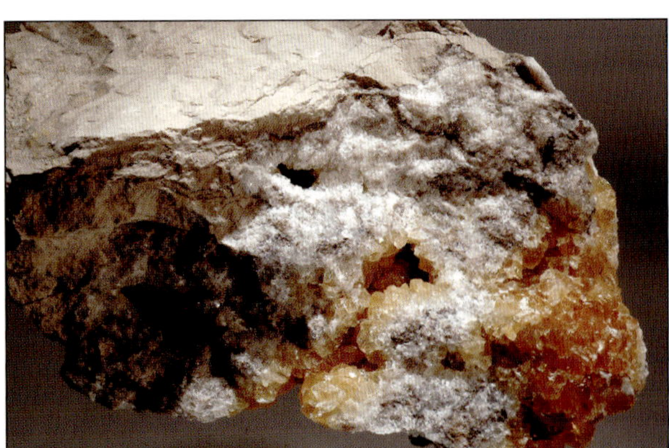

CALCITE. Calcite on limestone from near Short Gap, Mineral County. The calcite fluoresces yellow and blue LW and SW, better LW and has a long-lasting phosphorescense. The piece weighs 2 lb. 11.0 oz. and is 6.0 x 4.5 x 3.0 inches. Value $20-30

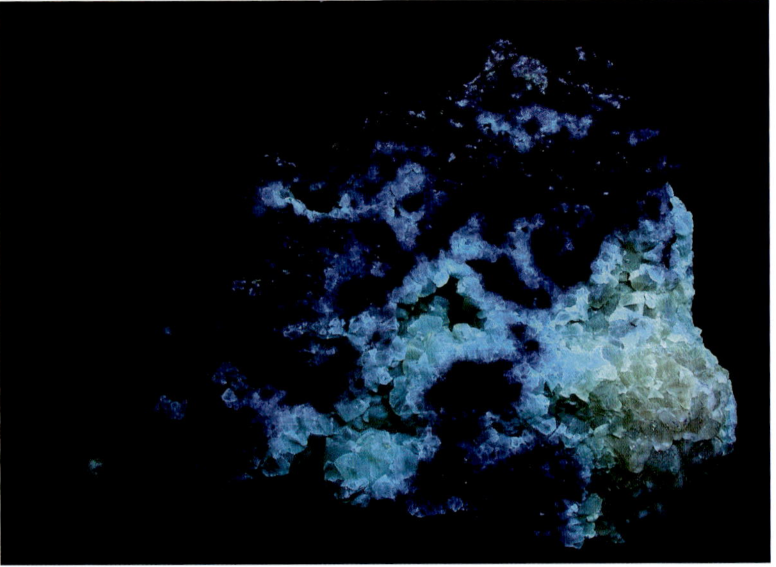

Same under SW

Western U.S.

Arizona

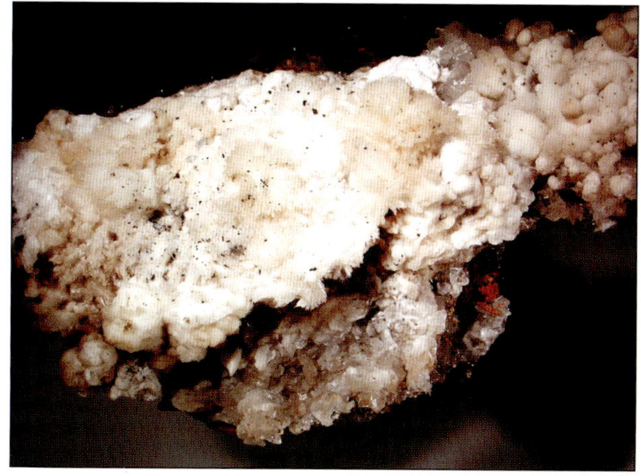

ARAGONITE. An unusual aragonite from the Glove mine, Santa Rita Mountains, Santa Cruz County. It forms needles, blades, and boytryoidal crusts. It fluoresces bright green SW. The piece weighs 1 lb. 6.0 oz. and is 5.5 x 3.0 x 1.0 inches. Value $50-60

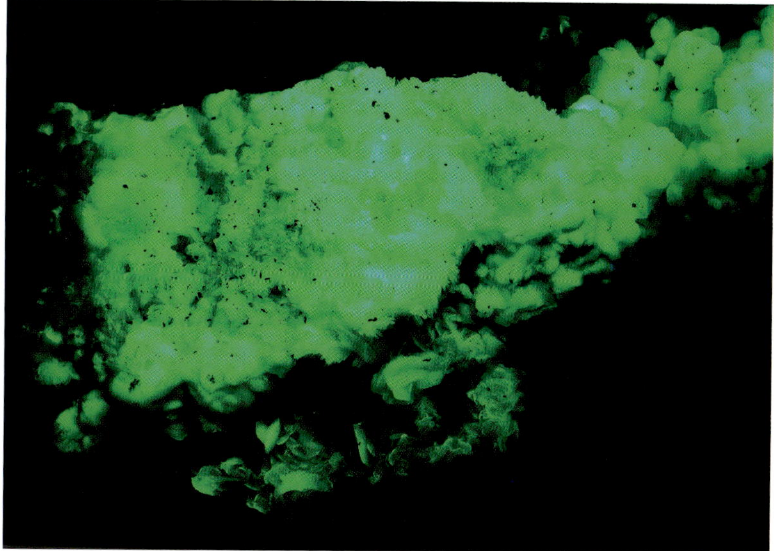

Same under SW

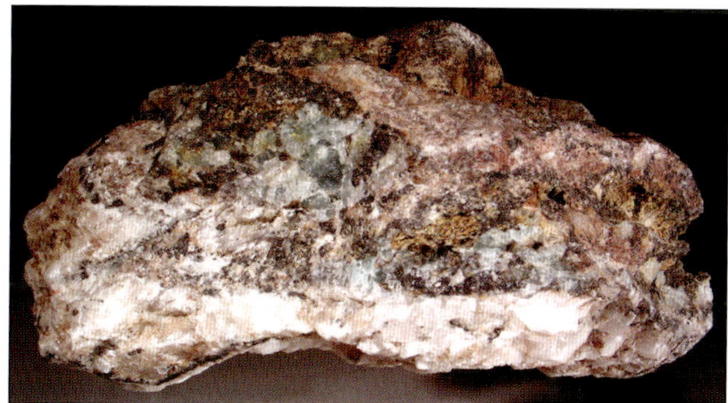

CALCITE, WILLEMITE, FLUORITE. Calcite, willemite and fluorite from the Hogan claim, near Wickenburg, Yavapai County. The calcite fluoresces a bright orange-red SW, the willemite fluoresces green SW and has a long-lasting phosphorescence, and the fluorite fluoresces violet SW and LW. These minerals except the fluorite, show up best under SW. The piece weighs 14.0 oz. and is 4.3 x 2.5 x 2.0 inches. Value $35-40

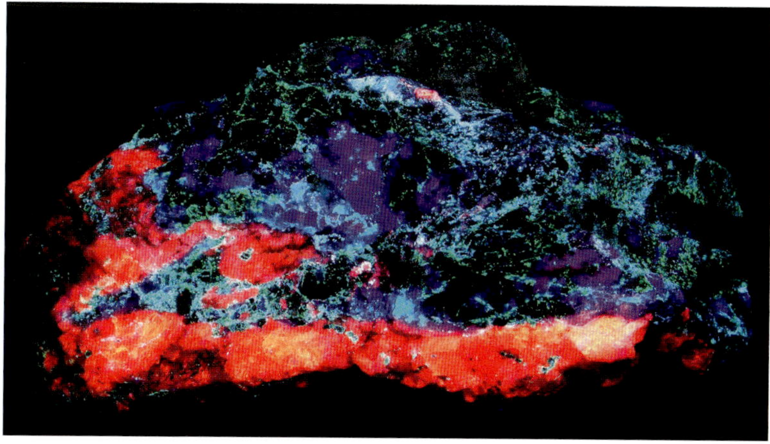

Same under SW

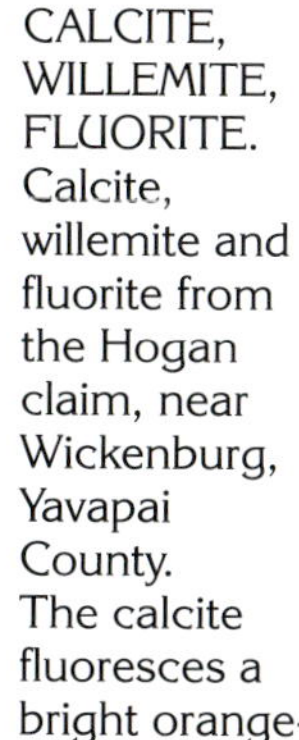

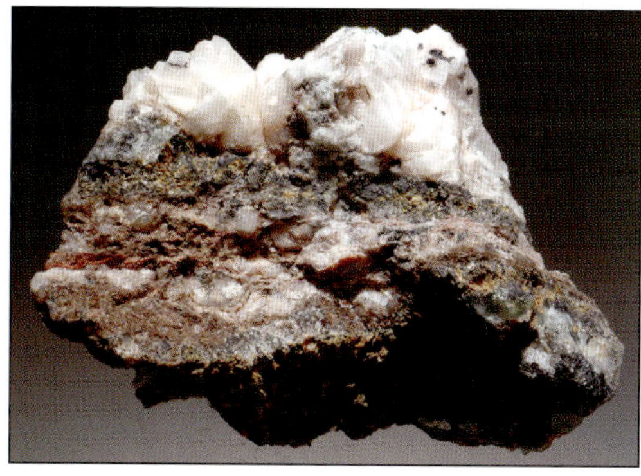

CALCITE, WILLEMITE, FLUORITE. Calcite, willemite and fluorite from the Hogan claim, near Wickenburg, Yavapai County. The calcite fluoresces a bright orange-red SW, the willemite fluoresces green SW, and the fluorite fluoresces violet SW and LW. These minerals except the fluorite, show up best under SW. The piece weighs 7.5 oz. and is 3.5 x 3.0 x 2.0 inches. Value $30-35

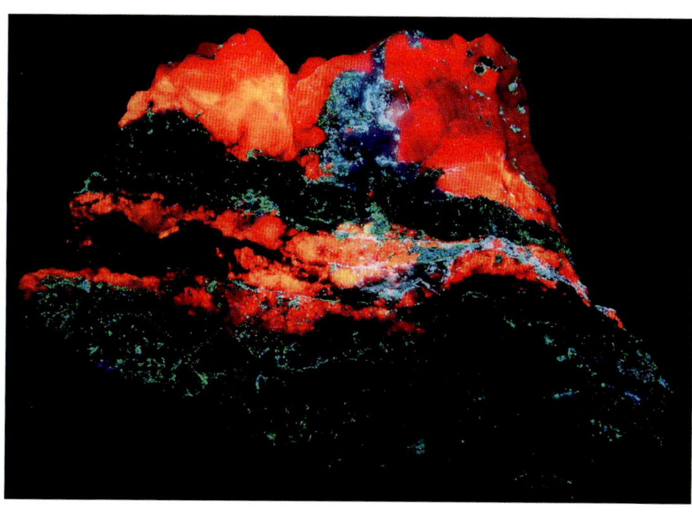

Same under SW

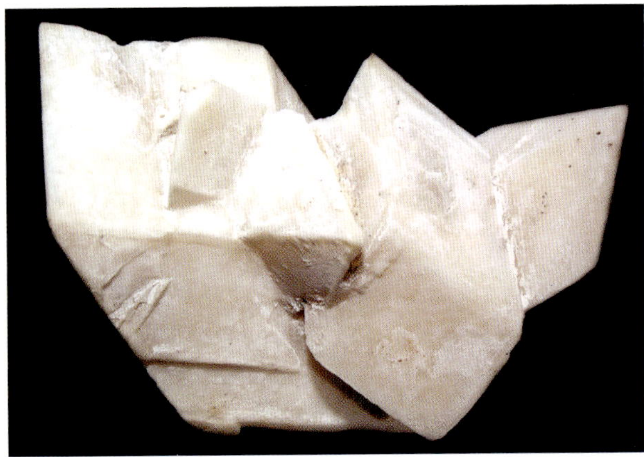

CALCITE. Calcite pseudomorph after glauberite from Camp Verde, Yavapai County. The crystal fluoresces pale yellow SW. It weighs 1.0 oz. and is 2.0 x 1.5 x 1.0 inches. Value $12-16

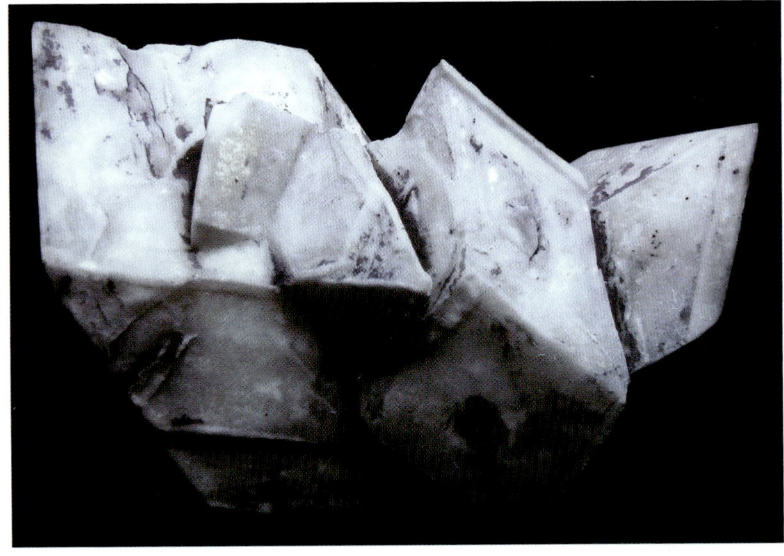

Same under SW

EUCRYPTITE. Eucryptite from the Midnight Owl mine, Yavapai County. Eucryptite fluoresces cherry red SW. The piece weighs 3.8 oz. and is 1.8 x 1.8 x 1.5 inches. Value $20-25

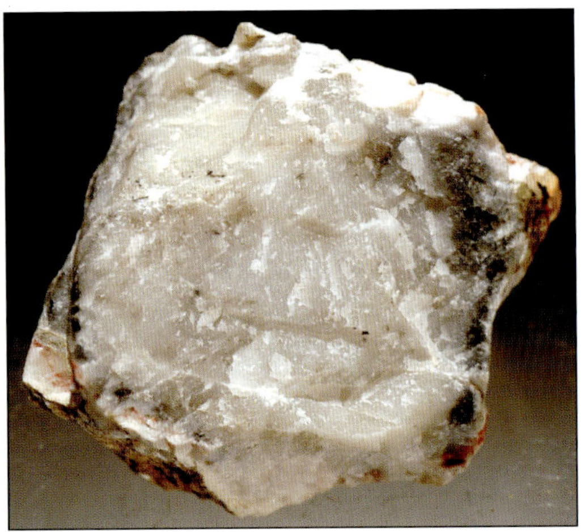

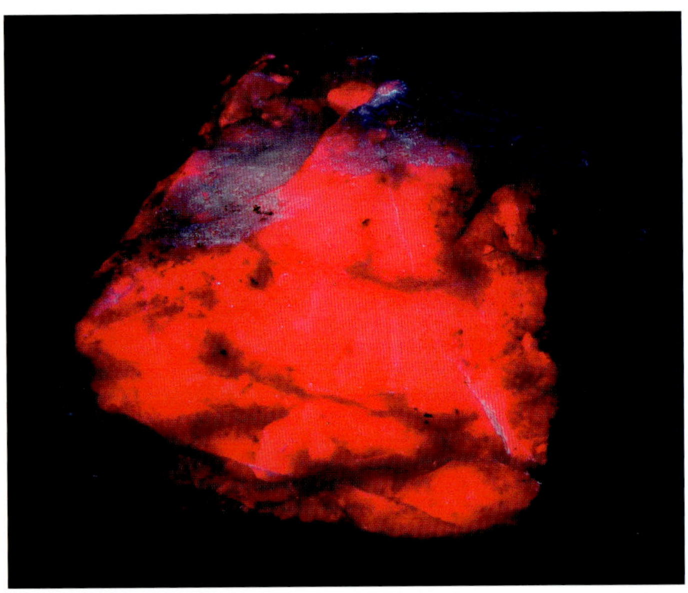

Same under SW

EUCRYPTITE. Eucryptite from the Midnight Owl mine, Yavapai County. Eucryptite fluoresces cherry red SW. The piece weighs 1.1 oz. and is 1.8 x 1.5 x 0.5 inches. Value $15-20

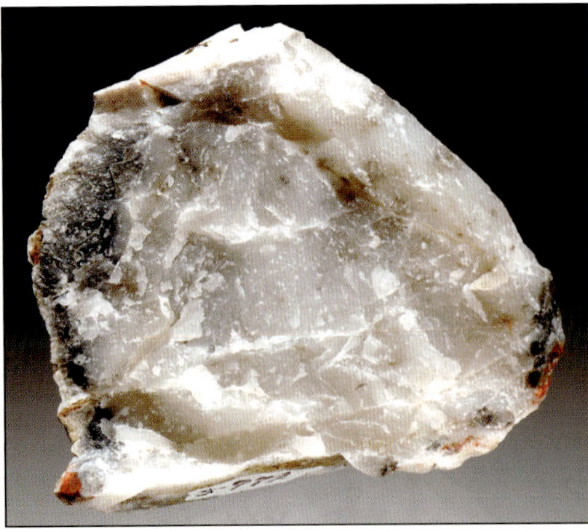

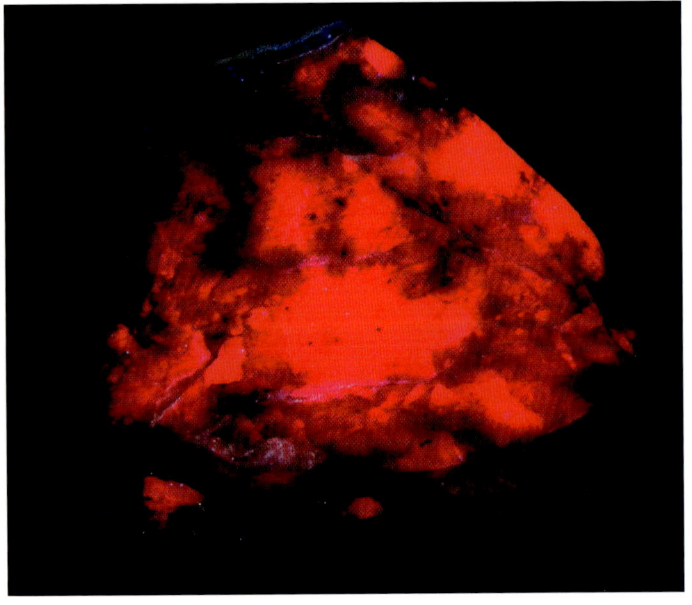

Same under SW

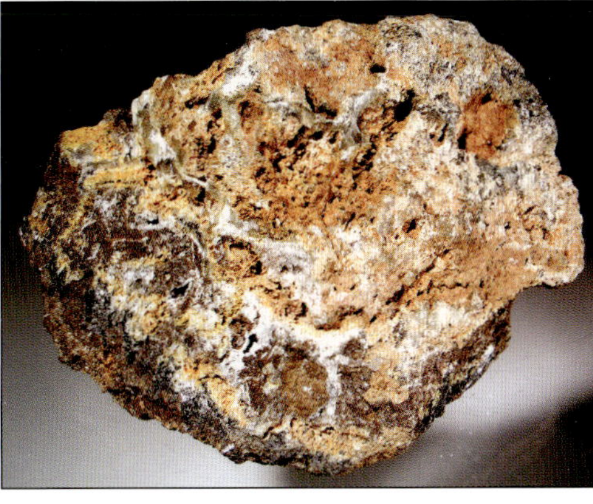

FLUORITE, CALICHE. A four-color fluorescent specimen containing fluorite, caliche (a general term for any secondary calcium carbonate that forms in sediments or in voids in bedrock just below the surface in semiarid regions), and two other unusual minerals from the Hogan claim, near Wickenburg, Yavapai County. One of the unidentified minerals (smithsonite?) fluoresces a bright bubblegum pink SW and the other fluoresces pale blue SW. The caliche fluoresces yellow SW, and the fluorite fluoresces violet SW and LW. All these minerals except the fluorite, show up best under SW. The piece weighs 6.3 oz. and is 3.0 x 2.0 x 2.0 inches. Value $45-55

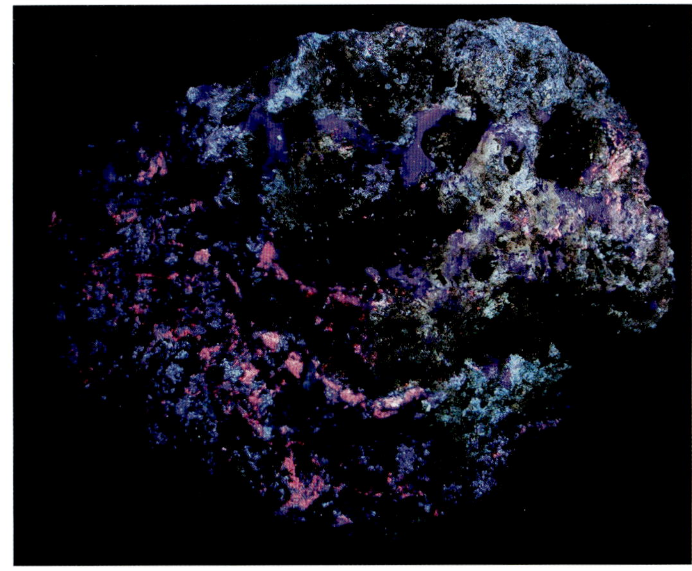

Same under SW

POWELLITE. Small powellite crystals on quartz from the St. Louis claim, Yavapai County. The powellite fluoresces pale yellow SW and LW. The piece weighs 1.5 oz. and is 1.8 x 1.3 x 1.0 inches. Value $8-10

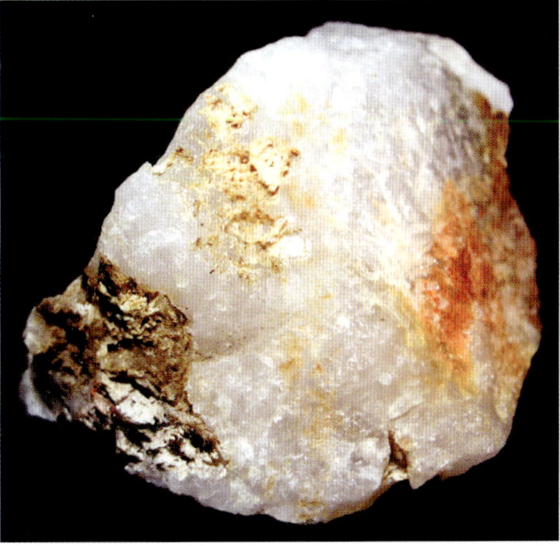

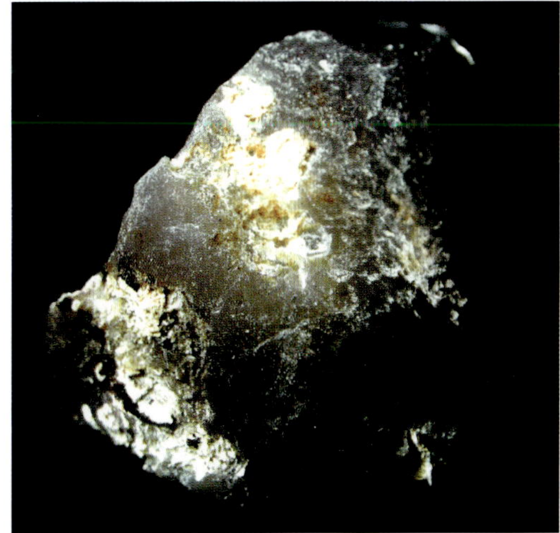

Same under SW

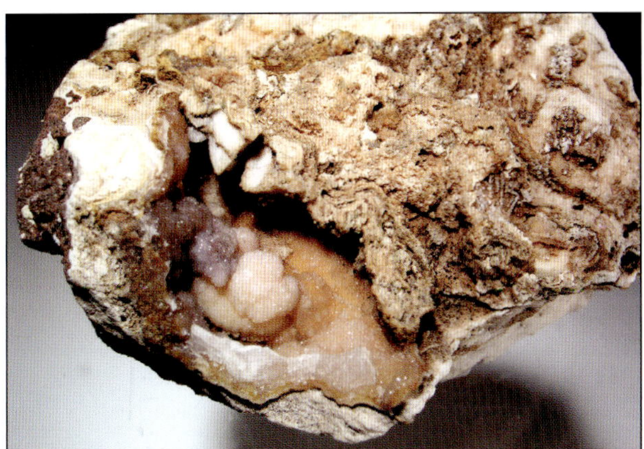

QUARTZ, CALICHE. Druzy quartz, var. chalcedony, with caliche (a general term for any secondary calcium carbonate that forms in sediments or in voids in bedrock just below the surface in semiarid regions) and 2 unidentified minerals from Cave Creek, Maricopa County. One unidentified mineral fluoresces a pale orange SW, the other fluoresces blue-grey SW, the caliche fluoresces a pale yellow SW, and the chalcedony fluoresces green SW. The piece weighs 3.0 oz. and is 3.0 x 2.5 x 1.0 inches. Value $20-25

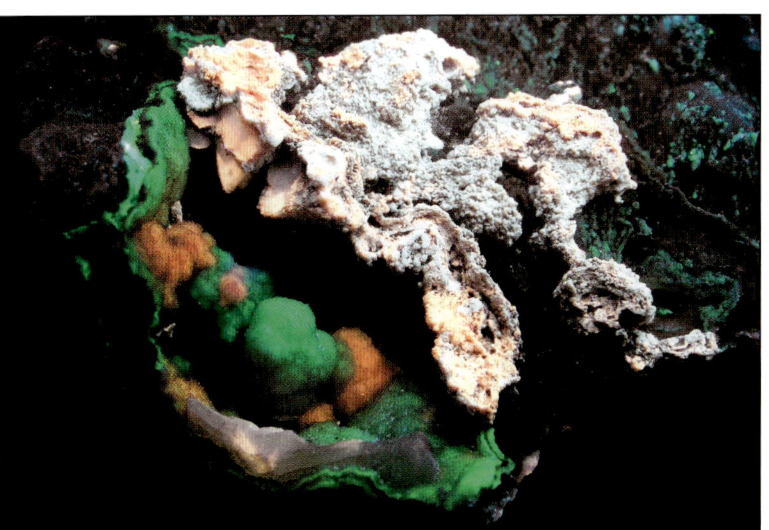

Same under SW

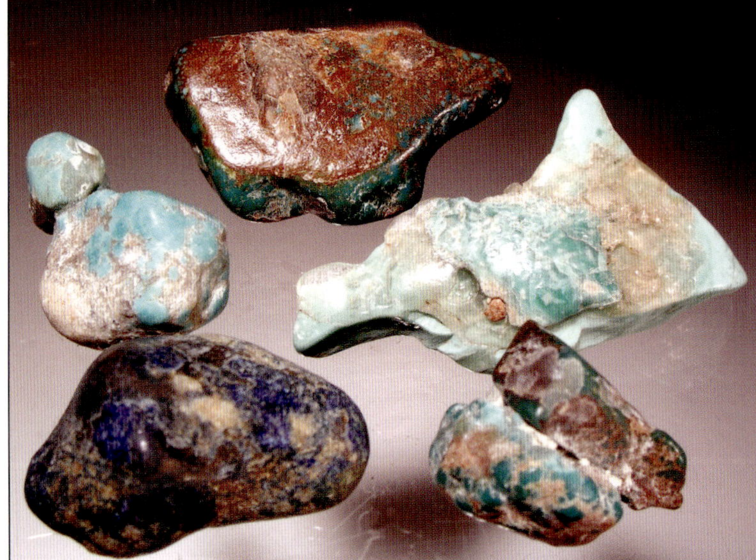

TURQUOISE. A group of turquoise nuggets. These were purchased in a small shop in Cave Creek, Maricopa County and were claimed to be local. They fluoresce brilliant blue to pale blue SW. The color is a result of the stabilizing material used to make low quality, porous turquoise stable enough for use in jewelry. The turquoise does not naturally fluoresce. About 0.5 inches each.

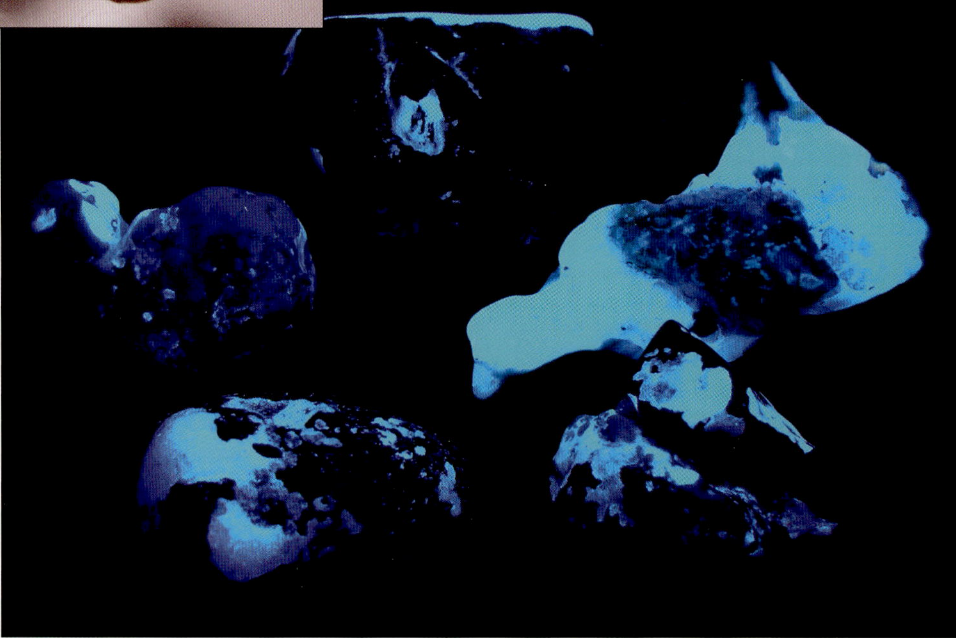

Same under SW

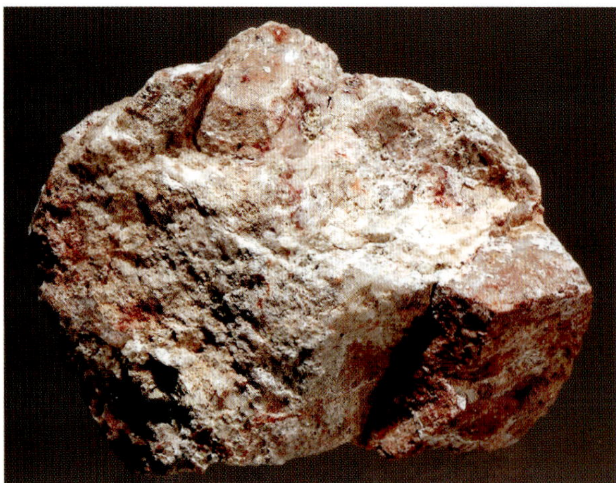

WICKENBURGITE. Wickenburgite with willemite and fluorite from near Wickenburg, Maricopa County. The wickenburgite fluoresces pink to red SW, the fluorite fluoresces violet LW and the willemite fluoresces pale green SW. The piece weighs 4.5 oz. and is 2.5 x 1.9 x 1.8 inches. Courtesy of Jeanne's Rock Shop. Value $25-30

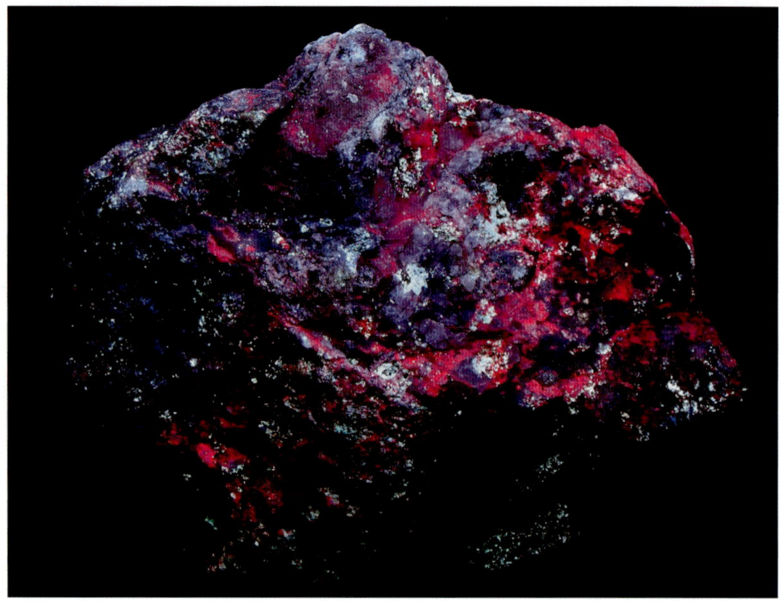

Same under SW

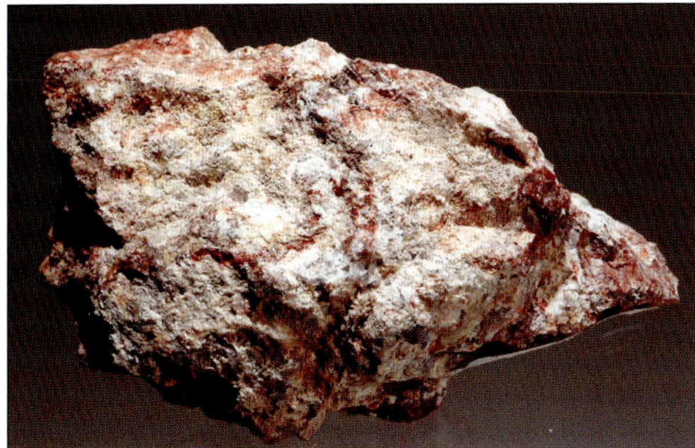

WICKENBURGITE. Wickenburgite with willemite and fluorite from near Wickenburg, Maricopa County. The wickenburgite fluoresces pink to red SW, the fluorite fluoresces violet LW and the willemite fluoresces pale green SW. The piece weighs 3.0 oz. and is 2.8 x 1.5 x 1.1 inches. Courtesy of Jeanne's Rock Shop. Value $25-30

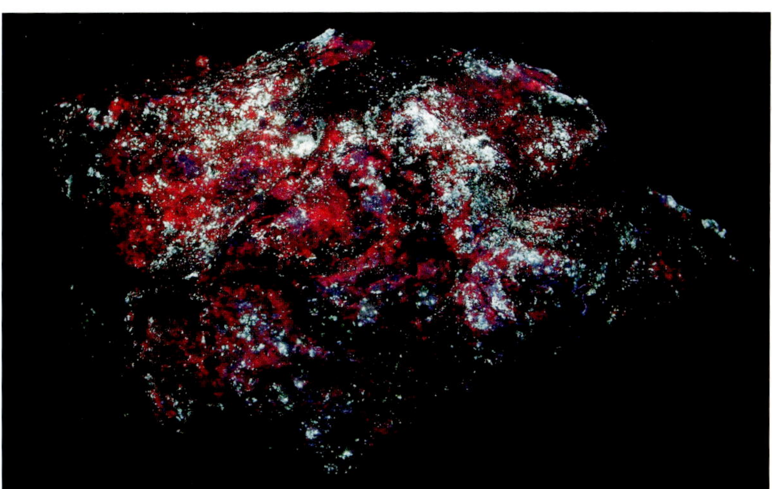

Same under SW

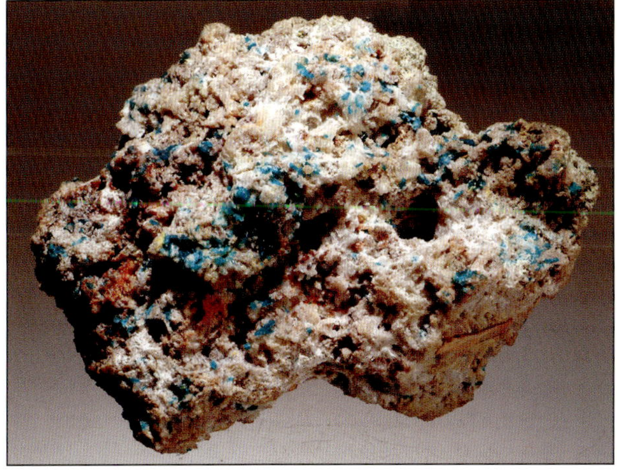

WILLEMITE. Willemite with green dioptase from the Mammoth mine, Tiger, Pinal County. The willemite fluoresces an interesting green-gray SW. The piece weighs 10.5 oz. and is 4.0 x 3.0 x 2.0 inches. From the James Zigras collection.

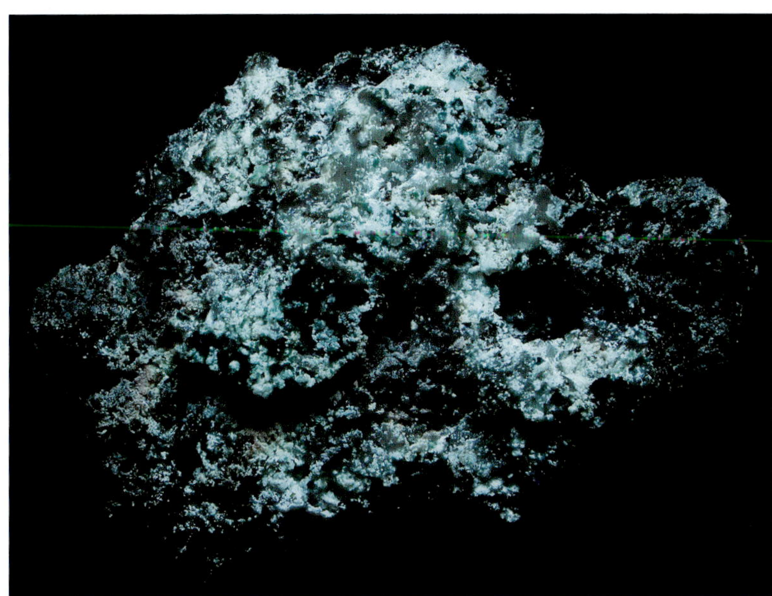

Same under SW

California

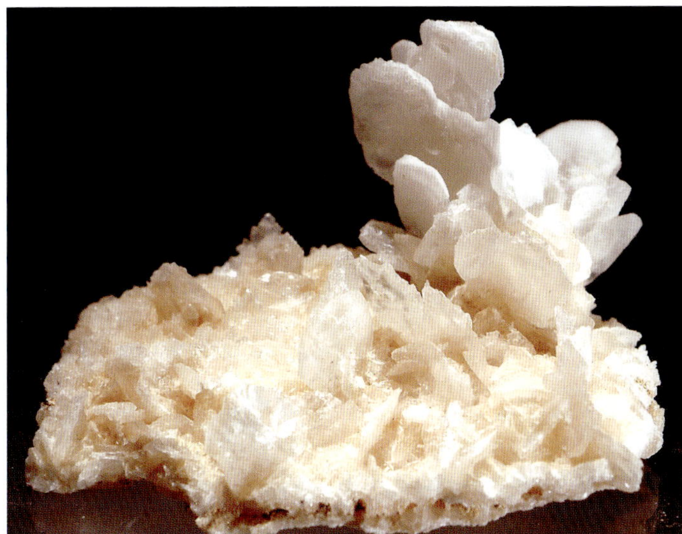

ARAGONITE. Beautiful bladed aragonite crystals from California (no further locality information). The aragonite fluoresces pale blue SW. The piece weighs 2.0 oz. and is 2.8 x 2.0 x 1.5 inches. Value $45-50

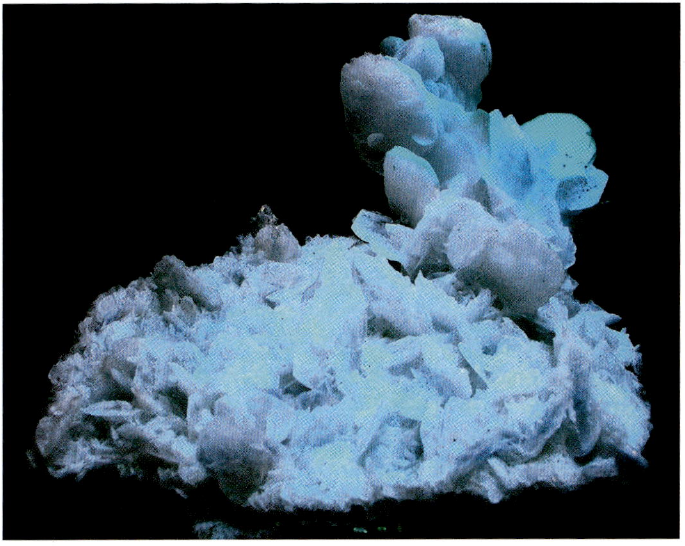

Same under SW

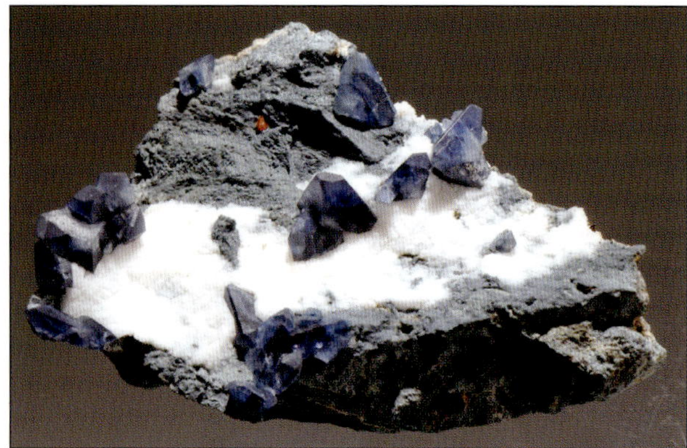

BENITOITE. Benitoite from the Benitoite Gem mine, San Benito County. Benitoite fluoresces a striking bright blue SW. Benitoite is only found in San Benito County. This piece has well defined crystals and an unusual group of stacked crystals on the left side. It also has a nonfluorescent, honey-colored joaquinite crystal. The piece weighs 1.5 oz. and is 2.0 x 1.4 x 1.5 inches. Value $200-225

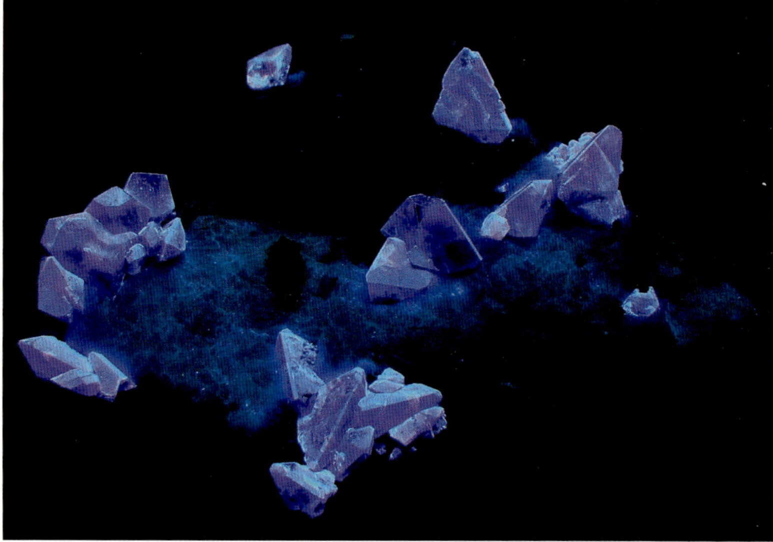

Same under SW

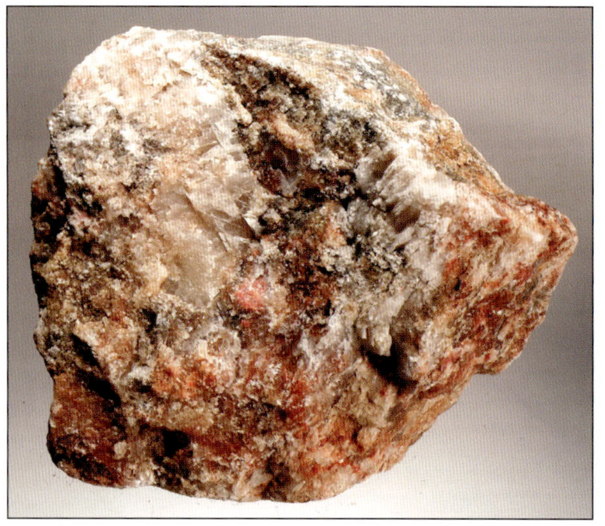

CALCITE. Calcite with a few dots of fluorapatite from the White Knob mine, Lucerne Valley Limestone Province, San Bernadino County. The calcite fluoresces orange-red SW and the fluorapatite fluoresces yellow SW. The piece weighs 3 lb. and is 3.5 x 3.5 x 3.0 inches. Value $10-15. Courtesy of Donald L. Fife.

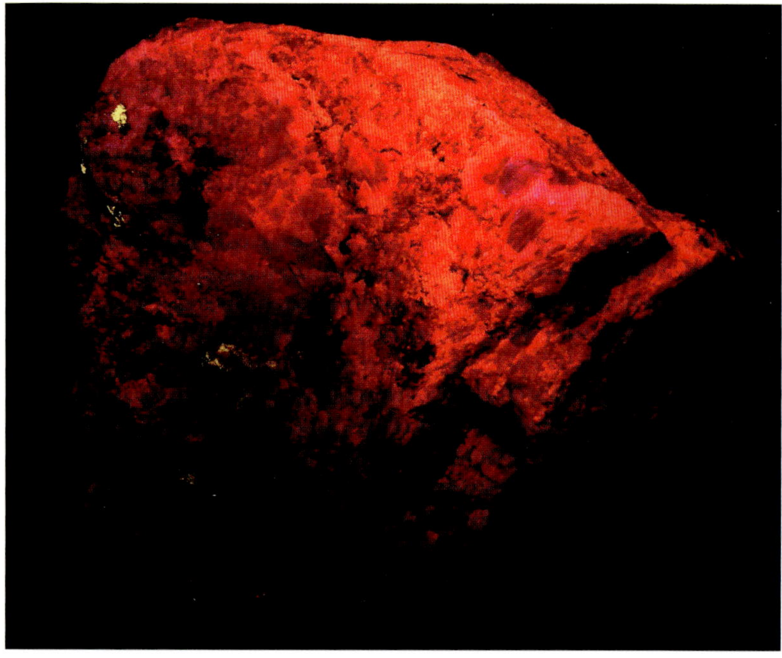

Same under SW

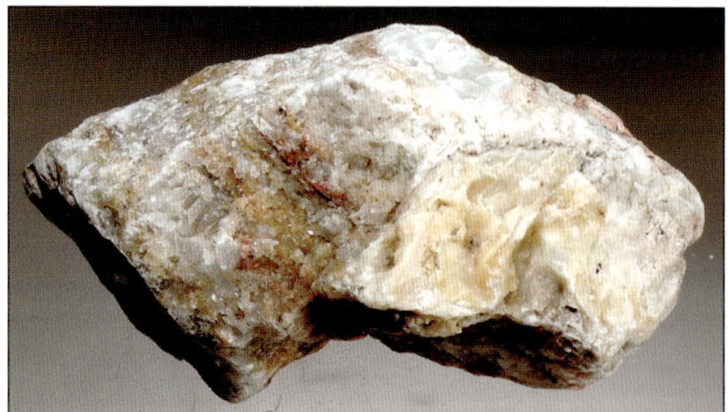

CALCITE, ARAGONITE. Calcite with aragonite from the White Knob mine, Lucerne Valley Limestone Province, San Bernadino County. The calcite fluoresces orange-red SW and the aragonite fluoresces pale blue SW. The piece weighs 2 lb. and is 3.5 x 2.5 x 2.0 inches. Value $10-15. Courtesy of Donald L. Fife.

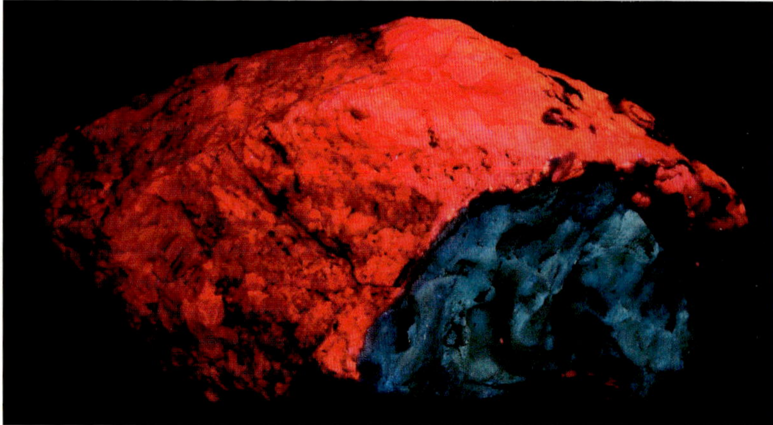

Same under SW

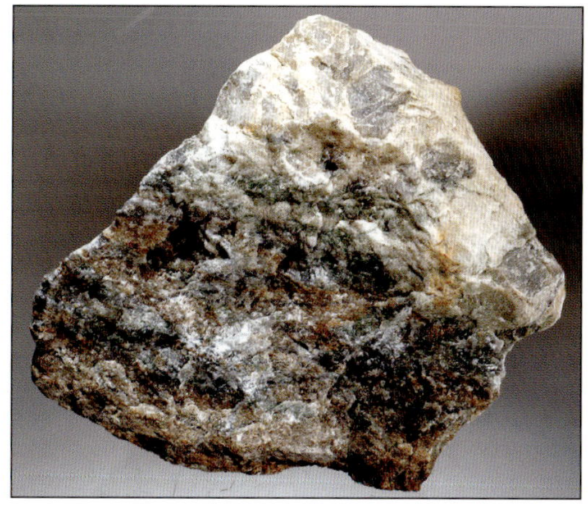

CALCITE, WOLLASTONITE. Calcite with wollastonite from the White Knob mine, Lucerne Valley Limestone Province, San Bernadino County. This mine is about 3 miles west of the Desert View mine. The calcite fluoresces weak orange-red SW and the wollastonite fluoresces orange-yellow SW. The piece weighs 3 lb. and is 3.5 x 3.5 x 3.0 inches. Value $40-45. Courtesy of Donald L. Fife.

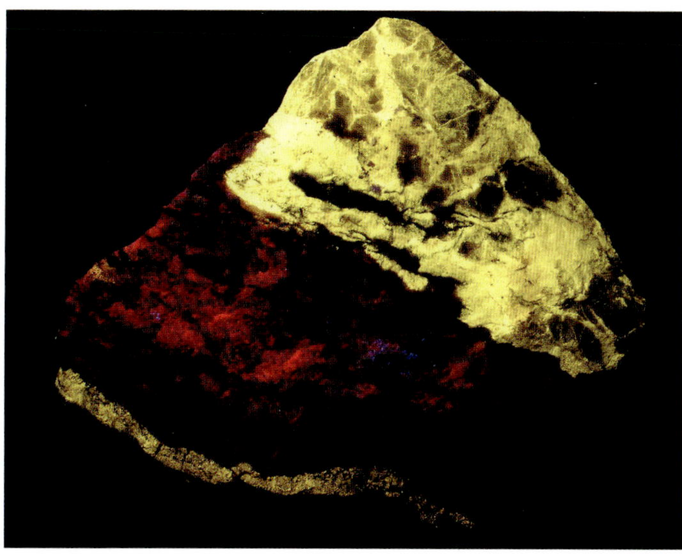

Same under SW

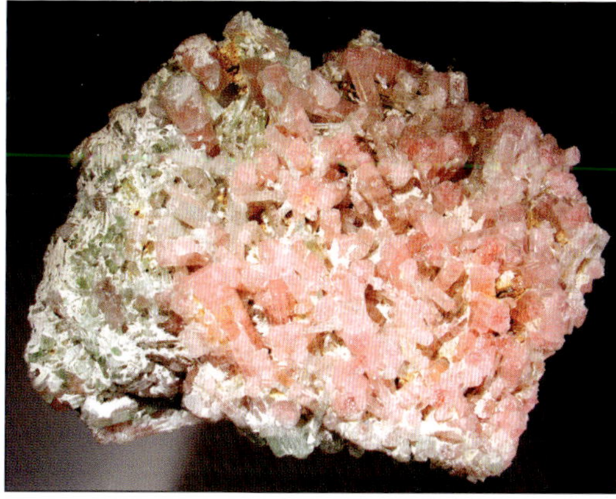

ELBAITE. Pink and blue elbaite (tourmaline group) from the Cryo-Genie mine, Warner Springs, San Diego County. The elbaite was discovered in September, 2001. The pink elbaite fluoresces pink and blue SW. The piece weighs 11.5 oz. and is 4.0 x 3.0 x 1.8 inches. Value $75-85

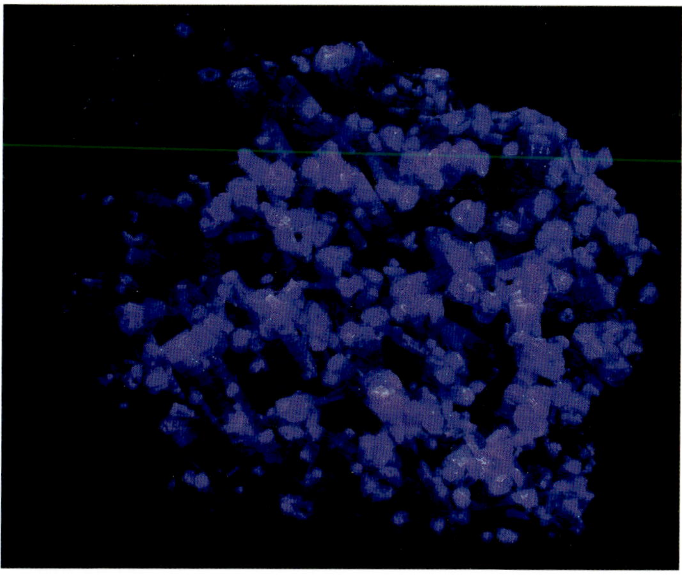

Same under SW

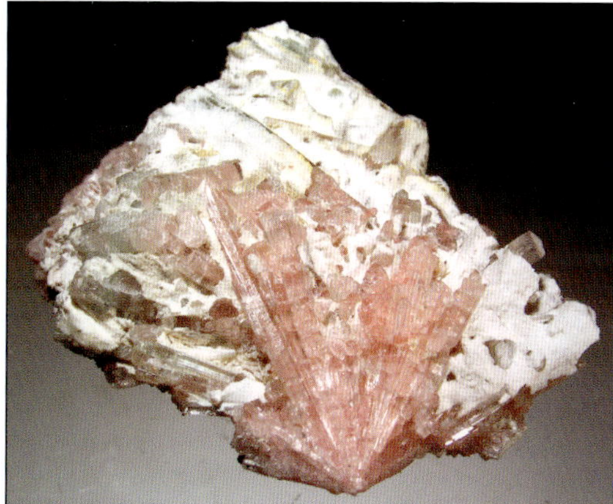

ELBAITE. Pink and blue elbaite (tourmaline group) from the Cryo-Genie mine, Warner Springs, San Diego County. The elbaite fluoresces pink and blue SW. The piece weighs 1.0 oz. and is 1.8 x 1.3 x 0.8 inches. Value $40-45

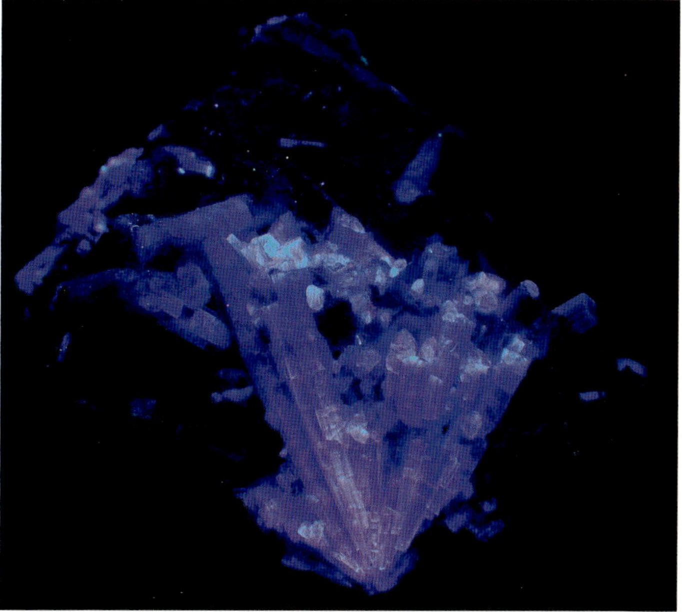

Same under SW

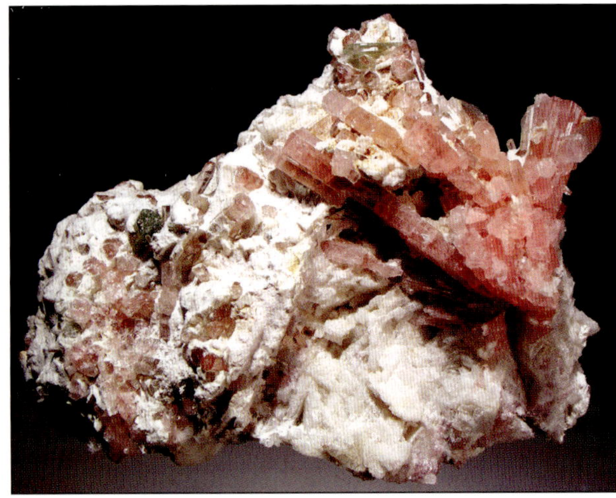

ELBAITE. Pink and blue elbaite (tourmaline group) from the Cryo-Genie mine, Warner Springs, San Diego County. The elbaite fluoresces pink and blue SW. The piece weighs 3.0 oz. and is 3.0 x 2.3 x 1.0 inches. Value $65-80

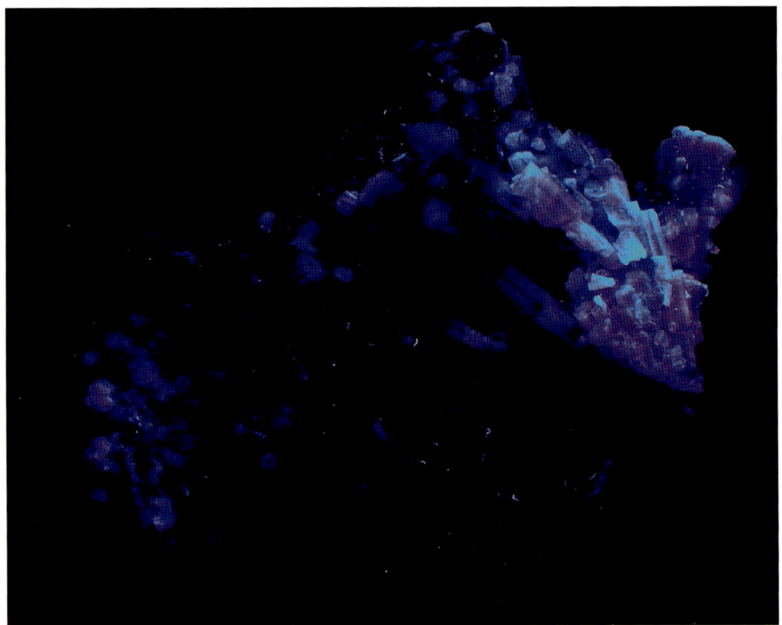

Same under SW

FRESNOITE. A fresnoite crystal from the Junilla mine, New Idria district, San Benito County. Collected by Scott Kleine in the summer of 1998. Fresnoite is rare and individual crystals are difficult to extract from their matrix. The fresnoite fluoresces pale yellow SW and weak yellow LW. It is 0.4 x 0.4 x 0.1 inches. Value $185-210

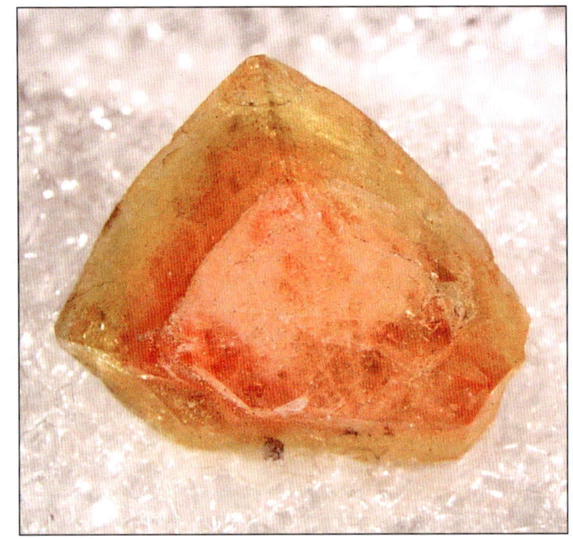

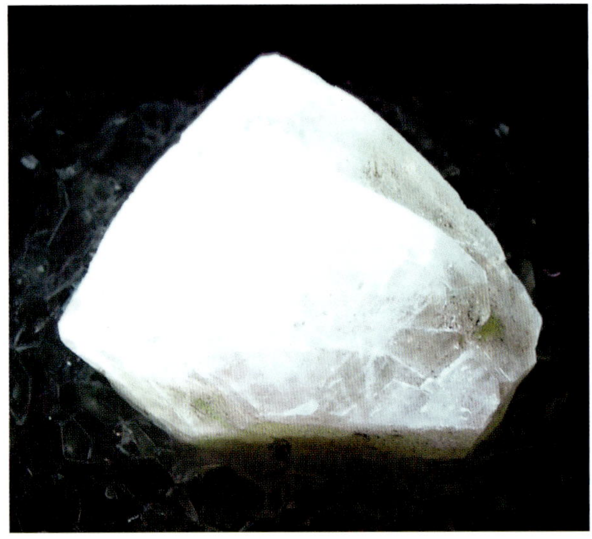

Same under SW

FRESNOITE. Fresnoite from the Junilla mine, New Idria district, San Benito County. The fresnoite fluoresces pale yellow SW and weak yellow LW. The piece weighs 0.2 oz. and is 1.4 x 0.5 X 0.3 inches. Value $15-17

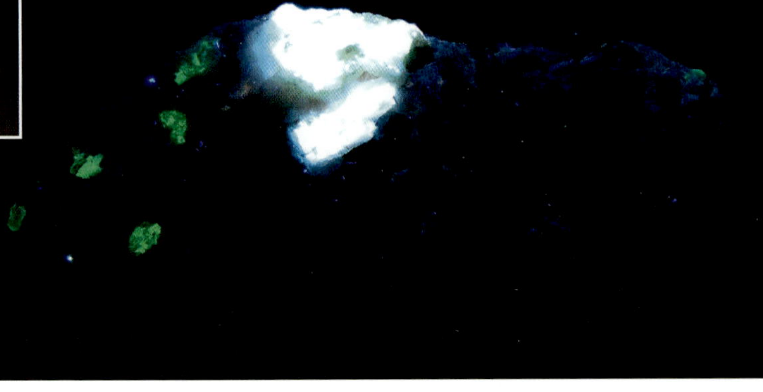

Same under SW

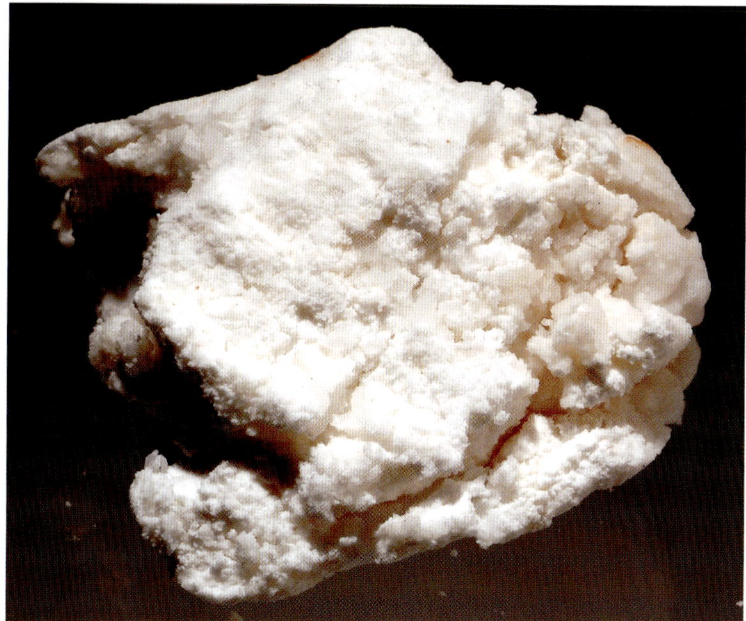

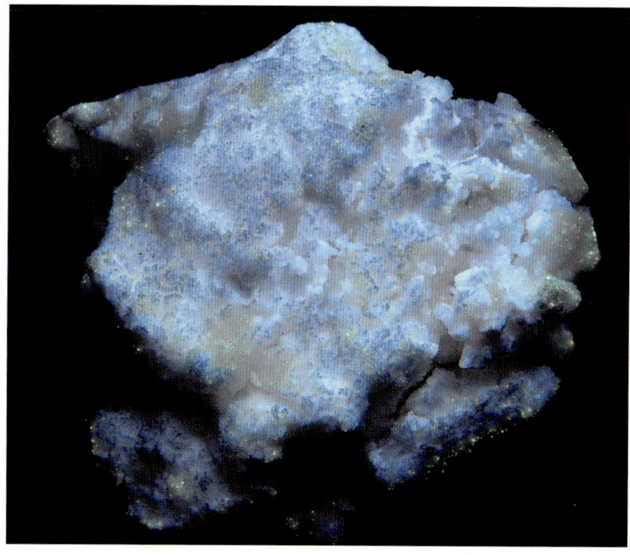

Same under SW

HECTORITE. Hectorite, from the Hector mine, near Hector, San Bernardino County, is a clay mineral from altered volcanic tuff ash with a high silica content. It looks like white paper-like matted masses with about a 30% water content. The mineral is commercially used as a thickener in cosmetics and toothpaste (Toms of Maine). The hectorite fluoresces a pale yellow LW, pale blue SW, and has a pale blue phosphorescence. The piece weighs 3.0 oz. and is 1.5 x 1.0 X 1.3 inches. From the Kevin Brady collection.

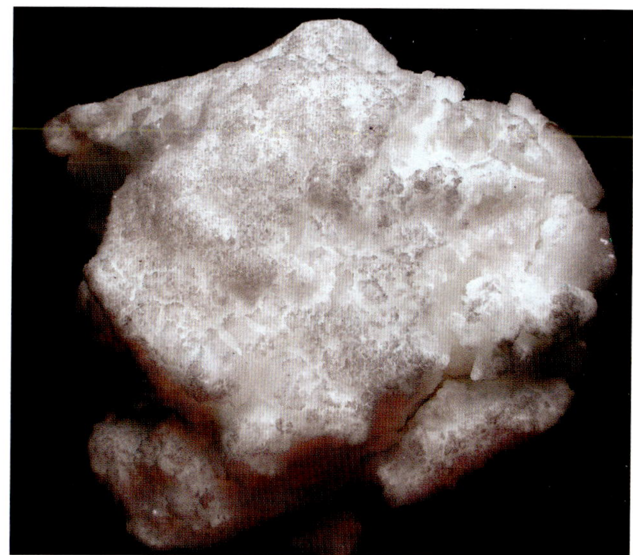

Same under LW

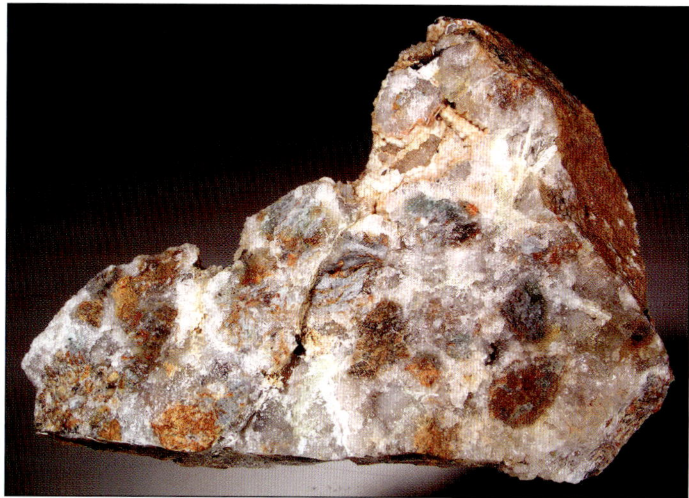

KARPATITE. Karpatite crystals from the Leza pit #1, 4th of July mine, New Idria district, San Benito County. Karpatite is a naturally occuring hydrocarbon mineral. It was first described in 1957 and was originally found in Russia. Karpatite fluoresces pale blue SW. The piece weighs 4.5 oz. and is 3.0 x 2.0 x 1.3 inches. Value $30-35.

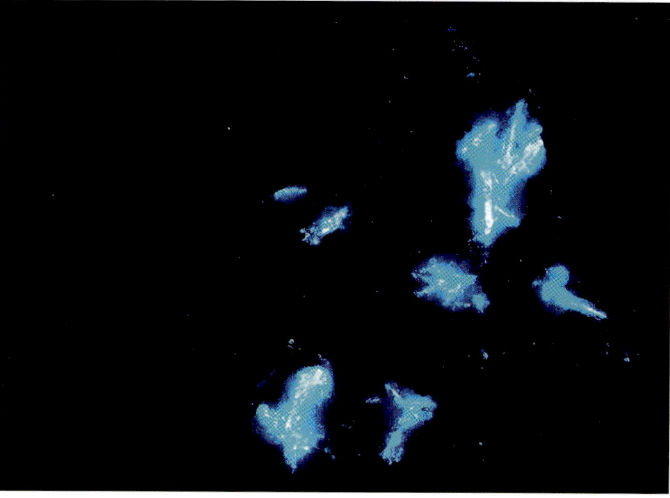

Same under SW

KARPATITE. Karpatite crystals from Picachos, New Idria district, San Benito County. The karpatite fluoresces two colors, pale blue and pale yellow SW and LW. The piece weighs 4.5 oz. and is 2.3 x 2.3 x 1.5 inches. Value $40-45

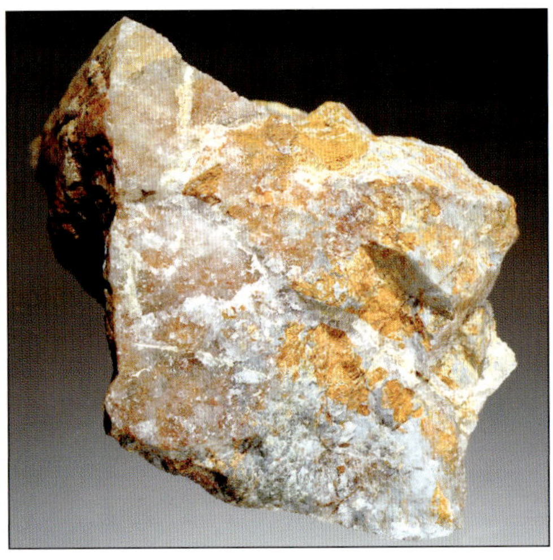

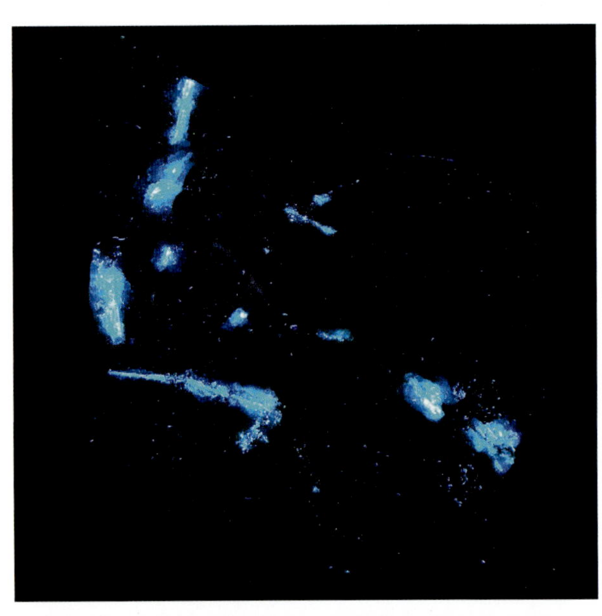

Same under SW

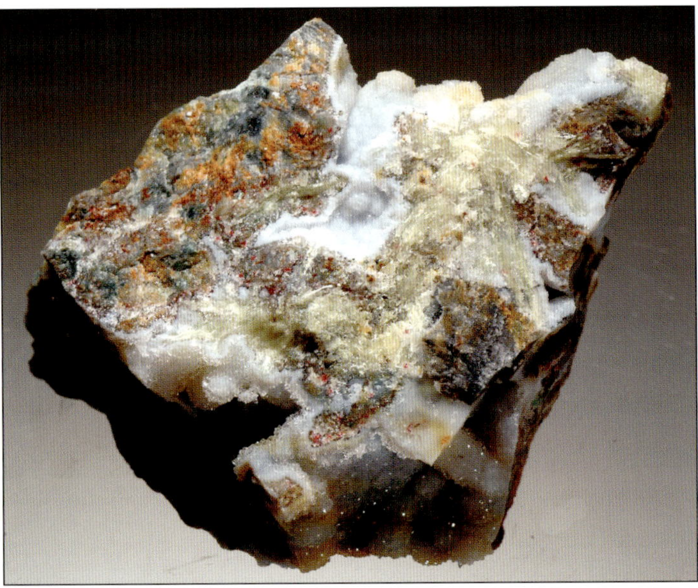

KARPATITE. Yellow karpatite crystals in a quartz coated piece from Picachos, New Idria district, San Benito County. This is a rich example. The karpatite fluoresces two colors, pale blue and pale yellow SW and LW. The piece weighs 6.5 oz. and is 3.3 x 2.5 x 1.5 inches. Value $90-110

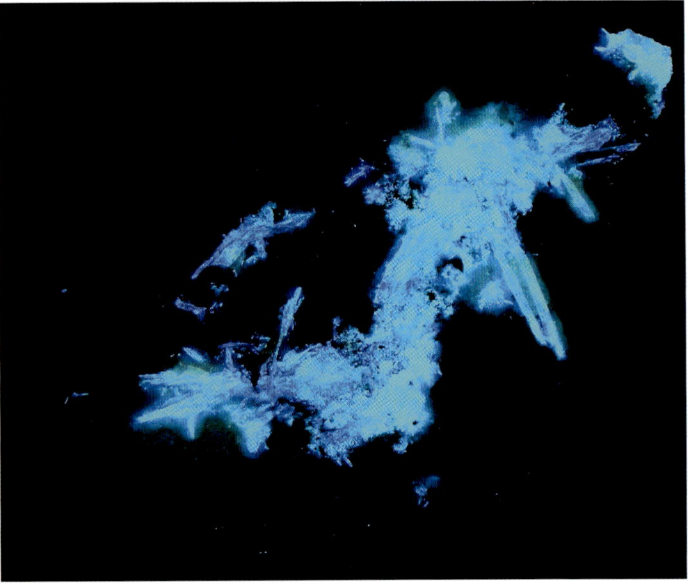

Same under SW

KARPATITE. Yellow karpatite crystals in a quartz coated piece from Picachos, New Idria district, Benito County. Another rich example. The karpatite fluoresces two colors, pale blue and pale yellow SW and LW. The piece weighs 7.8 oz. and is 2.8 x 2.3 x 1.9 inches. Value $90-110

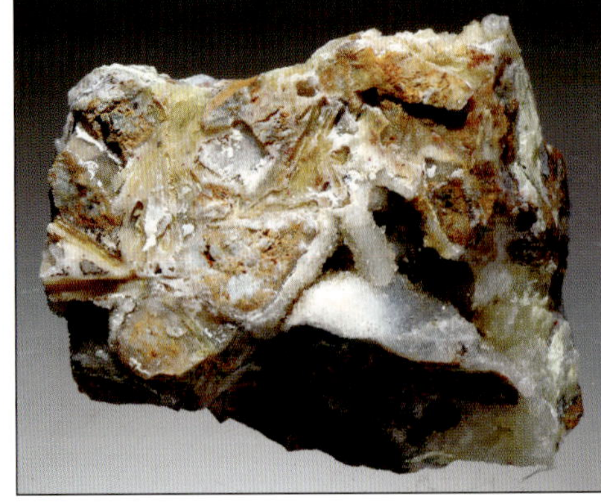

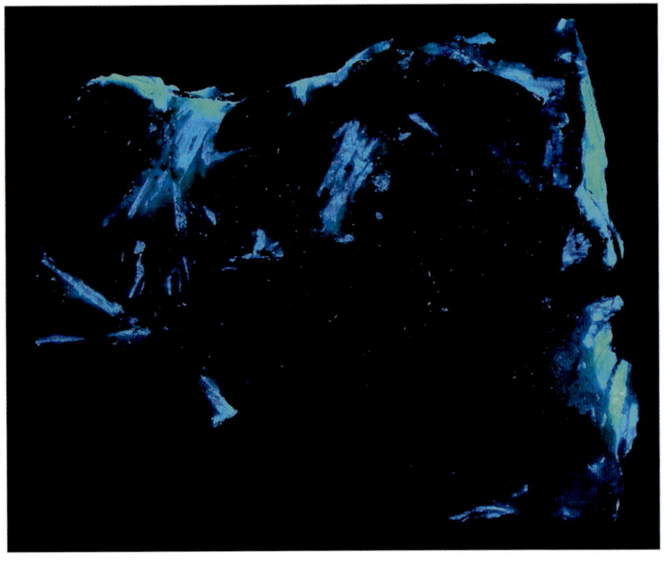

Same under SW

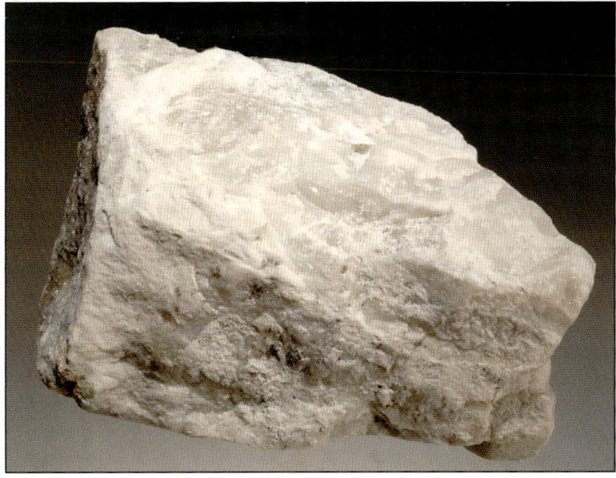

OPAL, CALCITE. Opal, var. hyalite, on calcite from the White Knob mine, Lucerne Valley Limestone Province, San Bernadino County. The opal fluoresces green SW and the calcite has a unusual bluish color SW. The piece weighs 2 lb. and is 3.3 x 2.0 x 2.0 inches. Value $10-15. Courtesy of Donald L. Fife.

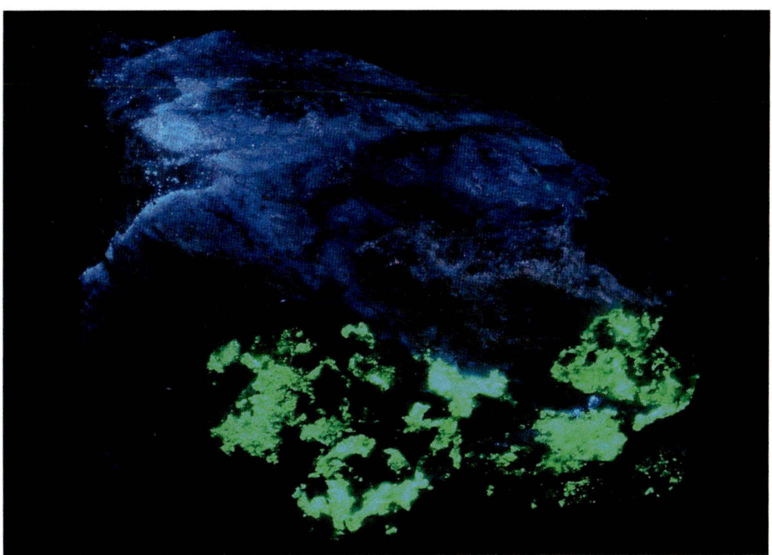

Same under SW

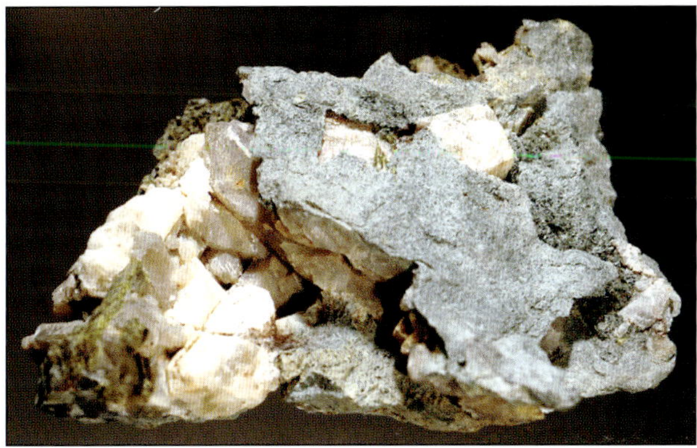

SCHEELITE. Crystals of scheelite with epidote on actinolite from the Sally Peterson mine, Inyo, Inyo County. Scheelite fluoresces pale blue SW. The piece weighs 5.0 oz. and is 3.0 x 2.5 x 1.0 inches. Value $20-25

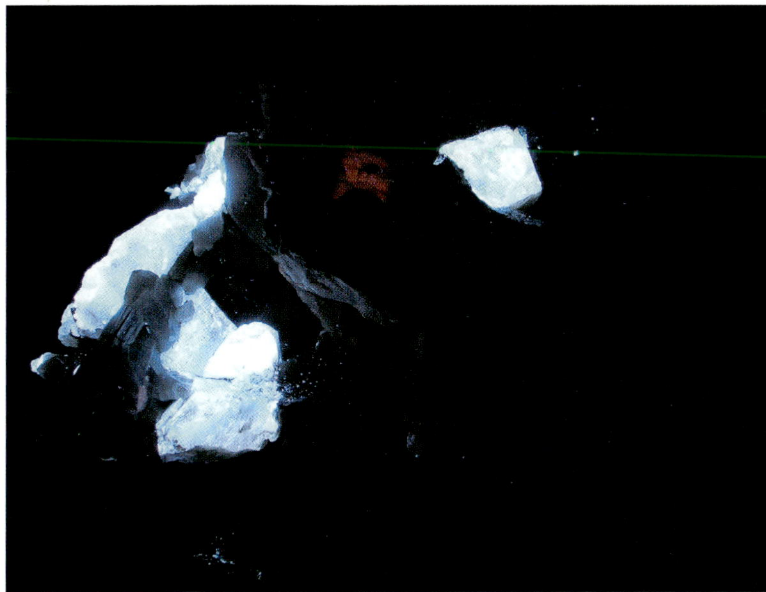

Same under SW

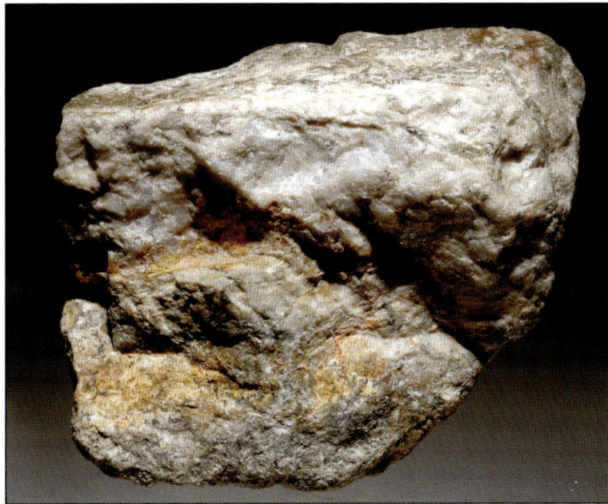

SCHEELITE. Scheelite from Atolia, San Bernardino County. This piece was sold by Tom Warren in 1945. It comes with his label and the original sales receipt. The scheelite fluoresces bright pale blue SW. This piece will sell for a premium since it comes with provenance from a leading fluorescent mineral dealer. The piece weighs 1 lb. 6.0 oz. and is 3.5 x 2.5 x 2.5 inches. Value $80-85

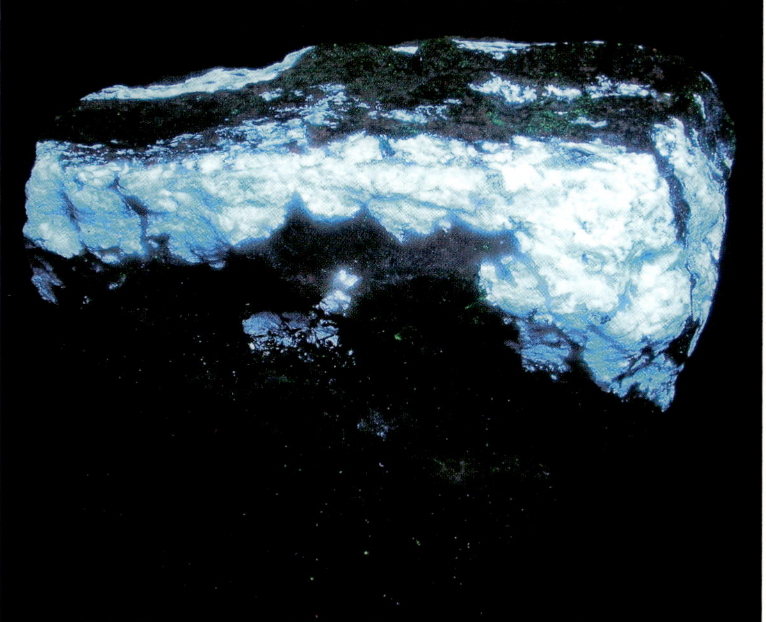

Same under SW

SCHEELITE, POWELLITE. Scheelite with powellite from Laurel Lakes, Mono County. The scheelite fluoresces pale blue and the powellite fluoresces yellow SW. The piece weighs 1.8 oz. and is 1.5 x 1.5 x 1.1 inches. Value $20-25

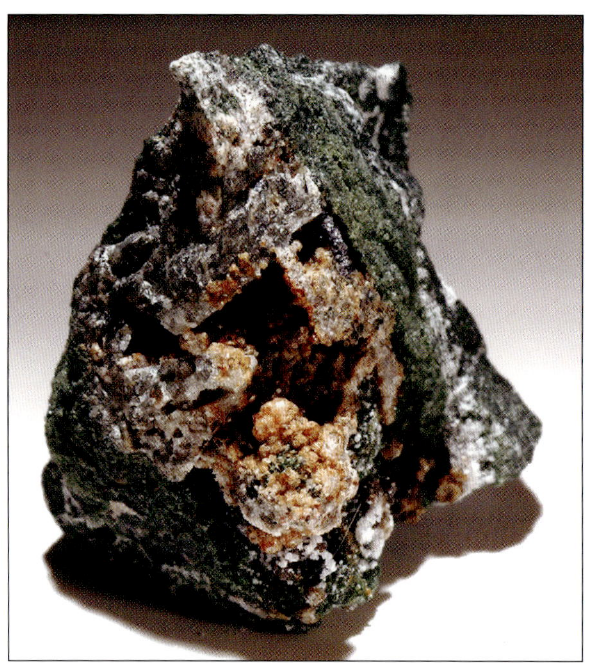

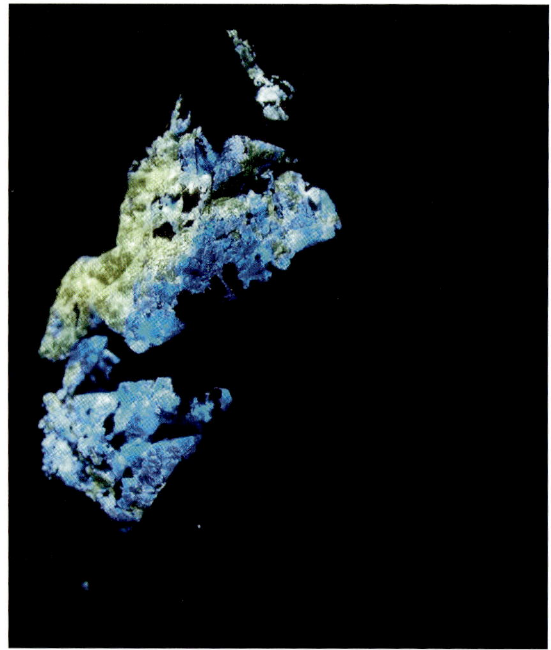

Same under SW

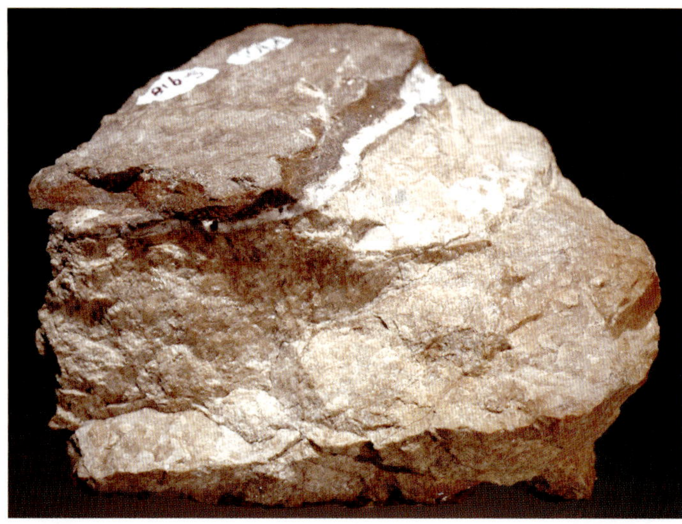

WOLLASTONITE, ARAGONITE. A brightly-fluorescing wollastonite with aragonite from the Desert View mine, Holcomb Valley, San Bernardino County. Collectors call this "Desert View wollastonite". The wollastonite fluoresces orange SW and the aragonite fluoresces pale blue SW and LW. The piece weighs 7.5 oz. and is 3.0 x 2.8 x 1.3 inches. Value $60-75

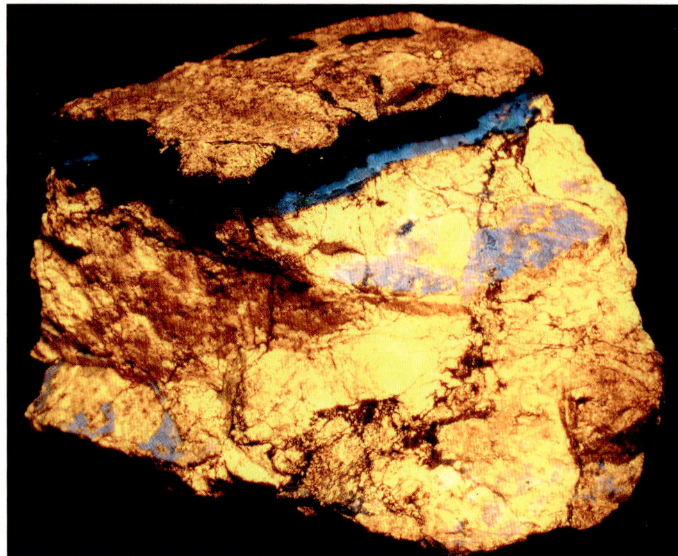

Same under SW

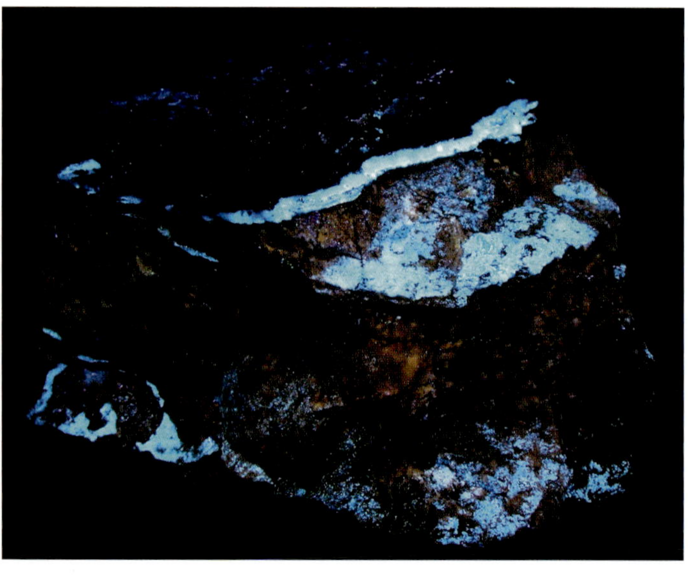

Same under LW

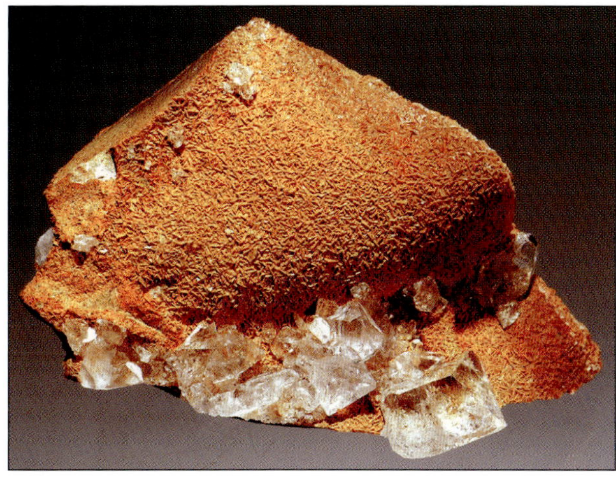

BARITE. Barite crystals from Brooks Cliff, Grand Junction, Mesa County. The crystals fluoresce in zones a pale yellow and white SW. The piece weighs 4.5 oz. and is 2.8 x 1.8 x 1.5 inches. Value $20-25

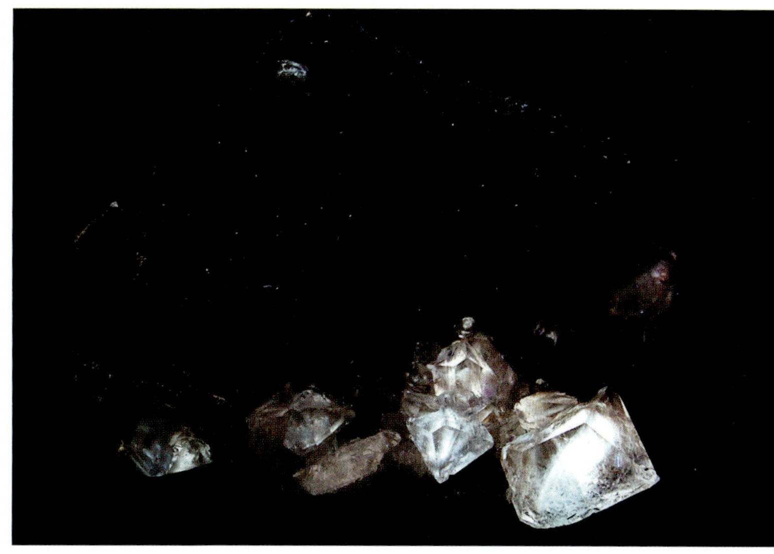

Same under SW

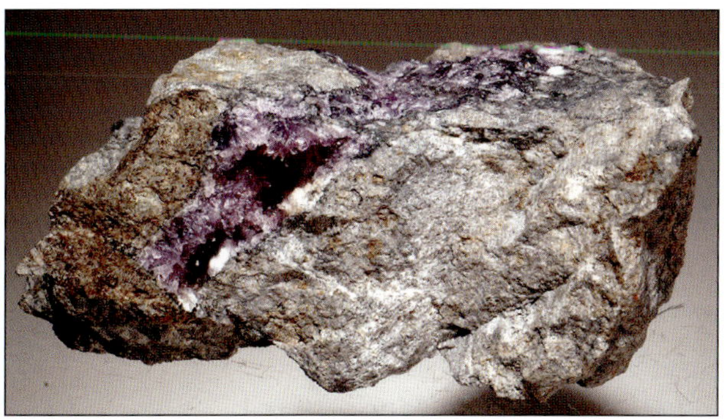

CREEDITE. Purple creedite crystals from the Cresson mine, Cripple Creek, Teller County. Creedite was first described in 1916 from the Creed Quadrangle in Mineral County, Colorado. The crystals fluoresce pale blue SW. The piece weighs 10.0 oz. and is 3.8 x 2.8 x 1.8 inches. Value $80-90

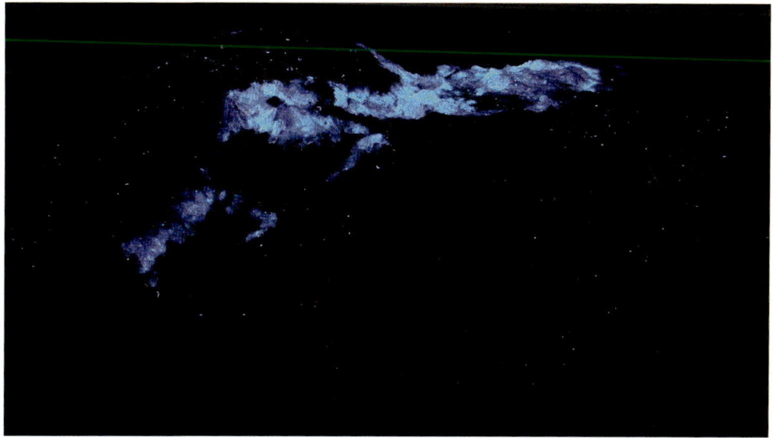

Same under SW

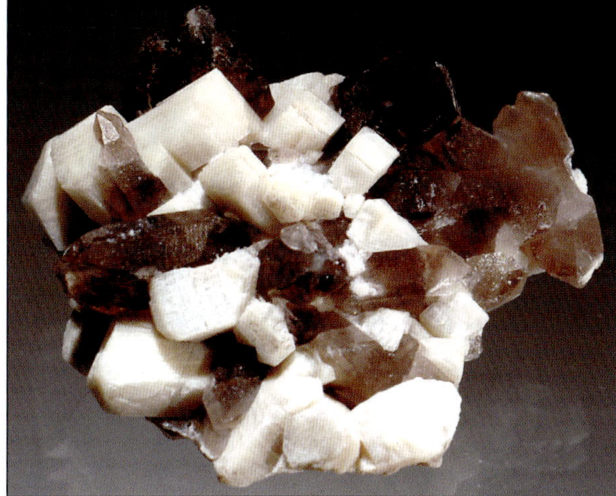

MICROCLINE. Microcline crystals with quartz from Crystal Park, El Paso County. The microcline fluoresces velvety red SW. The piece weighs 1.5 oz. and is 2.1 x 1.6 x 0.9 inches. Value $20-25

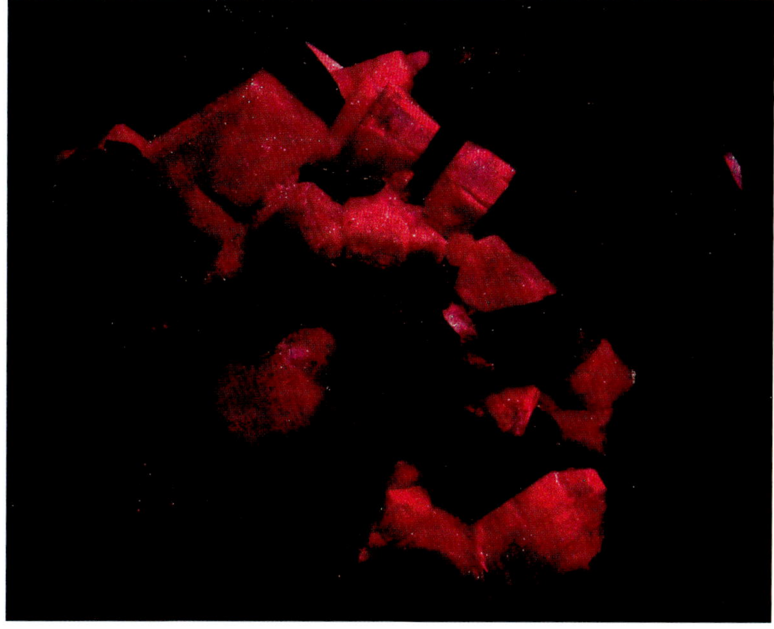

Same under SW

SPHALERITE, FLUORITE. Fluorite crystals on sphalerite with rhodochrosite, quartz and pyrite from the Sweet Home mine, Alma, Park County. The fluorite fluoresces violet LW and the sphalerite fluoresces yellow-orange LW and (not as brightly) SW. The piece weighs 4.8 oz. and is 3.5 x 1.5 x 1.0 inches. Value $50-55

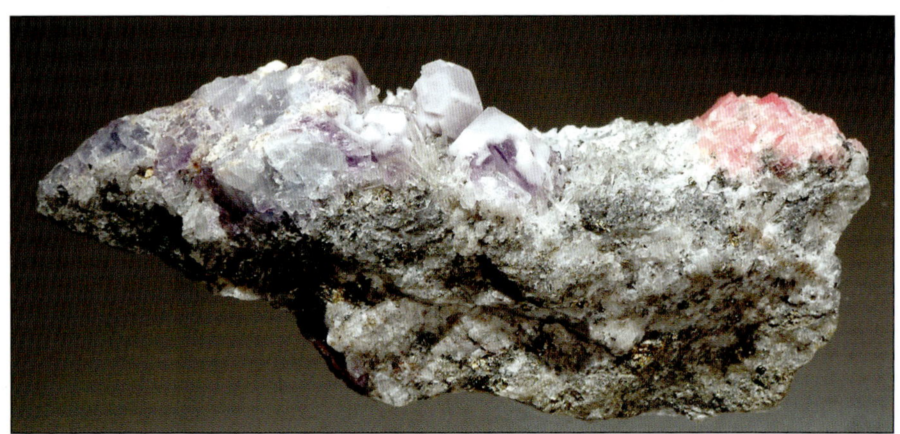

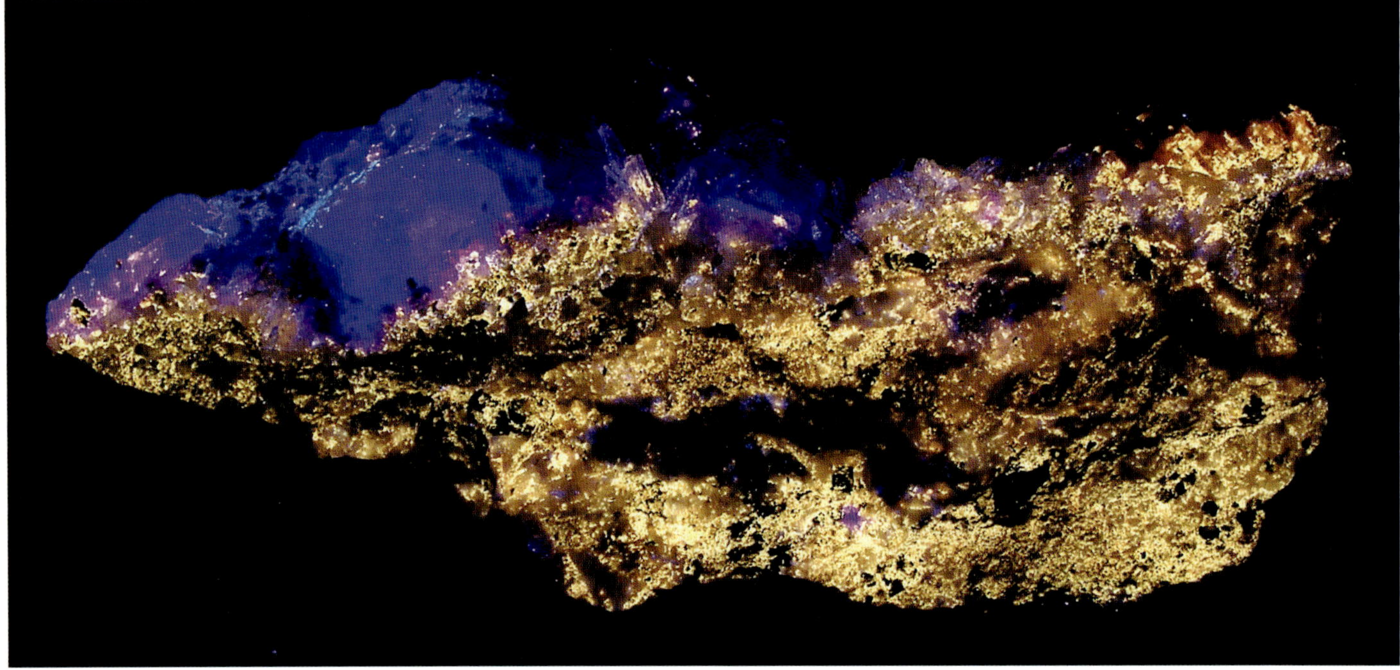

Same under LW

Missouri

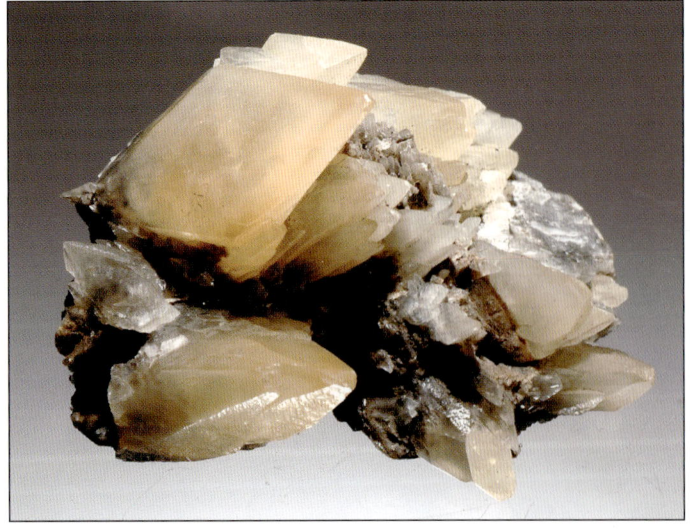

CALCITE. Dogtooth-shaped calcite crystals from Joplin, Jasper County. The calcite fluoresces pale blue SW. The piece weighs 4.5 oz. and is 2.5 x 2.0 x 1.5 inches. Value $20-25

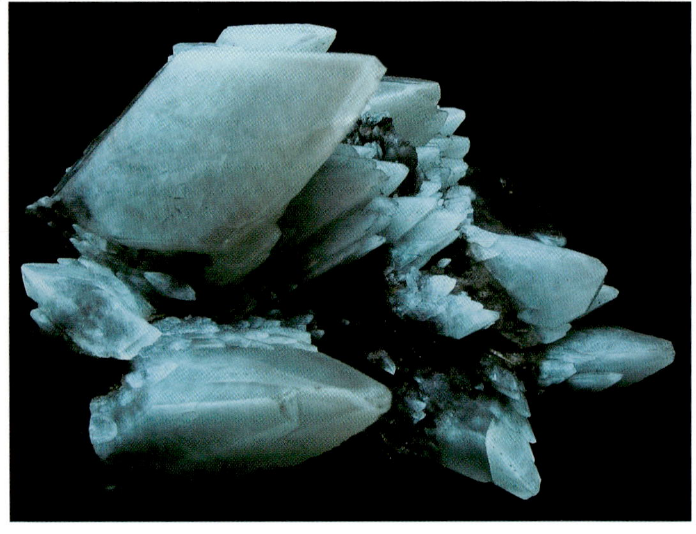

Same under SW

Nebraska

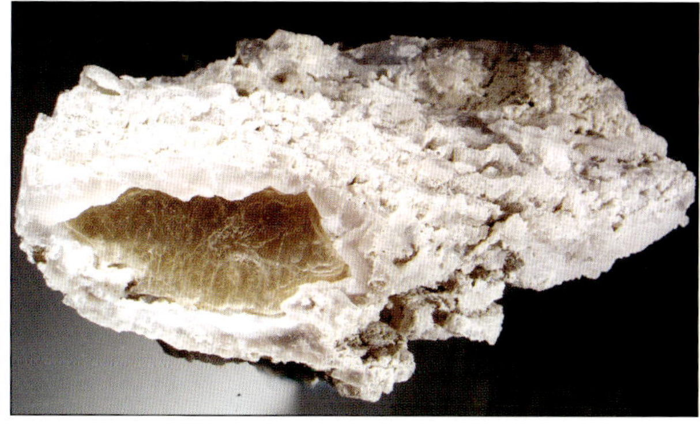

QUARTZ, GYPSUM. Quartz, var. chalcedony, and gypsum, var. selenite, from Crawford, Dawes County. The selenite is enveloped by the chalcedony. This is an exceptionally attractive display piece. The selenite fluoresces pale orange LW and SW, the chalcedony fluoresces green SW. The piece weighs 1 lb. 2.0 oz. and is 5.0 x 3.8 x 2.0 inches. Value $80-90

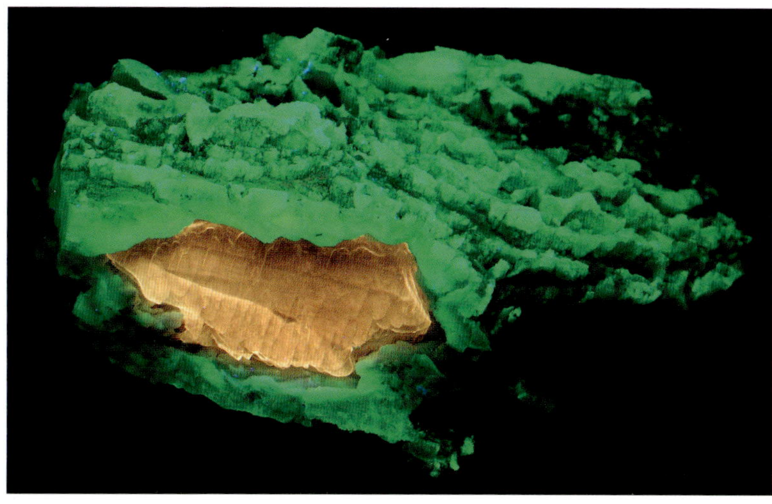

Same under SW

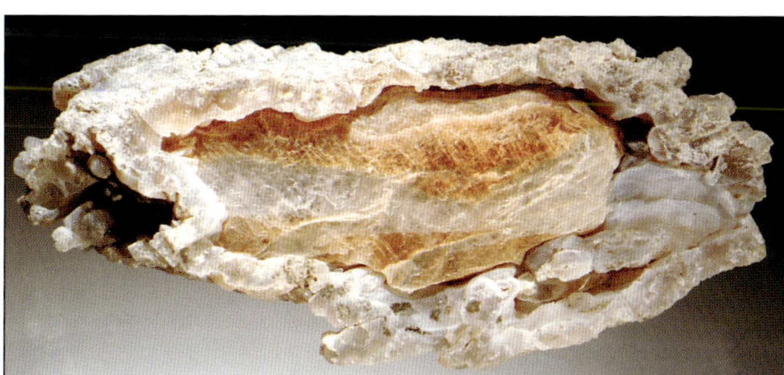

QUARTZ, GYPSUM. Quartz, var. chalcedony, and gypsum, var. selenite, from Crawford, Dawes County. The selenite fluoresces pale orange LW and SW, the chalcedony fluoresces green SW. The piece weighs 10.0 oz. and is 5.3 x 2.3 x 2.0 inches. Value $85-95

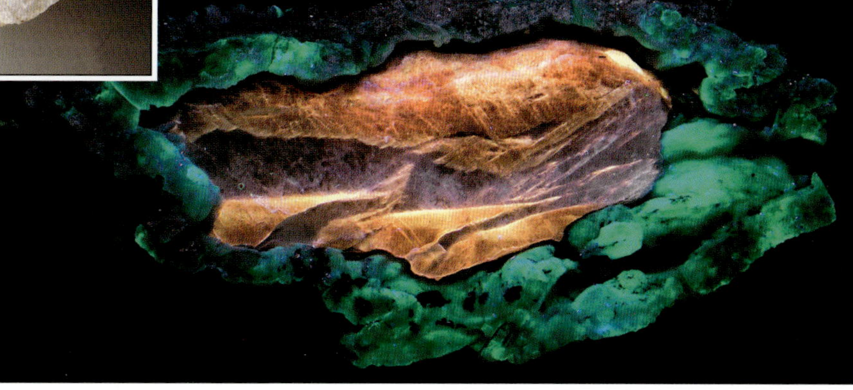

Same under SW

Nevada

ARAGONITE. A coral-like aragonite from the Northern Lights mine, near Yerington, Mineral County. The aragonite fluoresces white with green tips SW and bright pale yellow LW. The piece weighs 1.3 oz. and is 2.0 x 1.5 x 1.5 inches. Value $35-40

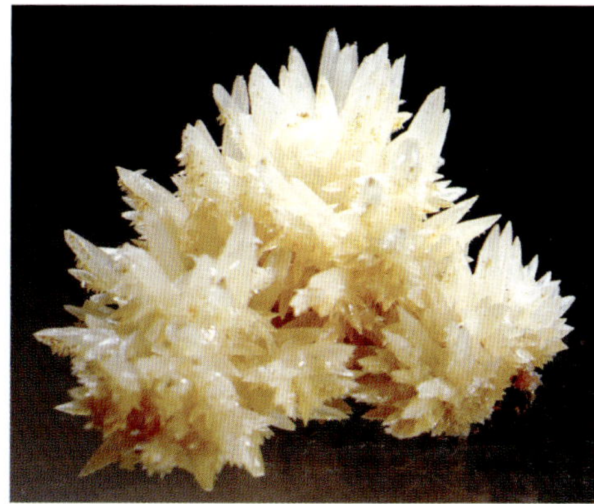

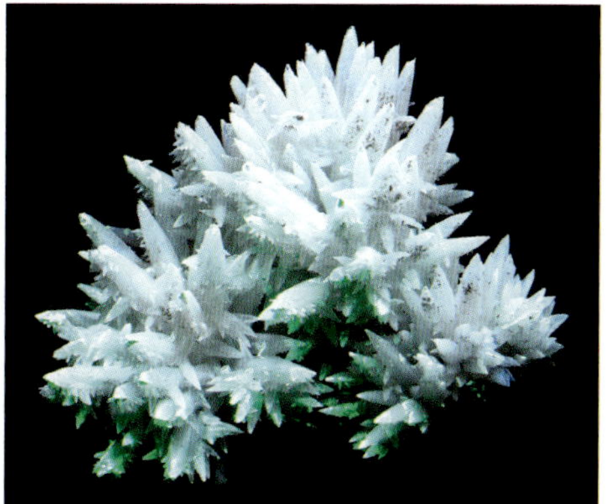

Same under SW

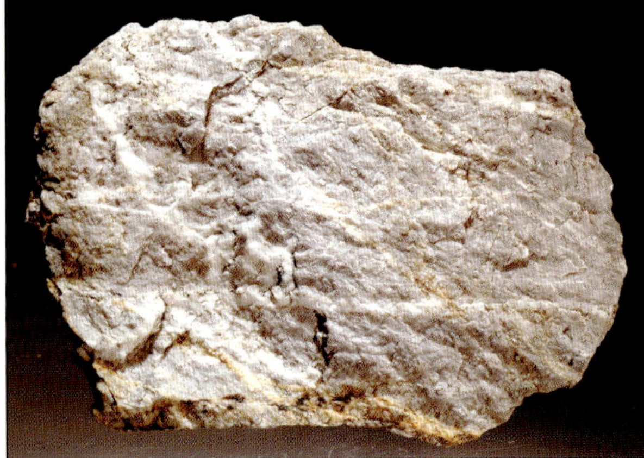

CALCITE. Calcite in matrix from Copper Canyon, Magdalena district, Socorro County. The calcite fluoresces orange SW in a lightning-like pattern across the piece. The piece weighs 5.0 oz. and is 3.8 x 2.5 x 0.8 inches. Value $55-60

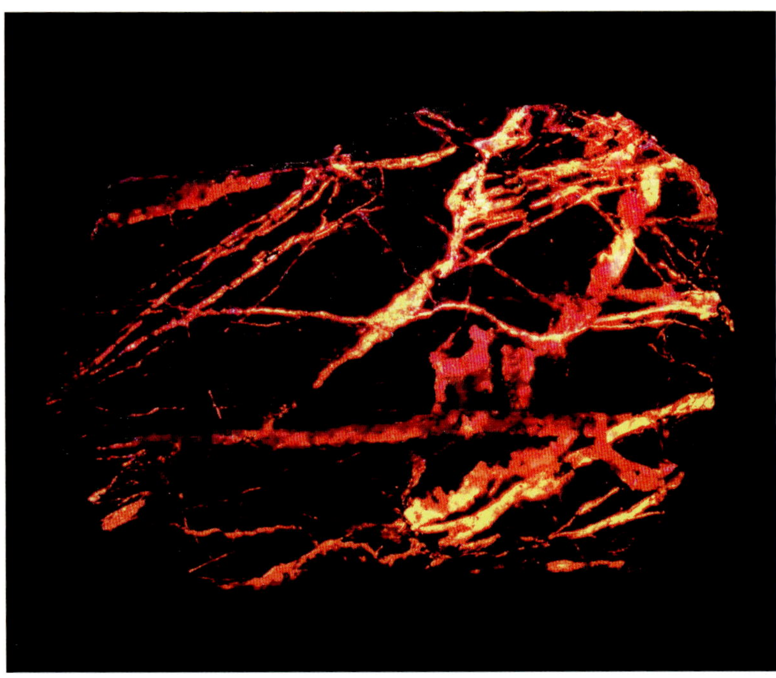

Same under SW

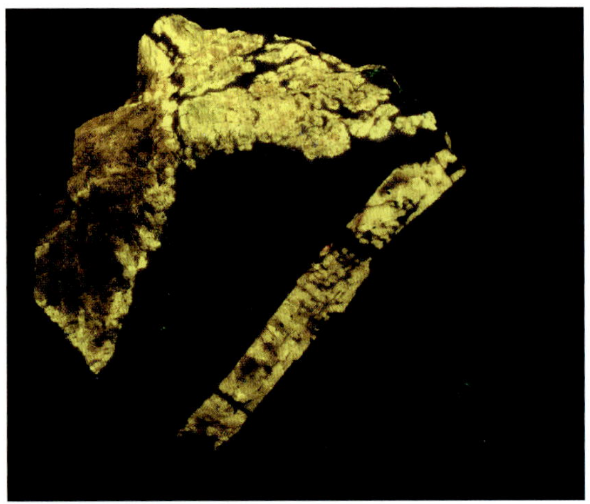

Same under SW

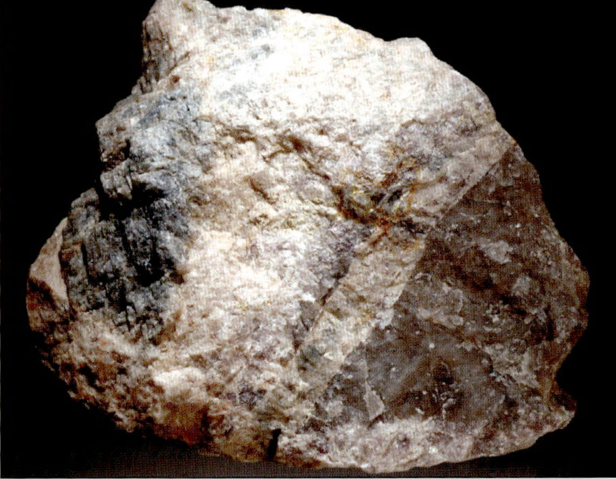

FLUORAPATITE. Fluorapatite on quartz, feldspar, and muscovite from the Harding mine, Taos County. This fluorapatite fluoresces yellow SW and brighter yellow MW. The piece weighs 2 lb. 3.0 oz. and is 4.5 x 4.0 x 2.5 inches. Value $25-30

WILLEMITE. Willemite from the Petroglyph mine, Hillsboro District, Sierra County. The willemite fluoresces pale blue SW. The piece weighs 7.0 oz. and is 3.3 x 2.3 x 1.8 inches. Value $30-35

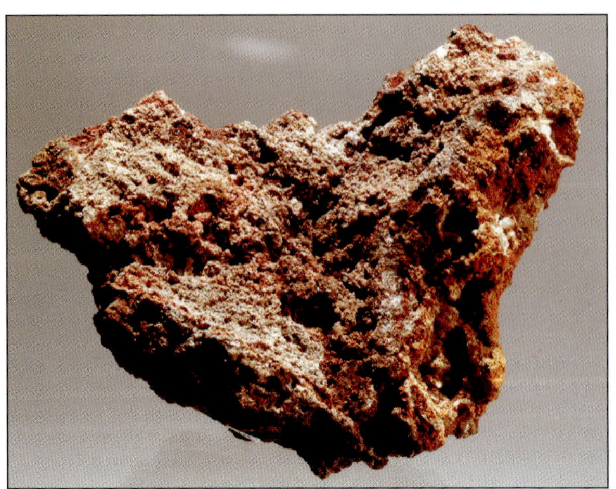

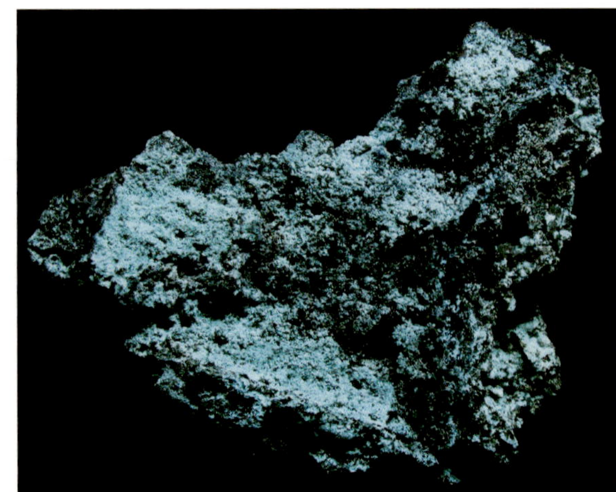

Same under SW

Oregon

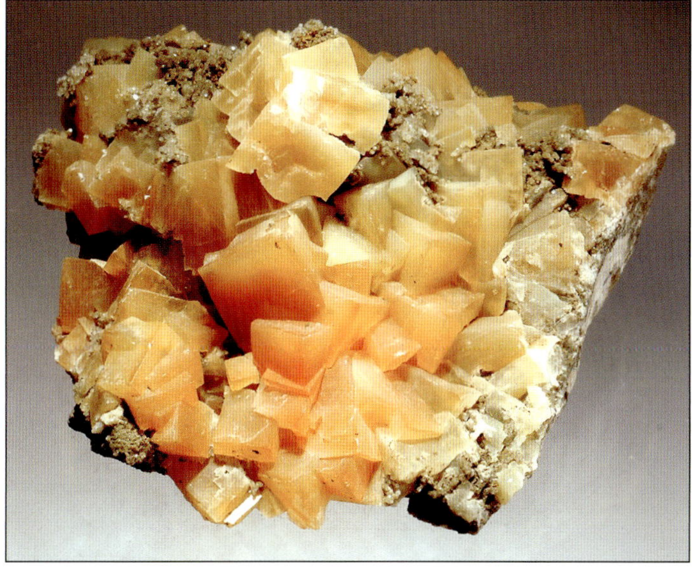

CALCITE. Stepped calcite crystals from the Gopher Valley quarry, Yamhill, Yamhill County. This calcite fluoresces bright pale yellow SW and LW. The piece weighs 7.0 oz. and is 3.0 x 3.0 x 1.8 inches. Value $55-65

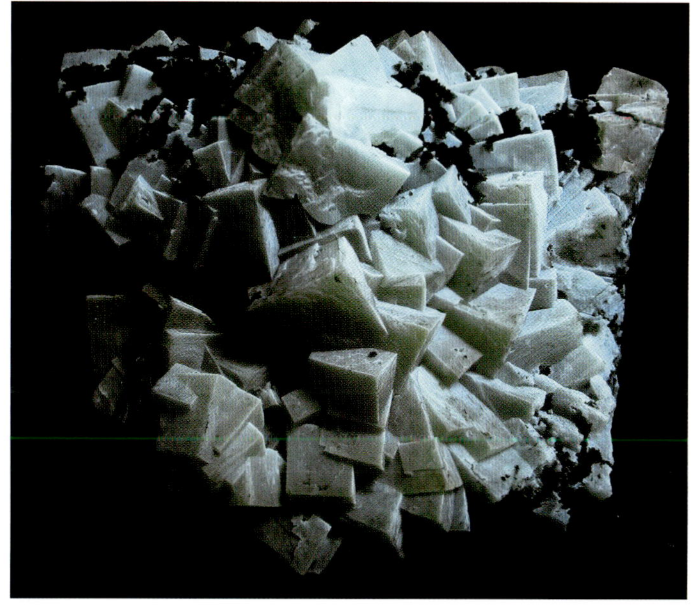

Same under SW

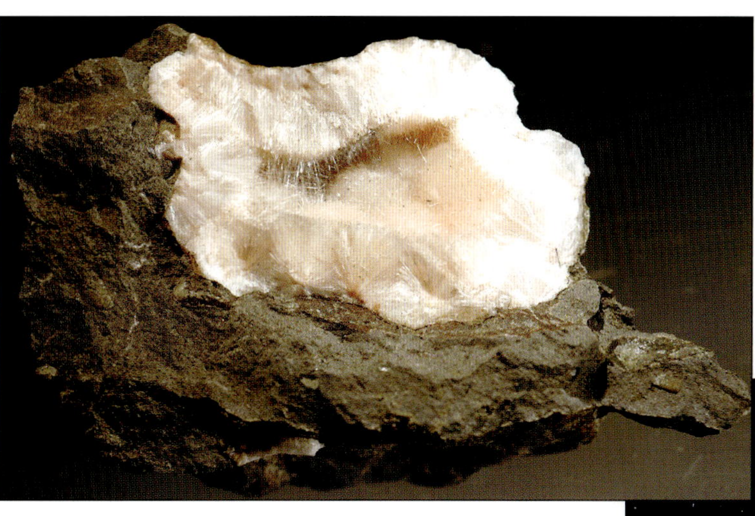

MESOLITE. Beautiful acicular mesolite crystals on basalt matrix from Goble, Columbia County. The mesolite fluoresces pale yellow SW. The piece weighs 2.0 oz. and is 2.5 x 1.5 x 1.0 inches. Value $30-35

Same under SW

ALBITE. Albite from the Tin Mountain mine, Custer district, Custer County. The albite fluoresces pale blue SW. The piece weighs 2 lb. 1.0 oz. and is 5.0 x 4.0 x 3.0 inches. Value $45-55

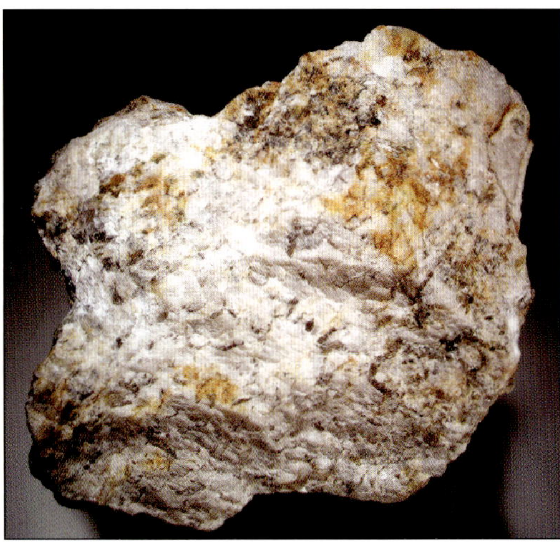

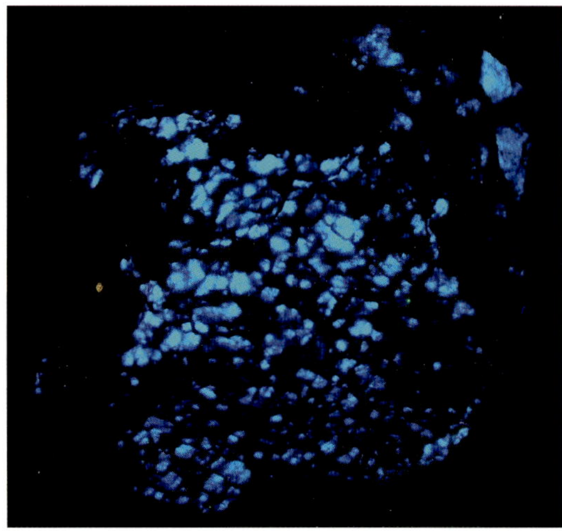

Same under SW

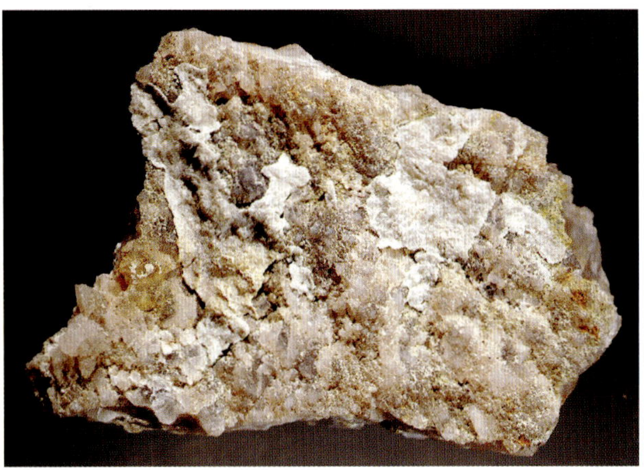

ARAGONITE, QUARTZ A very interesting and attractive specimen containing aragonite on quartz, var. chalcedony, from the Badlands, Jackson County. The thin layer of aragonite fluoresces pink SW and the chalcedony fluoresces green SW. The piece weighs 8.0 oz. and is 3.5 x 2.3 x 1.1 inches. $45-55

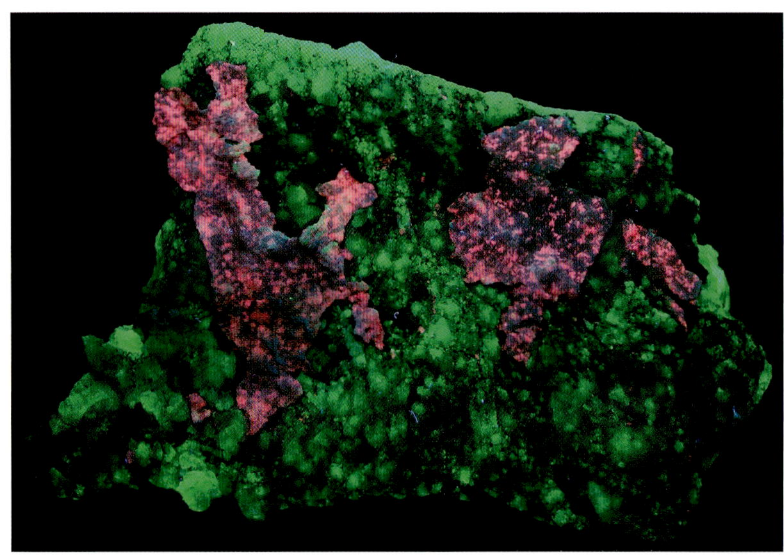

Same under SW

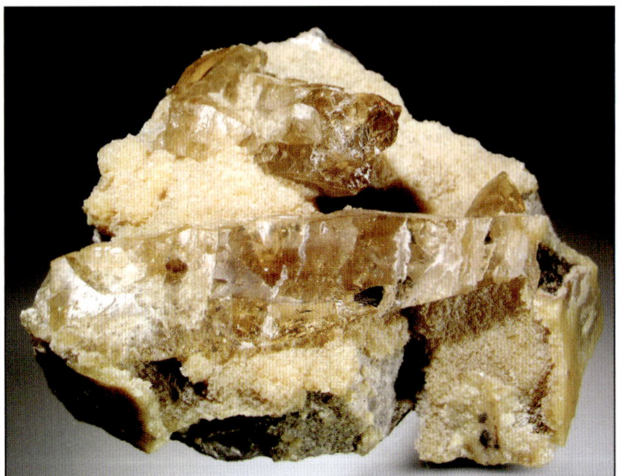

BARITE. Barite crystals from Butte County. The barite fluoresces pale yellow SW. The piece weighs 4.0 oz. and is 2.3 x 2.0 x 1.5 inches. Value $15-18.

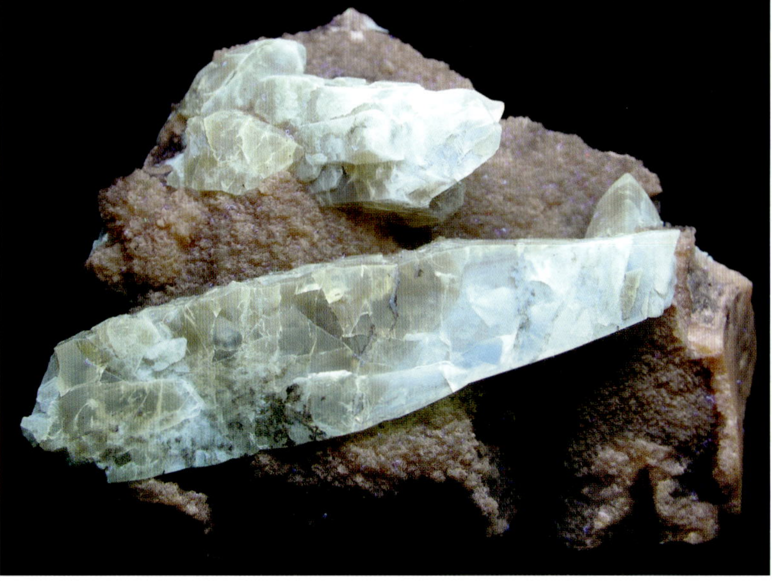

Same under SW

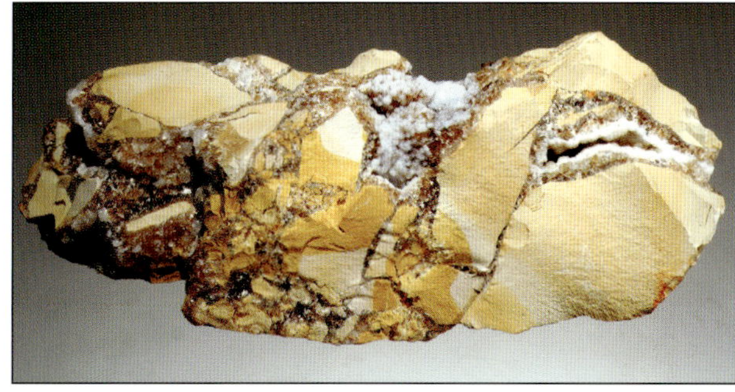

CALCITE. A piece from the Wright-McGrady mine, Karnes Uranium district, Karnes County. The calcite fluoresces green SW and bluish-green LW. There are veins of caliche (a secondary calcium carbonate that forms in sediments or in voids in bedrock just below the surface in semiarid regions) which fluoresce peach SW. The piece weighs 7.5 oz. and is 4.5 x 1.5 x 1.3 inches. Value $45-50

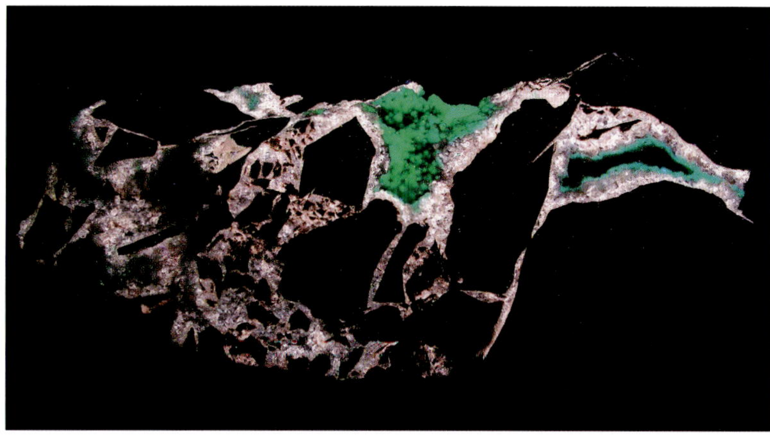

Same under SW

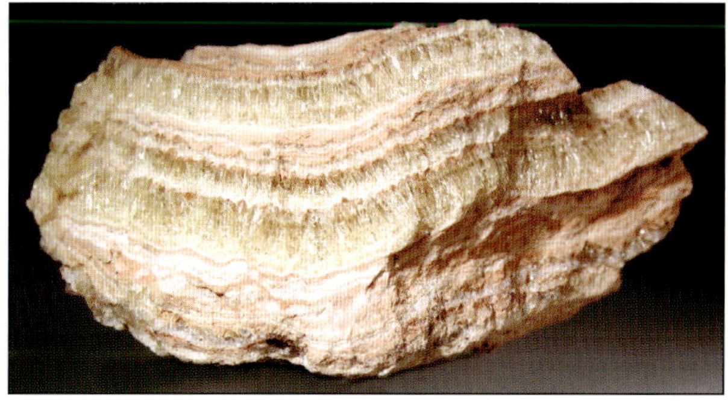

CALCITE. Calcite from a prospect pit near Study Butte in the Terlingua district, Brewster County. There are bands of calcite that fluoresce pale yellow equally well under SW, MW and LW. It phosphoresces after exposure to UV light. The piece weighs 5.3 oz. and is 2.3 x 2.3 x 1.0 inches. Value $45-55

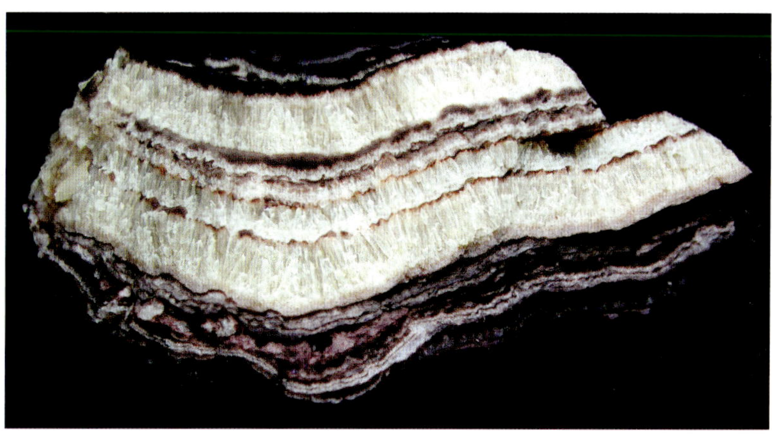

Same under SW

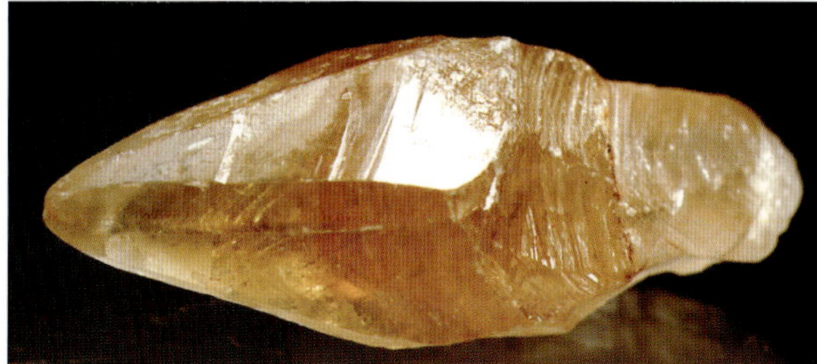

CALCITE. A spear-shaped crystal of honey-colored calcite from the Terlingua district, Brewster County. It fluoresces white and bright pale yellow under SW and LW. We are familiar with the calcite from the Terlingua district that fluoresces pink LW and blue SW, but there are other varieties of calcite and other fluorescent minerals found in this mercury-mining area. It weighs 1.0 oz. and is 2.0 x 0.8 x 0.6 inches. Value $15-17

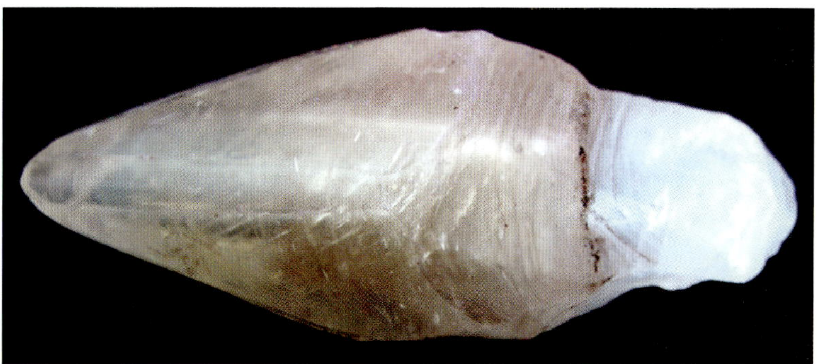

Same under SW

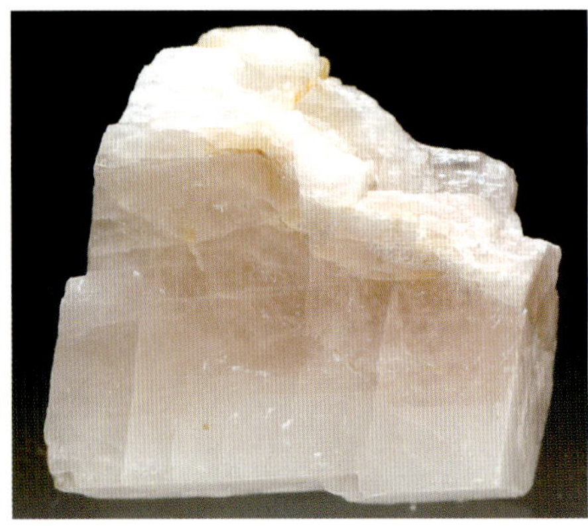

CALCITE. An unusual calcite rhomb from the Lampasas area, Lampasas County. Some sections fluoresce different colors in the calcite . It fluoresces pale blue SW, some areas fluoresce pink LW and it has a long-lasting blue phosphorescence. The piece weighs 2.5 oz. and is 1.9 x 1.8 x 0.8 inches. Value $25-30

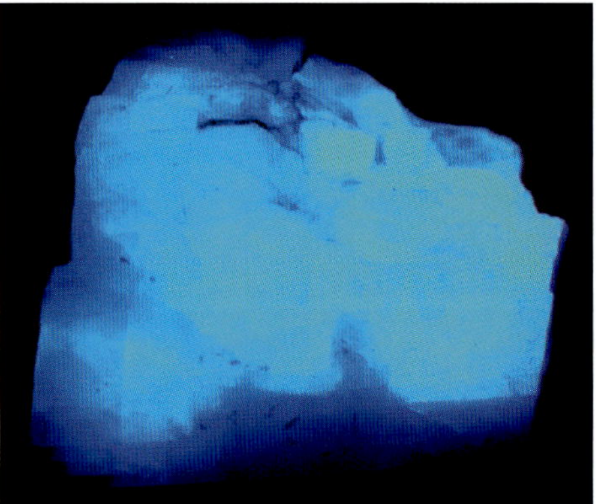

Same under SW

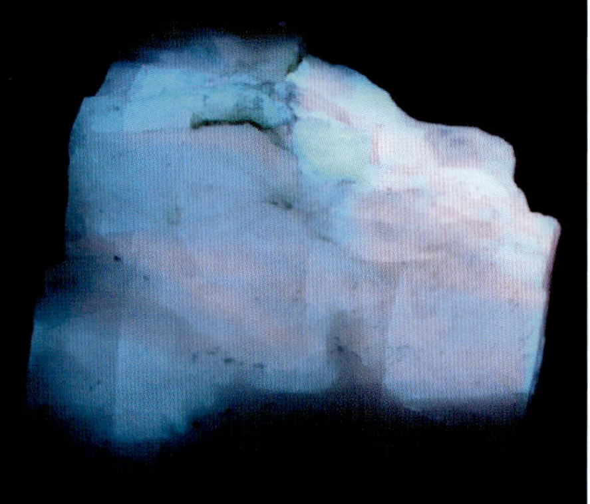

Same under LW

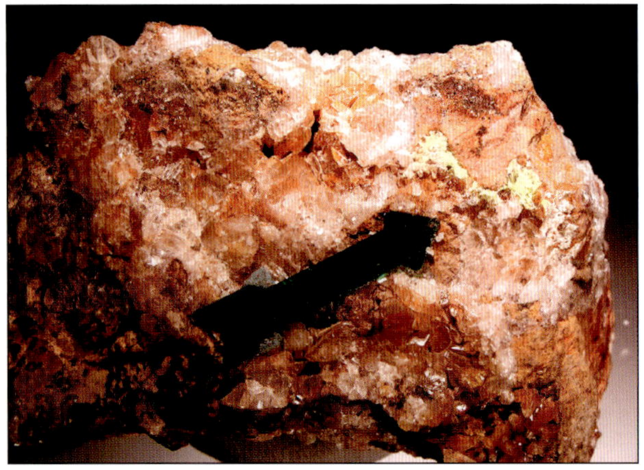

CALOMEL. Calomel with mosesite from the Mariposa mine, Terlingua district, Brewster County. Calomel, a mercury chloride mineral, fluoresces red-orange SW and LW. Mosesite is yellow in daylight and will change color to lime-green with exposure to sunlight. The piece weighs 6.0 oz. and is 2.5 x 2.0 x 1.8 inches. Value $30-35 (due to the small amount of calomel).

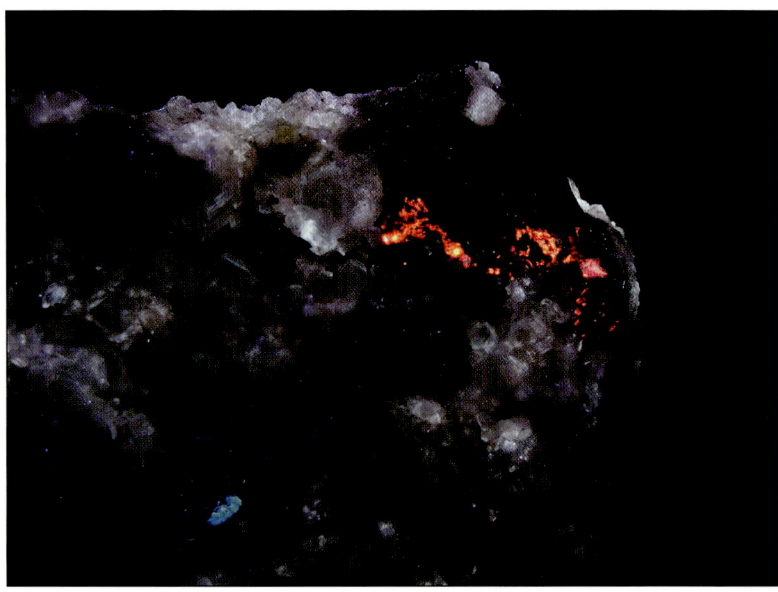

Same under SW

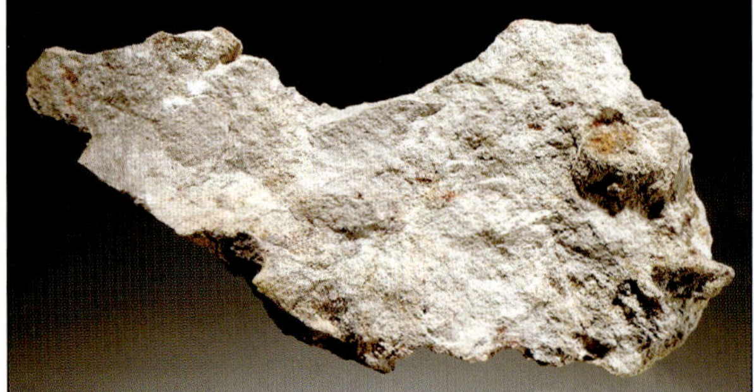

CALOMEL. Calomel with mosesite found in 1975 in the Perry pit of the Mariposa mine in the Terlingua district, Brewster County. Calomel fluoresces red-orange SW and LW. The piece weighs 2.8 oz. and is 3.3 x 1.8 x 1.3 inches. Value $60-65

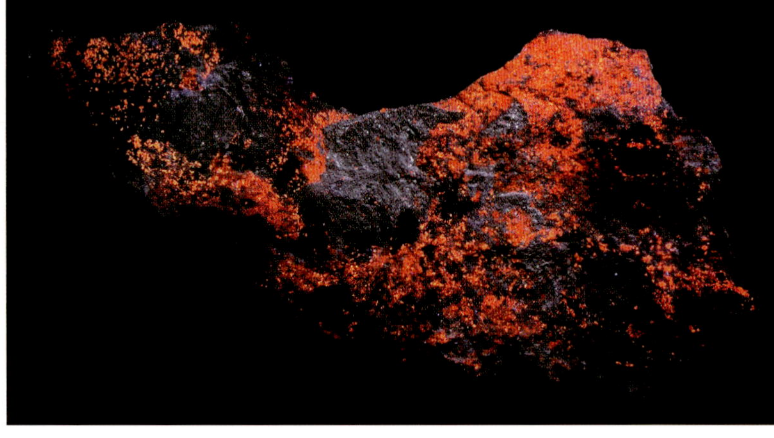

Same under SW

Utah

FLUORAPATITE. Fluorapatite crystals in magnetite from near Cedar City, Iron County. This fluorapatite fluoresces pink SW. The piece weighs 8.5 oz. and is 2.5 x 2.0 x 1.0 inches. Value $45-50

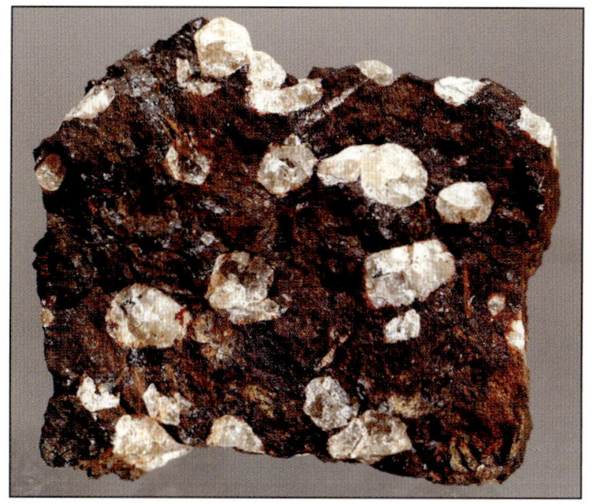

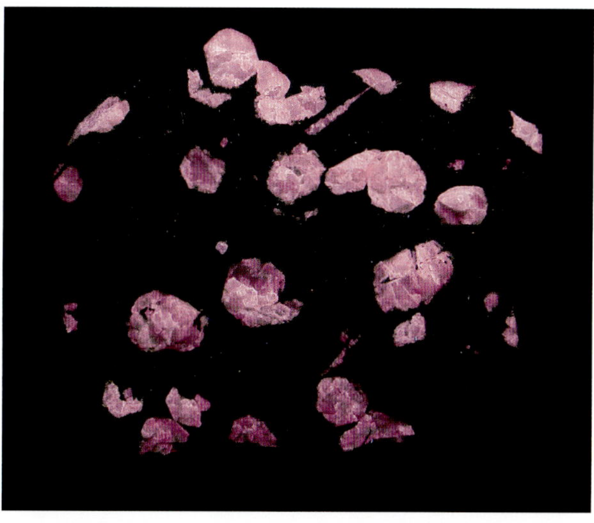

Same under SW

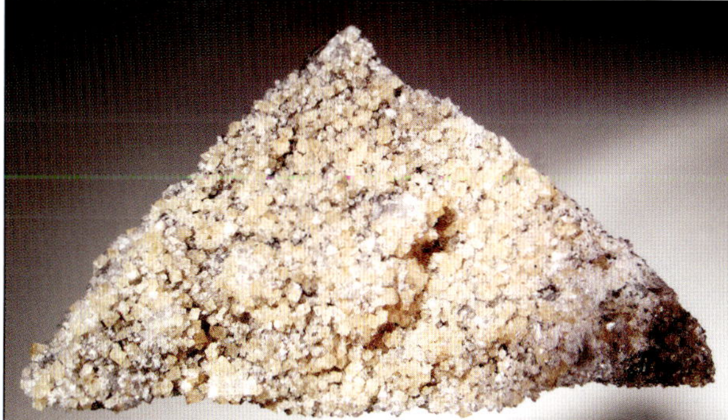

SCHEELITE. A specimen covered with small crystals of scheelite from the Ophir Hill mine, Ophir district, Tooele County. The scheelite fluoresces pale blue SW. The piece weighs 1.0 oz. and is 1.5 x 1.3 x 1.0 inches. Value $18-20

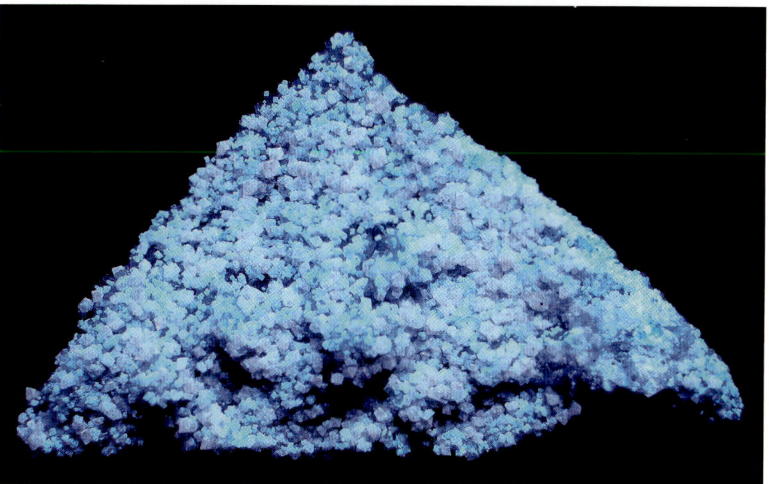

Same under SW

Washington

META-AUTUNITE. Meta-autunite crystals from the Daybreak mine, near Mt. Spokane, Spokane County. Meta-autunite is a radioactive mineral and fluoresces yellow-green SW and LW. The piece weighs 0.4 oz. and is 1.0 x 0.8 inches. Value $18-22

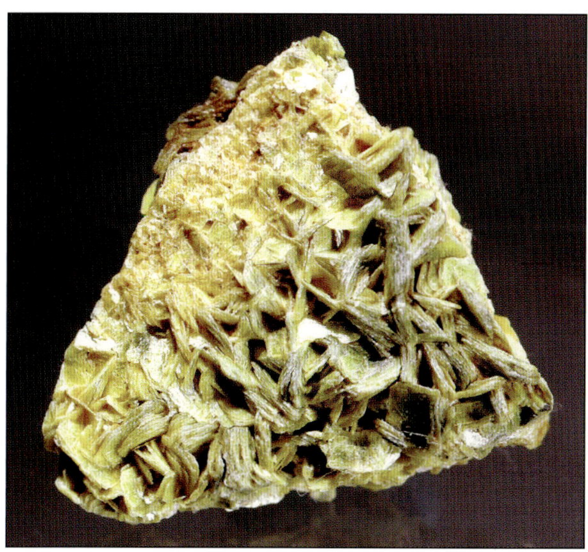

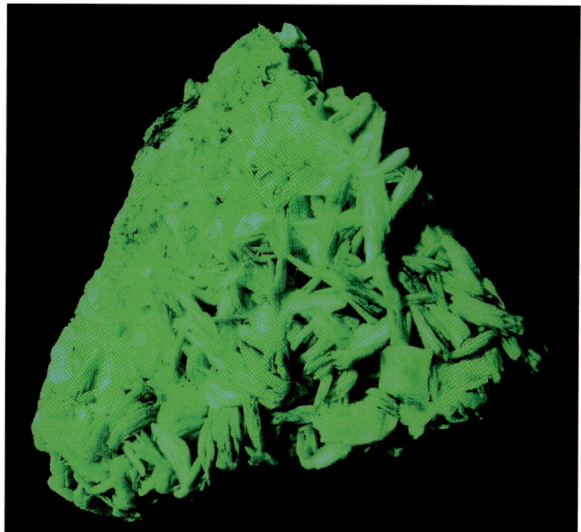

Same under SW

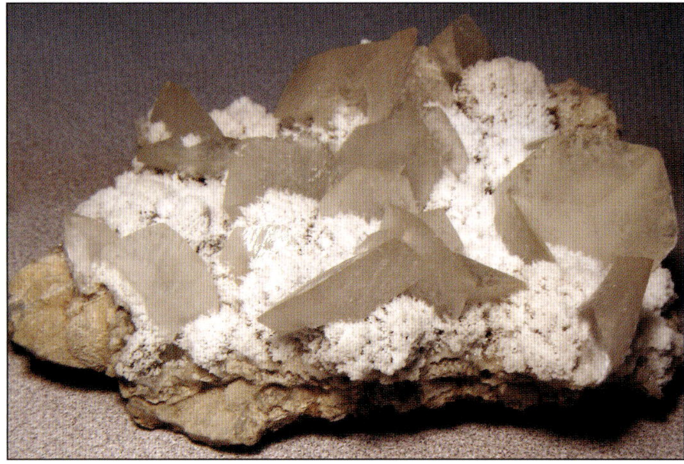

CALCITE. Calcite from the Cadman quarry, Monroe (about 30 miles northeast of Seattle), Snohomish County. This is an interesting and unusual calcite. It is almost nonfluorescent under SW, but under MW it lights up brightly. It is also rather nonreactive under LW 350 and LW 370. The piece weighs 18 lbs and is 4.5 x 8.5 x 11.5 inches. Photo courtesy of Don Newsome.

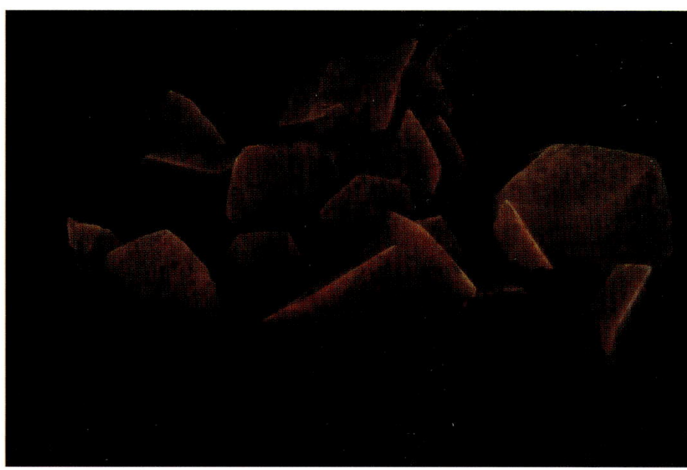

Same under SW

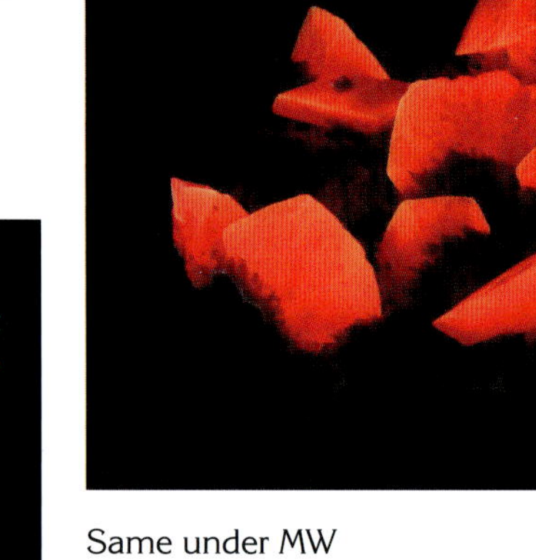

Same under MW

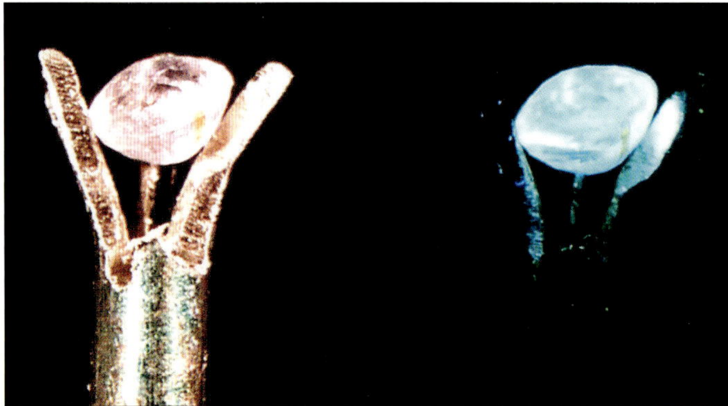

ZEKTZERITE. A zektzerite crystal from Kangaroo Ridge, near Washington Pass, Okanogan County that has been cut and faceted as a gem. Zektzerite fluoresces pale blue SW. It is 2.3 mm wide. Value $25-30

Wyoming

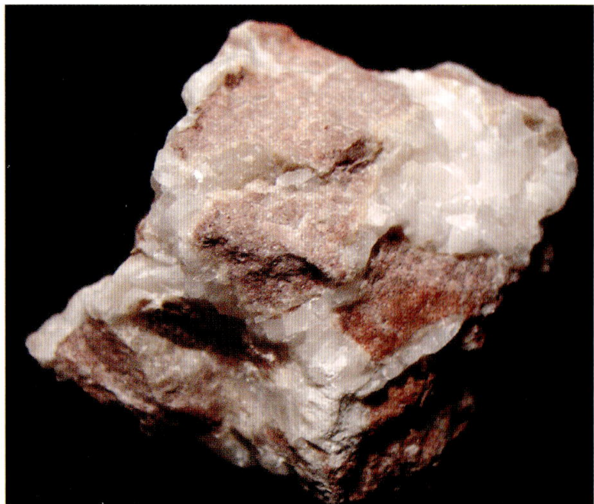

CALCITE. Calcite and aragonite from the Big Horn Mountains, Big Horn County. The calcite fluoresces dark red SW, and the aragonite fluoresces white SW. The piece weighs 8.0 oz. and is 3.3 x 2.5 x 2.0 inches. Photo courtesy of Dru Wilbur. Value $25-30

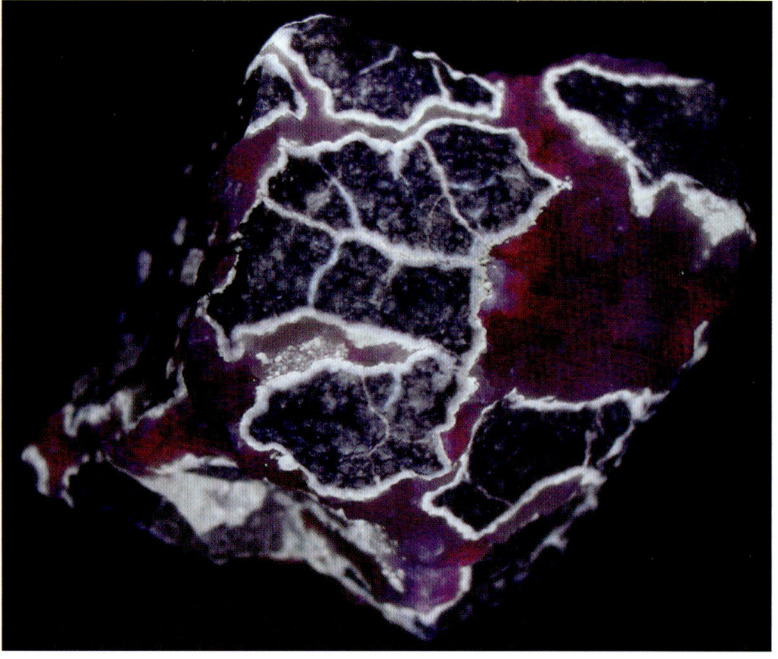

Same under SW

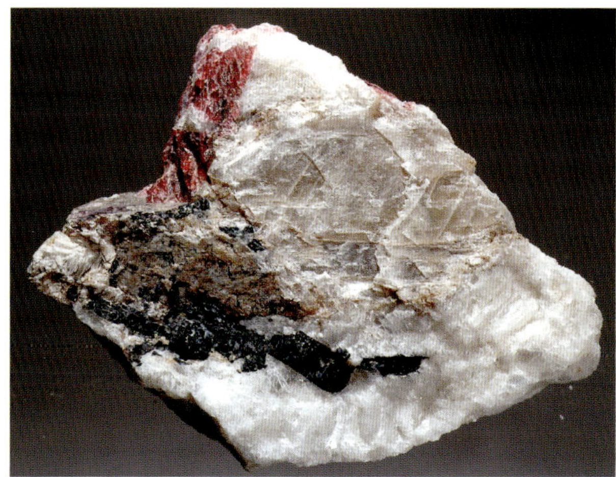

AGRELLITE. Agrellite with albite, eudialite, and calcite from the Mt. Kipawa complex, Sheffield Lake, Villedieu Township, Québec. Agrellite fluoresces pink SW, albite fluoresces cherry-red SW, and the calcite fluoresces green SW. The piece weighs 7.4 oz. and is 3.0 x 2.3 x 1.5 inches. Value $25-30

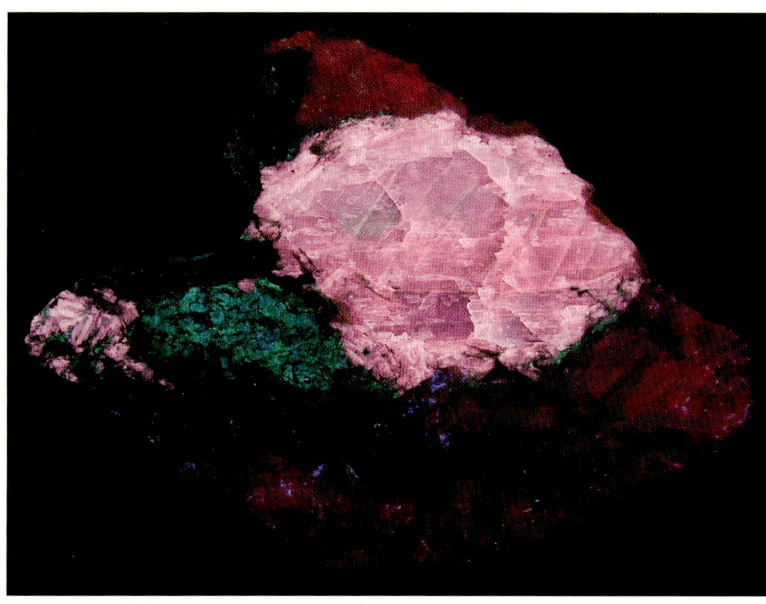

Same under SW

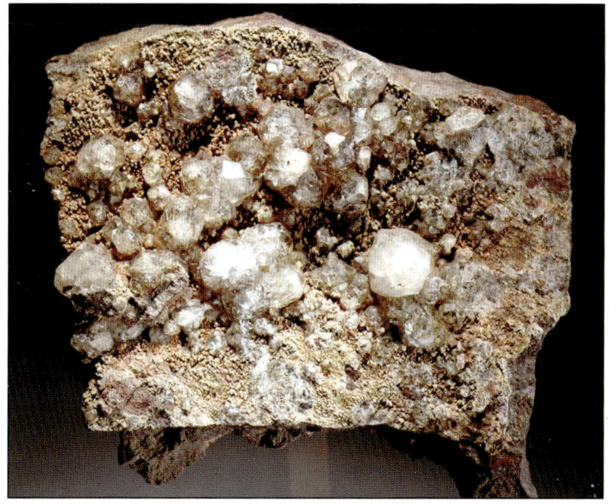

ANALCIME. Colorless crystals of analcime on a dark matrix from Parrsboro, Cumberland County, Nova Scotia. The crystals fluoresce electric blue SW in some areas. Under magnification, no visible cause was found to explain this electric blue fluorescence from some of the analcime crystals. The piece weighs 14.8 oz. and is 4.5 x 4 x 1.0 inches. Value $25-30

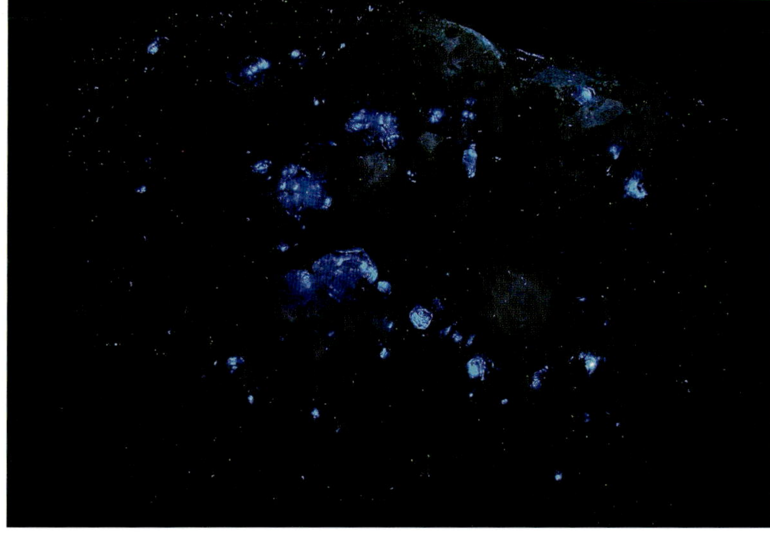

Same under SW

CALCITE. Dark tan dogtooth calcite crystals from Cyrille de Wendover, Québec. The calcite fluoresces blue-gray SW and green-tan LW. Other pieces from this location are reported to fluoresce and phosphoresce yellow. The piece weighs 5.0 oz. and is 3.0 x 2.5 x 1.5 inches. Value $40-45

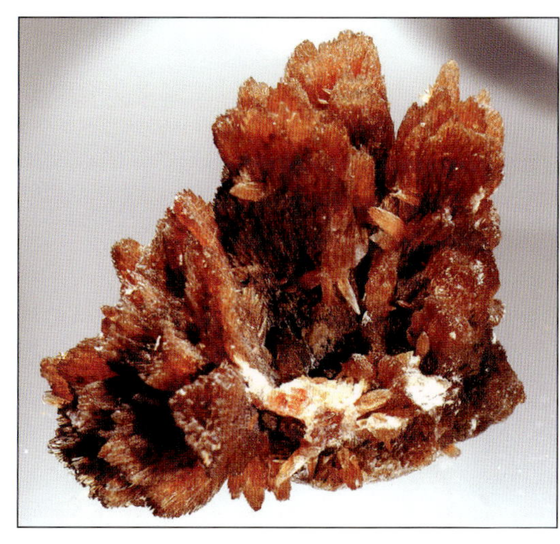

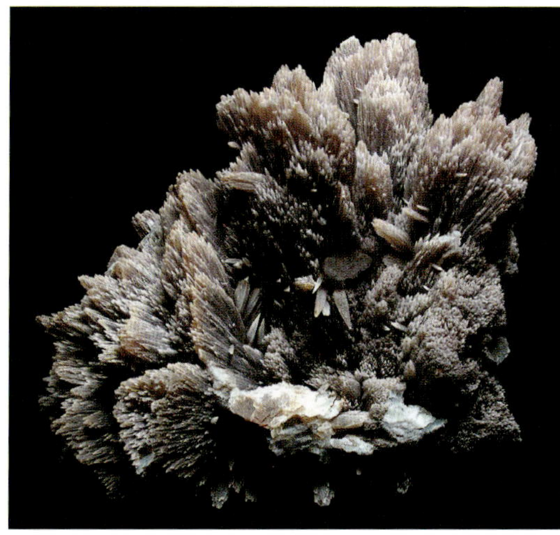

Same under SW

CATAPLEIITE, CALCITE. Calcite crystals with catapleiite, epididymite, and microcline from the Poudrette quarry, Mont Saint-Hilaire, Rouville County, Québec. It is rare at Mont Saint-Hilaire to find a group of fluorescent catapleiite crystals. Found in October, 2005, the catapleiite fluoresces green SW. Only the last crystallization layer of the calcite fluoresces dark orange-red (better MW), and the microcline fluoresces a cherry-red SW. The epididymite is nonfluorescent. The piece weighs 2.0 oz. and is 1.3 x 1.6 x 1.6 inches. Value $40-45

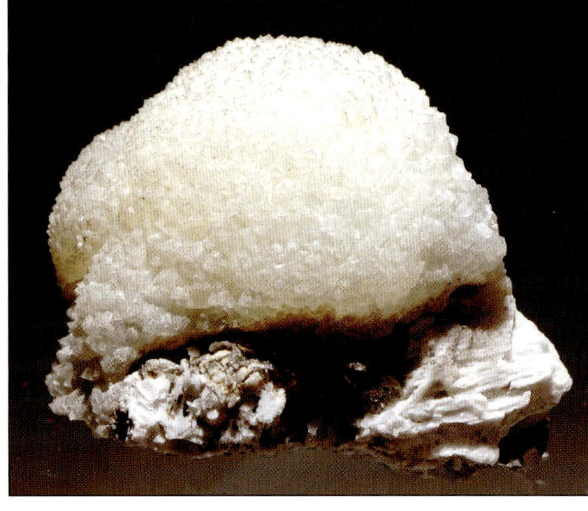

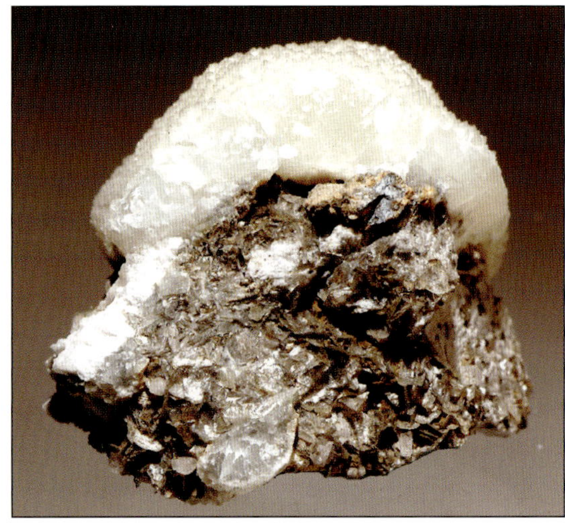

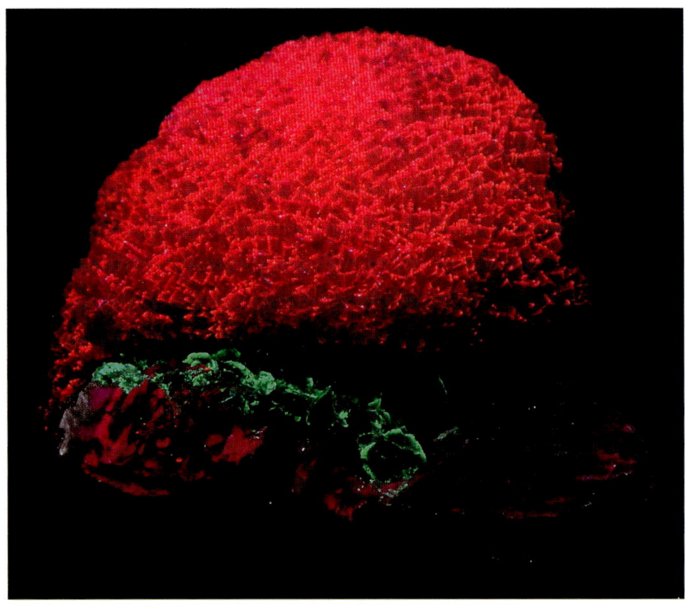

Same under SW

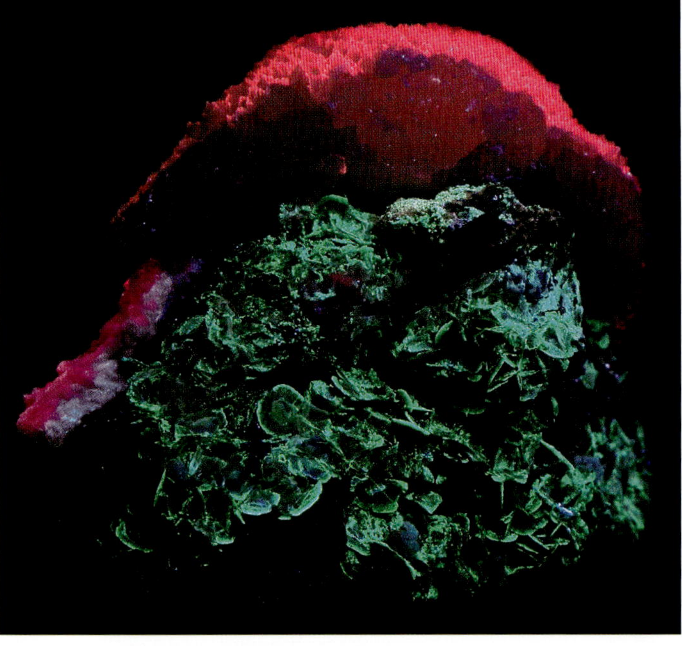

DIOPSIDE, VESUVIANITE. Crystals of diopside and manganoan vesuvianite from Asbestos, Richmond County, Québec. The diopside and the vesuvianite (most of it on the other side) both fluoresce pink-orange SW and LW. It appears that some coating on the minerals is fluorescing, but the vesuvianite crystals are gemmy and transparent and do not appear to have a coating. The piece weighs 8.0 oz. and is 3.0 x 2.8 x 1.5 inches. Value $40-45

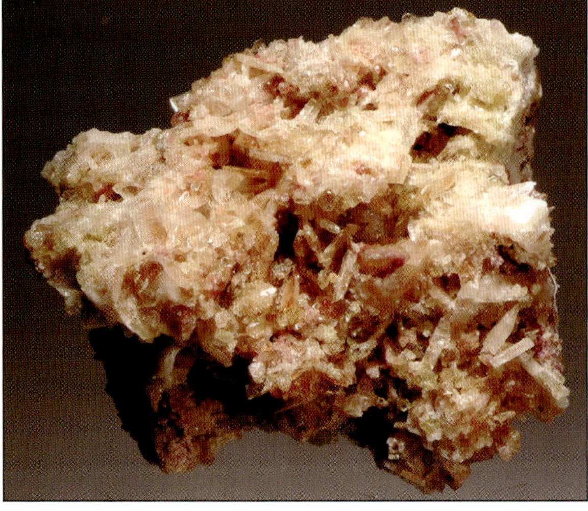

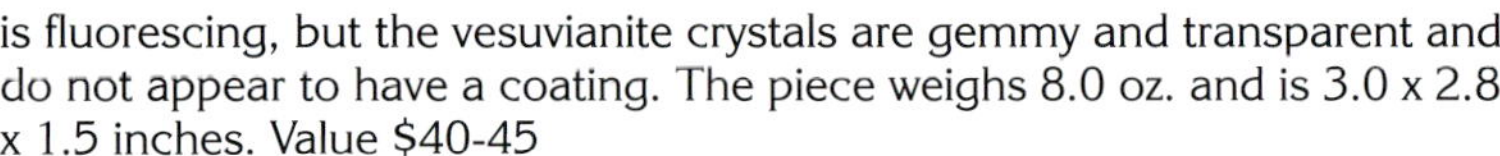

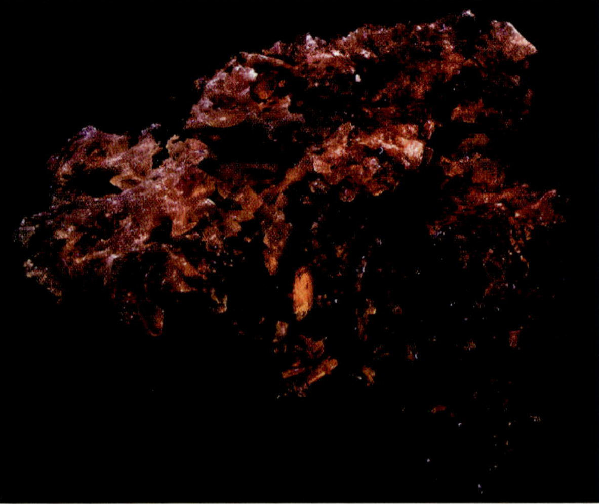

Same under SW

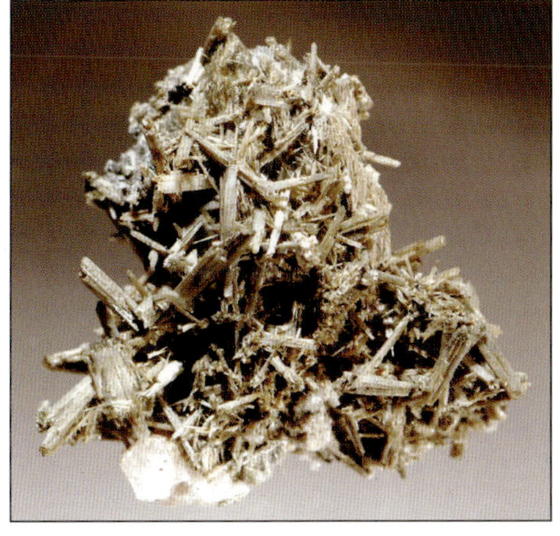

ELPIDITE. A group of elpidite crystals with albite from the Poudrette quarry, Mont Saint-Hilaire, Rouville County, Québec. Found in Fall 2003. Elpidite fluoresces pale green SW, the albite fluoresces cherry-red SW, and we are unsure what is causing the blue color. It could be an organic coating or contaminant or some unusual mineral reaction. The piece is 1.3 x 1.6 x 0.9 inches. Value $35-40

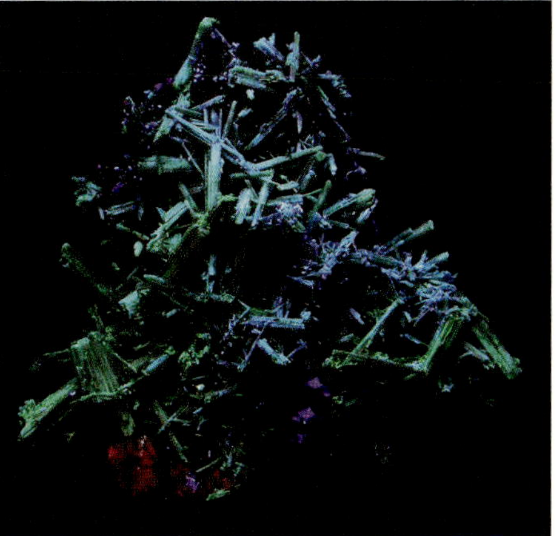

Same under SW

Same under LW

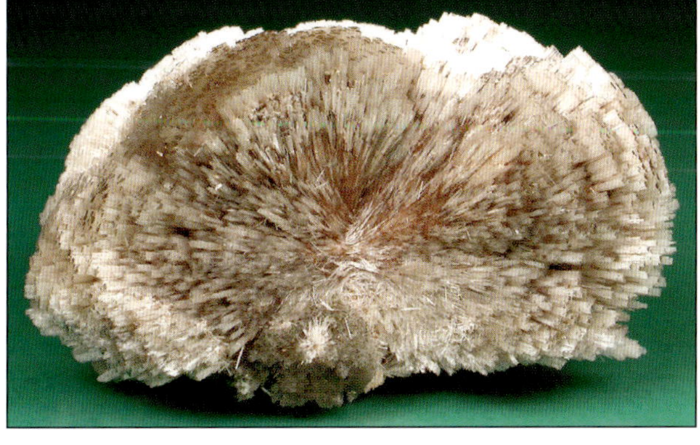

ELPIDITE. An outstanding group of elpidite crystals from the Poudrette quarry, Mont Saint-Hilaire, Rouville County, Québec. Elpidite fluoresces pale green SW. The piece is 2.5 x 1.5 inches. Specimen and photo courtesy of Gilles Haineault.

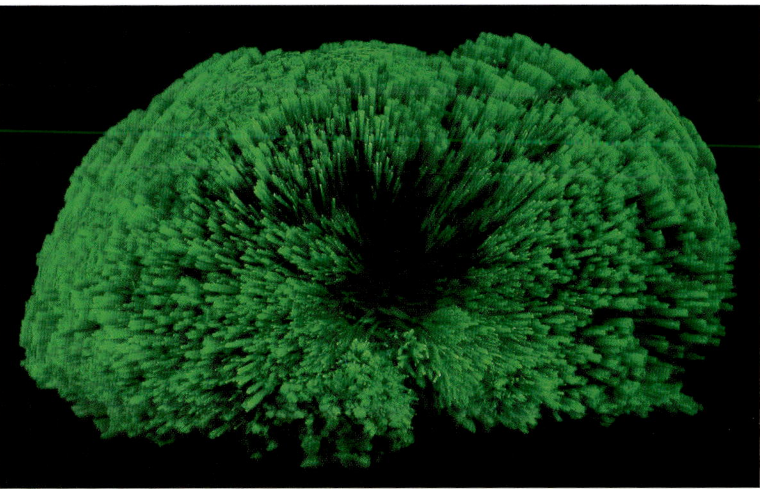

Same under SW

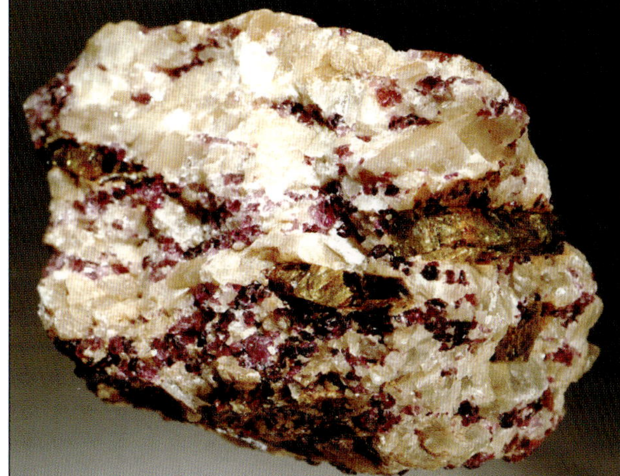

FLUORITE, CALCITE. Fluorapatite crystals and fluorite in calcite from the Fission mine in Wilberforce, Haliburton County, Ontario. The calcite fluoresces orange-red SW, the fluorite fluoresces violet LW, and the fluorapatite barely fluoresces weak tan SW. The piece weighs 1 lb. 12.0 oz. and is 4.8 x 3.3 x 2.5 inches. Value $35-40

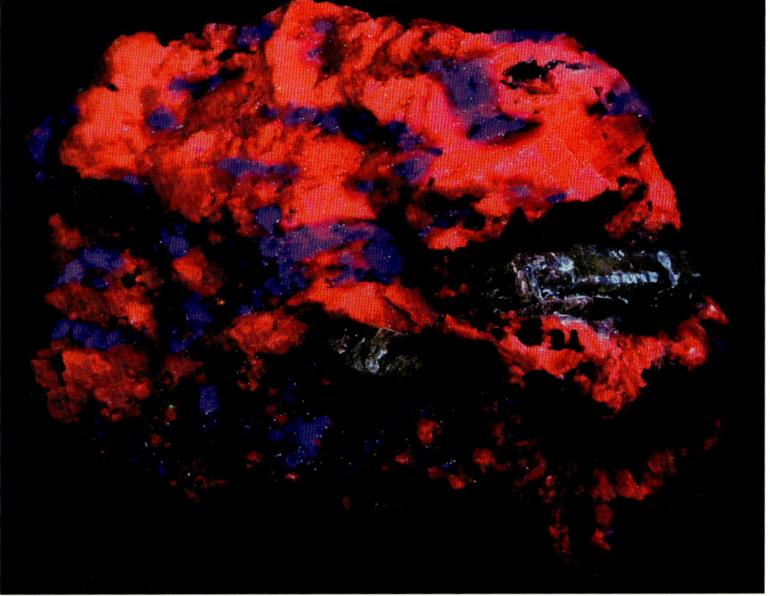

Same under combined SW and LW

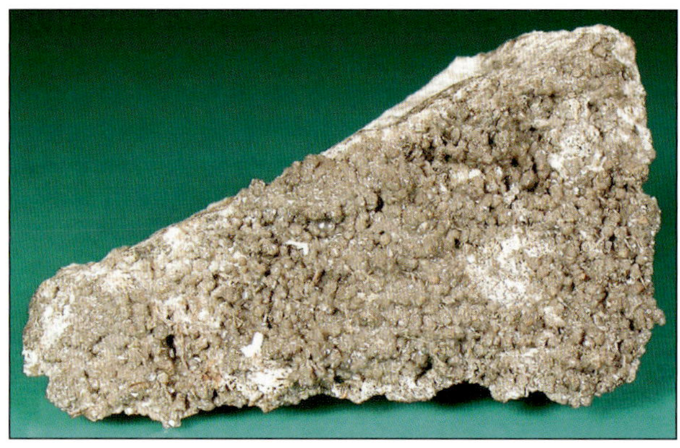

GAIDONNAYITE. A rare example of gaidonnayite from Mont Saint-Hilaire, Rouville County, Québec. The gaidonnayite fluoresces green SW. The piece is 2.3 x 1.8 inches. Photo by Gilles Haineault and courtesy of Jacques Poulin.

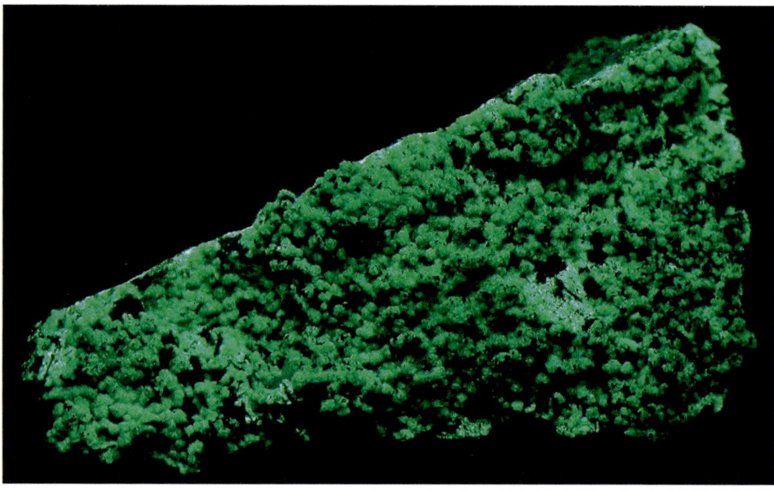

Same under combined SW and LW

GENTHELVITE. A thumbnail-sized mount of genthelvite yellow-tan, triangular crystals from the Poudrette quarry, Mont Saint-Hilaire, Rouville County, Québec. Genthelvite fluoresces green SW. Under MW it fluoresces an unusual combination of red, orange, and green. The piece weighs 0.4 oz. and is 0.6 x 0.8 inches. From the James Zigras collection.

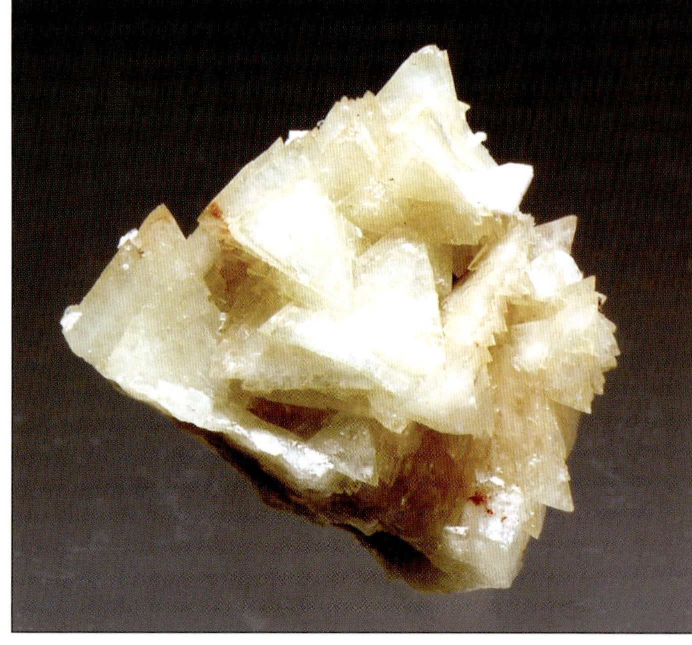

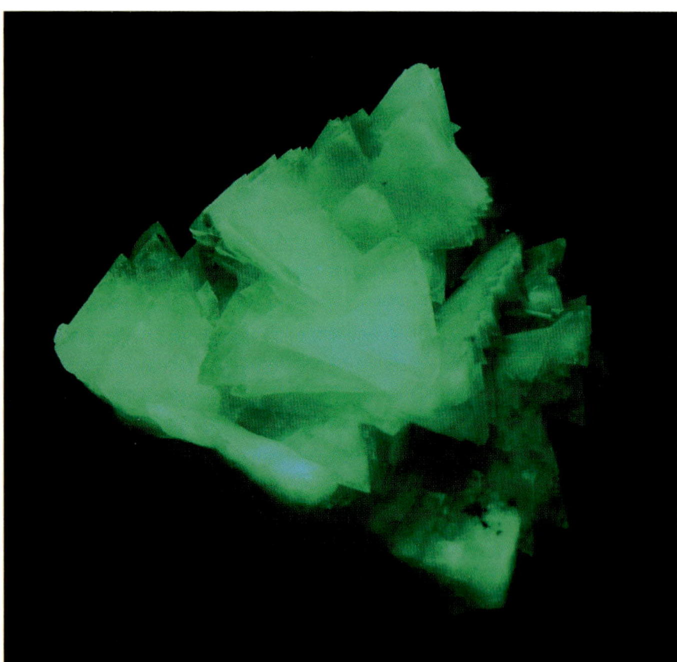

Same under SW

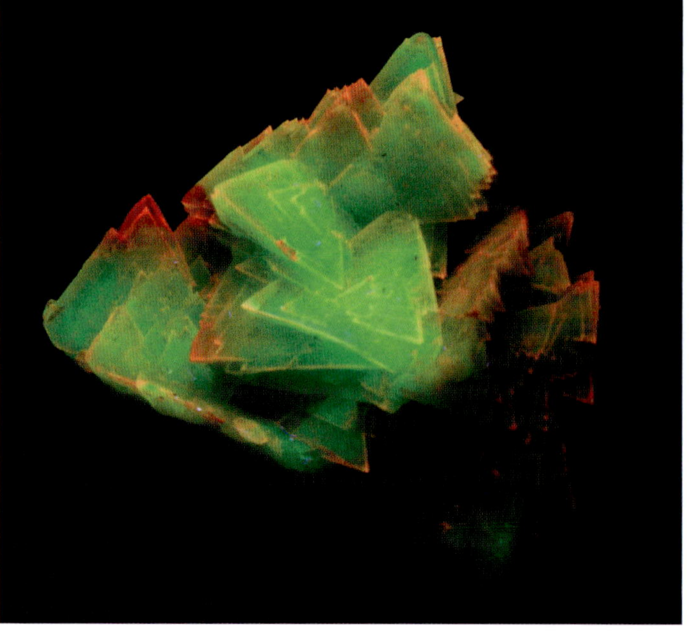

Same under MW. Note the orange and red colors that appear under MW.

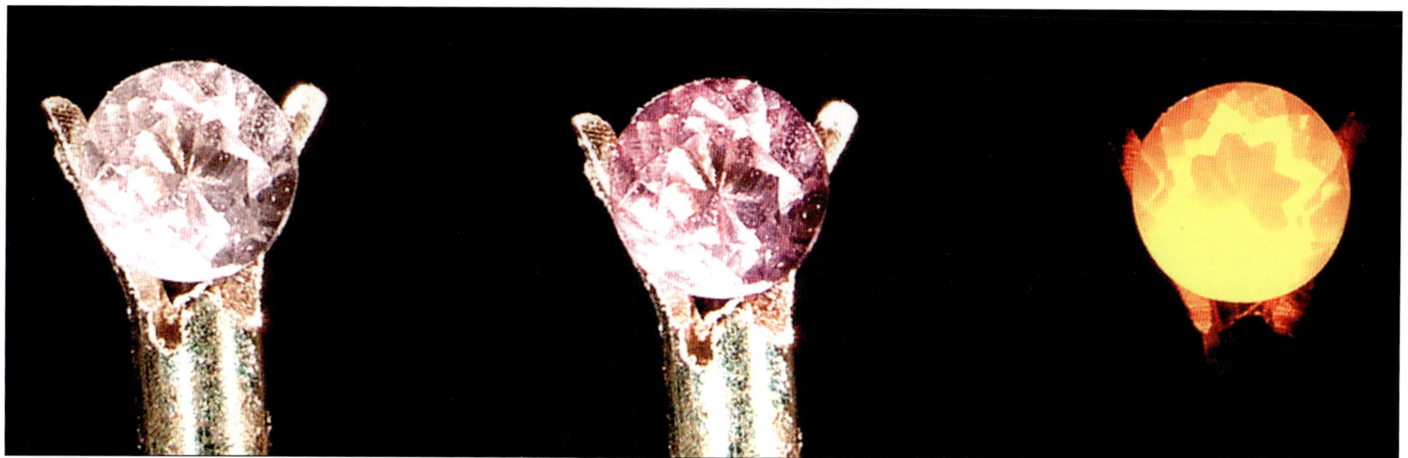

HACKMANITE. A color-changing sodalite, var. hackmanite, crystal that has been faceted into a gem from Mont Saint-Hilaire, Rouville County, Québec. Hackmanite fluoresces orange LW and is tenebrescent (darkens after exposure to SW). It is 3.0 mm wide. Value $55-70

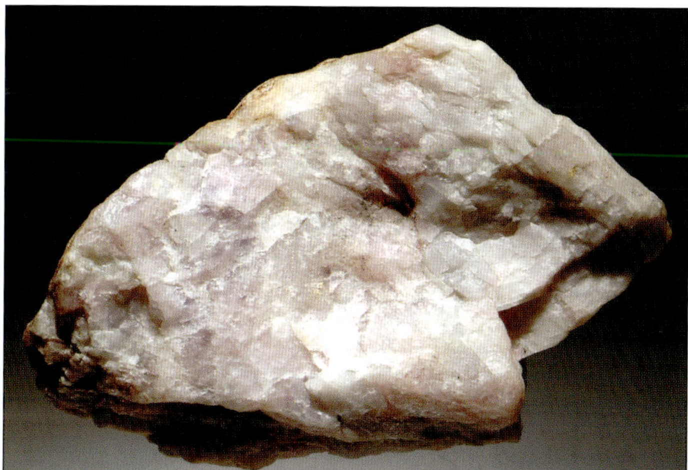

HACKMANITE. Sodalite, var. hackmanite, from Bancroft, Ontario. The hackmanite fluoresces a bright orange LW. After exposure to SW the hackmanite darkens to a raspberry color. The piece weighs 13.5 oz. and is 4.5 x 3.0 x 2.0 inches. Value $50-60

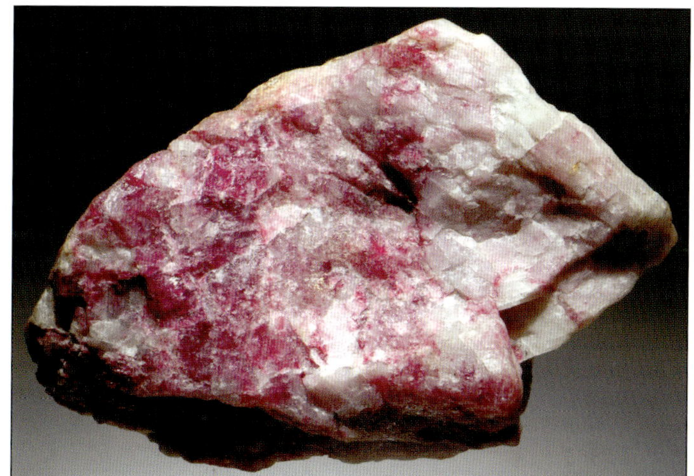

Same showing the tenebrescent effect after a few moments exposure to SW.

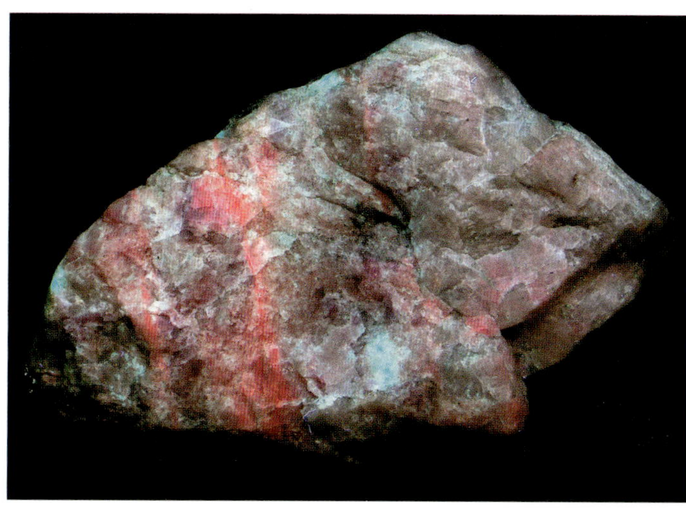

Same under SW

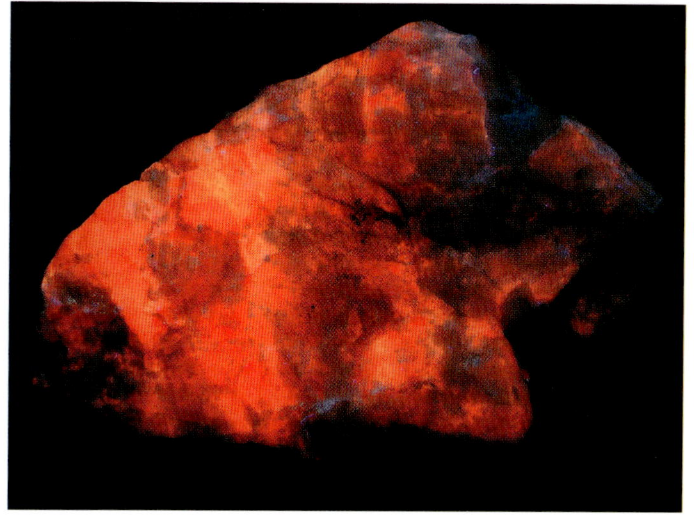

Same under LW

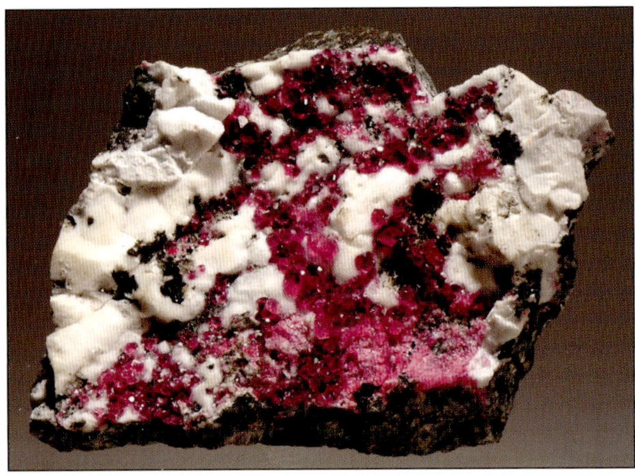

HACKMANITE. Sodalite, var. hackmanite, with analcime from Mont Saint-Hilaire, Rouville County, Québec. From a very small find, the hackmanite crystals, after exposure to SW, darken to a bright raspberry color and stay that way for weeks if kept out of direct sunlight. The hackmanite fluoresces a pale gray SW and barely fluoresces LW, and has a long-lasting phosphorescence. The analcime fluoresces an unusual pale blue LW. The piece weighs 5.0 oz. and is 3.3 x 2.5 x 0.8 inches. From the James Zigras collection.

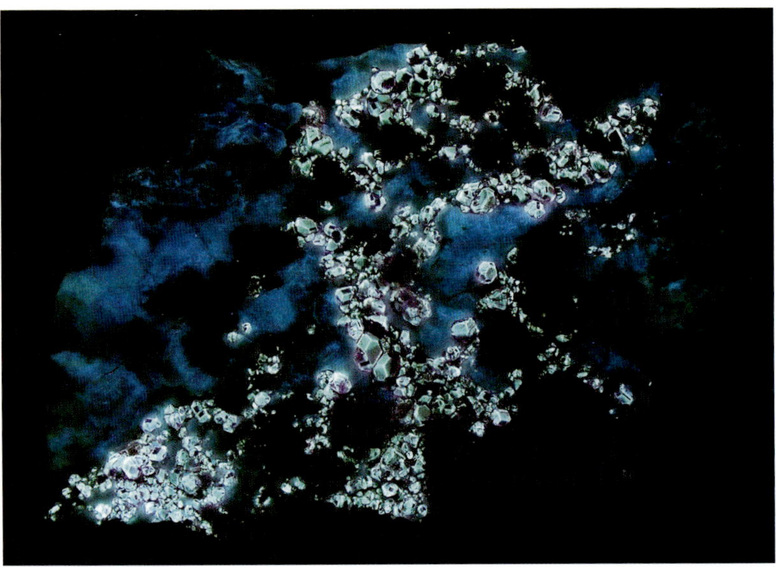

Same under SW

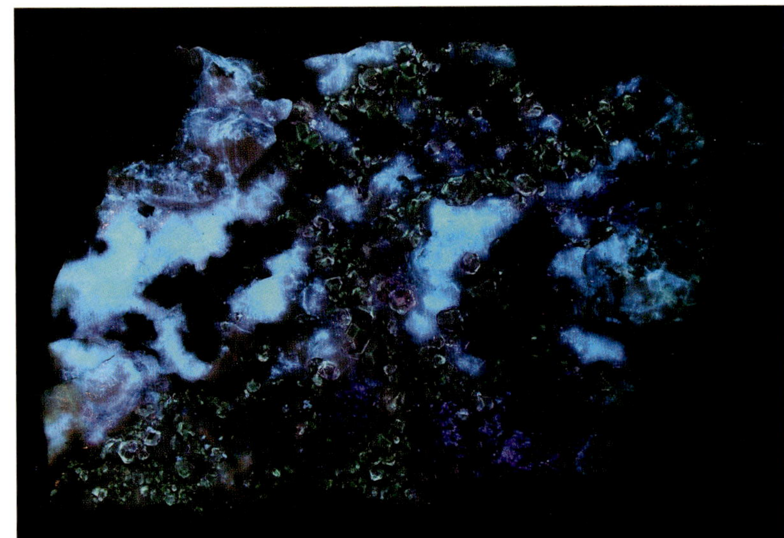

Same under LW

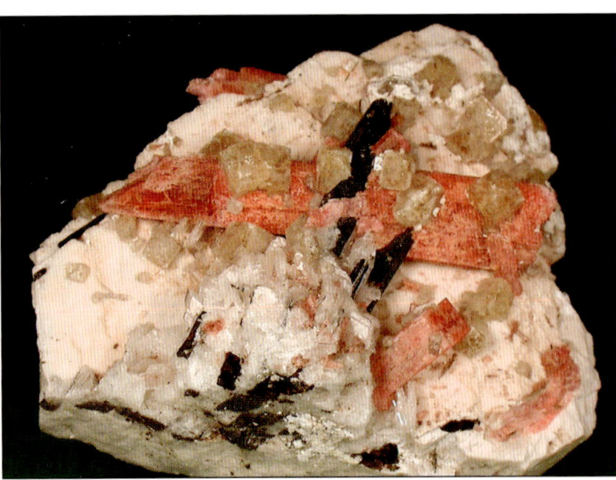

LEUCOPHANITE. Leucophanite on serandite from Mont Saint-Hilaire, Rouville County, Québec. This is an exceptionally rich example with good crystal structure. The leucophanite fluoresces lilac pink LW and (less brightly) SW. The piece is 3.0 x 3.0 inches. Photo courtesy of Jacques Poulin.

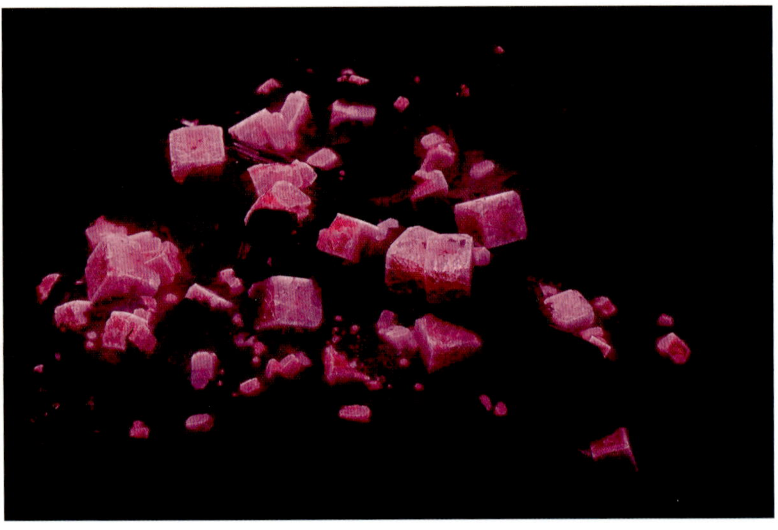

Same under SW

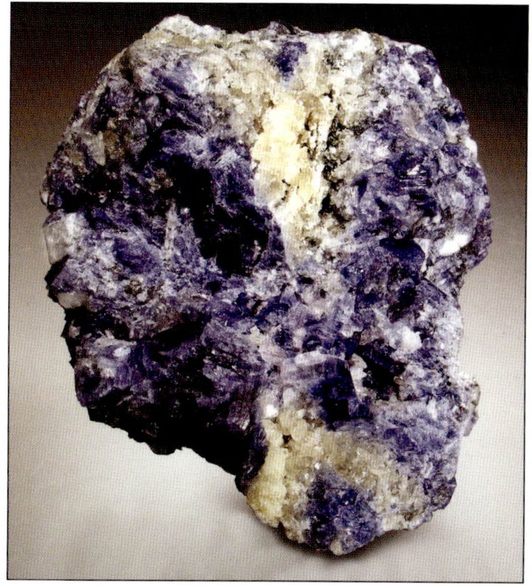

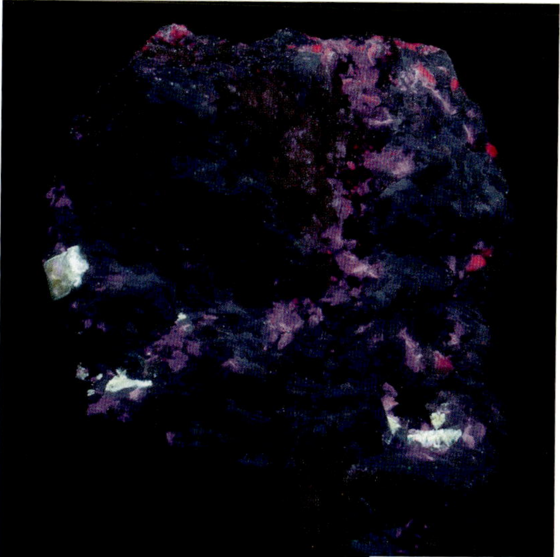

Same under SW

LEUCOPHANITE, LEUCOSPHENITE. Blue carletonite with leucophanite, albite, leucosphenite and an unidentified fluorescent mineral from Mont Saint-Hilaire, Rouville County, Québec. The leucophanite fluoresces lilac pink SW and better LW, the albite fluoresces cherry red SW, leucosphenite fluoresces yellow SW, and an unidentified mineral fluoresces pale red SW. The piece weighs 1.5 oz. and is 2.3 x 2.0 x 0.8 inches. Value $200-225. Note: about fifty percent of the value is attributable to the rare carletonite.

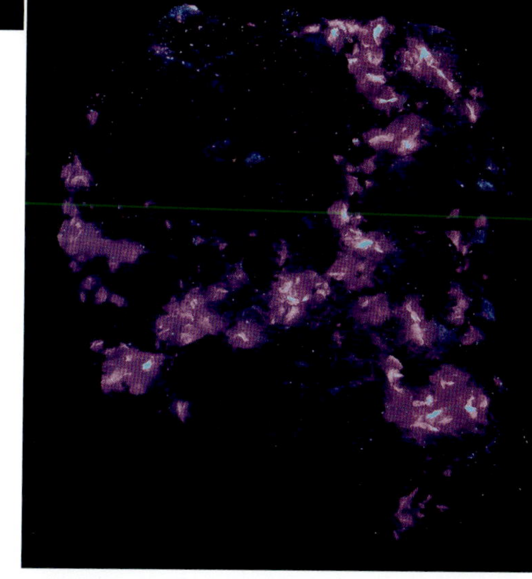

Same under LW

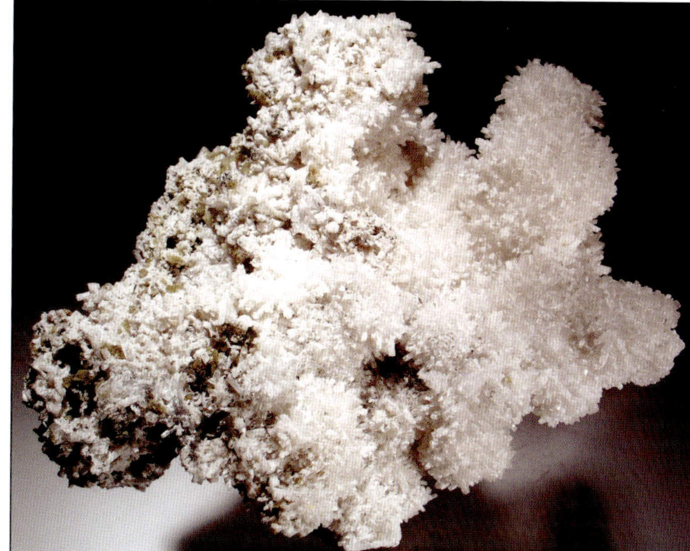

NATROLITE. Natrolite crystals with albite and an unidentified mineral from the Poudrette quarry, Mont Saint-Hilaire, Rouville County, Québec. The natrolite fluoresces pale green SW. The albite fluoresces cherry-red SW and the scattered crystals of the unidentified mineral fluoresce orange SW. The piece weighs 14.0 oz. and is 5.3 x 4.5 x 3.5 inches. Value $70-80

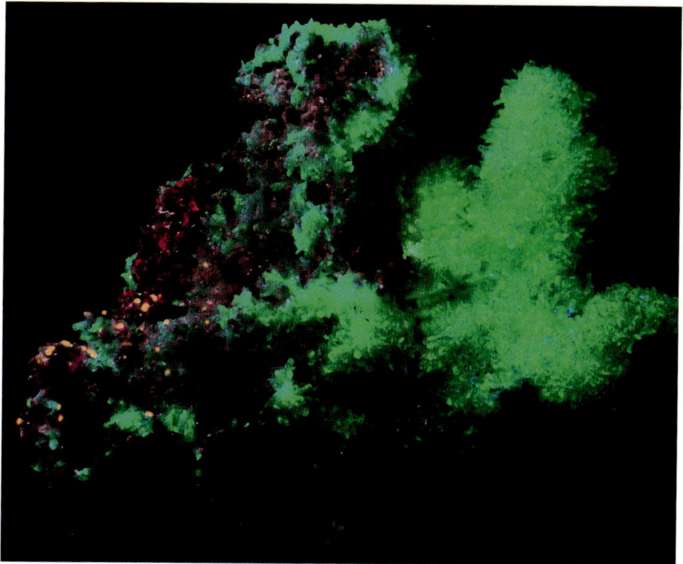

Same under SW

NATROLITE, CALCITE. This is a very aesthetically pleasing piece. Crystals of natrolite with a cluster of calcite crystals from the Poudrette quarry, Mont Saint-Hilaire, Rouville County, Québec. Found in June, 2004. The natrolite fluoresces green SW and the calcite fluoresces violet-pink SW. The piece weighs 0.3 oz. and is 1.0 x 1.0 x 0.9 inches. Value $40-45

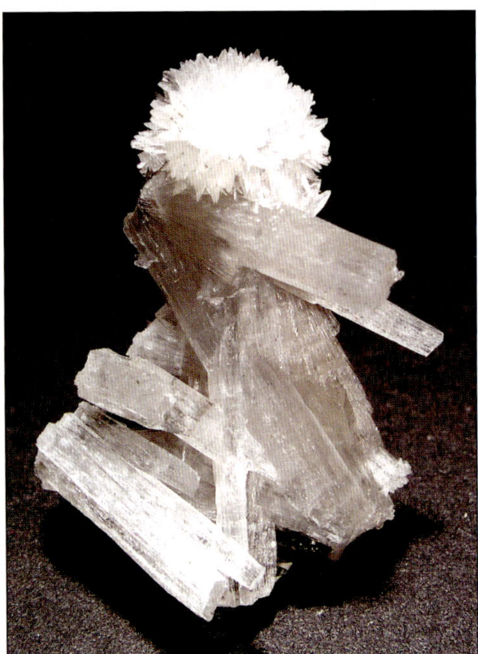

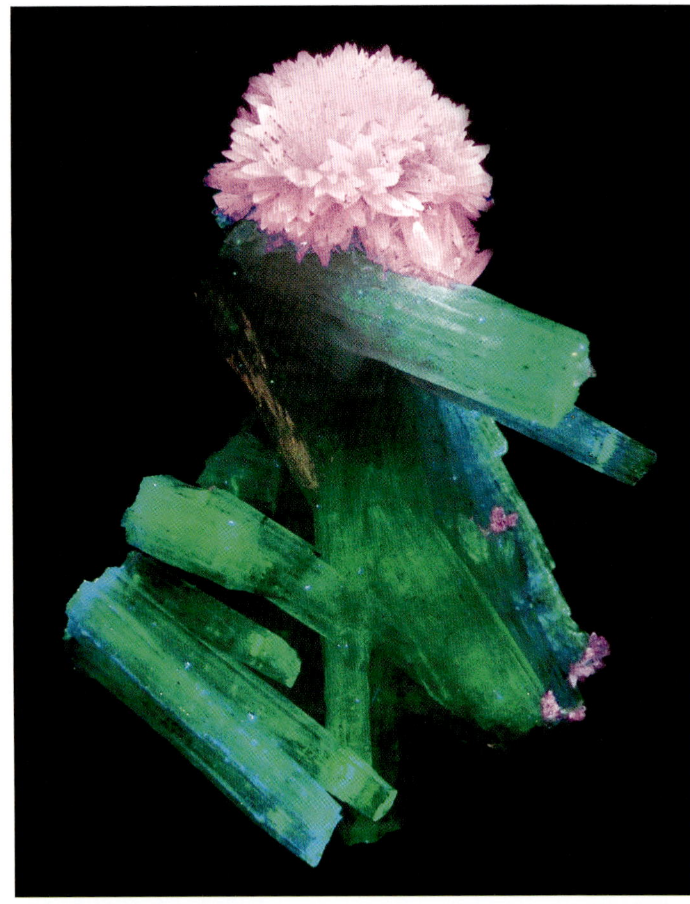

Same under SW

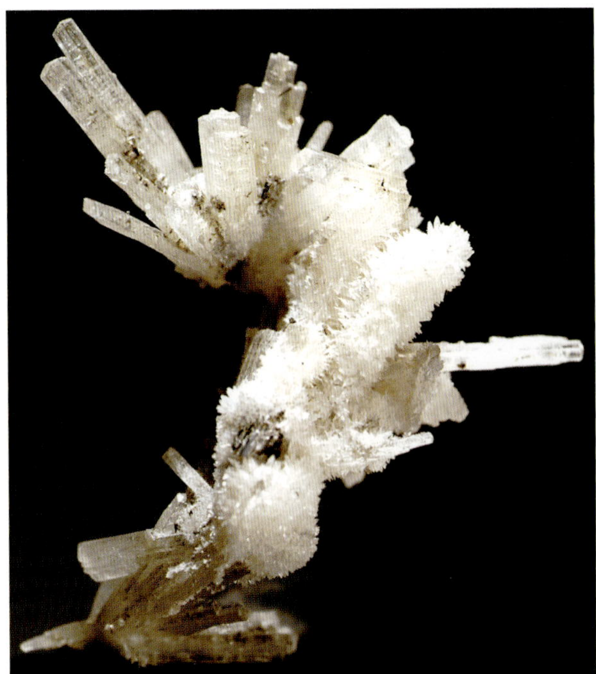

NATROLITE, CALCITE. Crystals of natrolite with clusters of calcite crystals from the Poudrette quarry, Mont Saint-Hilaire, Rouville County, Québec. Found in June, 2004, the natrolite fluoresces green SW and the calcite fluoresces violet-pink SW. The piece weighs 0.5 oz. and is 2.1 x 1.4 x 1.3 inches. Value $50-55

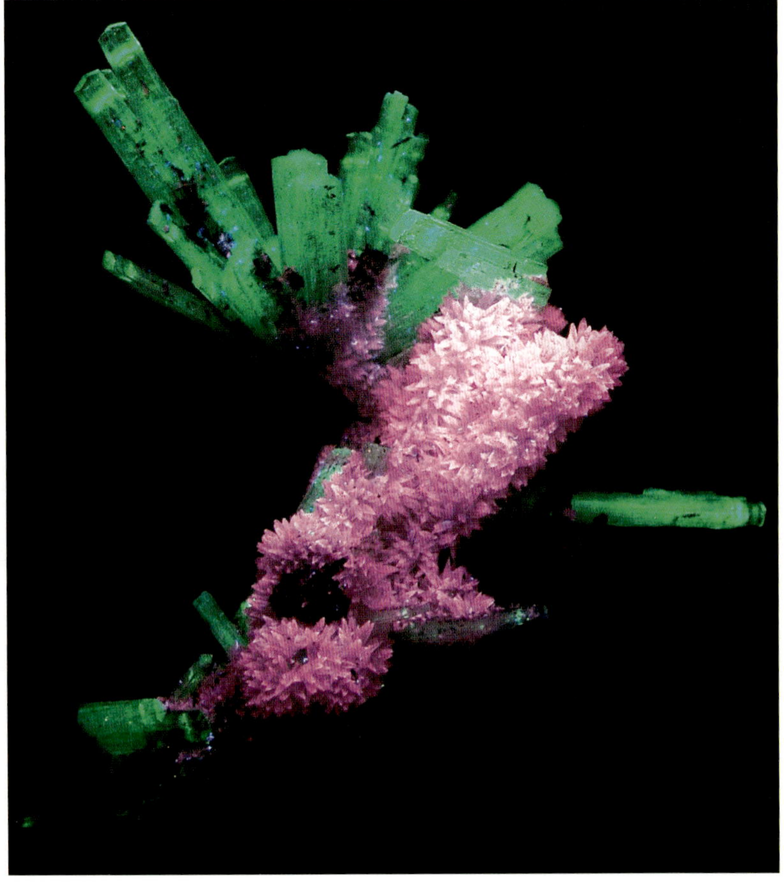

Same under SW

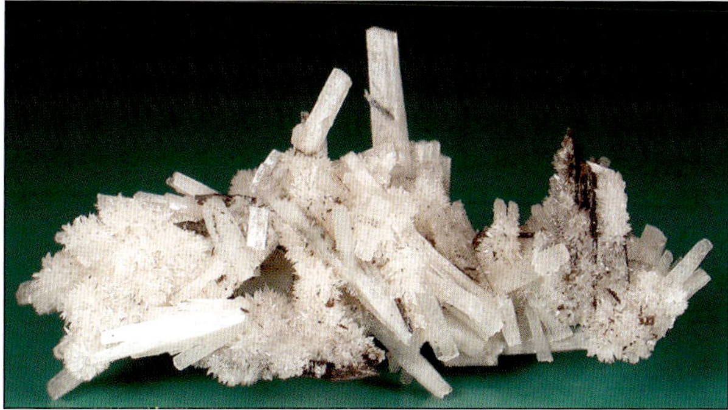

NATROLITE, CALCITE. Beautiful crystals of natrolite with clusters of calcite crystals, found in June, 2004, at the Poudrette quarry, Mont Saint-Hilaire, Rouville County, Québec. The natrolite fluoresces green SW and the calcite fluoresces violet-pink SW. The piece is 3.3 x 2.3 inches. Photo by Gilles Haineault and courtesy of Jacques Poulin.

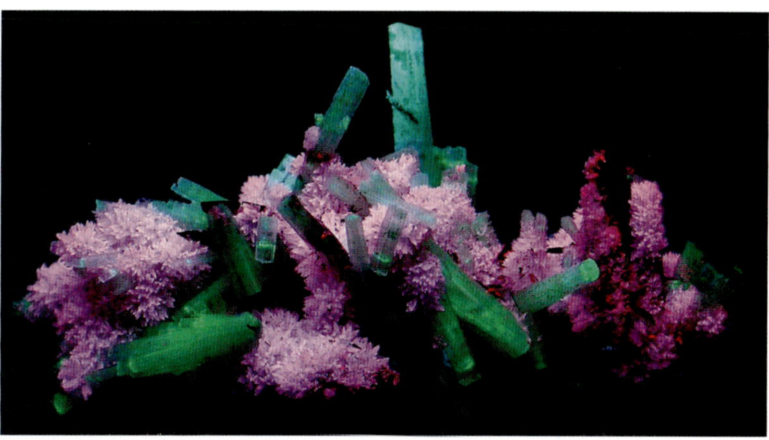

Same under SW

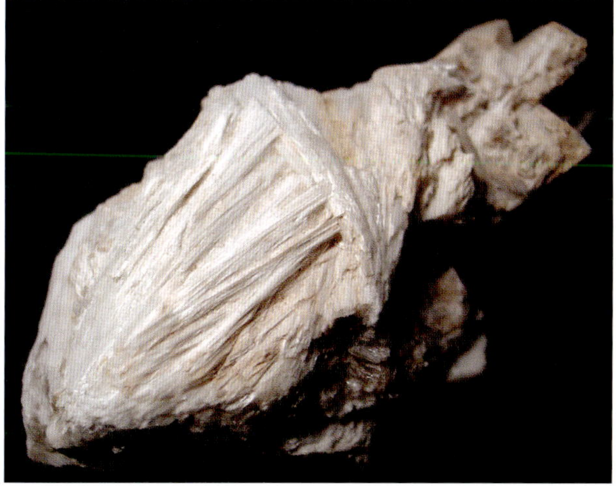

PECTOLITE. Needle-like pectolite crystals with albite from Asbestos, Richmond County, Québec. Pectolite fluoresces pink-violet LW and albite fluoresces cherry-red SW. The piece is 3.0 x 3.0 x 2.0 inches. Photo courtesy of Dru Wilbur. Value $25 -30

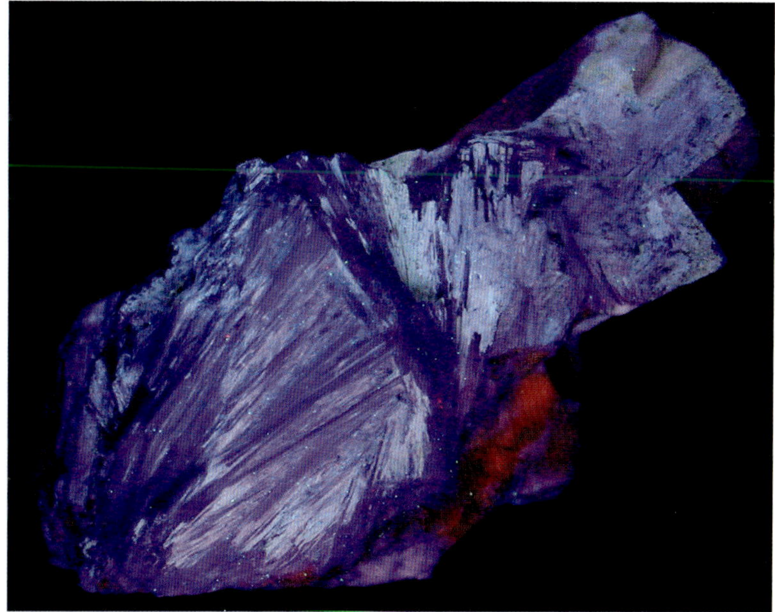

Same under LW

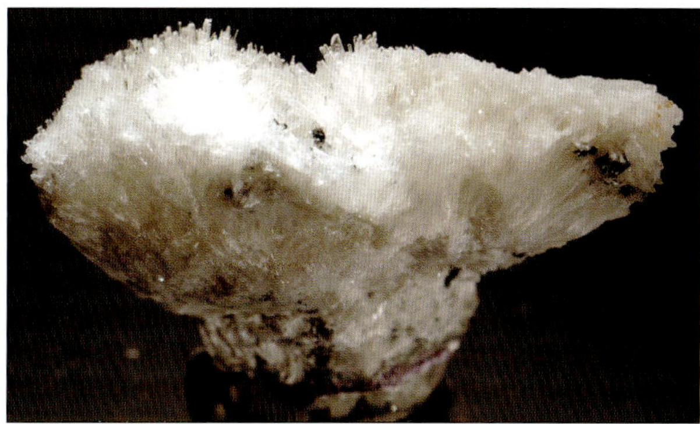

PECTOLITE. Needle-like pectolite crystals from the Poudrette quarry, Mont Saint-Hilaire, Rouville County, Québec. Found in fall 2003, the pectolite fluoresces pink-violet LW. The piece weighs 1.0 oz. and is 2.3 x 0.6 x 1.8 inches. Value $30-35

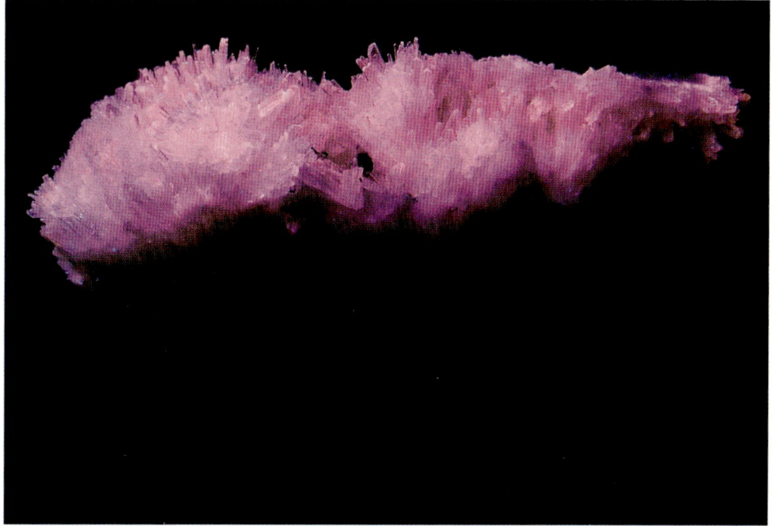

Same under LW

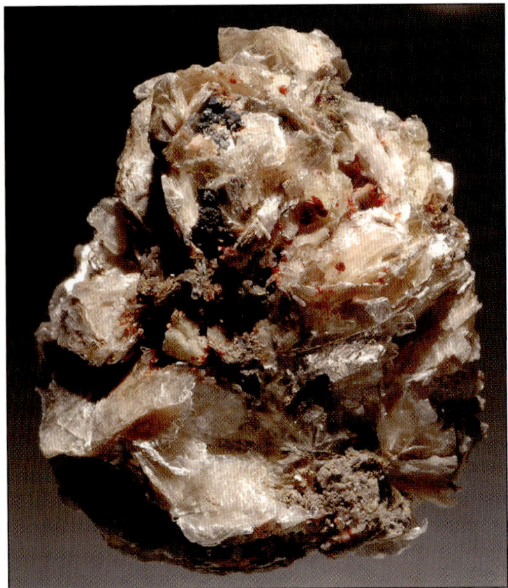

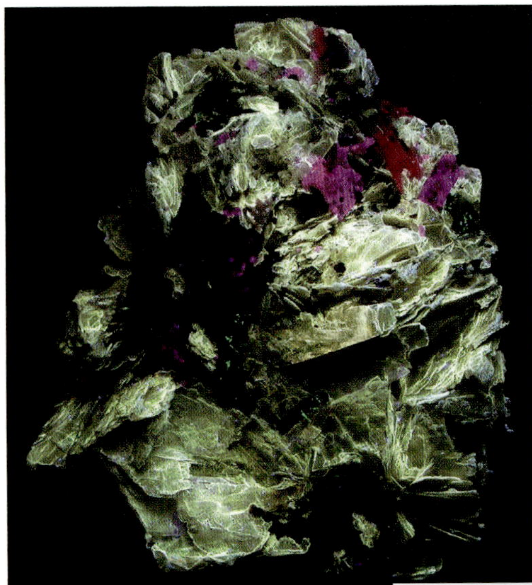

Same under SW

POLYLITHIONITE. A grouping of polylithionite crystals with leucophanite crystals and albite from the Poudrette quarry, Mont Saint-Hilaire, Rouville County, Québec. Polylithionite fluoresces yellow SW, albite fluoresces cherry-red SW, and leucophanite fluoresces lilac-pink LW and (less brightly) SW. The piece weighs 5.3 oz. and is 2.8 x 2.5 x 2.0 inches. Value $40-45

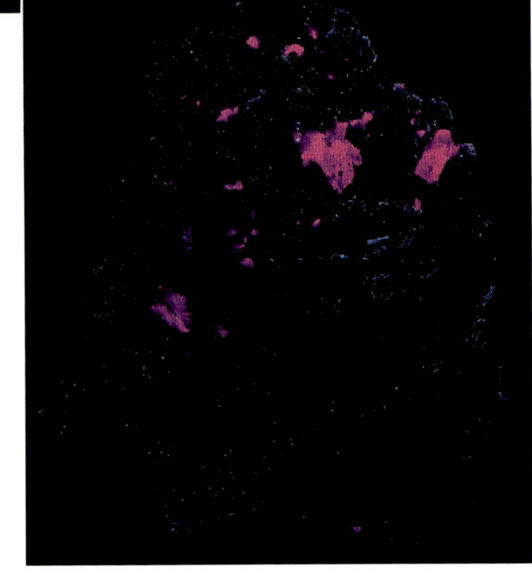

Same under LW

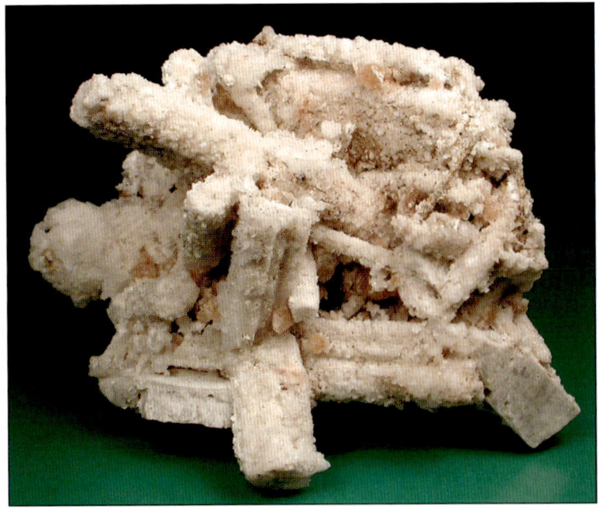

QUARTZ. An exceptionally rare piece of fluorescent quartz on analcime with microcline from Mont Saint-Hilaire, Rouville County, Québec. The quartz fluoresces yellow SW and the microcline fluoresces cherry-red SW. The piece is 4.0 x 5.0 inches. Courtesy of Jacques Poulin.

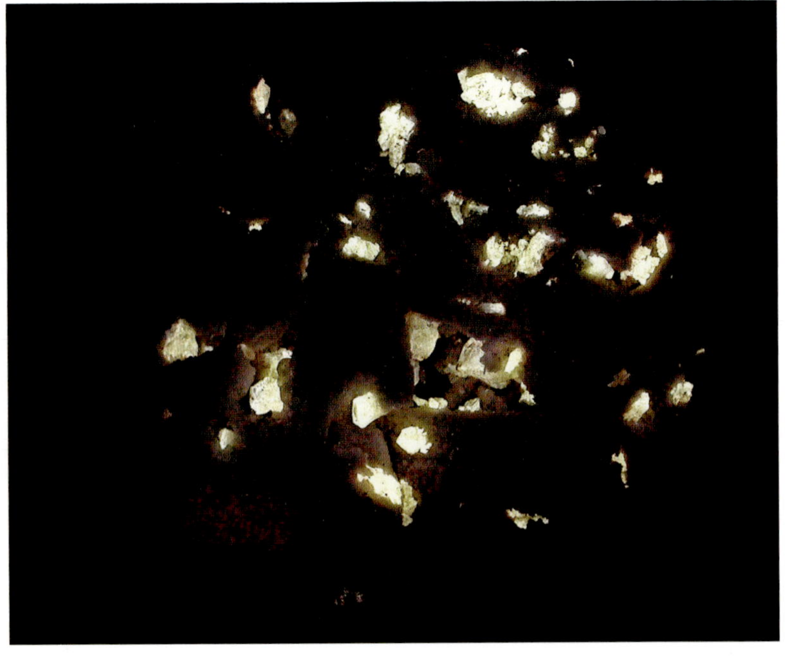

Same under SW

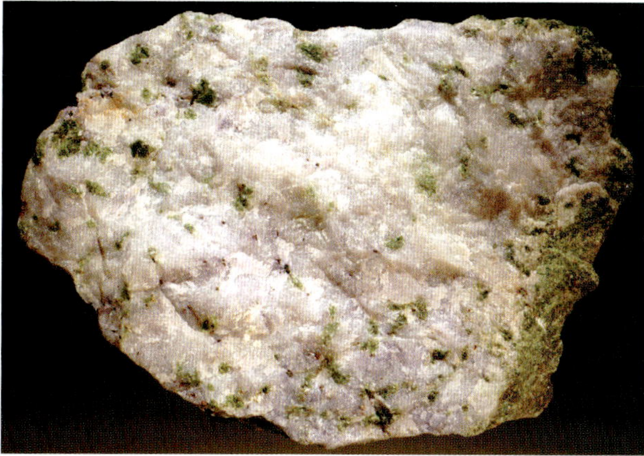

SCAPOLITE. Scapolite, var. marialite or meionite, from Frontenac County, Ontario. Scapolite often fluoresces velvety red, but this fluoresces a beautiful pale yellow and pink SW. The piece weighs 12.0 oz. and is 4.5 x 3.3 x 1.5 inches. Value $45-50

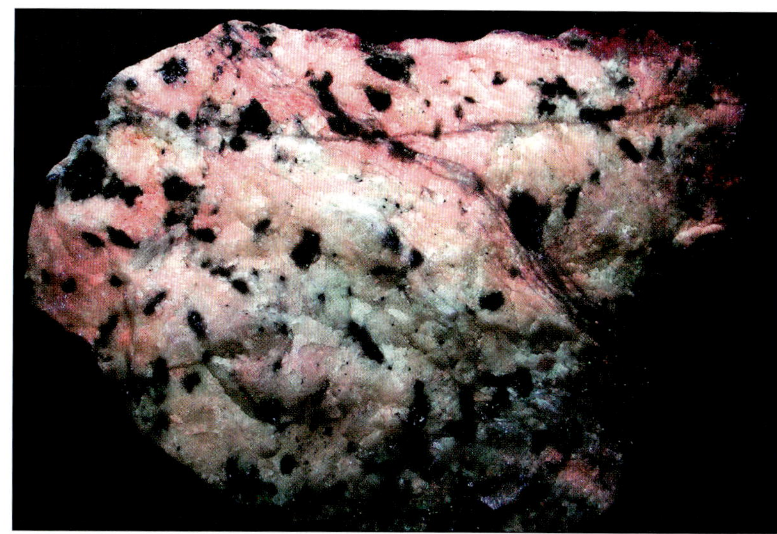

Same under SW

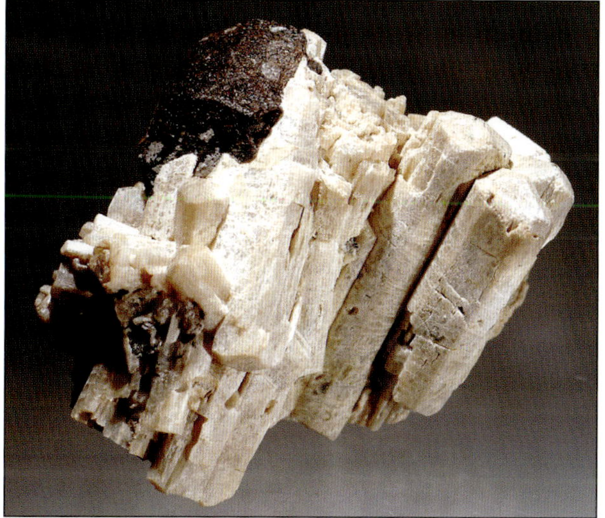

SCAPOLITE. Crystals of scapolite, var. marialite or meionite, with titanite from Sandy Creek, Otter Lake, Pontiac County, Québec. This scapolite fluoresces dark pink SW. The piece weighs 8.5 oz. and is 3.3 x 2.5 x 2.0 inches. Value $30-35

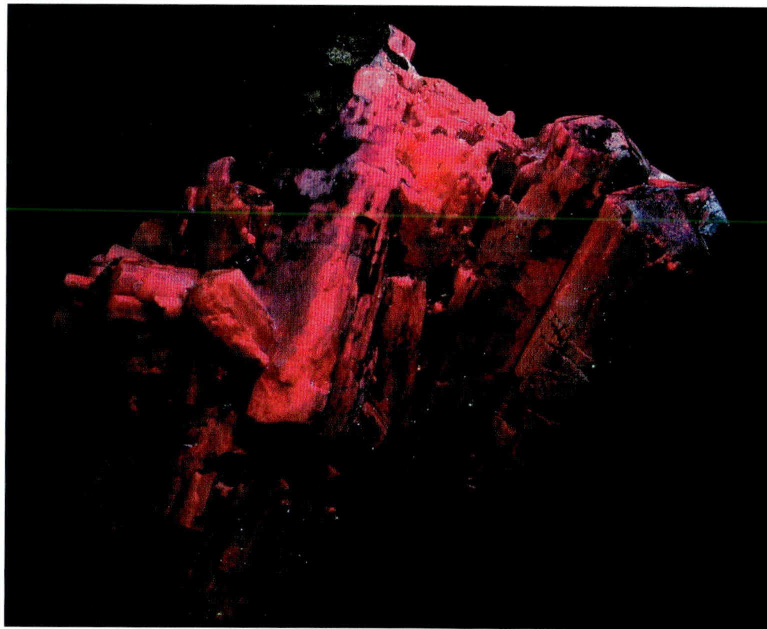

Same under SW

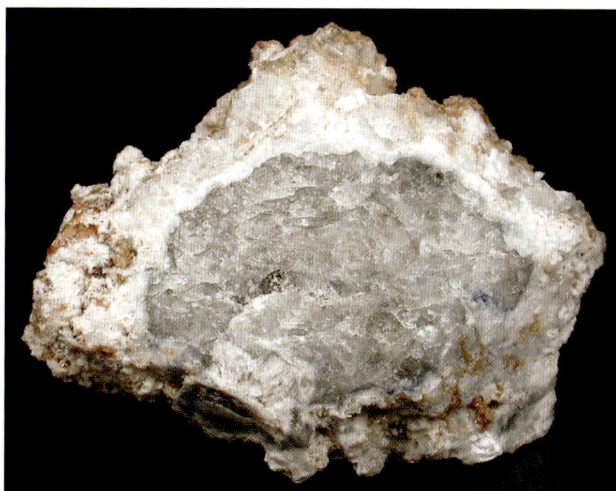

SODALITE. A rare, large crystal of sodalite from Mont Saint-Hilaire, Rouville County, Québec. The sodalite fluoresces orange LW. The piece is 4.0 x 3.0 inches. Courtesy of Jacques Poulin.

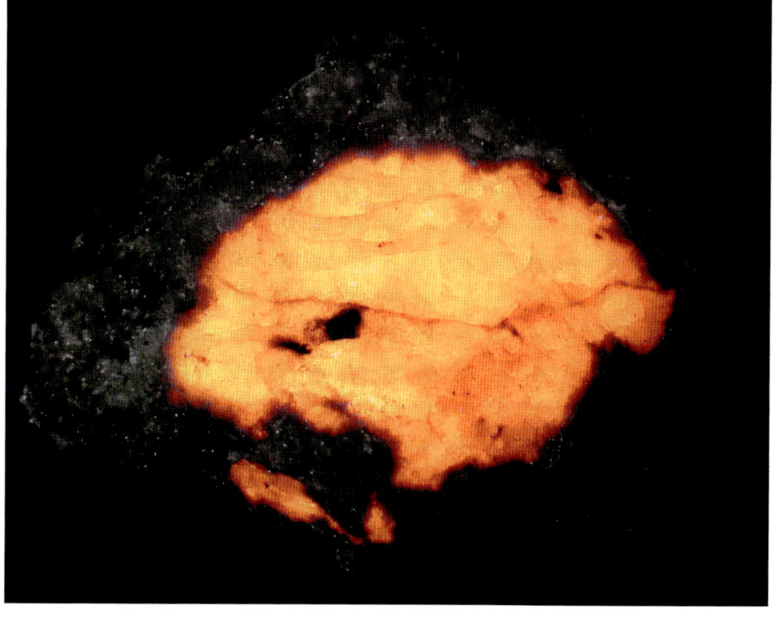

Same under LW

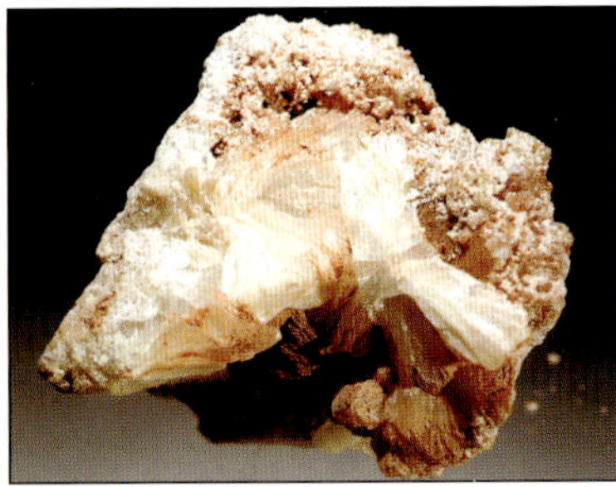

STILBITE. Stilbite crystals from Horseshoe Cove, Bay of Fundy, Nova Scotia. The stilbite fluoresces a weak pale blue LW and (less brightly) SW. This is an unusual reaction as most stilbite from Nova Scotia is not fluorescent. The piece weighs 3.0 oz. and is 2.3 x 2.0 x 1.5 inches. Value $25-30

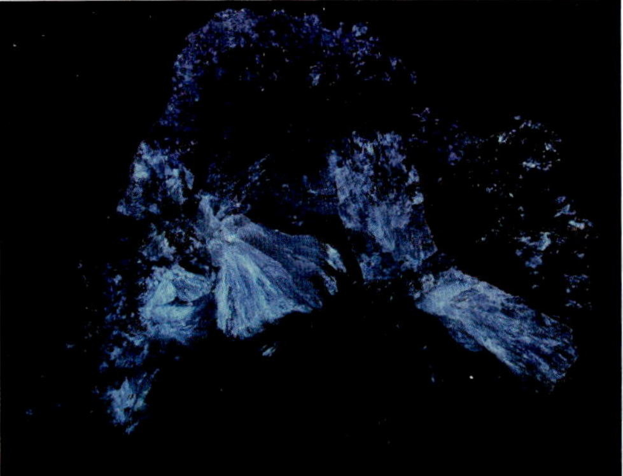

Same under SW

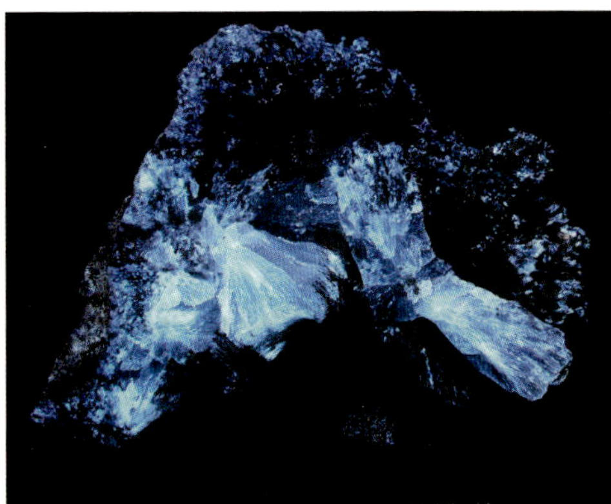

Same under LW

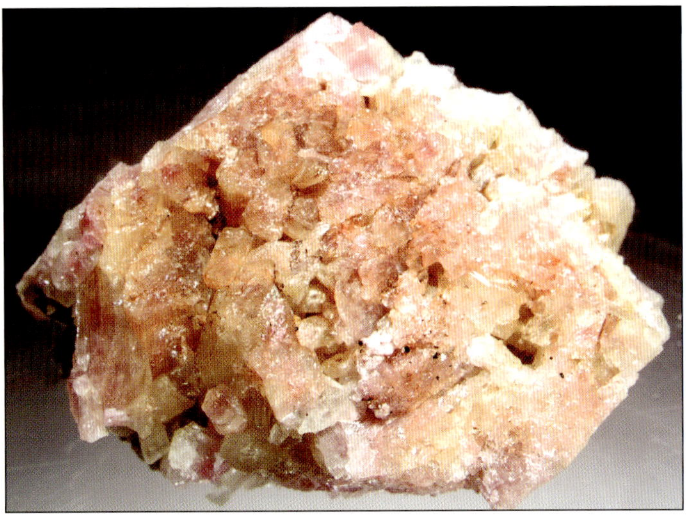

VESUVIANITE. Beautiful crystals of vesuvianite from the closed Jeffrey mine, Asbestos, Richmond County, Québec. The vesuvianite fluoresces orange SW and has a long-lasting phosphorescence. The piece weighs 1.1 oz. and is 1.8 x 1.3 x 0.8 inches. Value $20-25

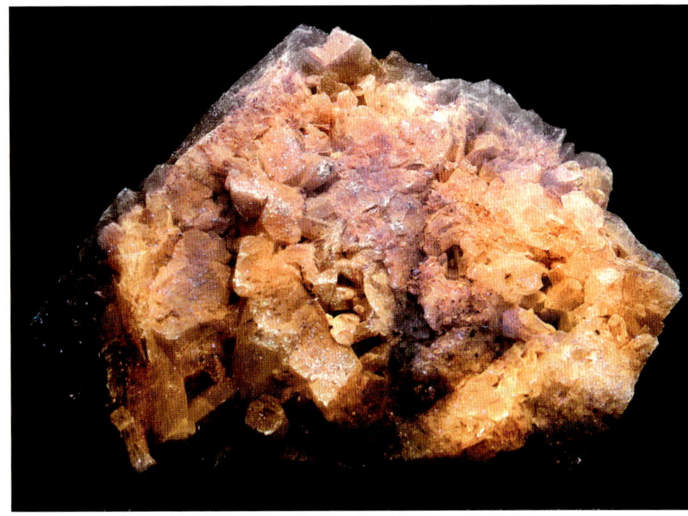

Same under SW

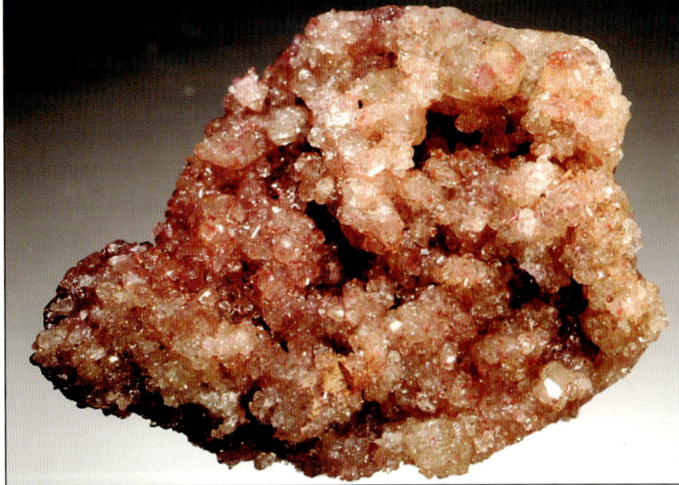

VESUVIANITE. Crystals of vesuvianite from the Jeffrey mine, Asbestos, Richmond County, Québec. The vesuvianite fluoresces orange SW and has a long-lasting phosphorescence. The piece weighs 1.8 oz. and is 2.3 x 1.6 x 0.8 inches. Value $40-45

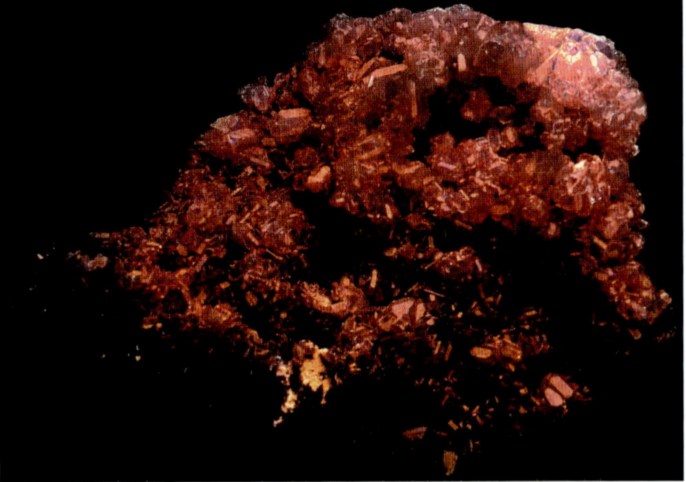

Same under SW

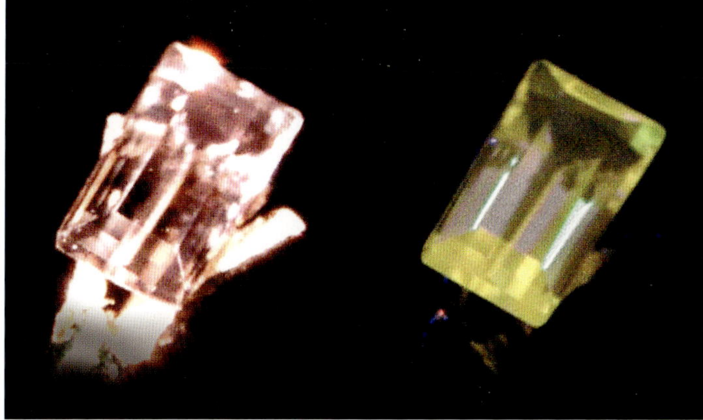

VLASOVITE. Vlasovite that has been faceted as an emerald-cut gem. It is from the Kipawa complex, Sheffield Lake, Témiscamingue County, Québec. Vlasovite, a rare mineral, fluoresces pale yellow SW. It is 2.5 mm by 4.5 mm. Value $45-55

Right:
WELOGANITE. A group of barrel-shaped weloganite crystals from the Francon quarry, Montréal, Jacques Cartier County, Québec. The crystals are almost 0.8 inches tall. The weloganite fluoresces pale yellow with green highlights SW. The green color is unusual. The piece weighs 1.8 oz. and is 1.8 x 1.5 x 1.7 inches. Value $60-65

Far right:
Same under SW

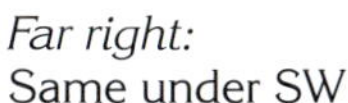

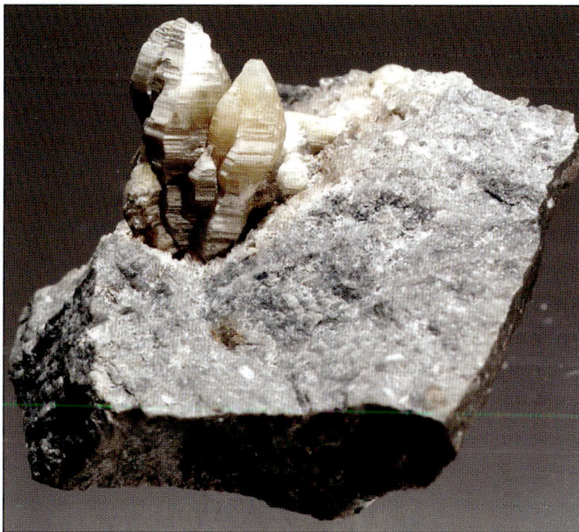

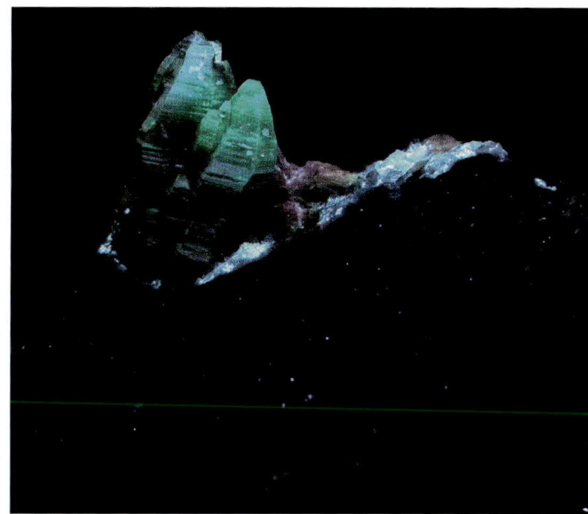

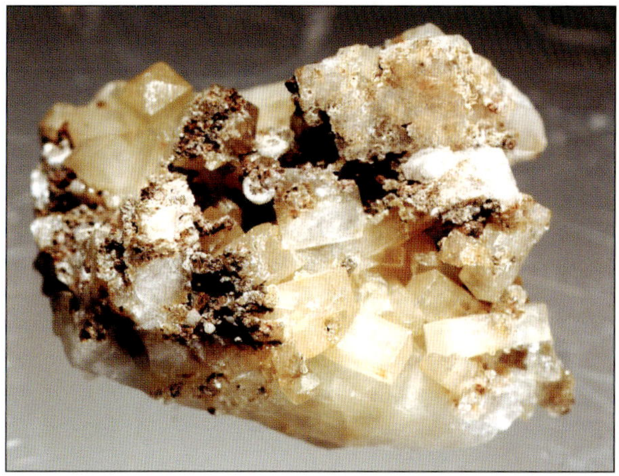

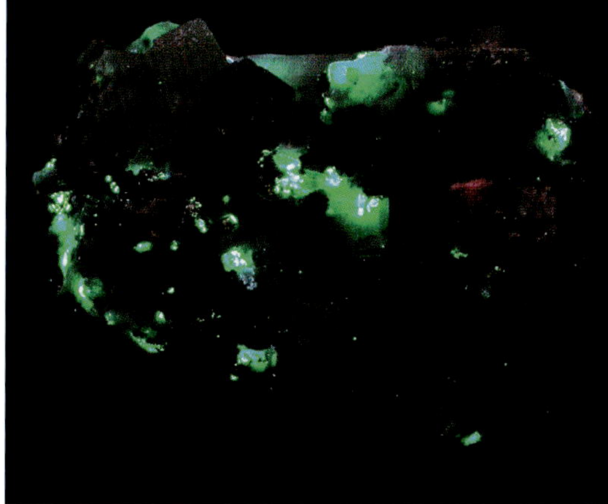

Far left:
WILLEMITE. Willemite with calcite from Mont Saint-Hilaire, Rouville County, Québec. Willemite fluoresces green SW and has a long-lasting green phosphorescence. The piece weighs 0.3 oz. and is 1.1 x 0.8 x 0.8 inches. Value $20-25

Left:
Same under SW

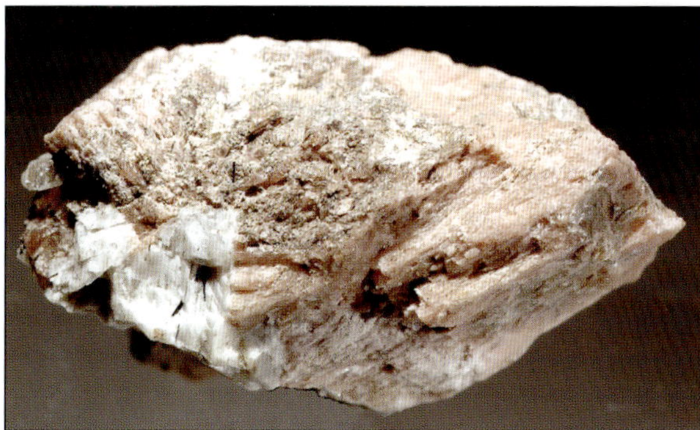

WILLEMITE. Willemite with polylithionite and albite from Mont Saint-Hilaire, Rouville County, Québec. Willemite fluoresces green SW and has a long-lasting green phosphorescence. Polylithionite fluoresces pale yellow SW and albite fluoresces cherry-red SW. The piece weighs 2.0 oz. and is 2.0 x 1.5 x 1.0 inches. Value $35-40

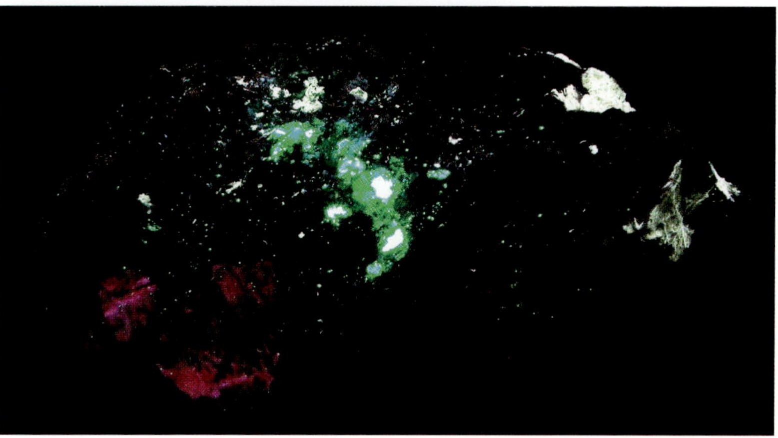

Same under SW

Mexico

ADAMITE. Beautiful adamite crystals with tiny quartz crystals from Mapimí, Durango County. The Adamite fluoresces bright green SW. The piece weighs 1.3 oz. and is 2.1 x 1.8 x 0.5 inches. Value $40-50

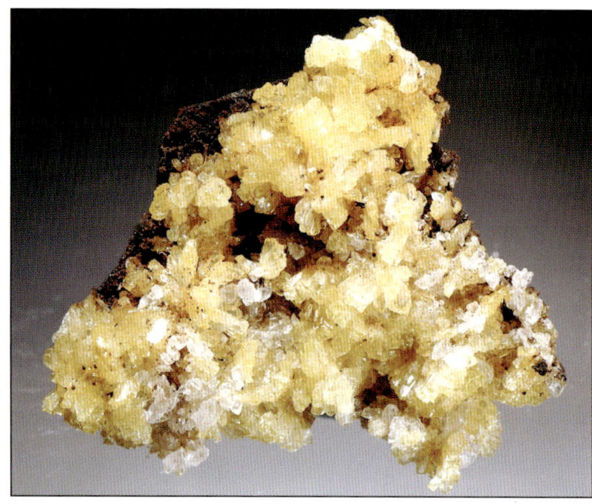

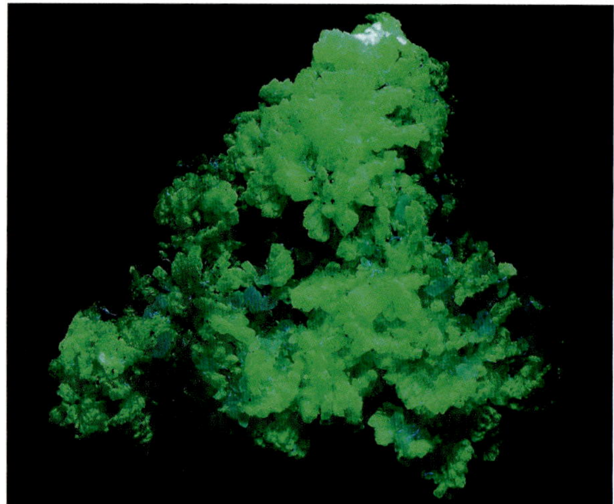

Same under SW

CALCITE. Fishscale-shaped calcite crystals with quartz from Guanajuato, Guanajuato County. The calcite fluoresces pink SW and LW. The piece weighs 1 lb. 3.0 oz. and is 8.3 x 2.5 x 1.3 inches. Value $55-60

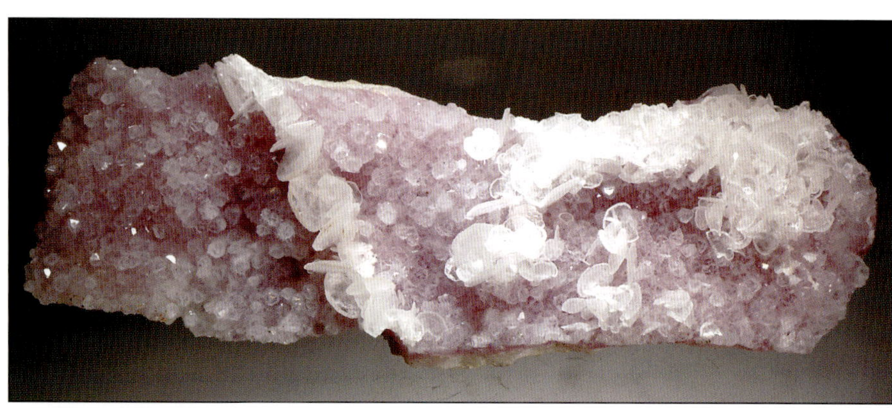

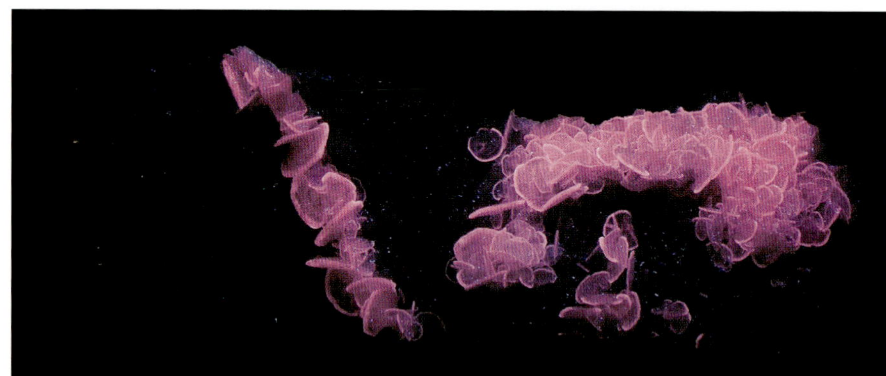

Same under SW

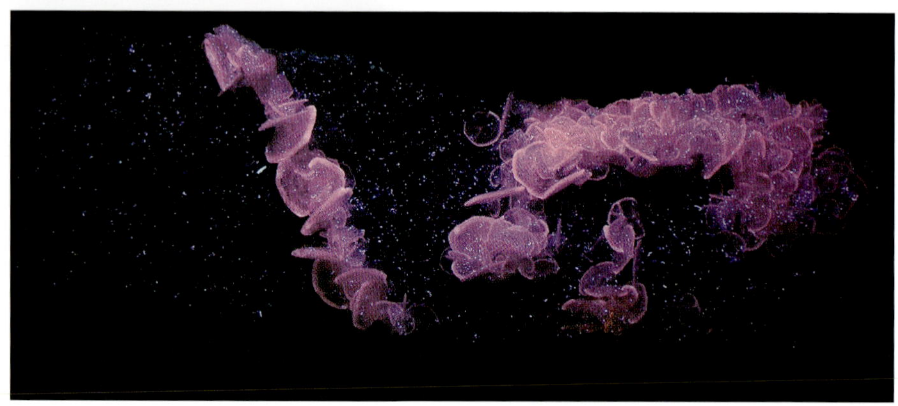

Same under LW

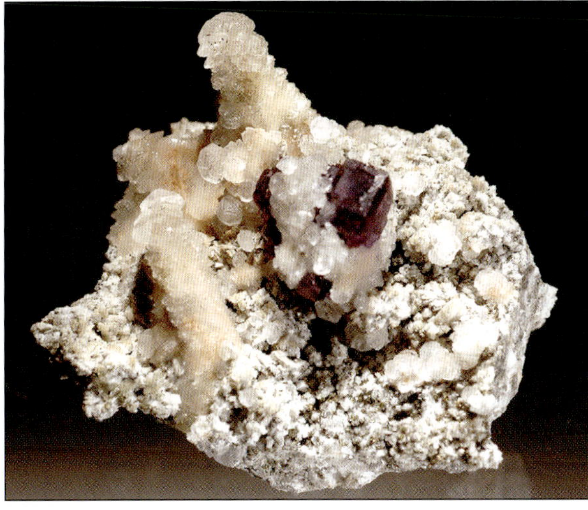

CALCITE. Calcite crystals with cubic fluorite crystals from San Martín, Zacatecas. The calcite crystals fluoresce orange-red SW and LW. The piece weighs 3.5 oz. and is 2.3 x 2.0 x 1.6 inches. Value $35-40

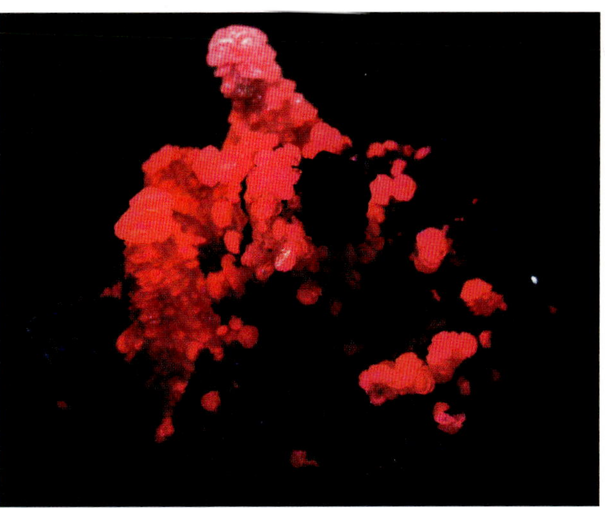

Same under SW

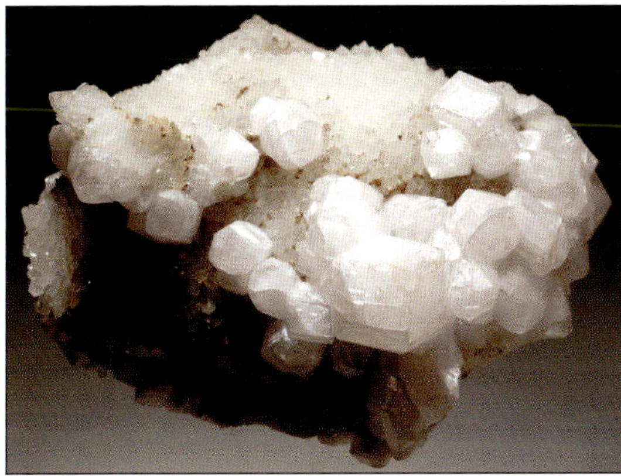

CALCITE. Crystals of calcite from the Valenciana mine, Guanajuato, Guanajuato County. The calcite fluoresces bright red SW. The piece weighs 9.5 oz. and is 3.8 x 3.3 x 1.3 inches. Value $40-50

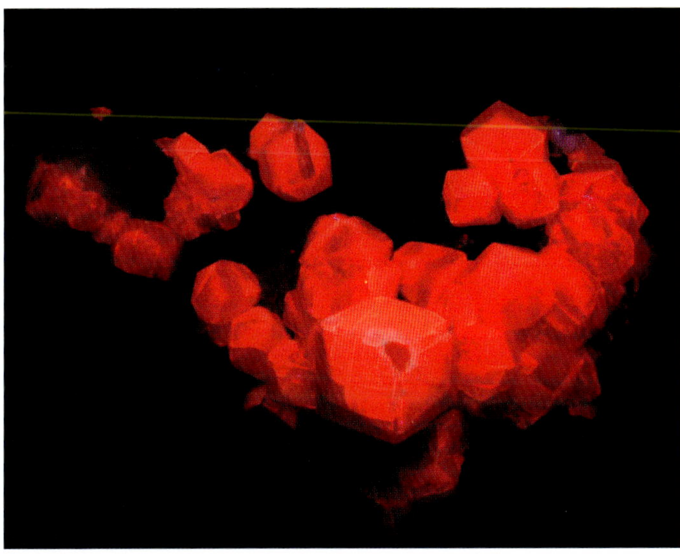

Same under SW

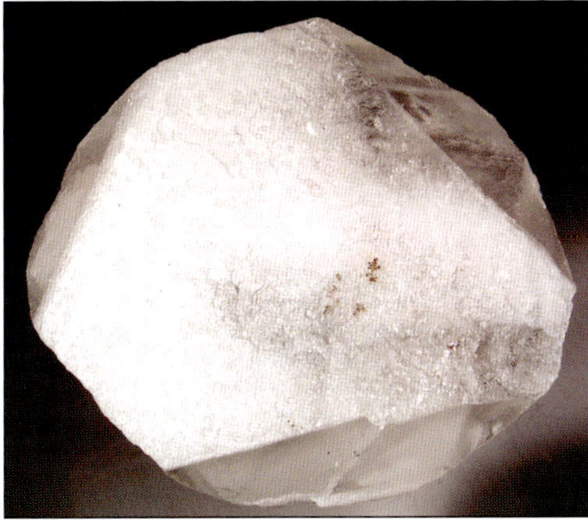

CALCITE. A calcite rhomb from the Amparo mine, San Antonio, Rodeo, Durango. Parts of the rhomb fluoresce bright green SW and (less brightly) LW. The piece weighs 0.8 oz. and is 1.8 x 1.5 x 0.3 inches. Value $10-12.

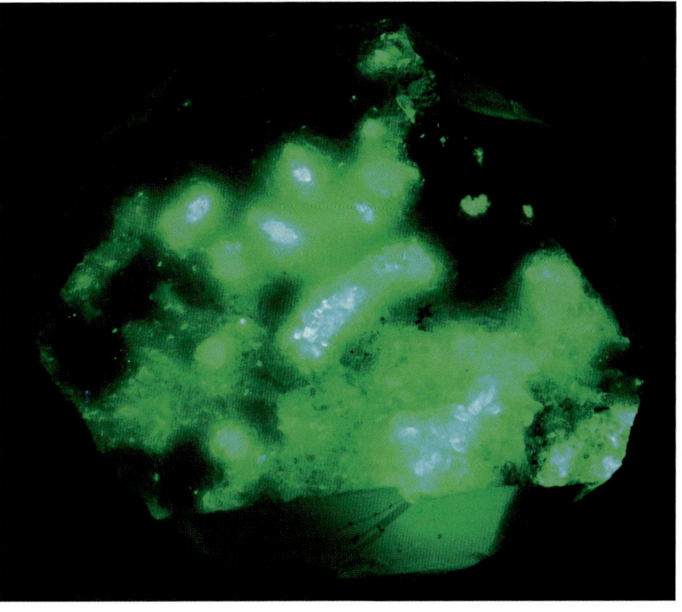

Same under SW

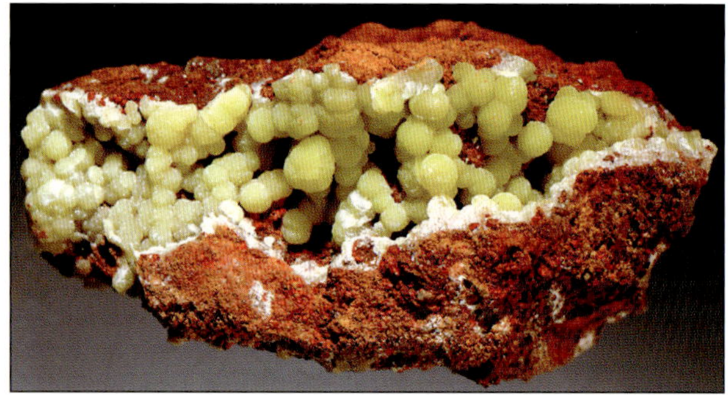

CUPROAUSTINITE. Tiny crystals of cuproaustinite in spheroidal groups on limonite gossan matrix from the Ojuela mine, Mapimí, Durango. Cuproaustinite fluoresces weak orange-tan SW and pale green with pale orange edges LW. The piece weighs 3.5 oz. and is 2.8 x 1.3 x 1.3 inches. Value $30-35

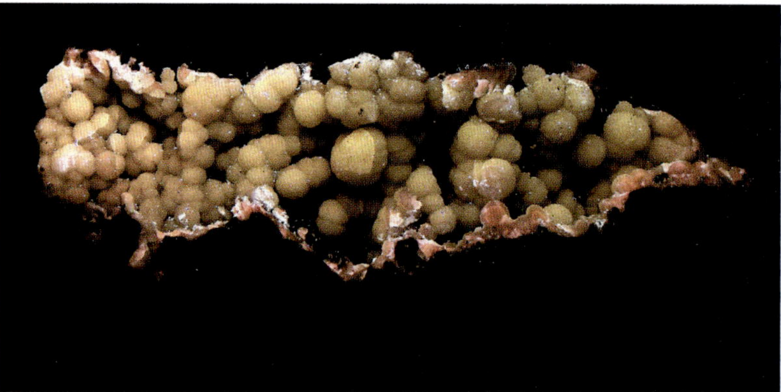

Same under SW

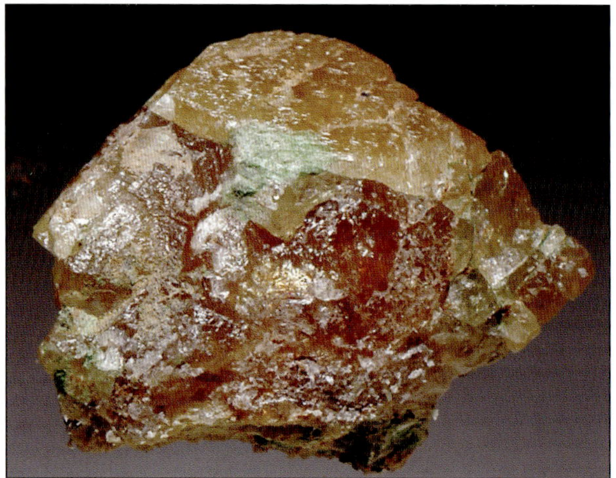

CUPROSCHEELITE. Cuproscheelite crystals from Sinoloa, Sonora. This scheelite fluoresces two colors, pale blue and pale yellow SW. From a group of material sold by Tom Warren in 1945 with an original label and sales receipt. There is a value premium for its provenance. The piece weighs 4.3 oz. and is 1.8 x 1.5 x 1.1 inches. Value $100-125

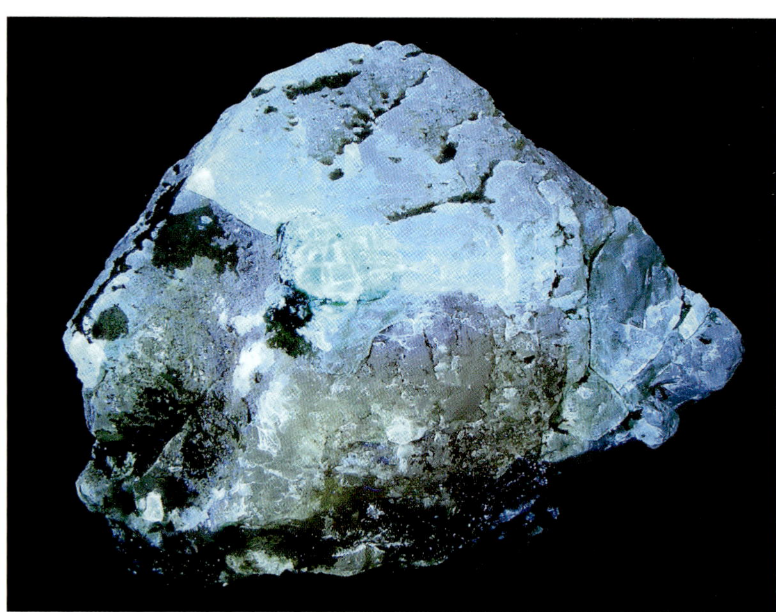

Same under SW

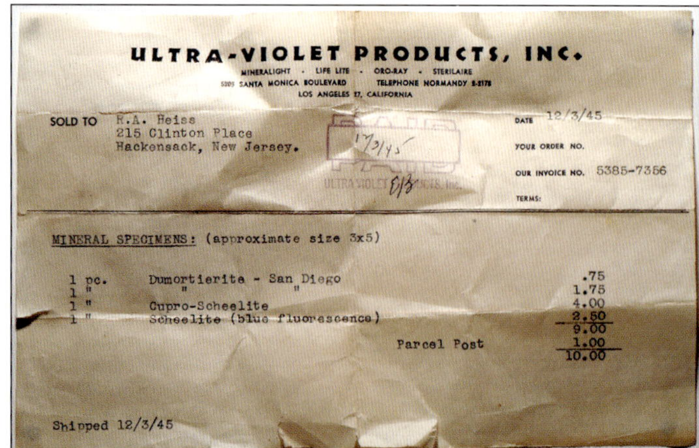

ULTRA-VIOLET PRODUCTS, INC.
MINERALIGHT · LIFE LITE · ORO-RAY · STERILAIRE
5205 SANTA MONICA BOULEVARD TELEPHONE NORMANDY 2-2175
LOS ANGELES 27, CALIFORNIA

SOLD TO R.A. Heiss
215 Clinton Place
Hackensack, New Jersey.

DATE 12/3/45
YOUR ORDER NO.
OUR INVOICE NO. 5385-7356
TERMS:

PAID 12/3/45 ULTRA VIOLET PRODUCTS, Inc.

MINERAL SPECIMENS: (approximate size 3x5)

1 pc.	Dumortierite - San Diego	.75
1 "	" "	1.75
1 "	Cupro-Scheelite	4.00
1 "	Scheelite (blue fluorescence)	2.50
		9.00
	Parcel Post	1.00
		10.00

Shipped 12/3/45

A 1945 invoice from UV Products of California selling the cuproscheelite above. The invoice also contains the reference to a piece of California scheelite that is shown under the California section. Tom Warren of UV Products is considered to be the "father" of the portable UV lamp. He was a noted collector of fluorescent minerals, maker of ultraviolet lamps, and an early dealer in fluorescent minerals.

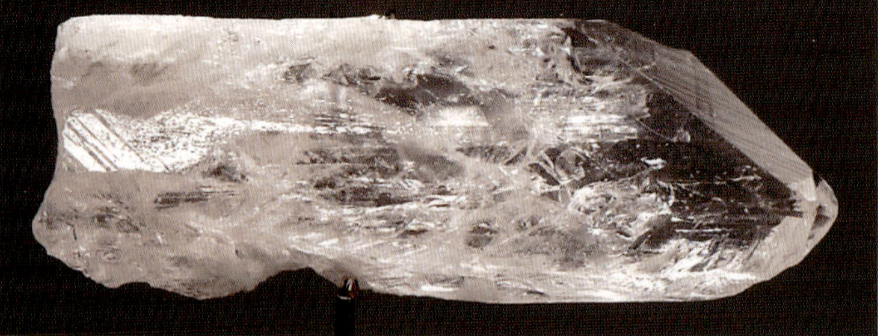

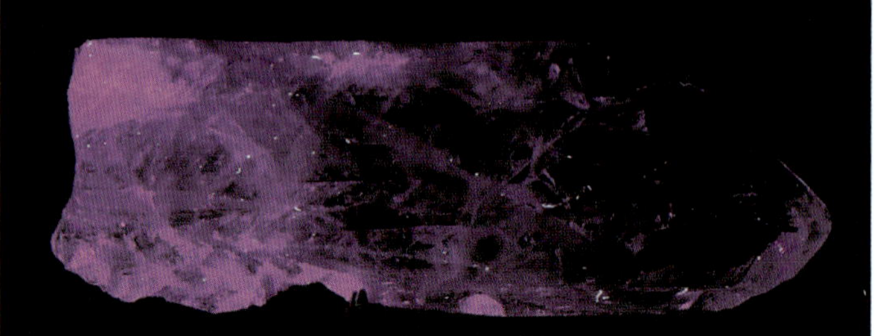

DANBURITE. Danburite from Charcas Mineral district, San Luis Potosi. It fluoresces weak violet SW. Danburite can be cut and faceted into gems. The piece weighs 0.4 oz. and is 1.5 x 0.6 x 0.3 inches. Value $15-20

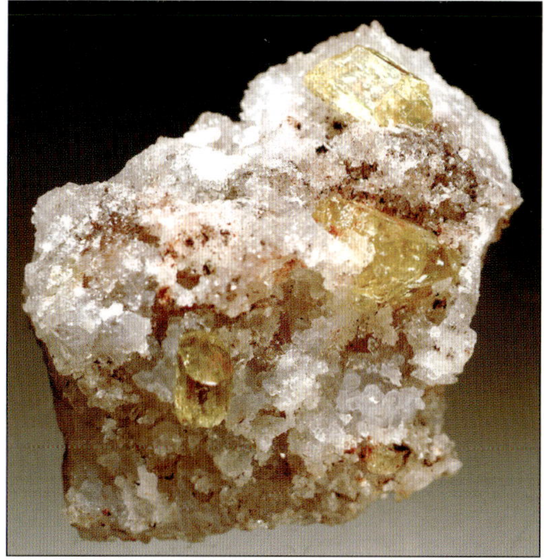

FLUORAPATITE. Beautiful gemmy crystals of fluorapatite from Cerro de Mercado, Durango. The fluorapatite fluoresces pink SW and calcite fluoresces green SW. The piece weighs 1.5 oz. and is 2.0 x 1.5 x 1.3 inches. Value $25-30

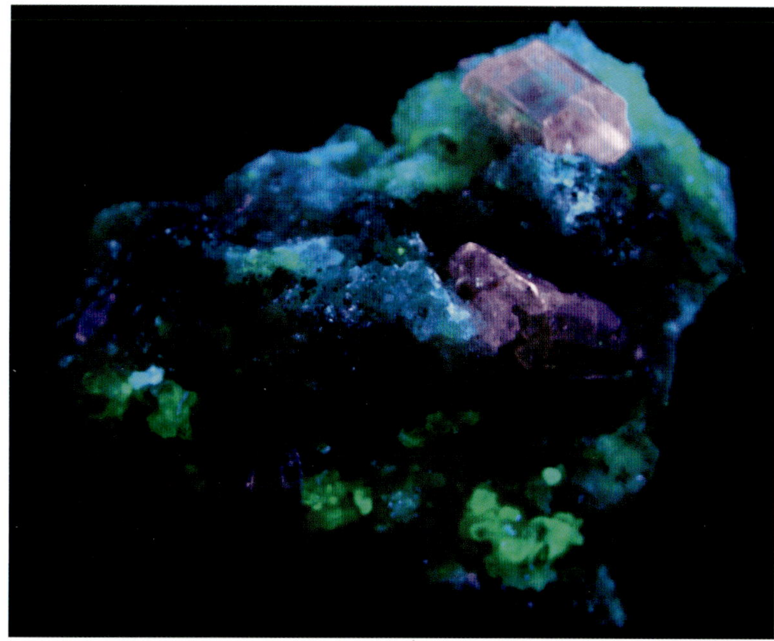

Same under SW

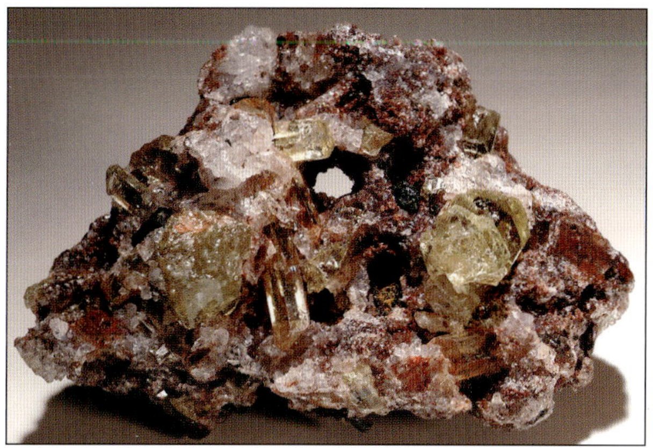

FLUORAPATITE. Crystals of fluorapatite with calcite from the Ojuela Mine, Mapimí, Durango. The largest crystal is 0.5 inches. The fluorapatite fluoresces pink SW and calcite fluoresces green SW in some parts of the piece. The piece weighs 2.5 oz. and is 2.5 x 1.8 x 1.3 inches. Value $25-30

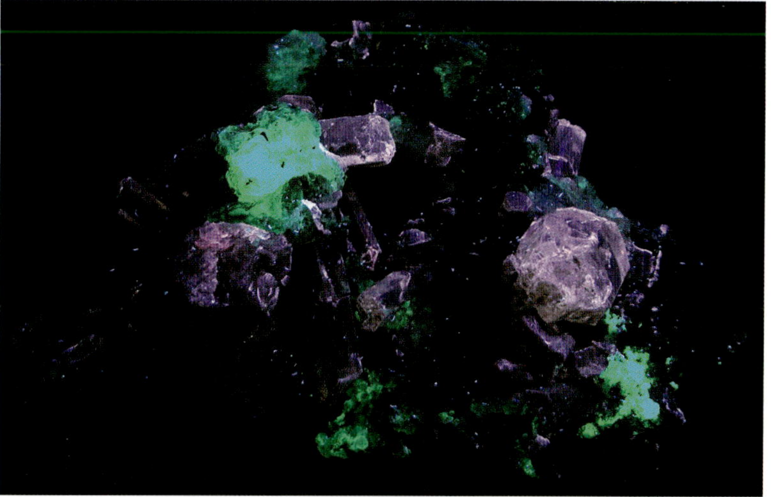

Same under SW

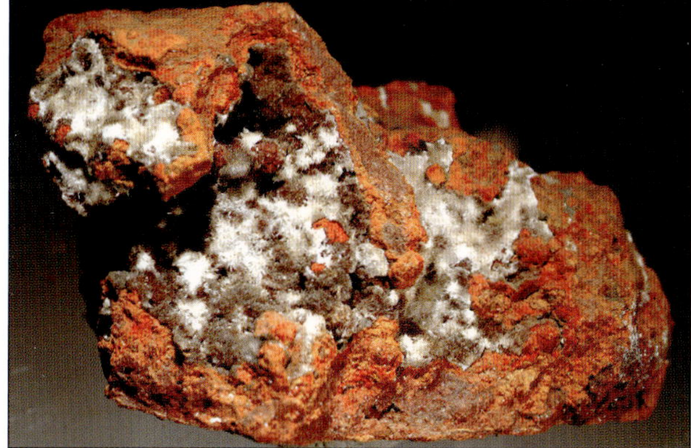

HYDROZINCITE. Well-crystallized hydrozincite in limonite gossan matrix from the Ojuela mine, Mapimí, Durango. Hydrozincite fluoresces bright pale blue SW. The piece weighs 8.0 oz. and is 2.5 x 2.0 x 2.0 inches. Value $30-35

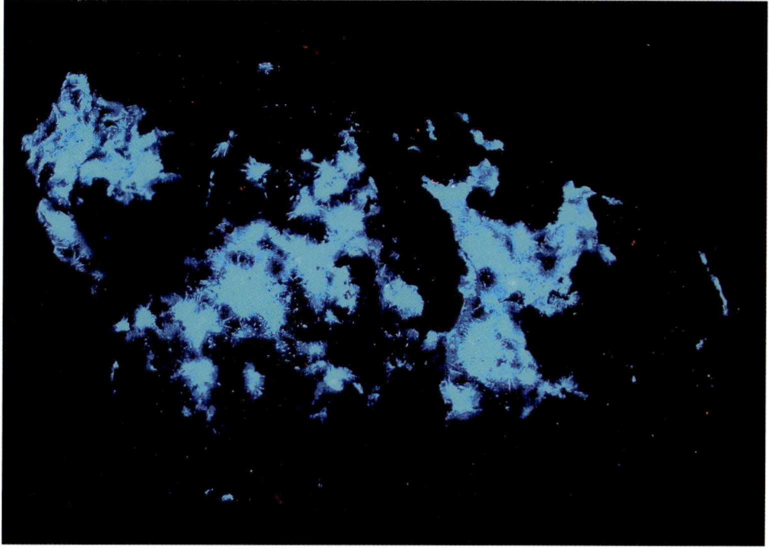

Same under SW

POWELLITE. Calcite crystals on powellite from San Javier, Sonora. The powellite fluoresces yellow SW and LW and the calcite fluoresces pale grey LW and SW. The piece weighs 10.3 oz. and is 3.0 x 2.5 x 1.8 inches. Value $25-35

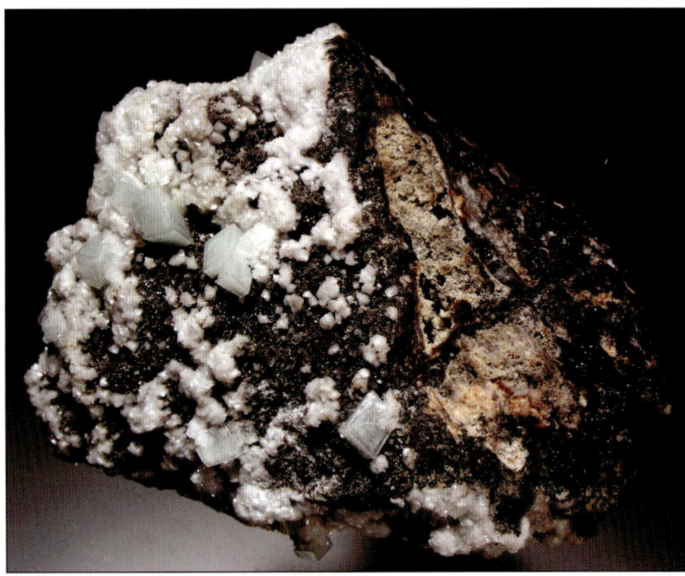

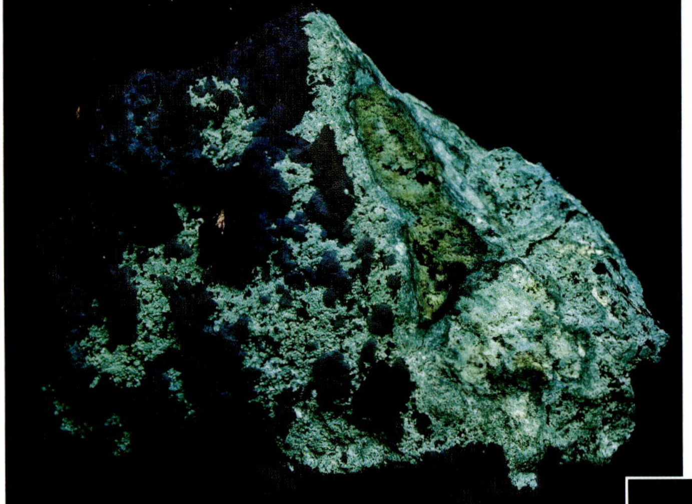

Same under SW

Same under LW

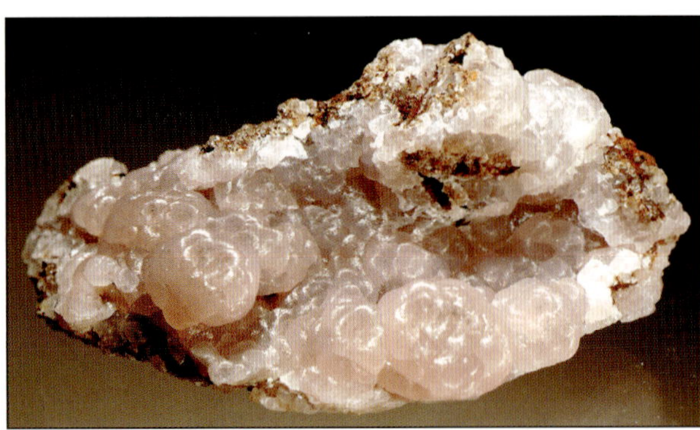

SMITHSONITE. Botryoidal-surfaced smithsonite from the Santa Anita mine, Choix, Sinaloa. The smithsonite has areas that fluoresce pink and blue SW. The piece weighs 5.0 oz. and is 3.0 x 1.8 x 1.3 inches. Value $20-25

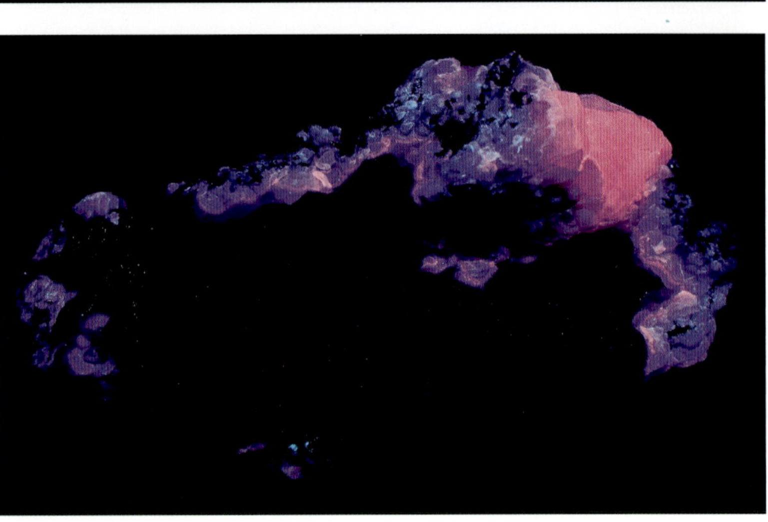

Same under SW

Other Countries (alphabetically)

Afghanistan

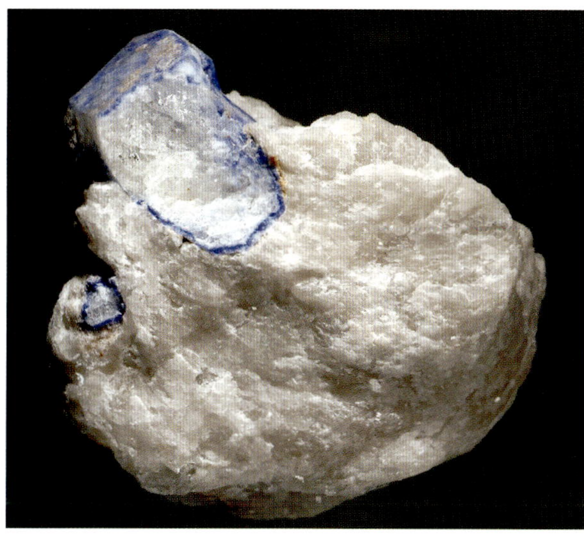

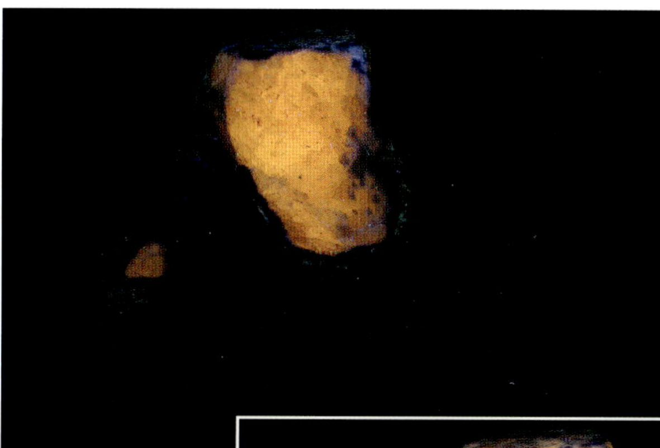

Same under SW

AFGHANITE. Unusual afghanite crystals that have been broken and show a white interior. They are from Koksha Valley, Sar-e-Sang, Badakhshan Province. The interior of the crystal fluoresces bright yellow LW and (less brightly) SW. I believe that the yellow-fluorescing mineral is scapolite, var. meionite. The piece weighs 8.8 oz. and is 2.8 x 2.5 x 1.5 inches. Value $75-85

Same under LW

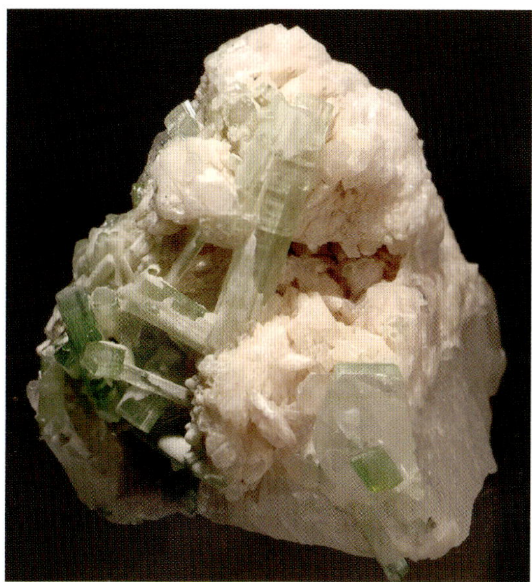

ELBAITE. Green to colorless tourmaline, var. elbaite, crystals from Paprook, Kunar Province. The elbaite fluoresces bright pale blue SW. The piece weighs 3.8 oz. and is 2.8 x 2.1 x 1.0 inches. Value $45-55

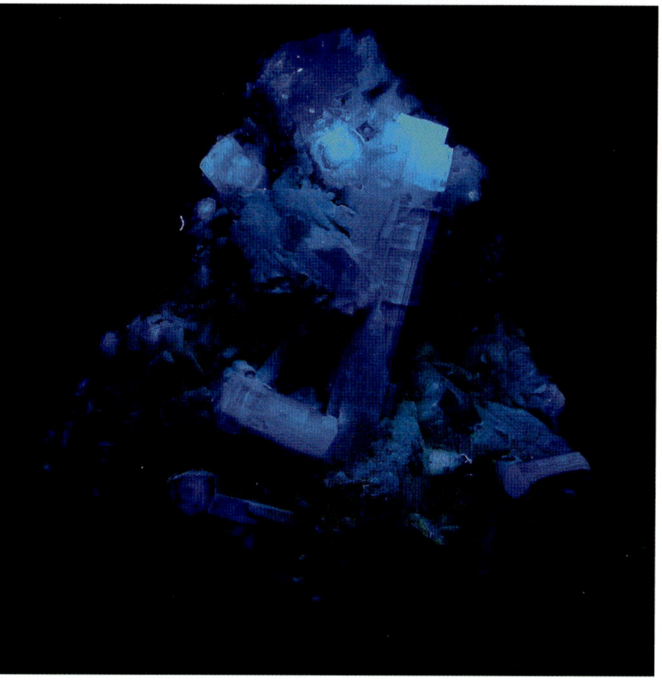

Same under SW

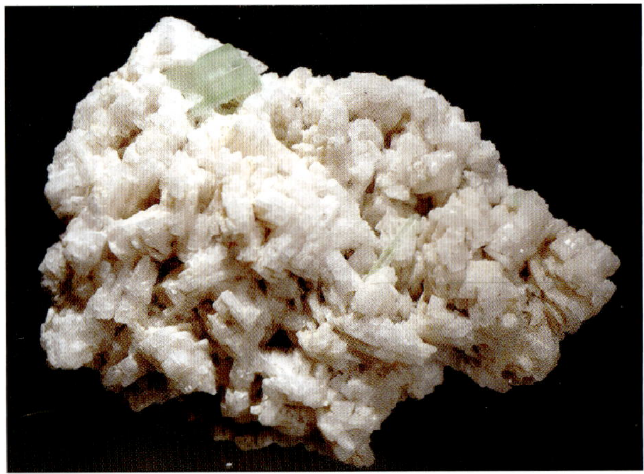

ELBAITE. Green tourmaline, var. elbaite (tourmaline group), crystals from Paprook, Kunar Province. The elbaite fluoresces bright pale blue SW. The piece weighs 3.8 oz. and is 1.9 x 2.0 x 2.0 inches. Value $40-45

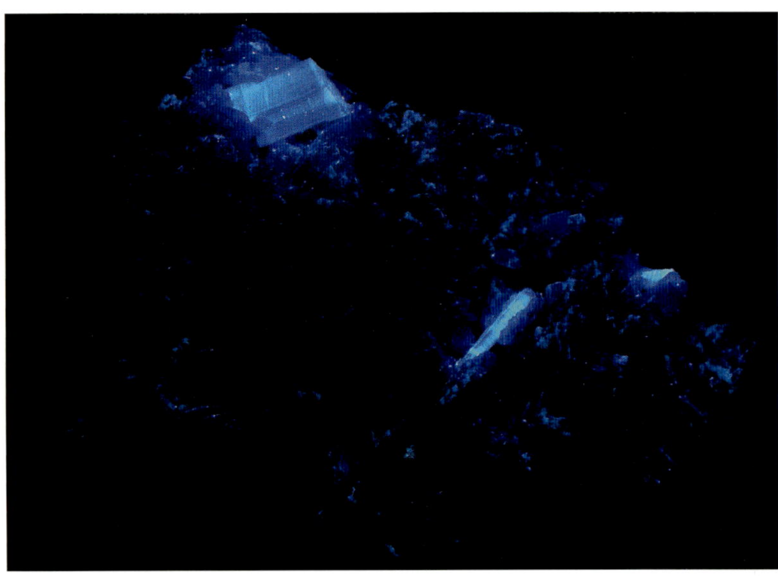

Same under SW

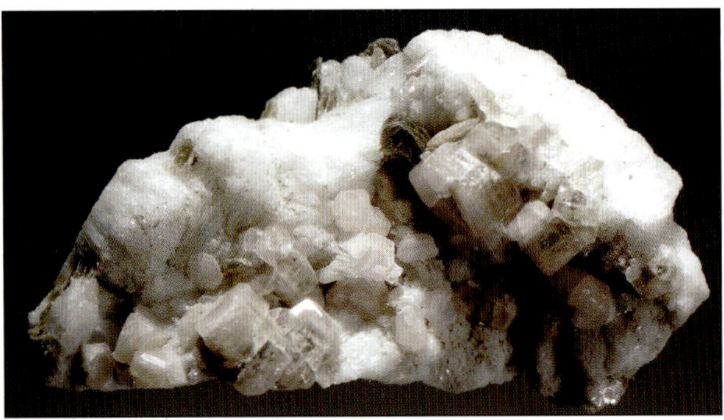

FLUORAPATITE. Fluorapatite crystals with mica from Paprook, Kunar Province. The fluorapatite fluoresces yellow SW. The piece weighs 5.5 oz. and is 3.3 x 2.0 x 1.5 inches. Value $65-75

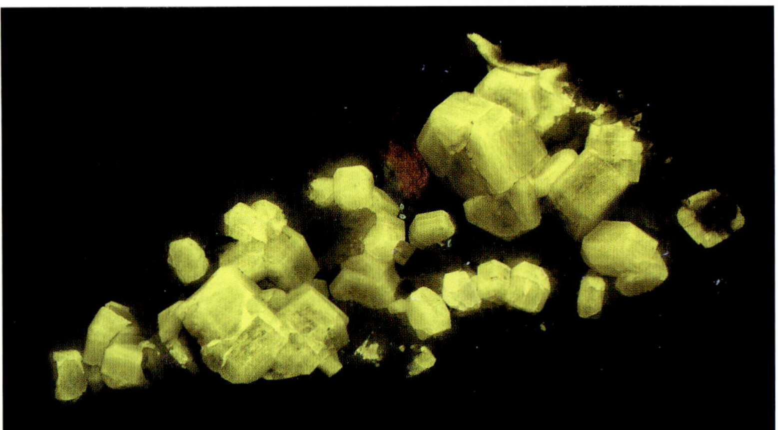

Same under SW

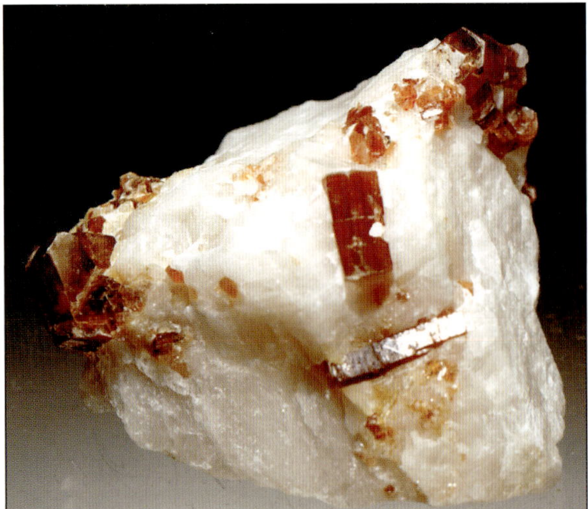

PHLOGOPITE. Calcite with hexagonal crystals of phlogopite from Koksha Valley, Sar-e-Sang, Badakhshan Province. The phlogopite fluoresces yellow SW and the calcite fluoresces a weak pink SW. The piece weighs 4.0 oz. and is 2.5 x 1.8 x 1.5 inches. Value $35-45

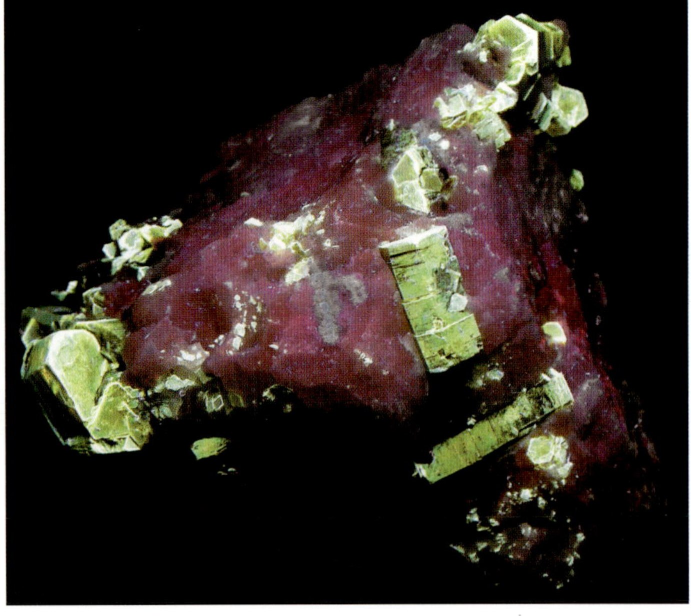

Same under SW

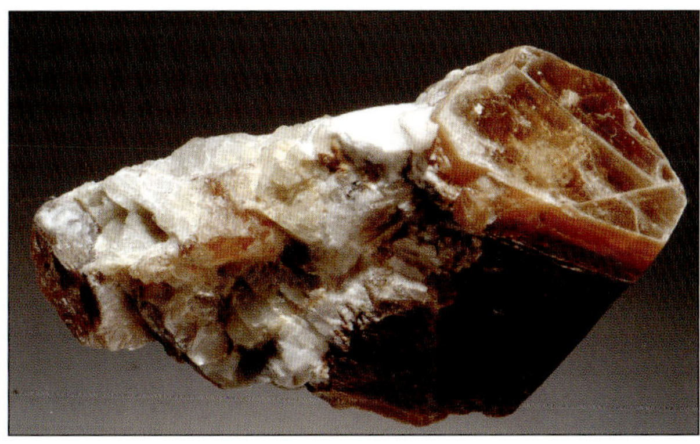

PHLOGOPITE. Phlogopite crystals on calcite from Koksha Valley, Sar-e-Sang, Badakhshan Province. The crystal fluoresces yellow SW. The piece weighs 4.3 oz. and is 2.8 x 1.8 x 1.1 inches. Value $35-40

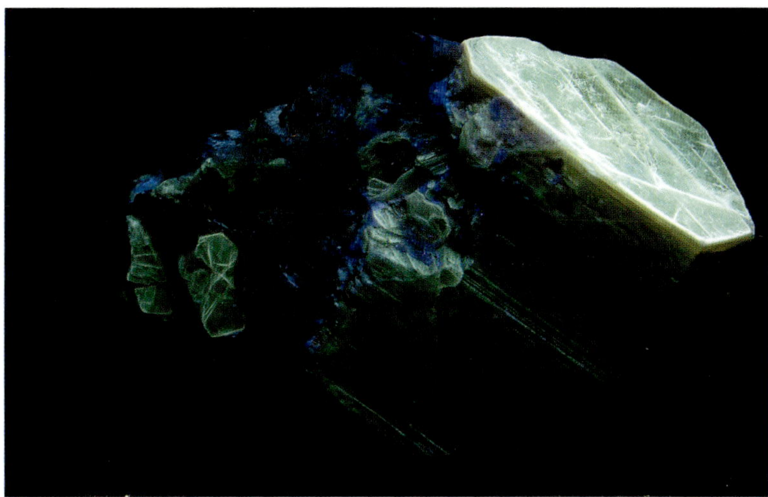

Same under SW

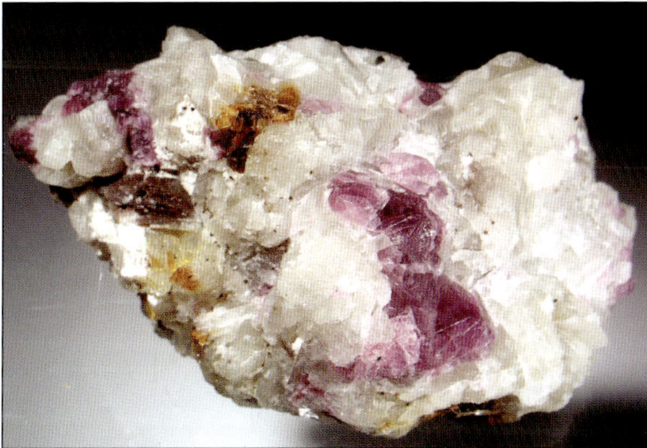

SODALITE. Sodalite, var. hackmanite, crystals on winchite from Koksha Valley, Sar-e-Sang, Badakhshan Province. The hackmanite fluoresces orange LW, has a long-lasting blue phosphorescence, and is tenebrescent. The piece weighs 3.5 oz. and is 3.8 x 2.3 x 2.0 inches. Value $65-75

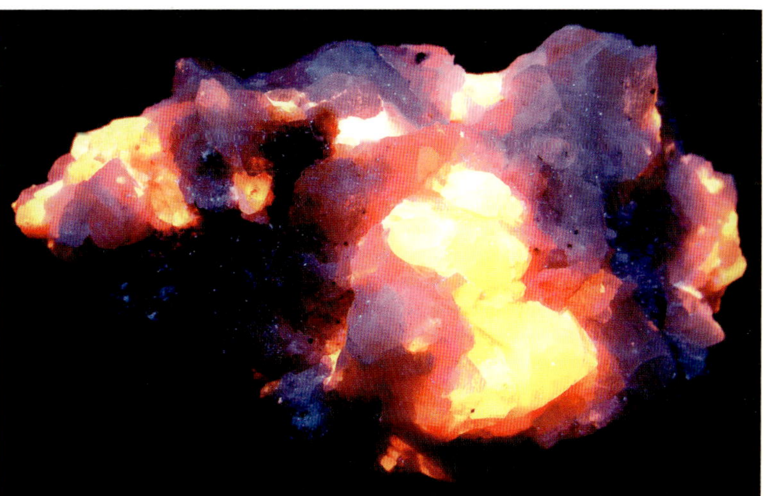

Same under LW

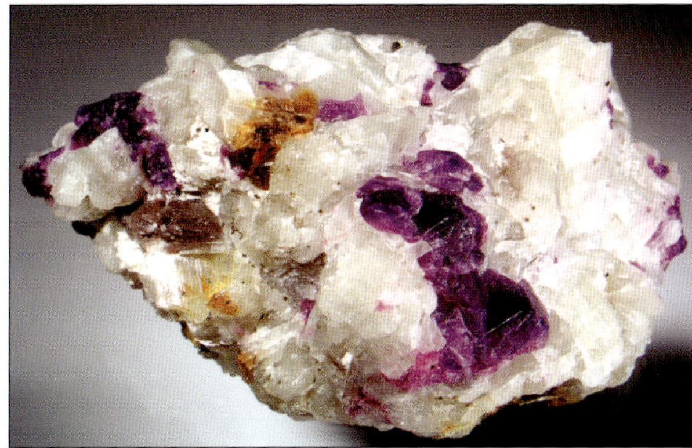

Showing tenebrescencex

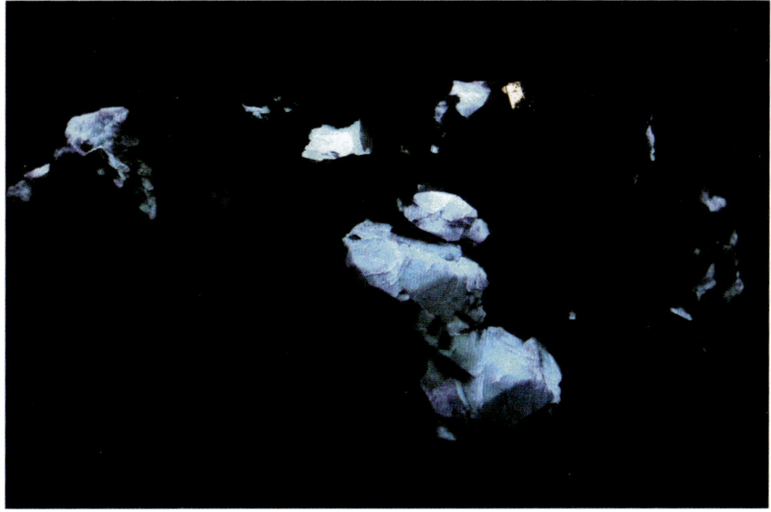

Same under SW. The photo shows the crystals beginning to phosphoresce white, After the SW lamp is removed, they will continue to phosphoresce.

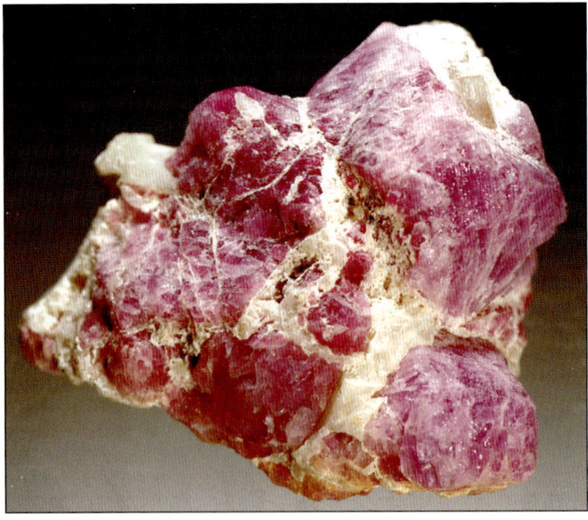

SODALITE. A piece containing a large crystal of sodalite, var. hackmanite, from Koksha Valley, Sar-e-Sang, Badakhshan Province. The hackmanite fluoresces orange LW, has a long-lasting blue phosphorescence and is tenebrescent. The piece weighs 1.0 oz. and is 1.8 x 1.4 x 1.0 inches. Value $30-35

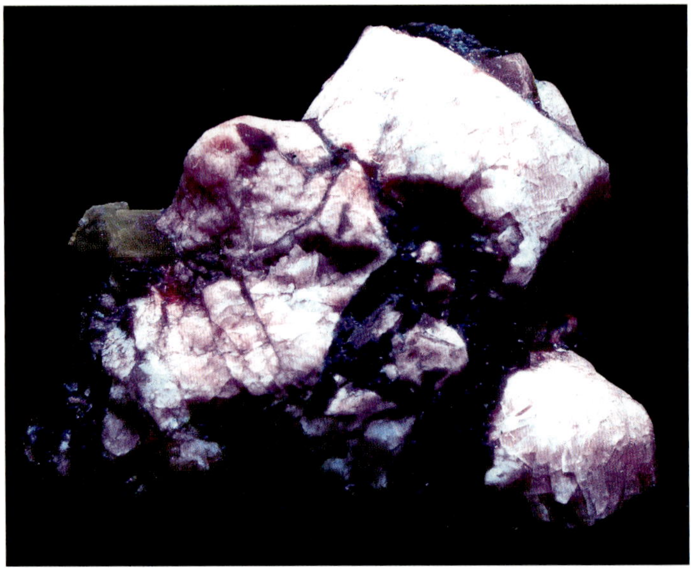

Same under SW. In the photo the hackmanite is beginning to phosphoresce white and then continues to phosphoresce pale blue.

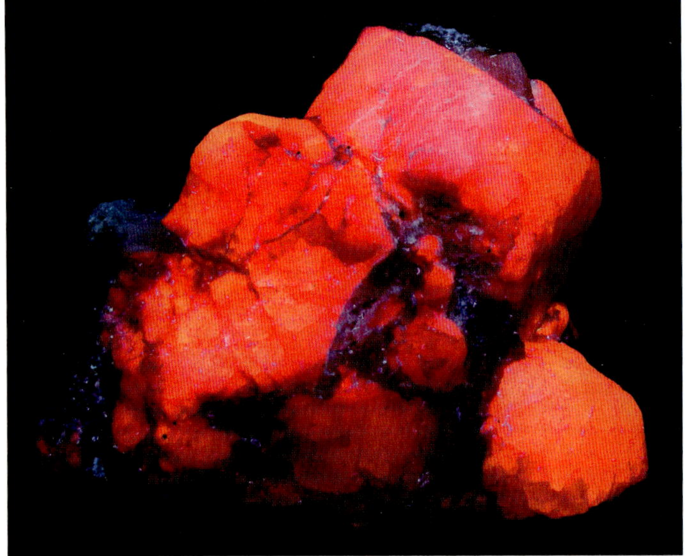

Same under LW

SODALITE. Sodalite from Koksha Valley, Sar-e-Sang, Badakhshan Province faceted into a gem. The sodalite fluoresces orange LW and white SW, and has a pale blue phosphorescence. It is 6.0 x 9.0 mm. Value $60-75

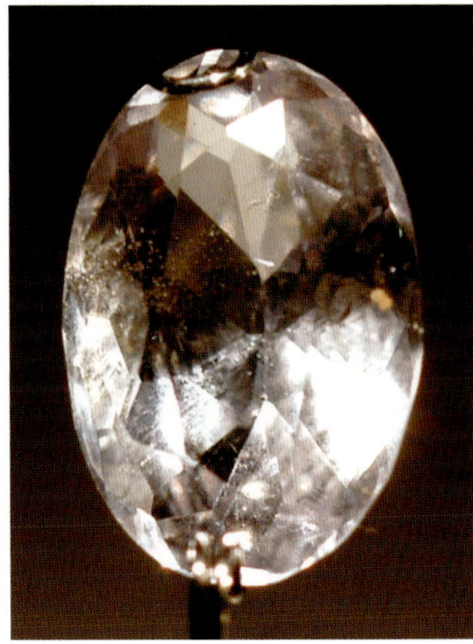

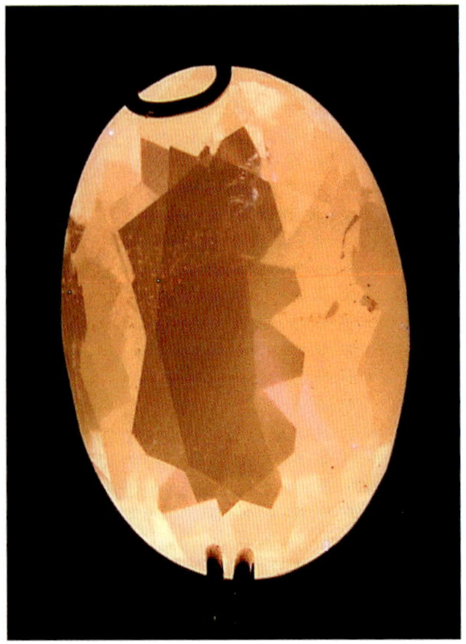

Same under LW

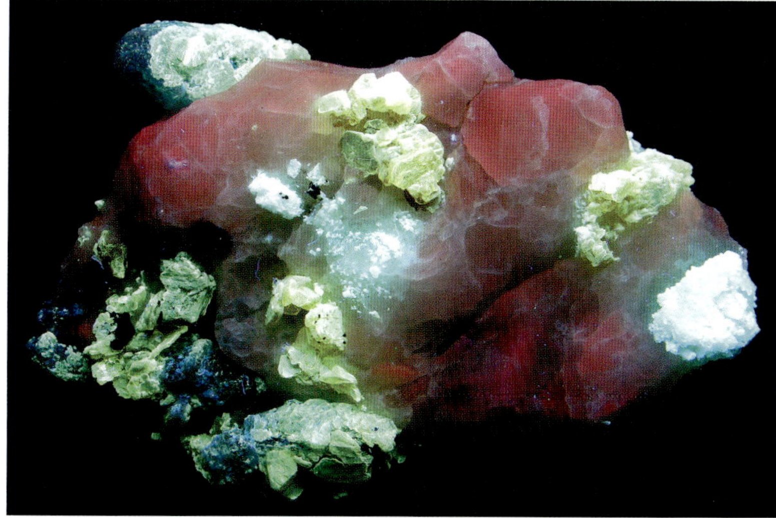

Same under SW

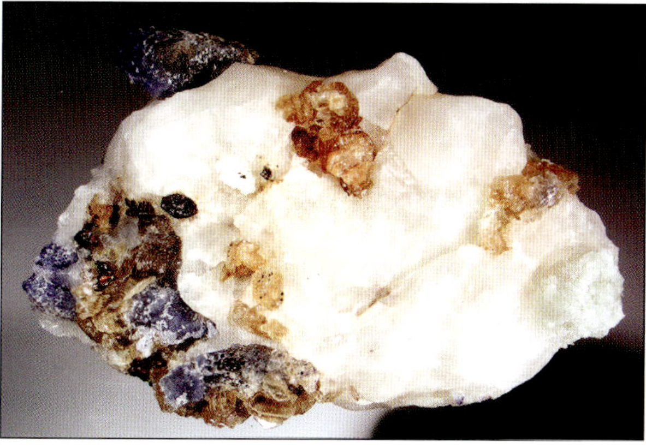

SODALITE. Blue sodalite crystals with phlogopite, calcite, and feldspar from Koksha Valley, Sar-e-Sang, Badakhshan Province. If you find a mineral dealer selling blue afghanite, check it with your longwave lamp. Sometimes the material sold as afghanite is occasionally blue sodalite. The sodalite fluoresces orange LW, the phlogopite fluoresces yellow SW, the feldspar fluoresces white LW and SW, and the calcite fluoresces pale red-orange SW. The piece weighs 1.0 oz. and is 2.3 x 1.5 x 0.5 inches. Value $175-200

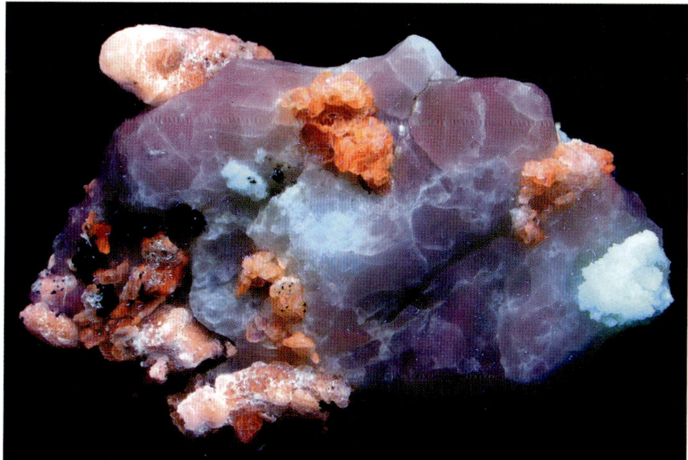

Same under LW

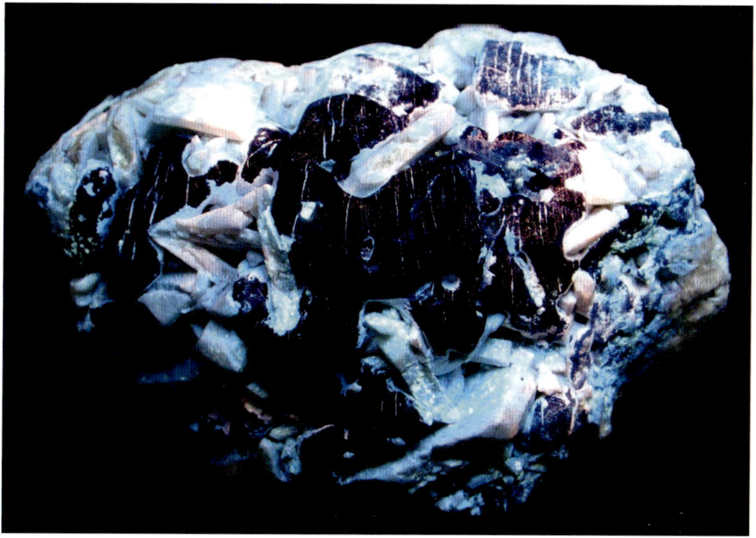

Same under SW

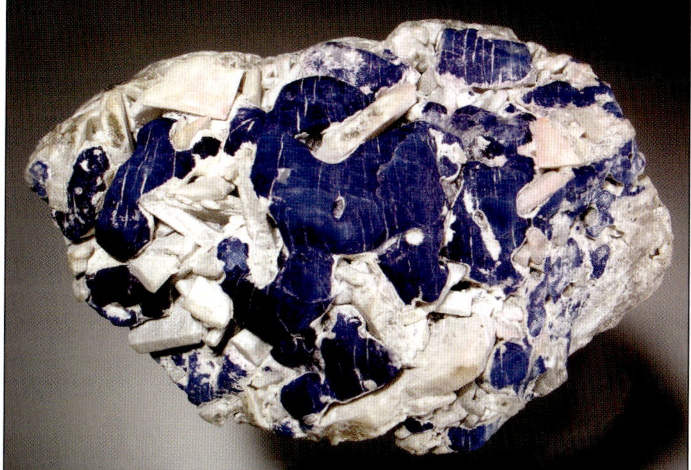

SODALITE. Blue sodalite, afghanite, and feldspar from Koksha Valley, Sar-e-Sang, Badakhshan Province. The sodalite fluoresces orange LW and the feldspar fluoresces white SW and LW. The piece weighs 6.5 oz. and is 3.0 x 2.0 x 1.3 inches. Value $200-225

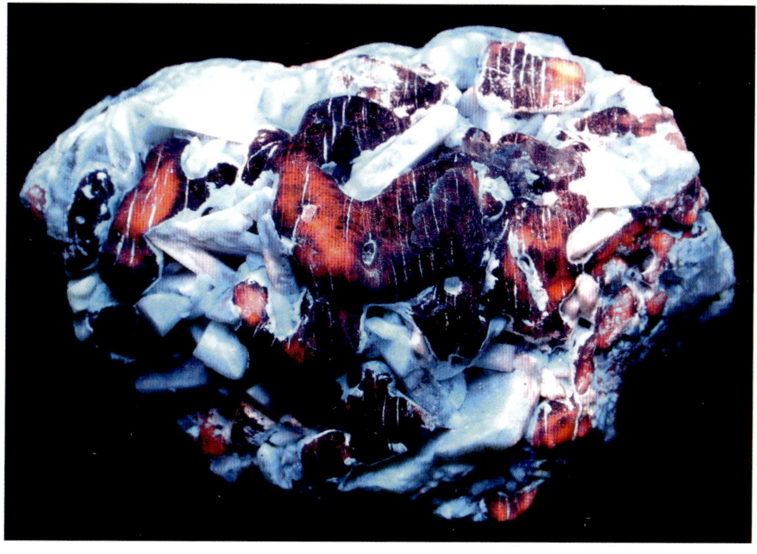

Same under LW

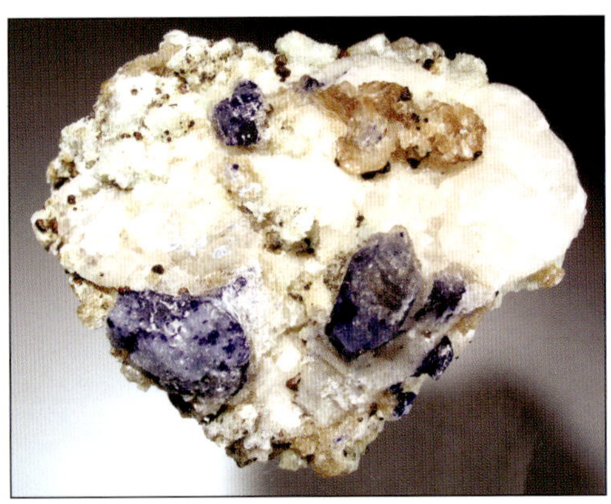

SODALITE. Blue sodalite, phlogopite, calcite, feldspar, and an unidentified mineral, all from Koksha Valley, Sar-e-Sang, Badakhshan Province. The sodalite fluoresces orange LW. The phlogopite fluoresces yellow SW, the feldspar fluoresces white LW and SW, the calcite fluoresces dull red-orange SW, and the unidentified mineral, fluoresces yellow LW. The piece weighs 4.0 oz. and is 2.3 x 2.0 x 1.5 inches. Value $255-275

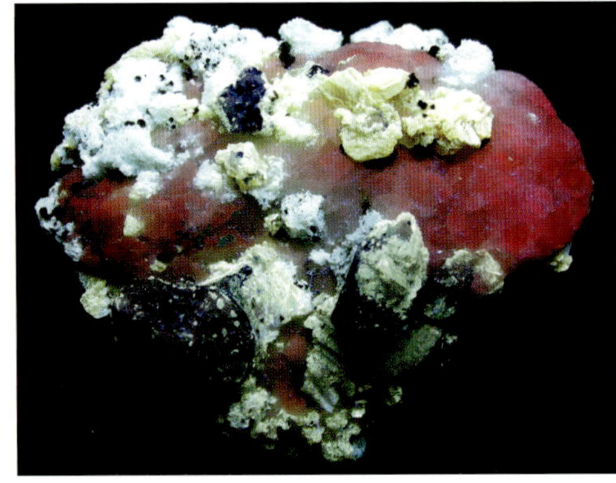

Same under SW

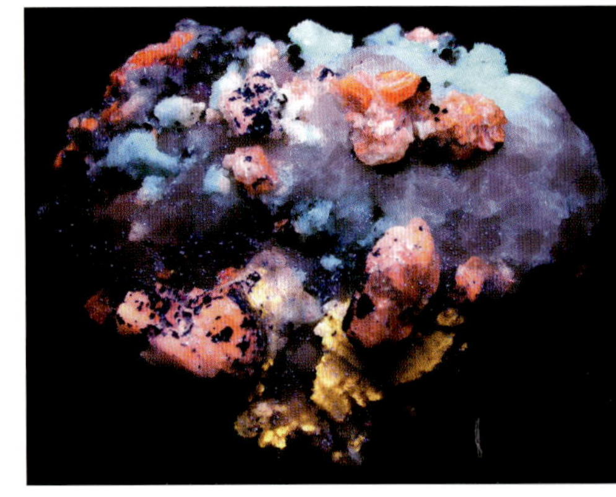

Same under LW

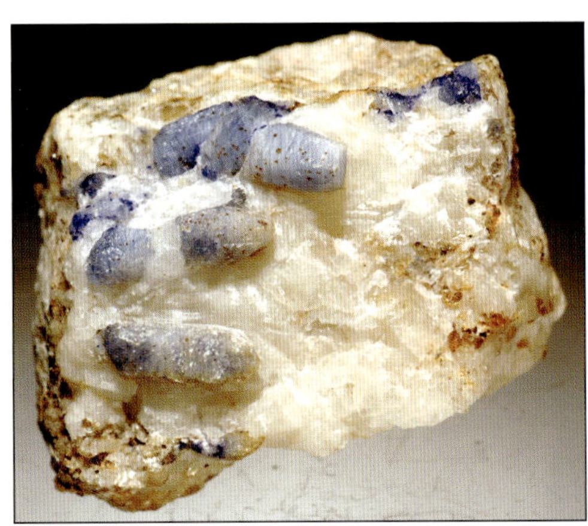

SODALITE. A specimen with crystals of blue sodalite, phlogopite, and calcite from Koksha Valley, Sar-e-Sang, Badakhshan Province. The sodalite fluoresces orange LW. The calcite fluoresces pale red-orange SW and the phlogopite fluoresces bright yellow-orange SW. The piece weighs 3.5 oz. and is 2.0 x 2.0 x 1.5 inches. Value $200-235

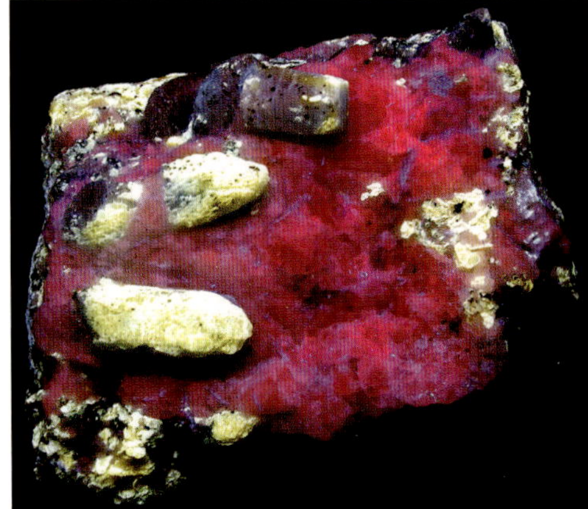

Same under SW

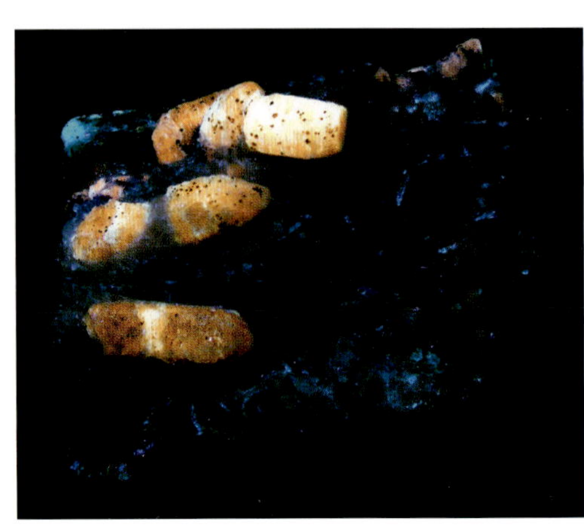

Same under LW

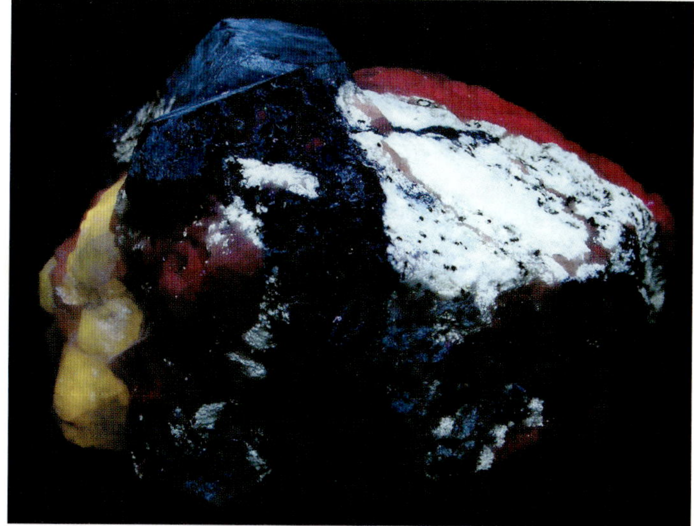

Same under SW

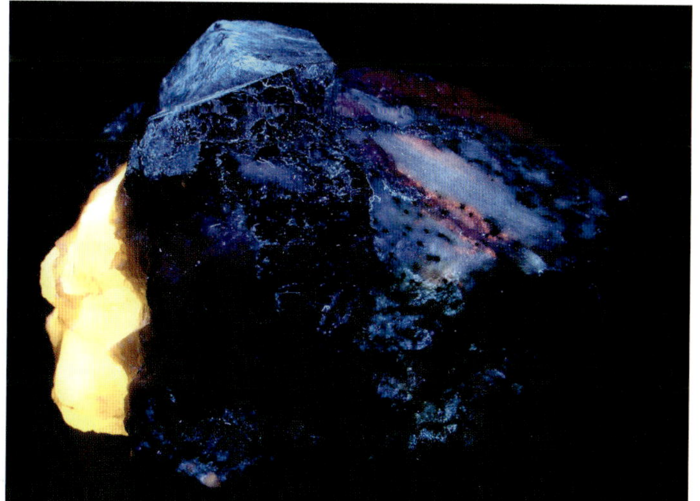

Same under LW

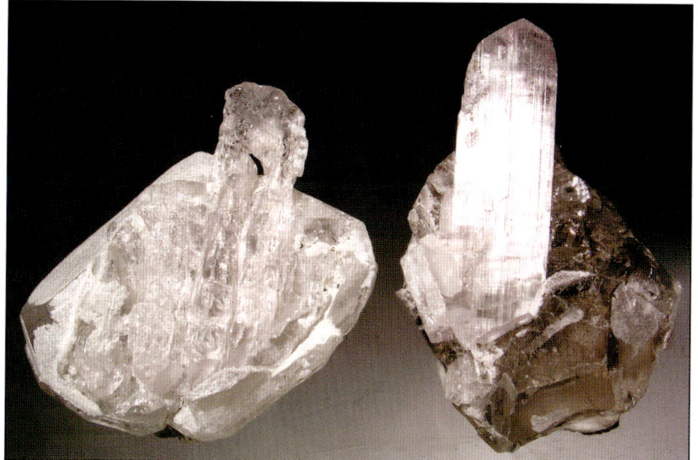

SPODUMENE, A crystal of spodumene, var. kunzite, from Kunar Province. The spodumene fluoresces bright lavender SW, pink to orange-pink LW, and blue MW. The piece weighs 0.3 oz. and is 1.3 x 1.0 x 0.5 inches. Value $20-25.

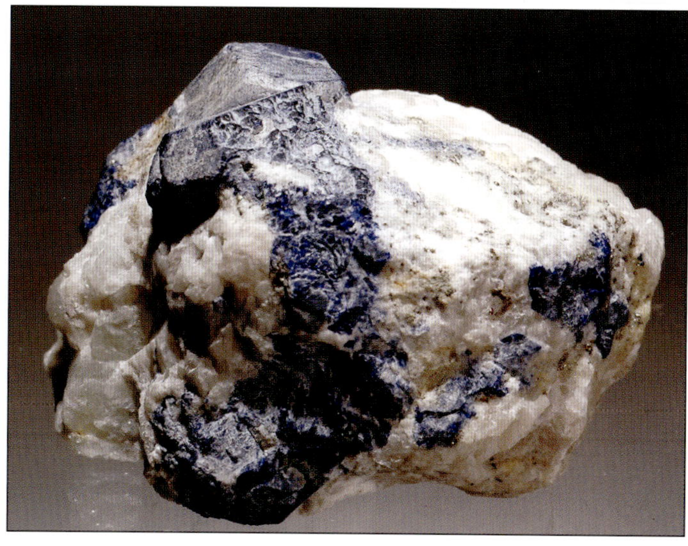

SCAPOLITE, FELDSPAR, An interesting piece with calcite, scapolite, feldspar, and afghanite crystals, from Koksha Valley, Sar-e-Sang, Badakhshan Province. The scapolite (believed to be var. meionite) fluoresces bright yellow LW, the calcite fluoresces pale red SW, and the feldspar fluoresces white SW. The piece weighs 7.0 oz. and is 2.5 x 2.0 x 1.7 inches. Value $60-65

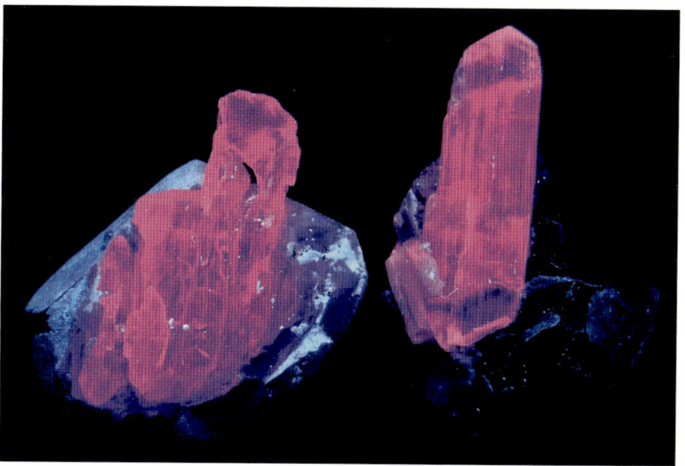

Same under SW

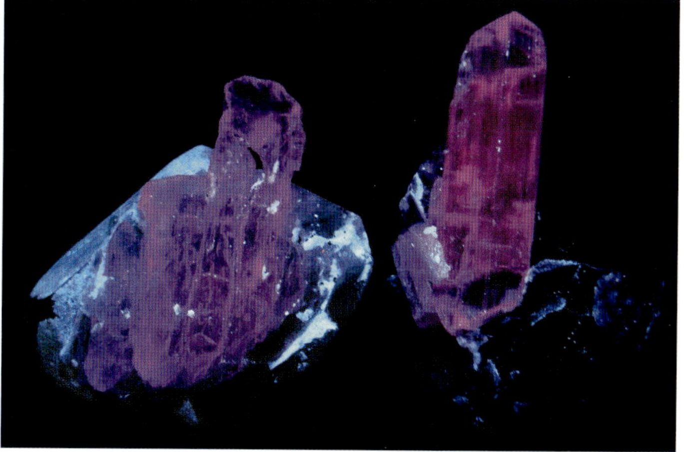

Same under LW

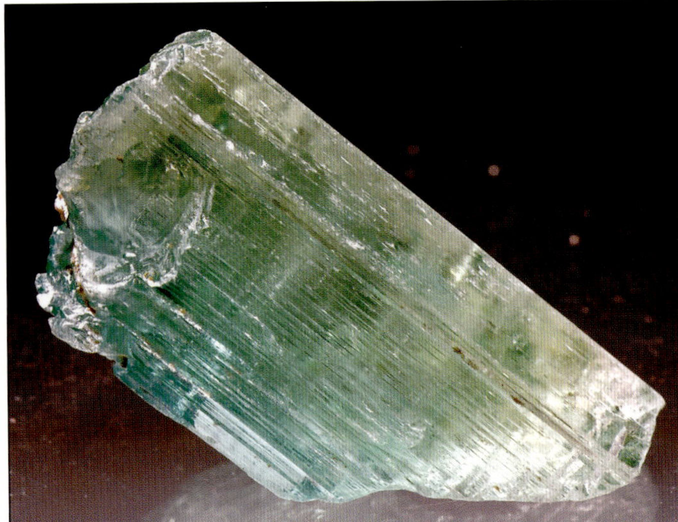

SPODUMENE, Gemmy green spodumene, var. kunzite, from Kolum district, Nuristan region. It pays to examine all spodumene from Afghanistan and Pakistan under both SW and LW. Very few pieces react like this one. The spodumene fluoresces pink SW and orange LW. The piece weighs 1.5 oz. and is 1.9 x 1.1 x 0.8 inches. Value $125-140

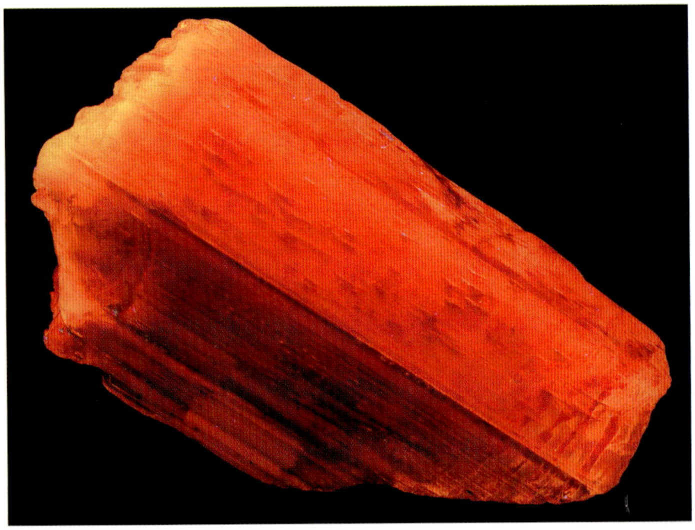

Same under LW

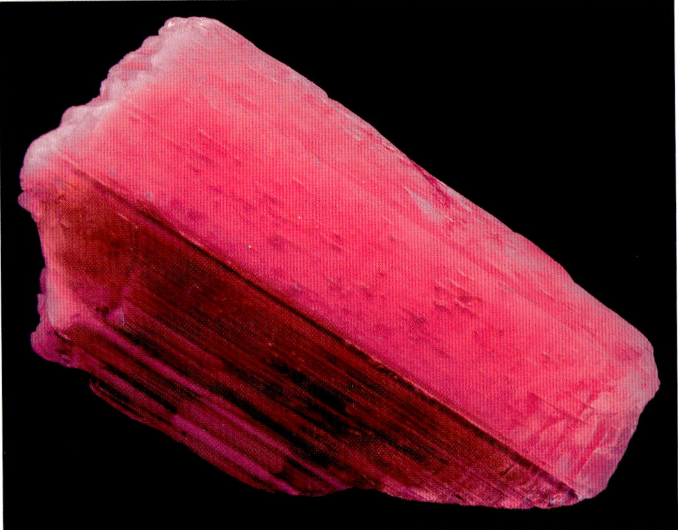

Same under SW

Australia

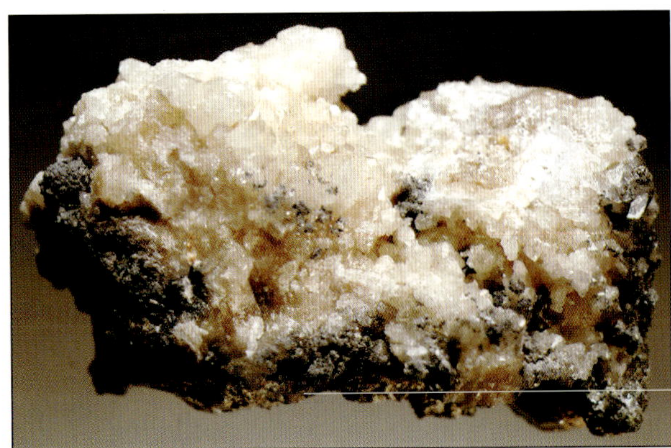

ANGLESITE. Anglesite crystals from Broken Hill, New South Wales. The anglesite fluoresces pale yellow SW. The piece weighs 5.0 oz. and is 2.3 x 1.3 x 1.1 inches. Value $50-60

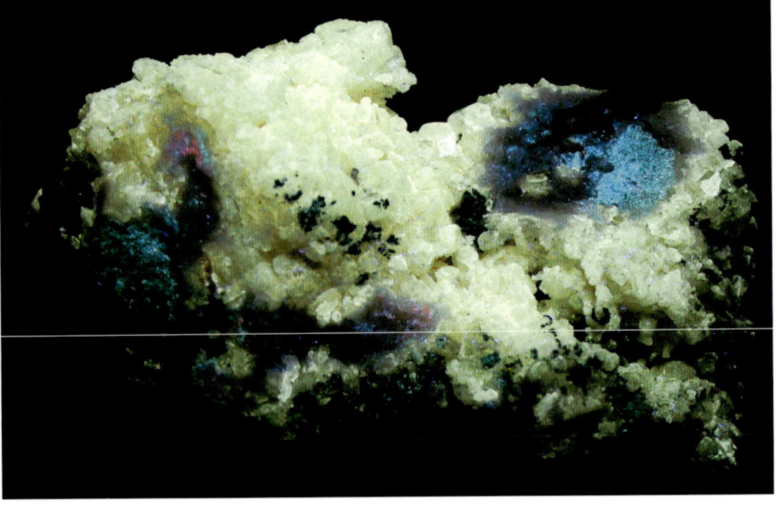

Same under SW

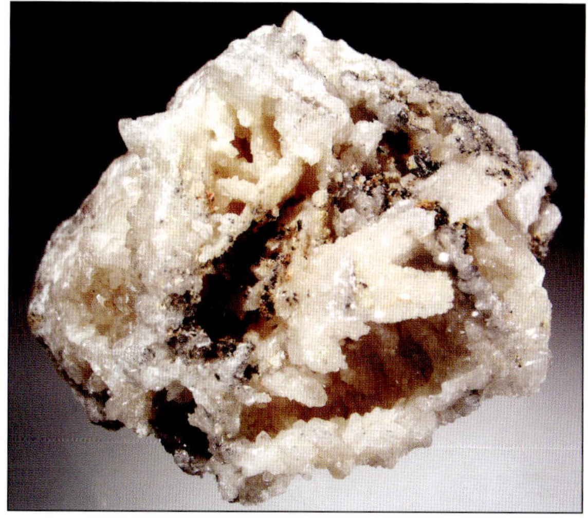

CERUSSITE. Cerussite crystals from Broken Hill, New South Wales. The cerussite fluoresces bright yellow SW. The piece weighs 4.8 oz. and is 2.0 x 1.5 x 1.5 inches. Value $20-25.

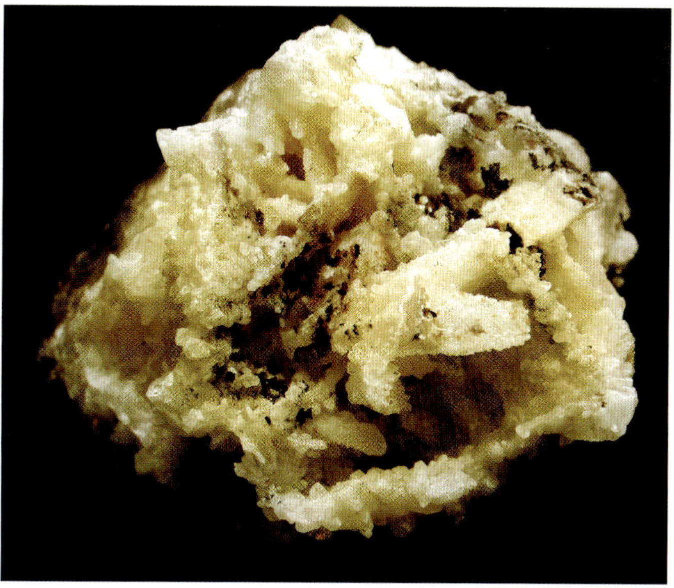

Same under SW

MICROCLINE. Blue-green microcline, var. amazonite, from the Kintore Open Cut, Broken Hill, New South Wales. The amazonite fluoresces bright pale blue SW. The piece weighs 6.0 oz. and is 2.8 x 2.0 x 1.8 inches. Value $55-60

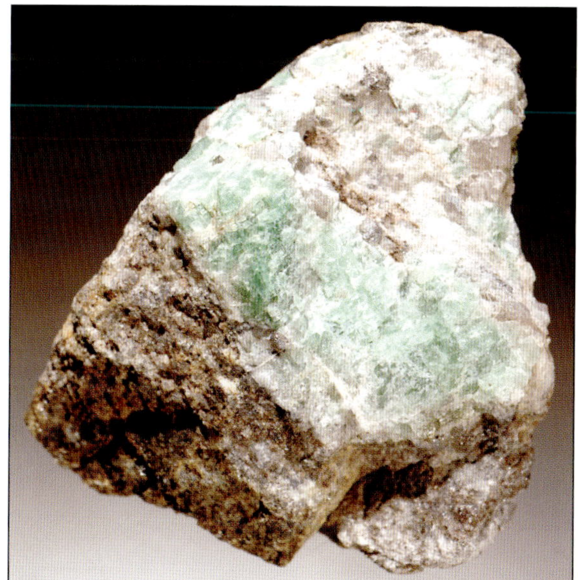

Same under SW

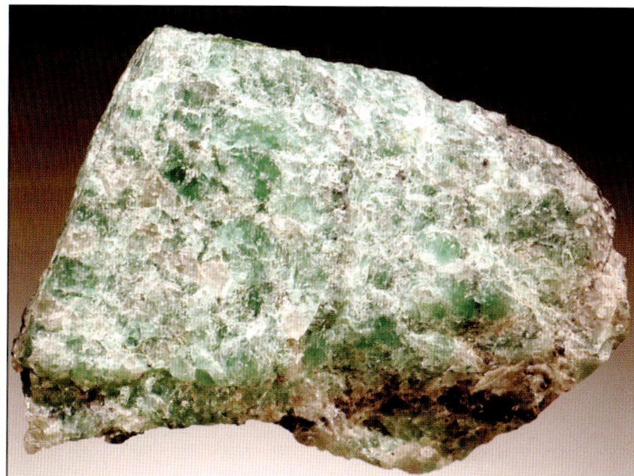

MICROCLINE. Blue-green microcline, var. amazonite, from the Kintore Open Cut, Broken Hill, New South Wales. The amazonite fluoresces bright pale blue SW. The piece weighs 4.8 oz. and is 2.5 x 1.8 x 1.8 inches. Value $55-60

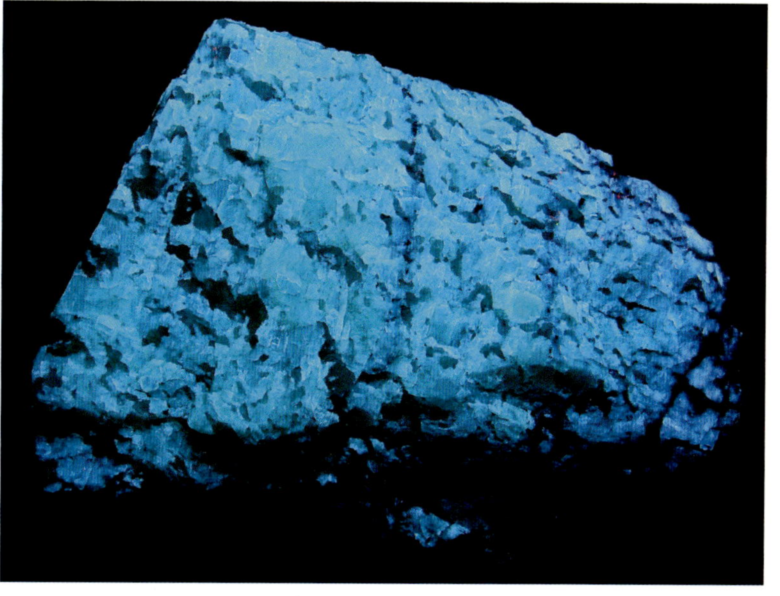

Same under SW

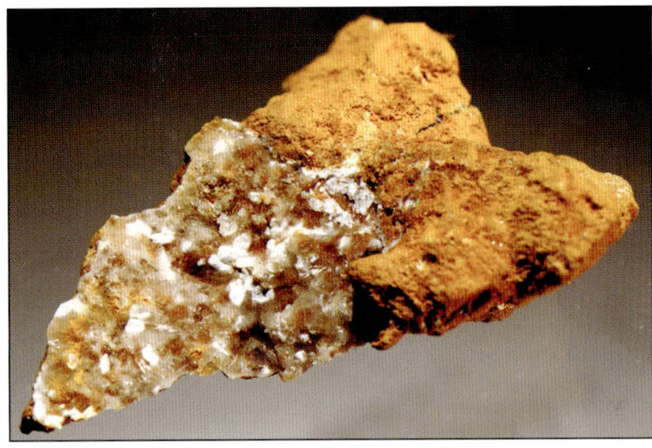

MINYULITE, CRANDALLITE. Minyulite in radiating crystals with white puffs of crandallite from Kapunda, South Australia. The crandallite fluoresces green SW and the minyulite fluoresces pale tan LW and pale blue SW. The piece weighs 1.0 oz. and is 2.3 x 1.5 x 1.0 inches. Value $40-50

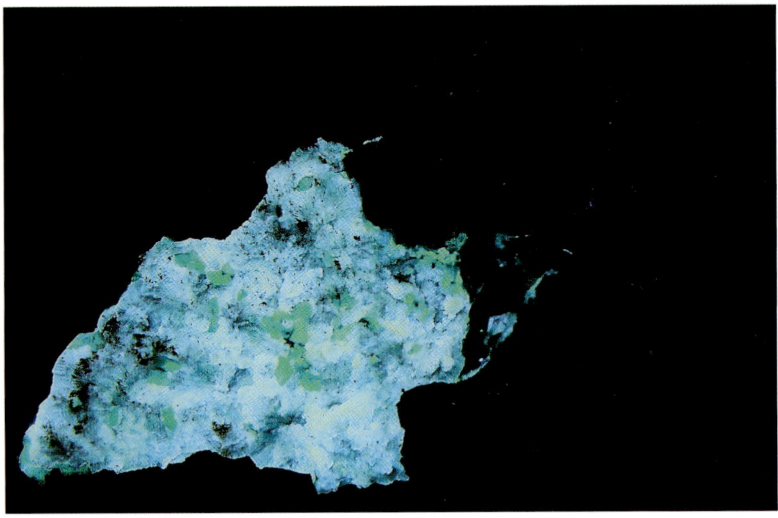

Same under SW

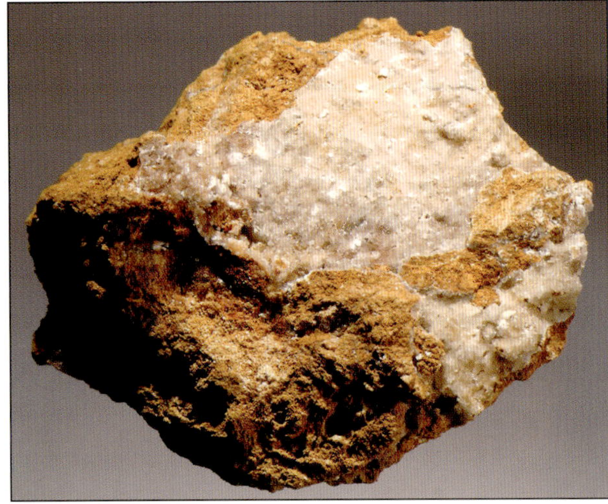

MINYULITE, CRANDALLITE. Small radiating minyulite crystals with white puffs of crandallite from Kapunda, South Australia. The minyulite fluoresces pale tan LW and pale blue SW and the crandallite fluoresces green SW. The piece weighs 7.0 oz. and is 3.3 x 3.0 x 2.0 inches. Value $70-80

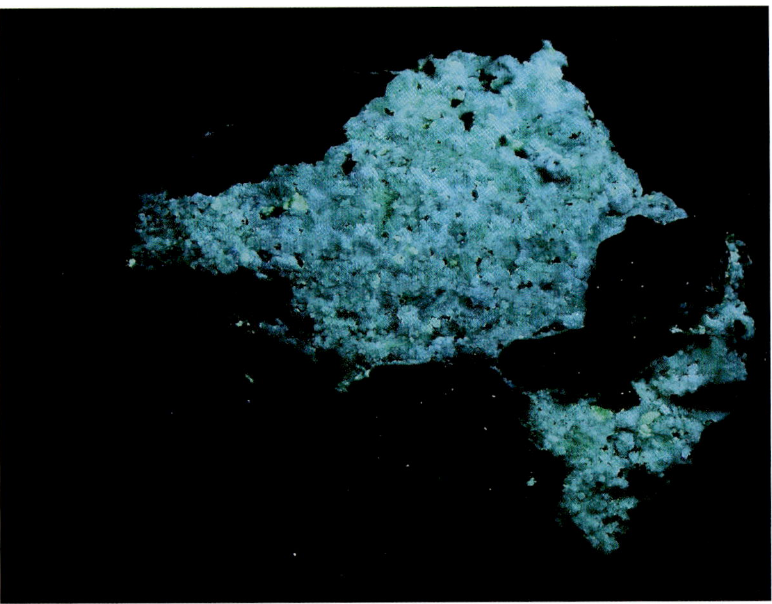

Same under SW

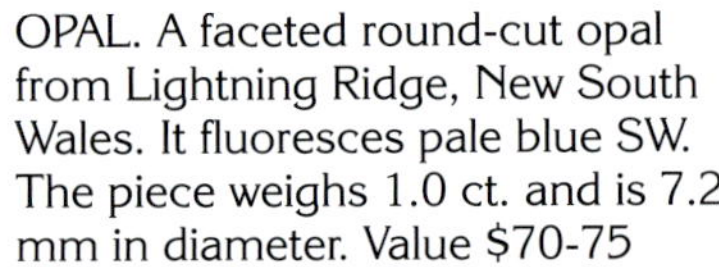

OPAL. A faceted round-cut opal from Lightning Ridge, New South Wales. It fluoresces pale blue SW. The piece weighs 1.0 ct. and is 7.2 mm in diameter. Value $70-75

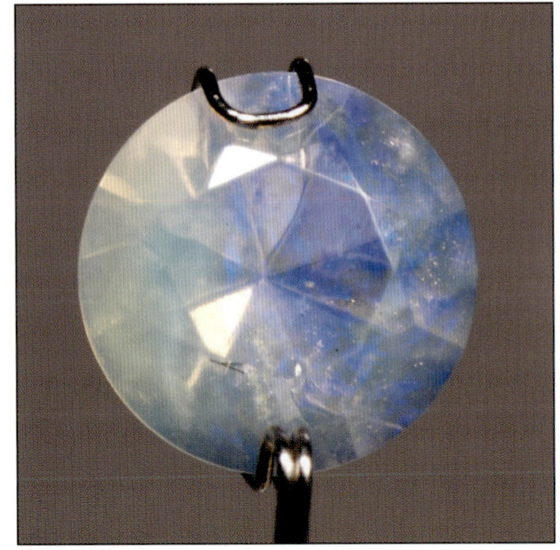

Same under SW

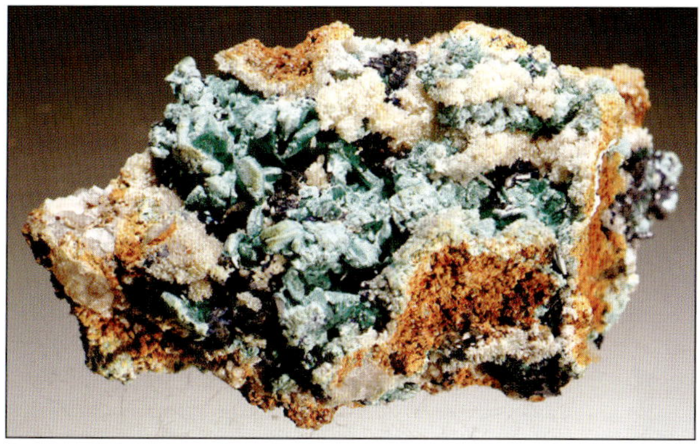

SMITHSONITE. A specimen containing smithsonite with azurite and malachite from Broken Hill, New South Wales. The smithsonite fluoresces hot pink SW and mustard yellow LW. There is an unidentified crystalline mineral that fluoresces pale green SW. The piece weighs 3.8 oz. and is 2.8 x 1.5 x 1.5 inches. Value $50-60.

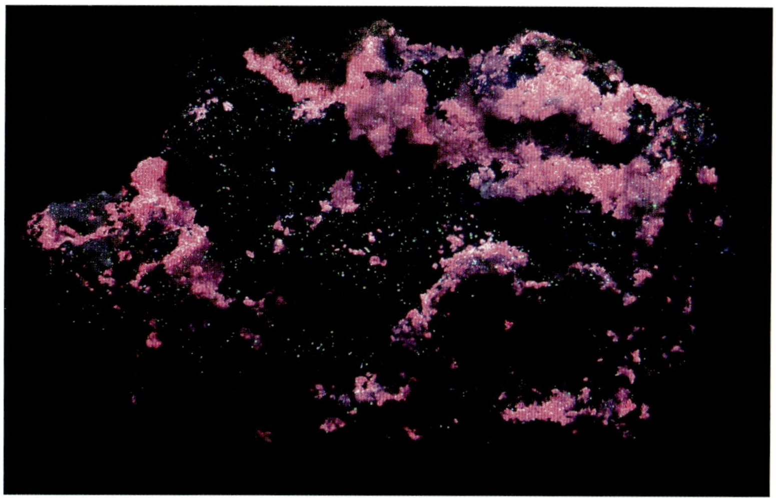

Same under SW

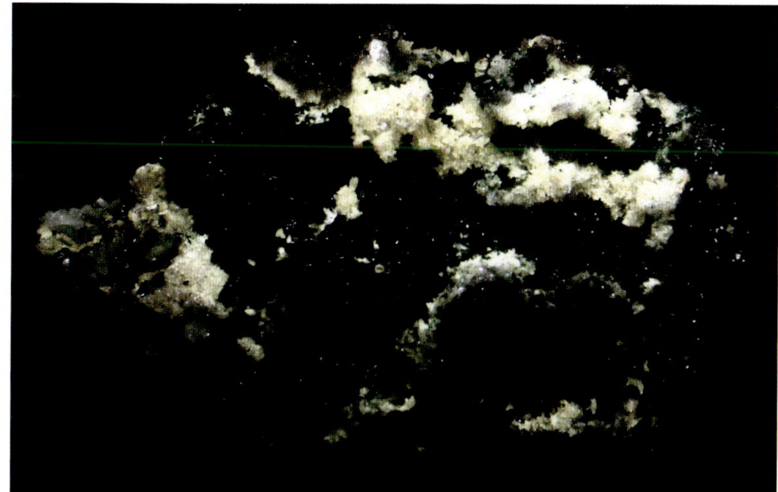

Same under LW

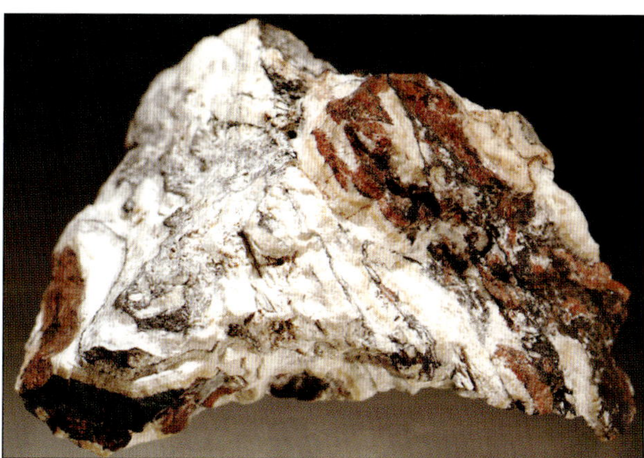

WILLEMITE. A folded hematitic ore containing willemite from the Puttapa deposit, Flinders Ranges, South Australia. The willemite fluoresces yellow and white SW. The piece weighs 5.5 oz. and is 2.8 x 2.3 x 1.3 inches. Value $30-40

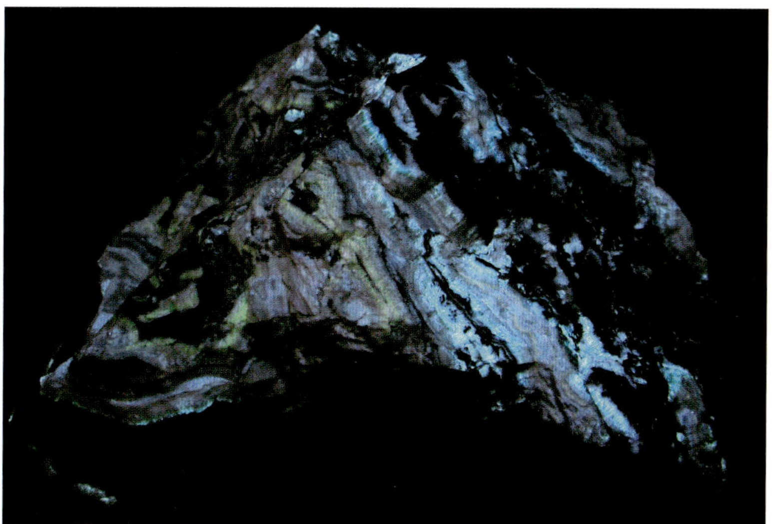

Same under SW

WILLEMITE. Willemite from the Puttapa deposit, Flinders Ranges, South Australia. The willemite fluoresces yellow and white SW. The piece weighs 5.5 oz. and is 2.3 x 2.3 x 1.5 inches. Value $30-40

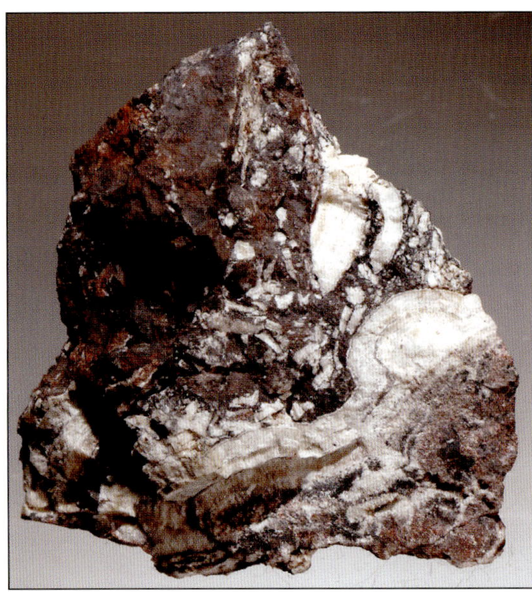

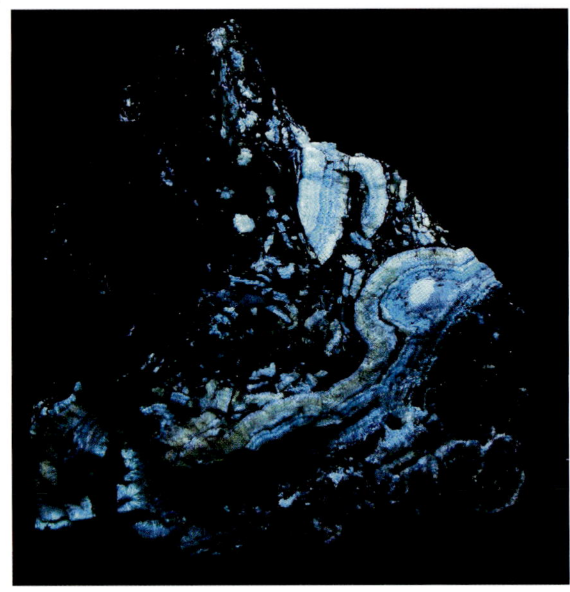

Same under SW

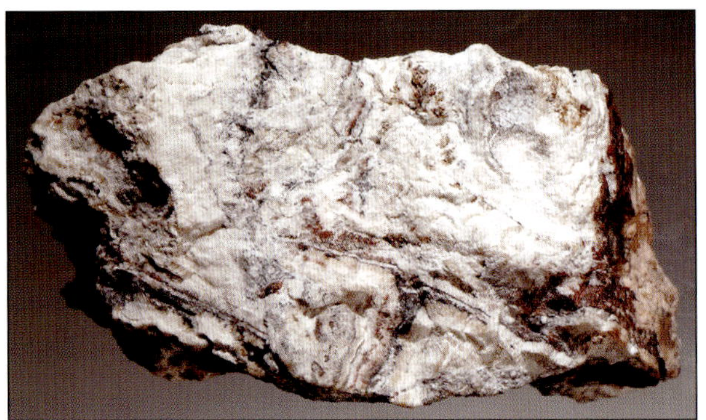

WILLEMITE. Willemite from the Puttapa deposit, Flinders Ranges, South Australia. The willemite fluoresces yellow and white SW. The piece weighs 5.0 oz. and is 2.8 x 1.5 x 1.3 inches. Value $30-40

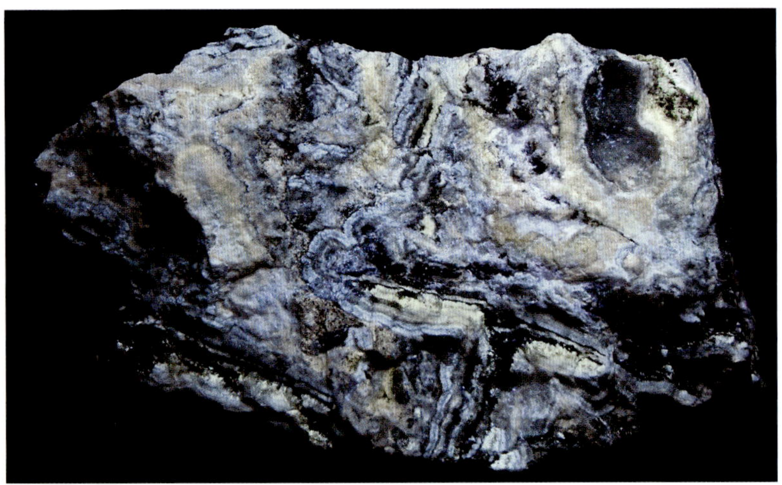

Same under SW

Bolivia

CREEDITE. Rare creedite crystals from Colquiri, Inquisivi Province, La Paz Department. The creedite fluoresces pale blue SW. The piece weighs 0.3 oz. and is 1.0 x 1.0 inches. Value $20-22

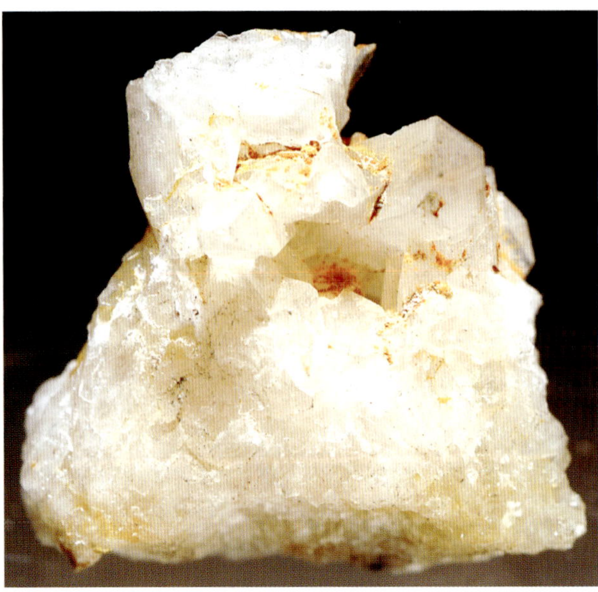

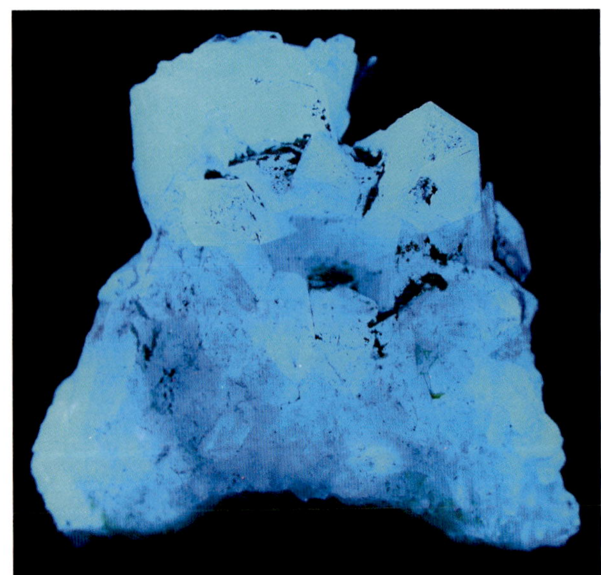

Same under SW

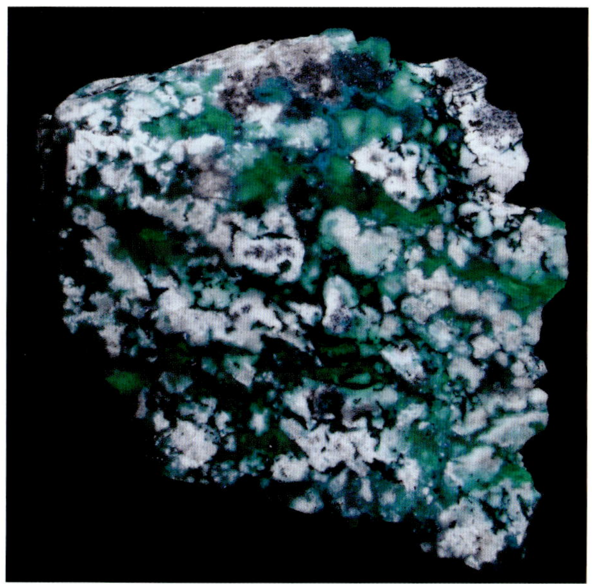

Same under SW

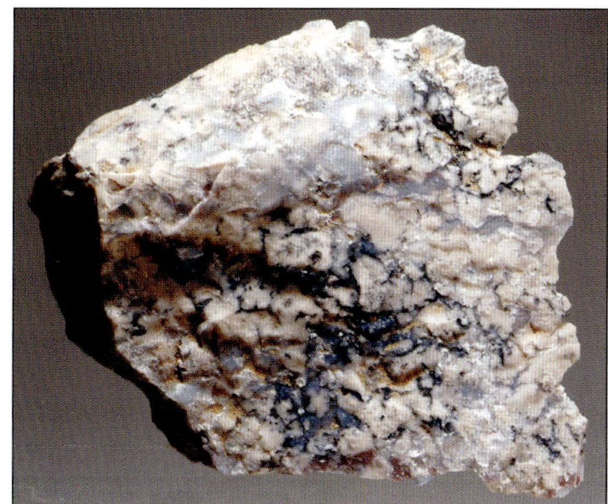

DOLOMITE, QUARTZ. Dolomite with quartz, var. chalcedony, from Cliza, Cochabamba Department. The dolomite fluoresces white SW and LW and the chalcedony fluoresces green SW. The piece weighs 1.8 oz. and is 2.8 x 2.3 x 0.5 inches. Value $35-40

Brazil

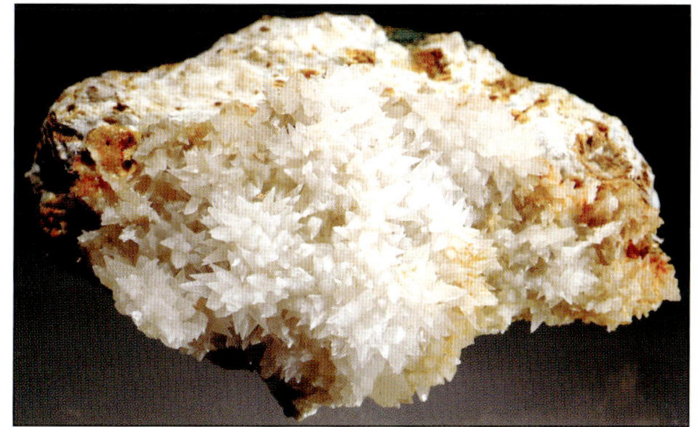

CALCITE. Calcite crystals on matrix from Rio Grande do Sul, South Region. The calcite fluoresces bright pink SW. The piece weighs 5.3 oz. and is 3.5 x 2.5 x 1.3 inches. Value $45-50

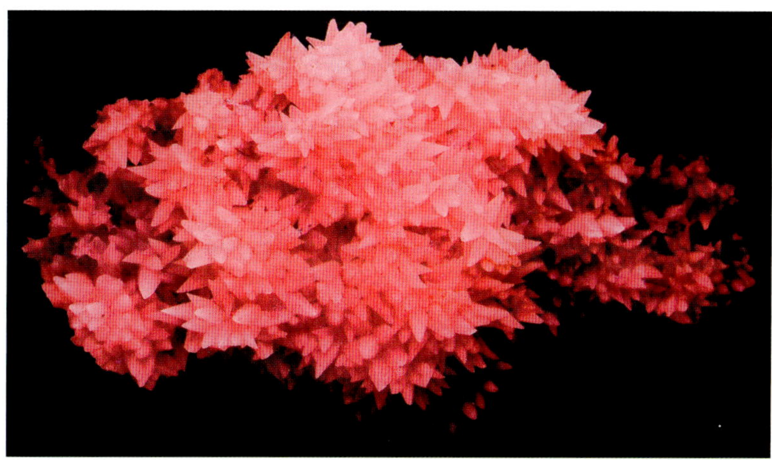

Same under SW

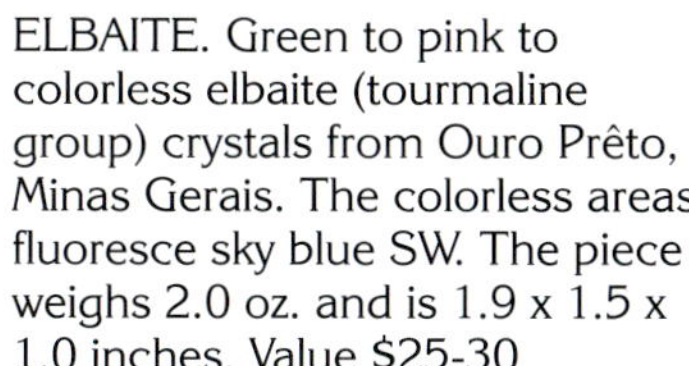

ELBAITE. Green to pink to colorless elbaite (tourmaline group) crystals from Ouro Prêto, Minas Gerais. The colorless areas fluoresce sky blue SW. The piece weighs 2.0 oz. and is 1.9 x 1.5 x 1.0 inches. Value $25-30

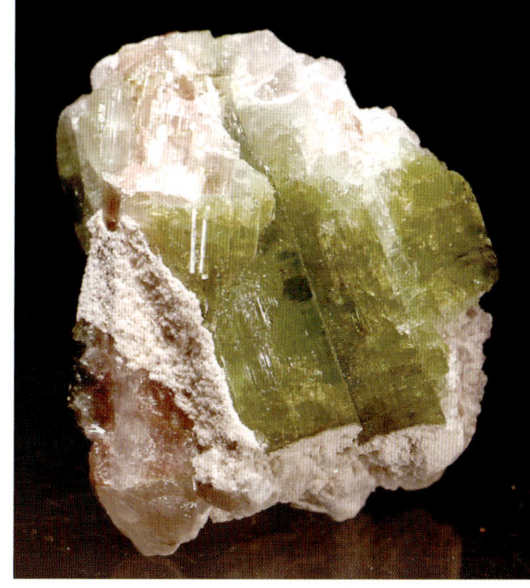

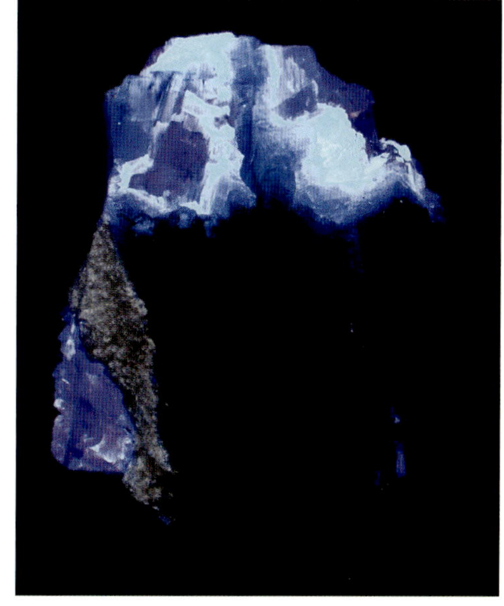

Same under SW

ELBAITE. Green and pink elbaite crystals on a tan matrix with aegirine crystals from Ouro Prêto, Minas Gerais. The pink area fluoresces sky blue SW. The piece weighs 4.0 oz. and is 2.3 x 2.3 x 1.5 inches. Value $45-50

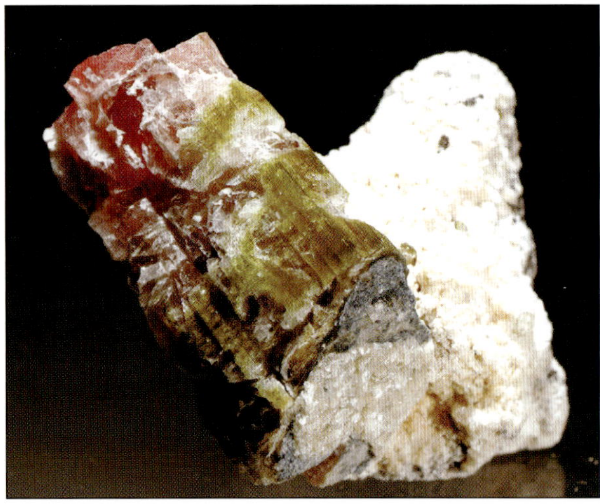

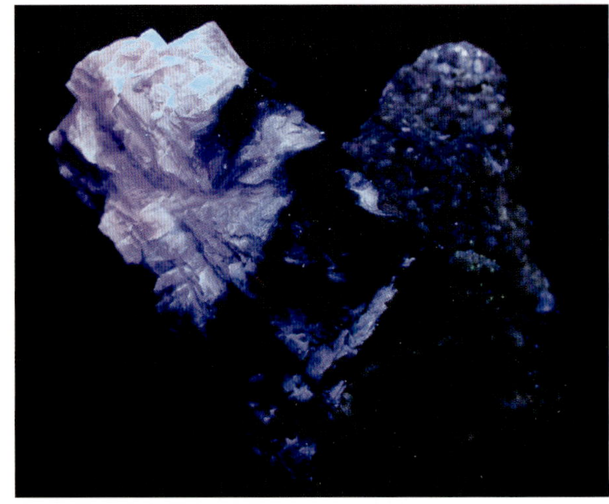

Same under SW

FLUORAPATITE. Fluorapatite crystals on albite crystals from Medina, Minas Gerais. The fluorapatite fluoresces yellow SW and the albite fluoresces purple-red SW. The piece weighs 1.8 oz. and is 1.9 x 1.9 x 1.3 inches. Value $50-55

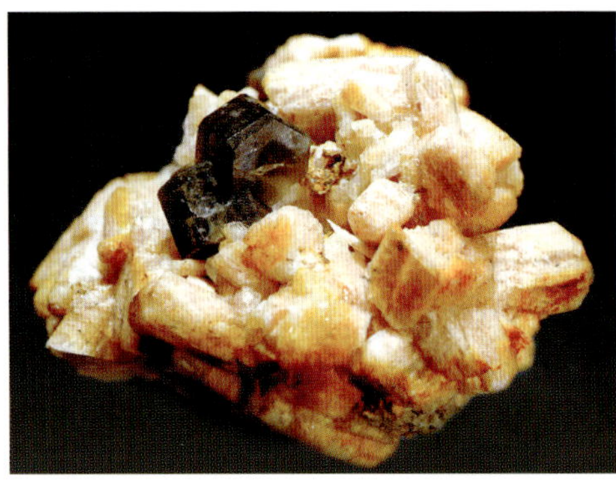

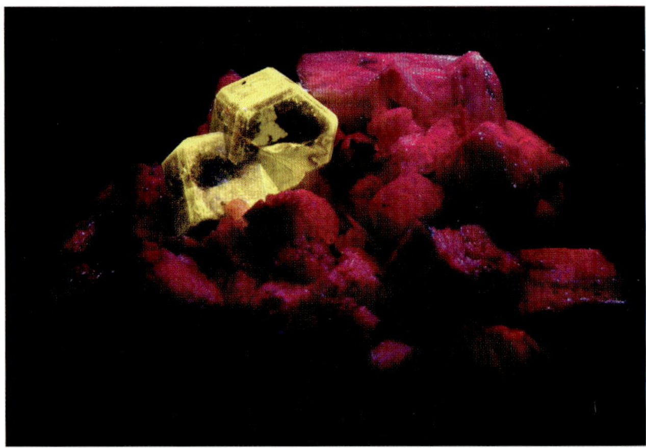

Same under SW

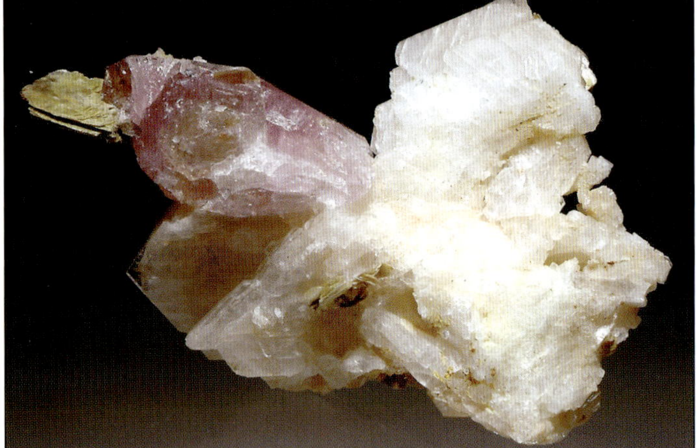

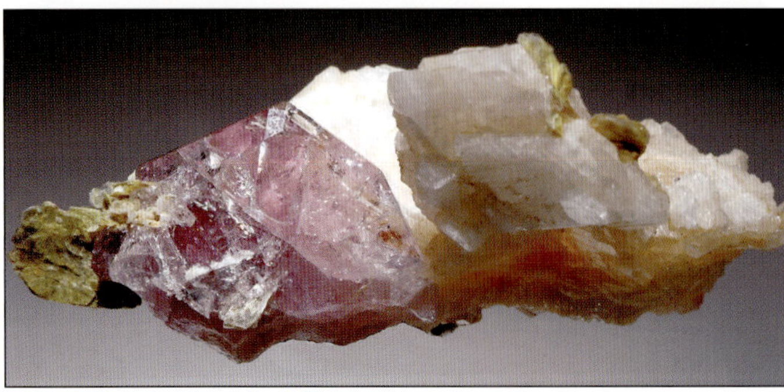

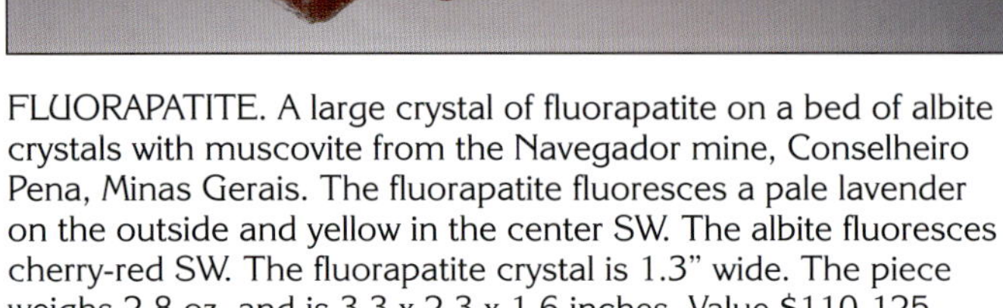

FLUORAPATITE. A large crystal of fluorapatite on a bed of albite crystals with muscovite from the Navegador mine, Conselheiro Pena, Minas Gerais. The fluorapatite fluoresces a pale lavender on the outside and yellow in the center SW. The albite fluoresces cherry-red SW. The fluorapatite crystal is 1.3" wide. The piece weighs 2.8 oz. and is 3.3 x 2.3 x 1.6 inches. Value $110-125

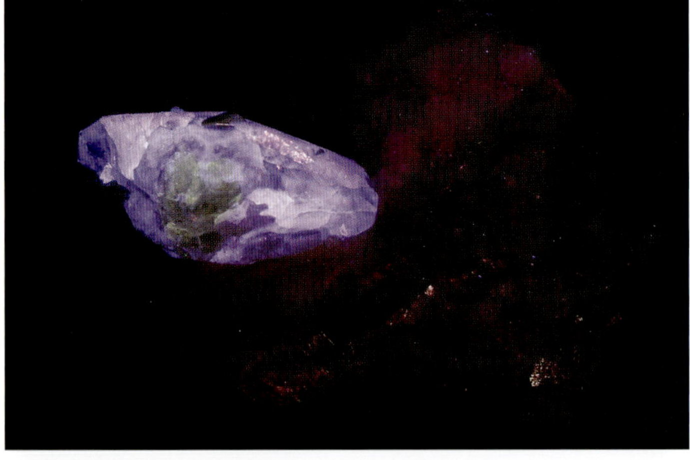

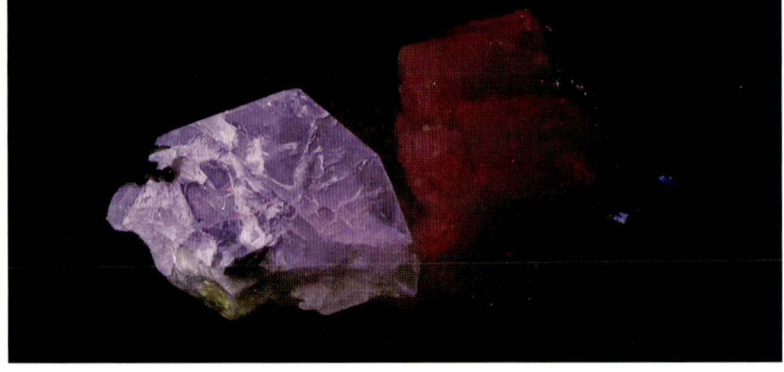

Same under SW

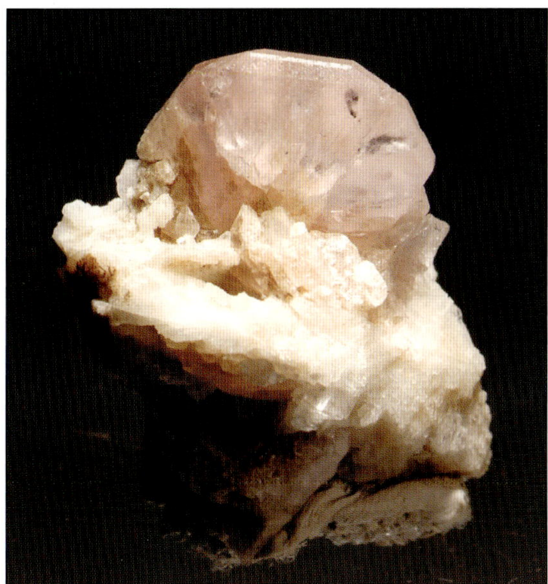

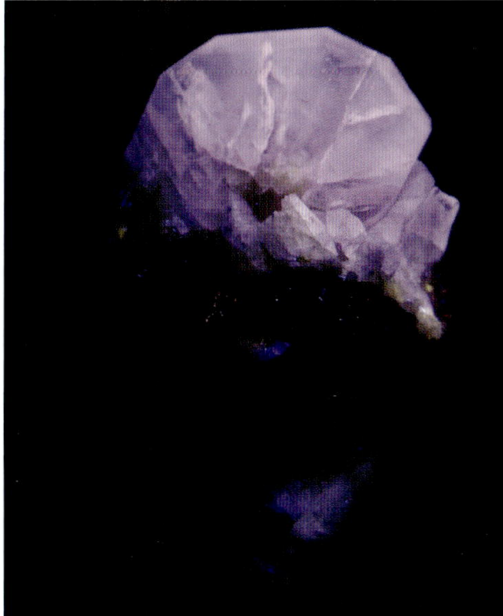

Far left:
FLUORAPATITE. A fluorapatite crystal from the Navegador mine, Conselheiro Pena, Minas Gerais. The fluorapatite fluoresces pale lavender SW. The piece weighs 0.4 oz. and is 1.1 x 0.9 x 0.9 inches. Value $40-50

Left:
Same under SW

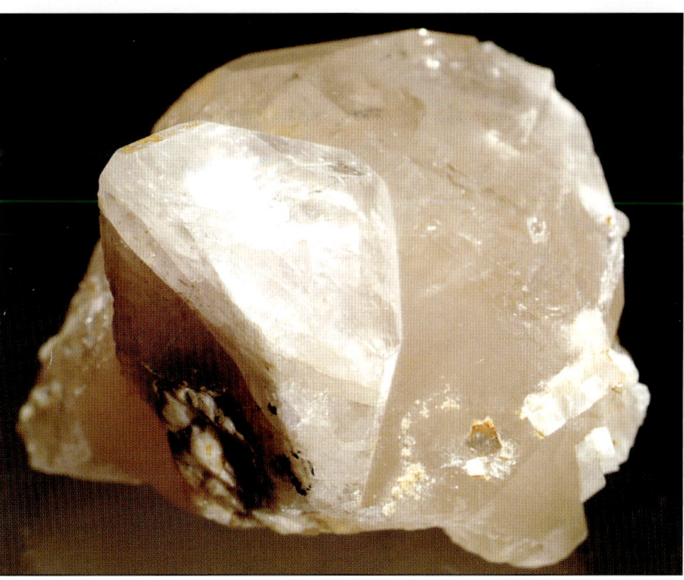

FLUORAPATITE. A crystal of fluorapatite on quartz, var. citrine, from Governador Valadares, Minas Gerais. The fluorapatite fluoresces pale lavender on the outside and yellow on the inside SW. The piece weighs 3.0 oz. and is 2.0 x 1.8 x 1.6 inches. Value $50-55

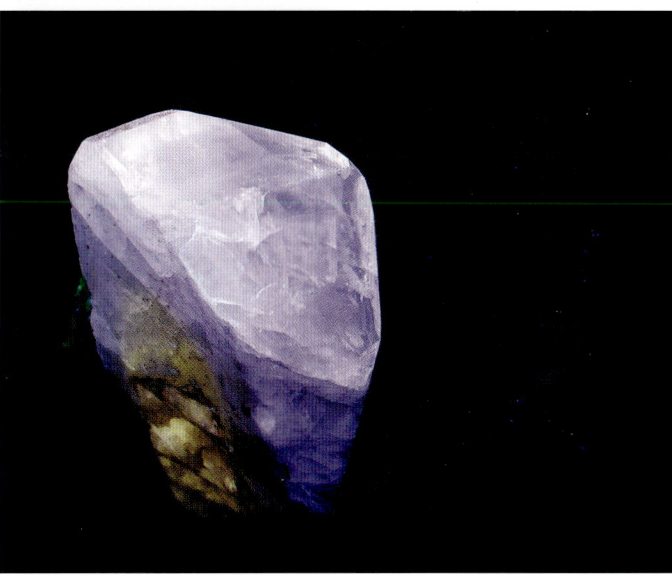

Same under SW

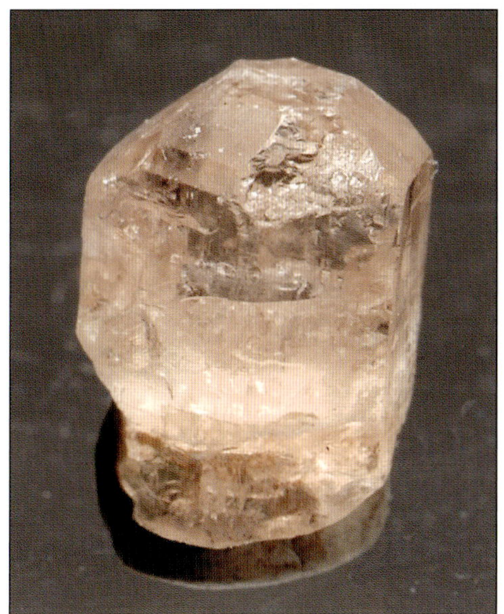

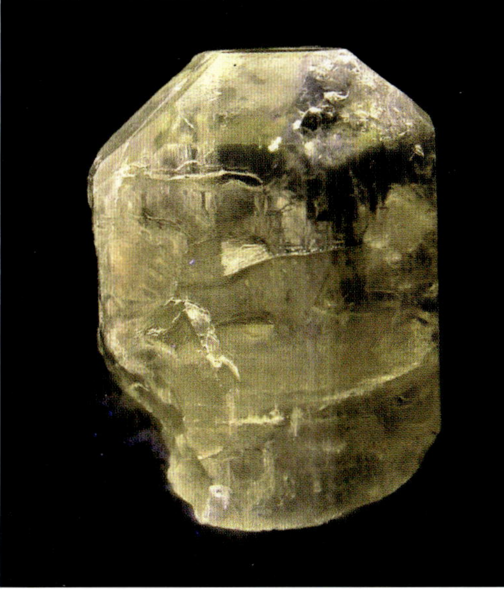

Far left:
FLUORAPATITE. A fluorapatite crystal from the Olhos D'Agua mine, Jequitinhonha River, Taquaral, Itinga, Minas Gerais. The fluorapatite fluoresces pale yellow SW. The piece weighs 0.1 oz. and is 0.5 x 0.3 x 0.3 inches. Value $12-15

Left:
Same under SW

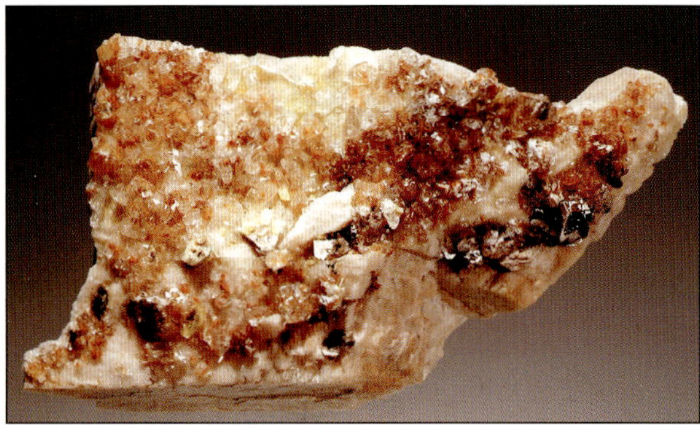

HAIWEEITE. Rare yellow crystals of haiweeite with uranophane on albite matrix from Teofilo Otoni, Minas Gerais. Haiweeite is radioactive and fluoresces green SW. The piece weighs 4.3 oz. and is 3.5 x 1.8 x 1.0 inches. Courtesy of Jeanne's Rock Shop. Value $125-130

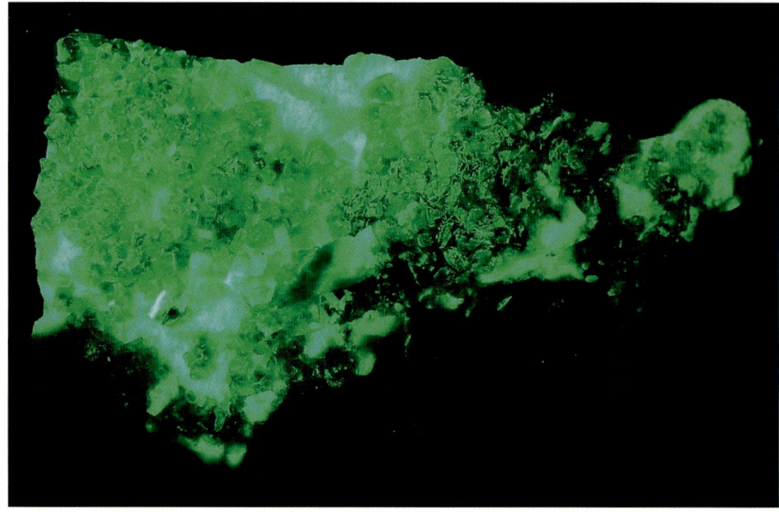

Same under SW

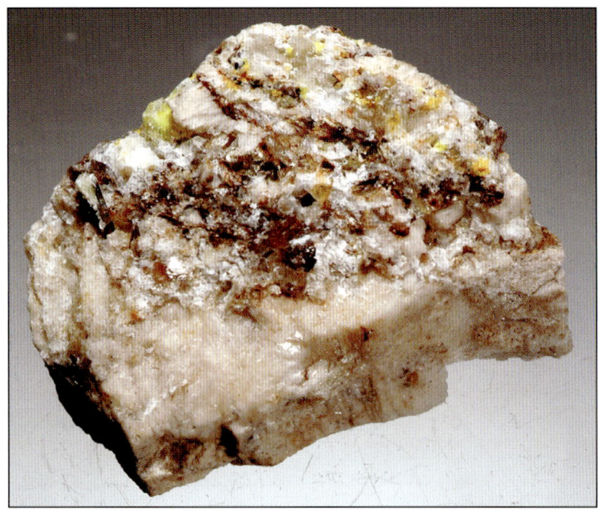

HAIWEEITE. Rare yellow crystals of haiweeite with uranophane on albite matrix from Teofilo Otoni, Minas Gerais. Haiweeite is radioactive and fluoresces green SW. The piece weighs 1.8 oz. and is 1.9 x 1.4 x 1.1 inches. Courtesy of Jeanne's Rock Shop. Value $30-35

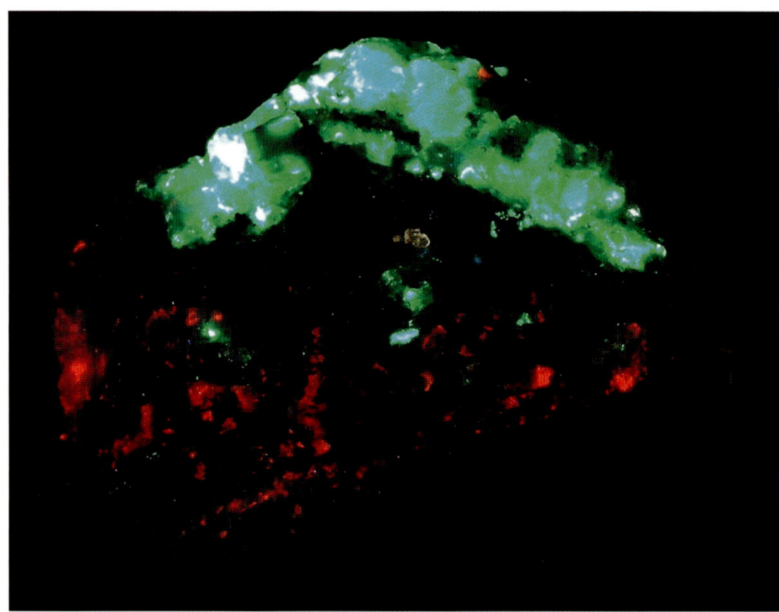

Same under SW

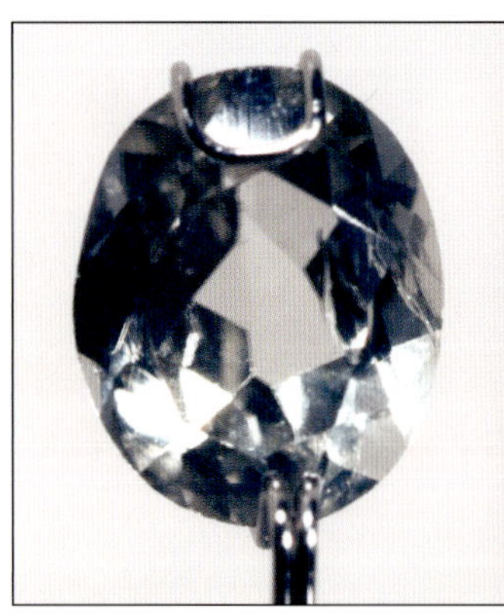

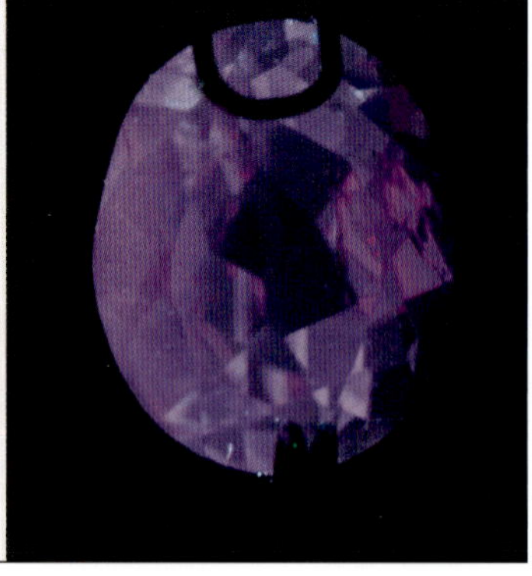

HERDERITE. Herderite from Minas Gerais that has been faceted as an oval cut gem. It fluoresces pale lavender SW. Hardness is 5.5. Herderite is named for Siegmund August Wolfgang von Herder who was a mining official in Germany. It weighs 0.8 ct. and is 6.0 x 5.0 mm. Value $35-45

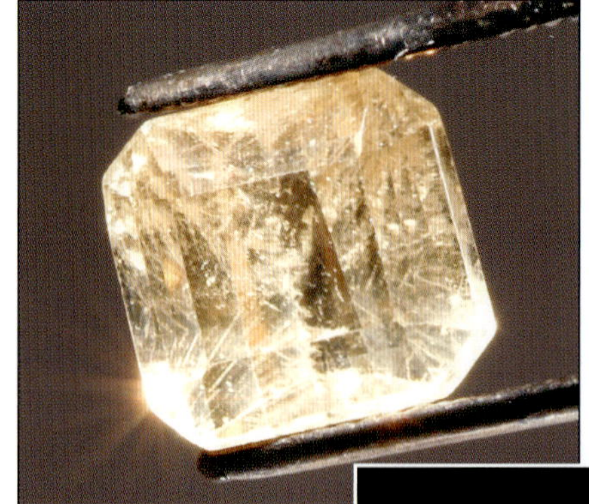

SCHEELITE. Scheelite from Currais Novos, Rio Grande do Norte that has been faceted as a modified emerald-cut gem. Transparent with slight inclusions. It fluoresces pale blue SW. It weighs 1.9 ct. and is 5.9 x 5.7 mm. Value $100-115

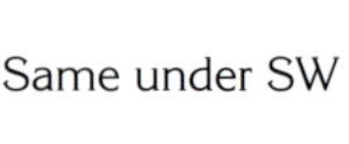

Same under SW

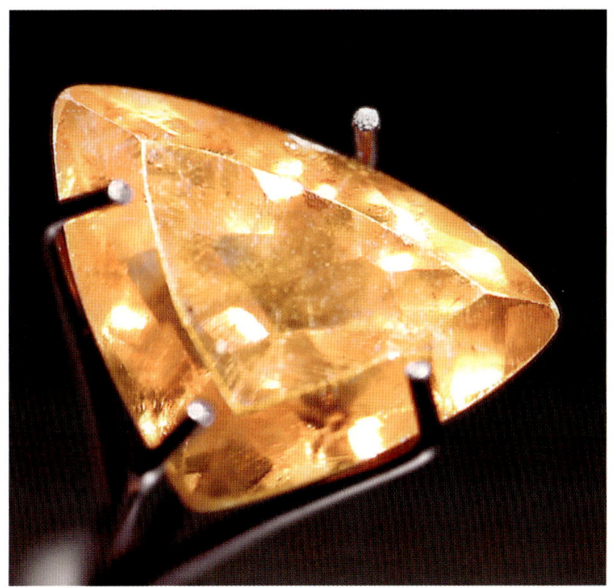

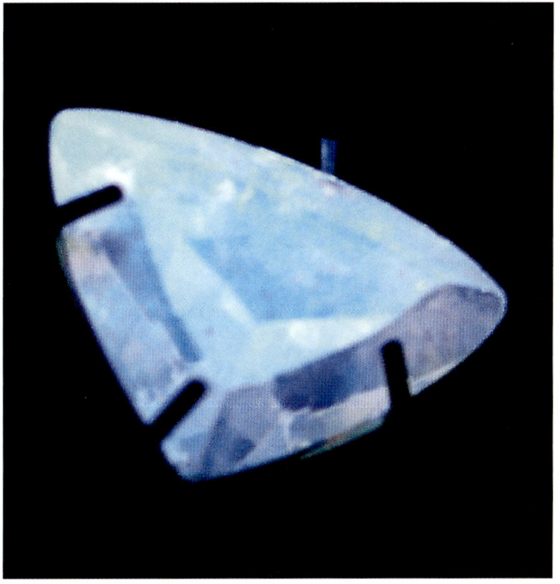

Far left:
SCHEELITE. Scheelite from Currais Novos, Rio Grande do Norte that has been faceted as a three-sided gem. It fluoresces pale blue SW. It weighs 2.4 ct. and is 8.5 x 6.5 mm. Value $150-175

Left:
Same under SW

Burma (Myanmar)

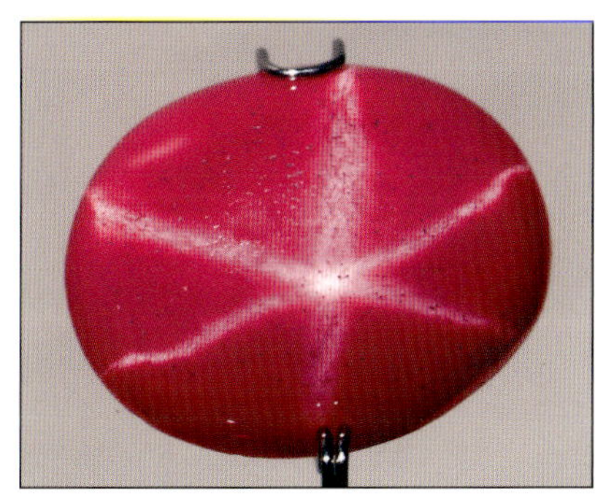

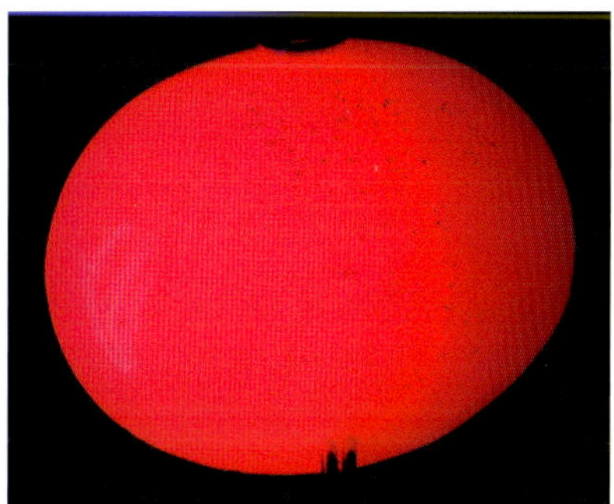

CORRUNDUM. Lab-made star ruby from Burma. It fluoresces red LW and SW. The piece weighs 3.6 ct. and is 11.0 x 9.0 mm. Value $30-35

Czech Republic

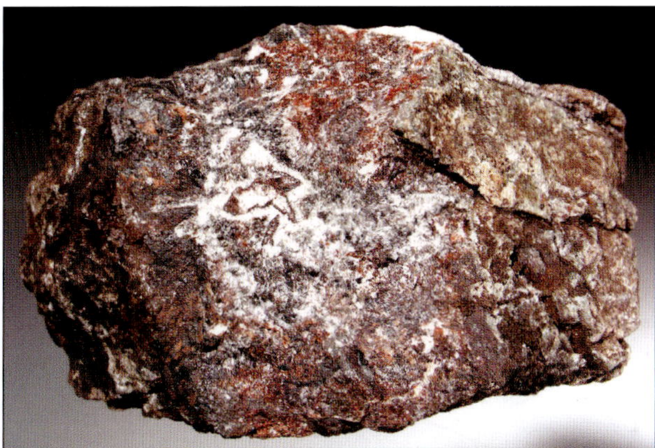

WILLEMITE. Willemite and calcite from Alexander Mine, Vrancice, Príbram, Bohemia. The willemite fluoresces green SW and the calcite fluoresces orange-red SW. The piece weighs 5.0 oz. and is 2.8 x 1.5 x 1.3 inches. Value $25-30.

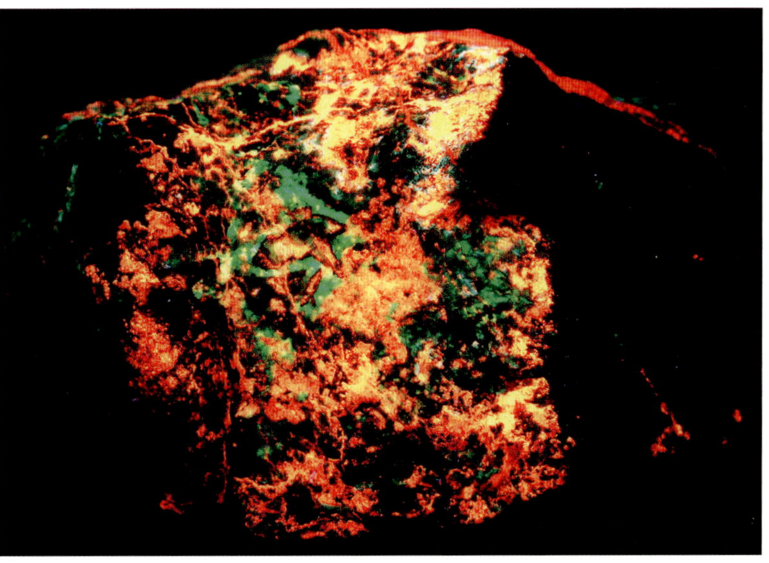

Same under SW

CARACOLITE, GLAUBERITE. Crystals of caracolite with glauberite from the Challacollo mine, Challacollo, Tarapacá Province. Caracolite fluoresces pale pink SW and the glauberite fluoresces blue-green SW. The piece weighs 3.5 oz. and is 2.0 x 2.0 x 1.5 inches. Value $40-45

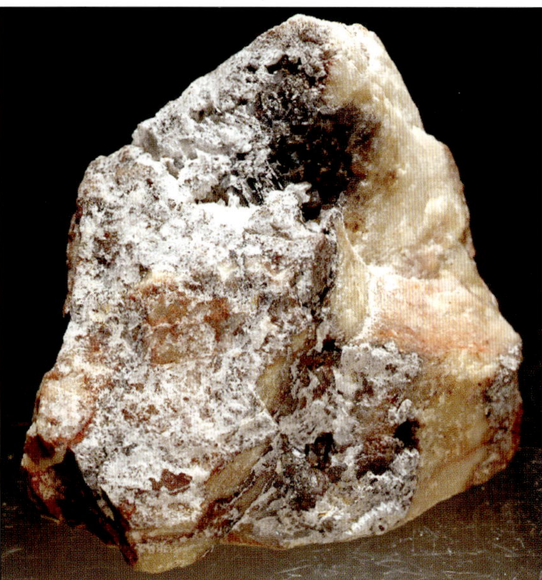

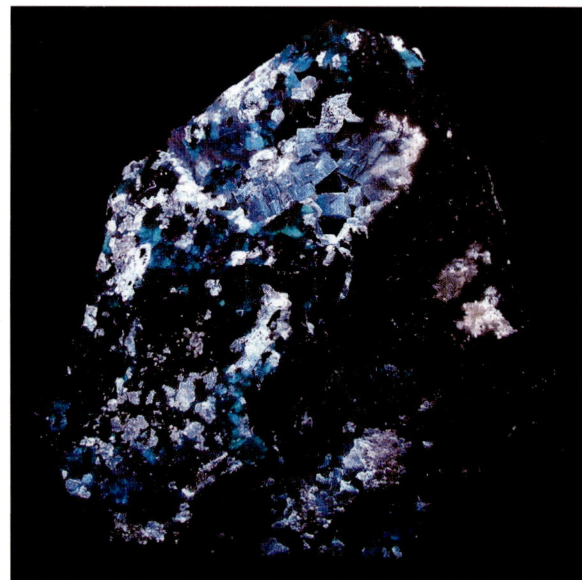

Same under SW

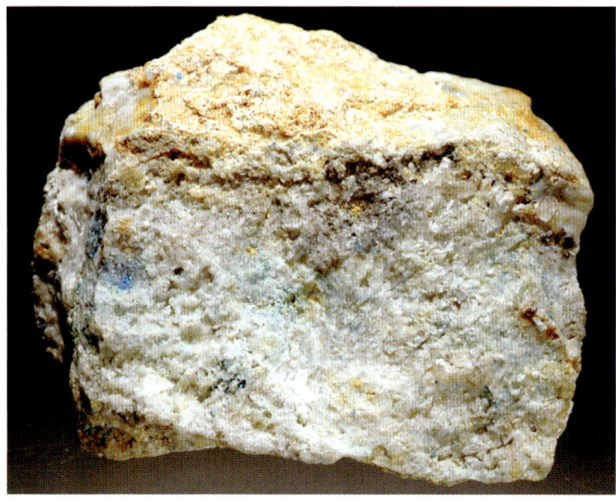

PENFIELDITE. Crystals of penfieldite with blue boleite from the Margarita mine, Sierra Gorda, Antofagasta Province. Penfieldite fluoresces mustard yellow SW. The piece weighs 5.5 oz. and is 2.8 x 2.3 x 1.1 inches. Value $50-65

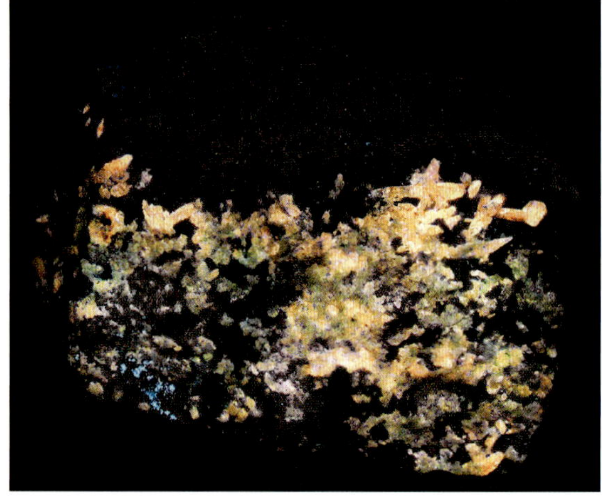

Same under SW

POWELLITE. Green crystals of cuprian powellite with crystals of molybdenite from the Jardinera #1 mine, Inca de Oro, Copiapó, Atacama Province. The powellite fluoresces yellow SW. The piece weighs 1 lb. 5.0 oz. and is 4.5 x 2.8 x 2.3 inches. Value $65-75

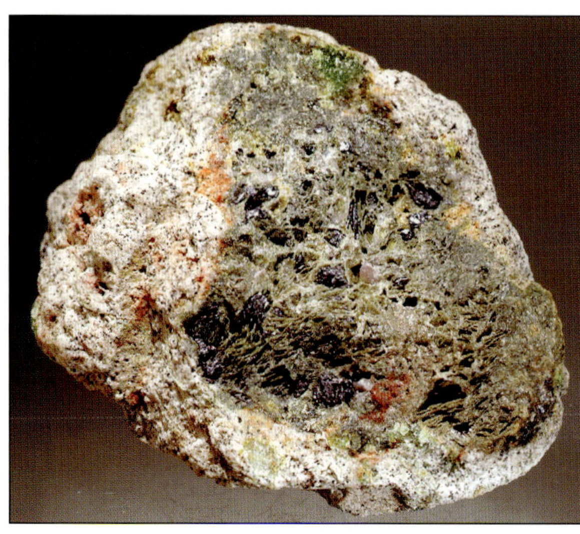

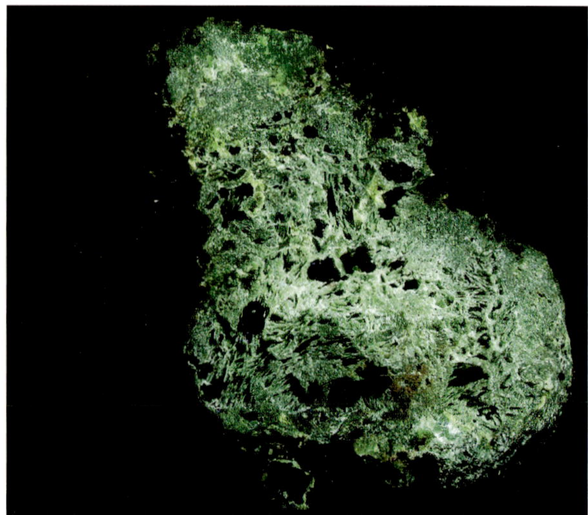

Same under SW

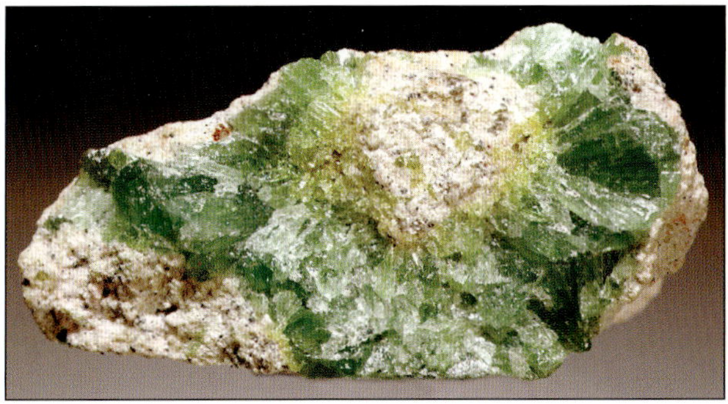

POWELLITE. Rich green crystalline mass of cuprian powellite from the Jardinera #1 mine, Inca de Oro, Copiapó, Atacama Province. The powellite fluoresces yellow SW. The piece weighs 1.5 oz. and is 2.0 x 0.9 x 0.9 inches. Value $45-50 (small but beautiful)

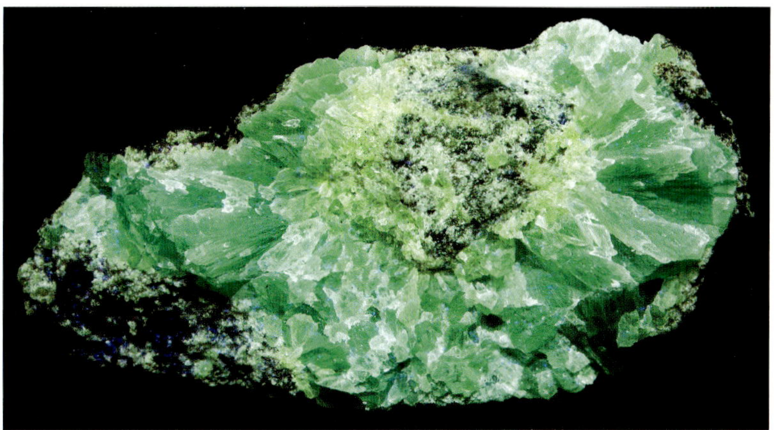

Same under SW

China

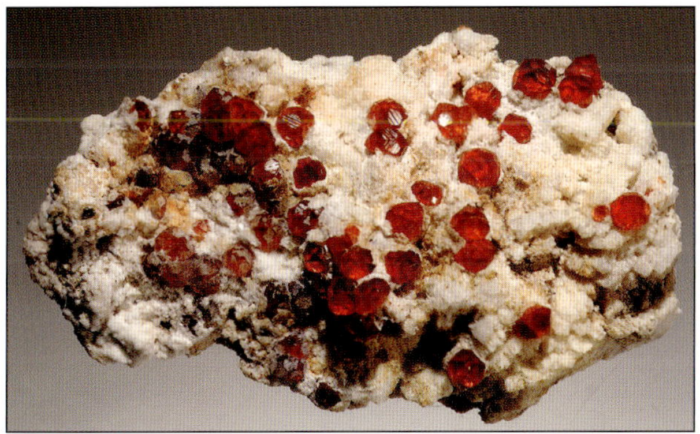

OPAL. A matrix of opal, var. hyalite, covered with brown-orange spessartine garnets from the Le Cheng mine, Guang Dong. The hyalite fluoresces bright green SW. The piece weighs 1.8 oz. and is 2.8 x 1.5 x 0.8 inches. Value $40-50.

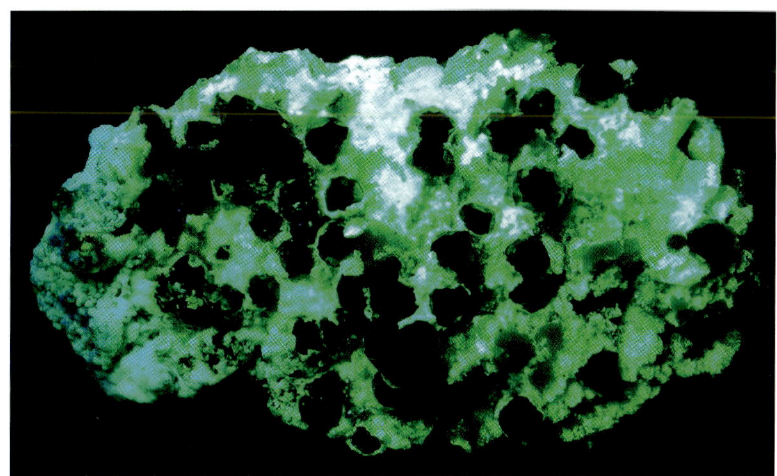

Same under SW

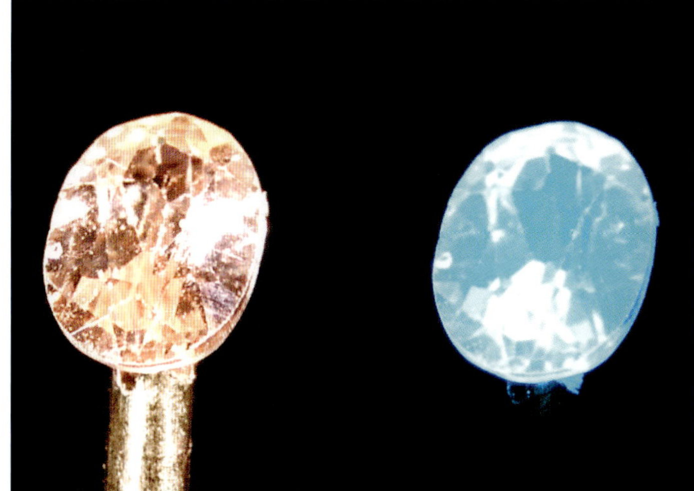

SCHEELITE. Scheelite faceted as an oval cut gem. Shown under daylight and SW. Scheelite fluoresces pale blue SW. It weighs 2.0 ct. and is 4.0 mm by 6.0 mm. Value $55-60

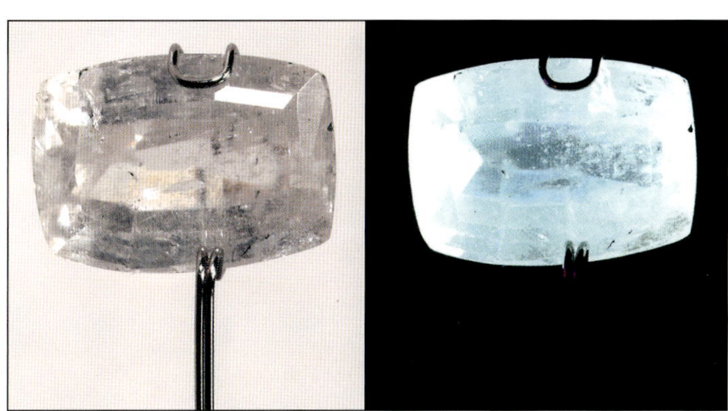

SCHEELITE. Scheelite faceted as an emerald cut gem. Hardness is 5.5. It fluoresces pale blue SW. It weighs 4.6 ct. and is 9.3 x 6.5 mm. Value $75-85

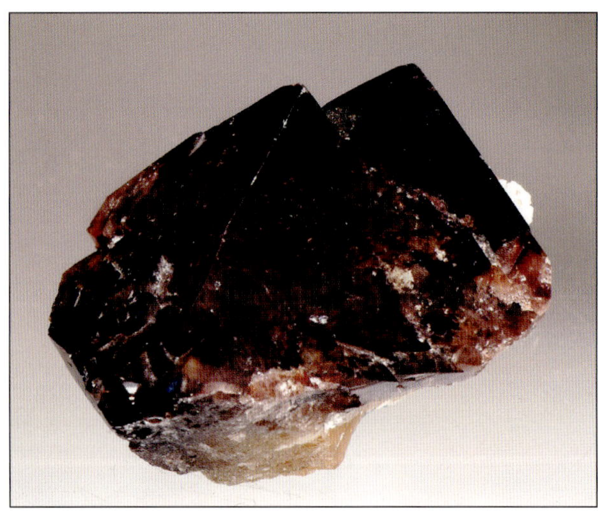

SCHEELITE. A very fine example of black scheelite from the Yaogangxian mine, Chenzhou, Hunan Province. The surface of the scheelite looks like obsidian. The scheelite fluoresces pale blue SW. The piece weighs 6.5 oz. and is 2.3 x 1.8 x 1.5 inches. Value $300-350

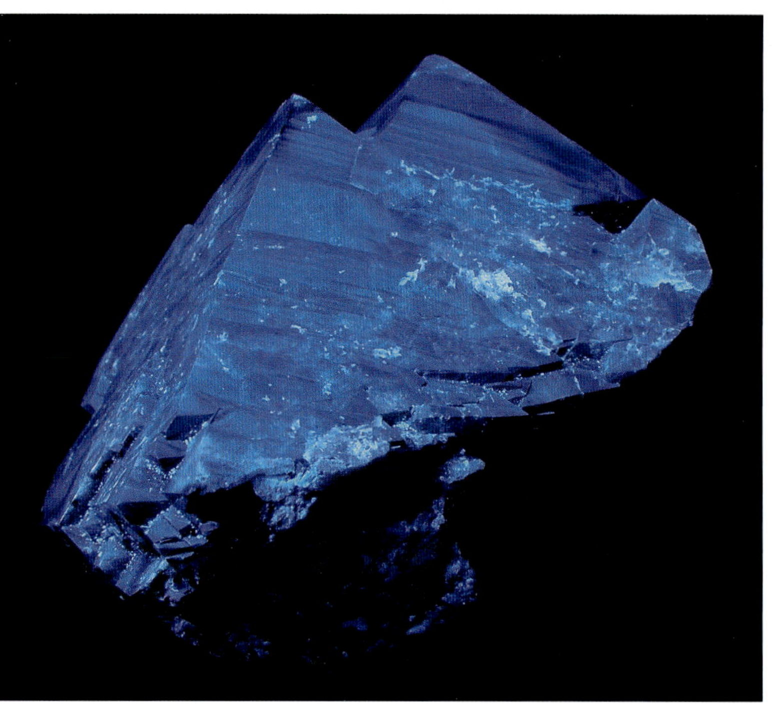

Same under SW

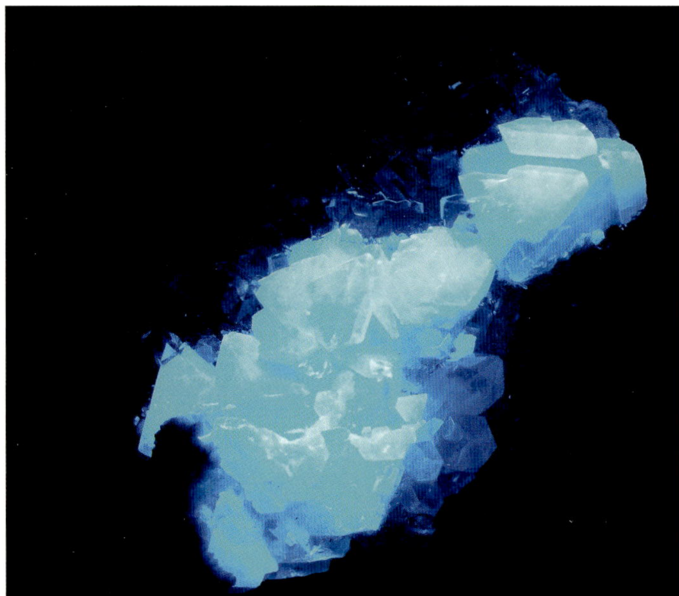

Same under SW

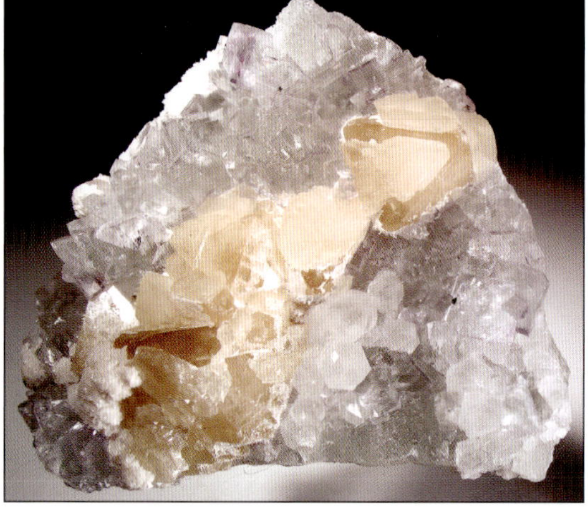

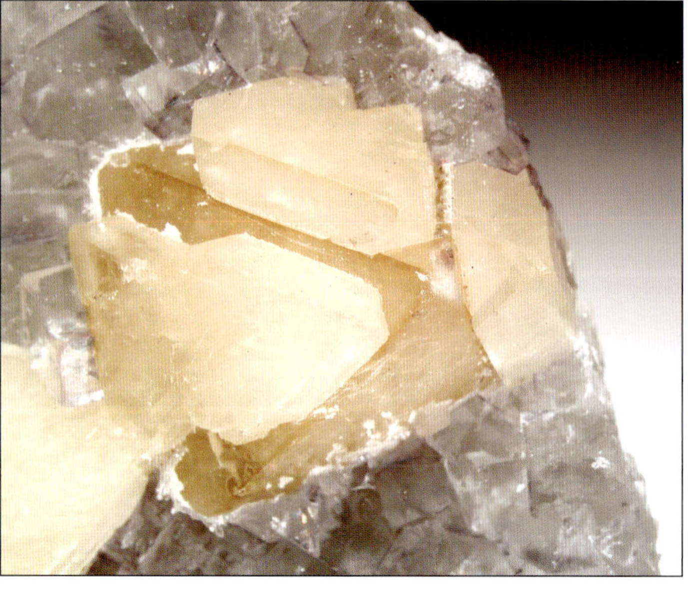

SCHEELITE. Scheelite crystals on fluorite with stilbite from the Shizhuyuan mine, Chenzhou, Hunan Province. The closeup shows crystals growing on crystals. Scheelite fluoresces pale blue SW and the fluorite fluoresces violet LW. The piece weighs 10.0 oz. and is 3.8 x 3.0 x 1.0 inches. Value $100-135

SCHEELITE. Amber scheelite crystals with pale blue beryl crystals on muscovite from the Xue Bao Ding mine, Pingwu, Sichuan Province. Scheelite fluoresces pale blue SW. The piece weighs 2 lb. 8.0 oz. and is 6.0 x 5.0 x 2.5 inches. Value $175-200

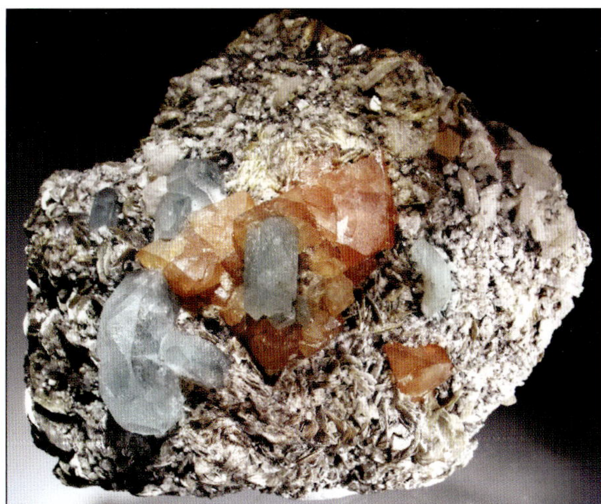

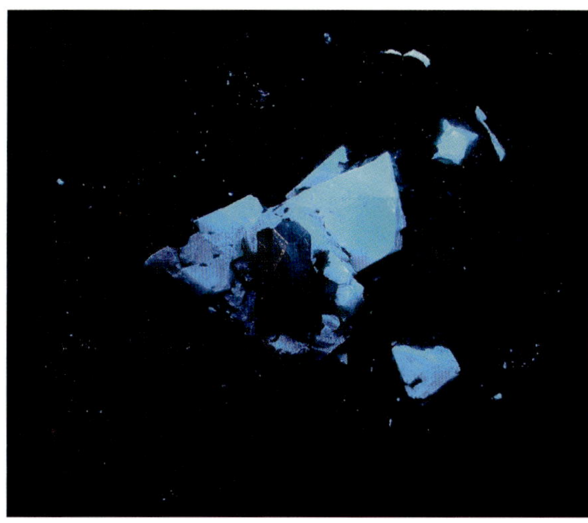

Same under SW

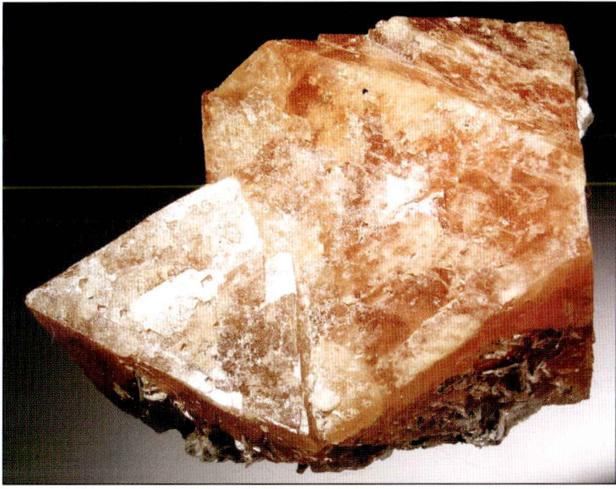

SCHEELITE. A pair of scheelite crystals from the Xue Bao Ding mine, Pingwu, Sichuan Province. Scheelite fluoresces pale blue SW. The piece weighs 10.0 oz. and is 2.5 x 1.8 x 1.5 inches. Value $95-100

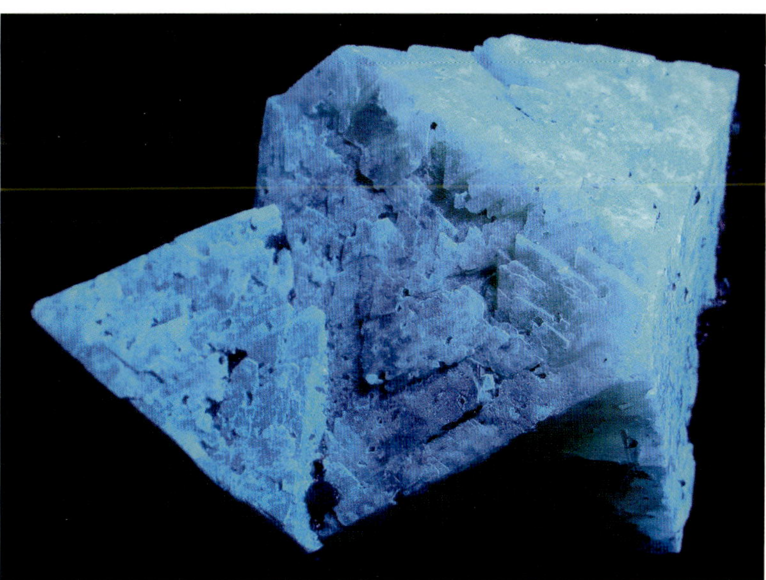

Same under SW

Greenland

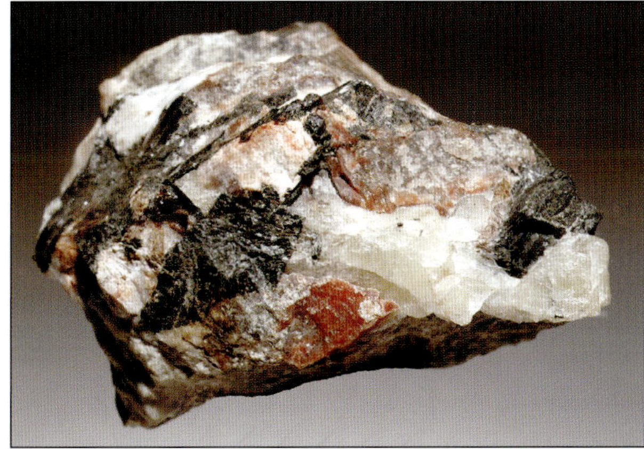

LEUCOPHANITE. Very rare leucophanite with albite and other unidentified minerals from Nakalaq, Ilimaussaq complex, Narsaq, Kitaa Province. Leucophanite fluoresces lavender-pink LW and orange-pink SW, the albite fluoresces cherry red SW. The piece weighs 5.8 oz. and is 2.8 x 2.0 x 2.0 inches. Value $250-300

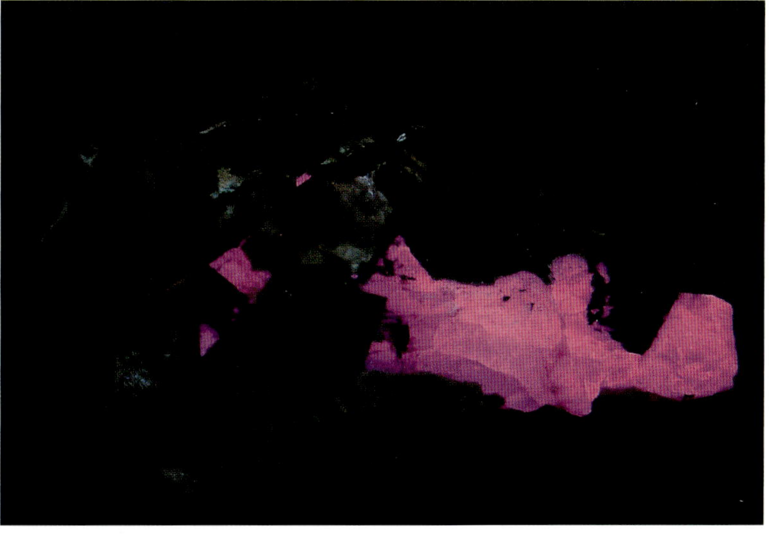

Same under LW

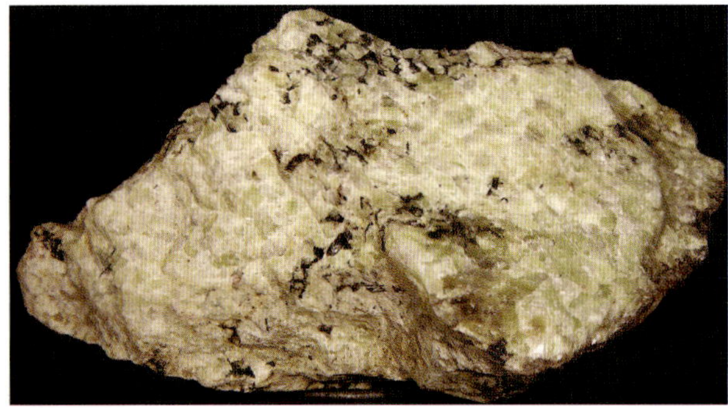

SODALITE. Pea-green sodalite from Tunuliarfik, Ilimaussaq complex, Narsaq, Kitaa Province. Sodalite fluoresces orange LW. The piece is 4.0 x 3.0 x 2.0 inches. Courtesy of Mark Cole and MinerShop.

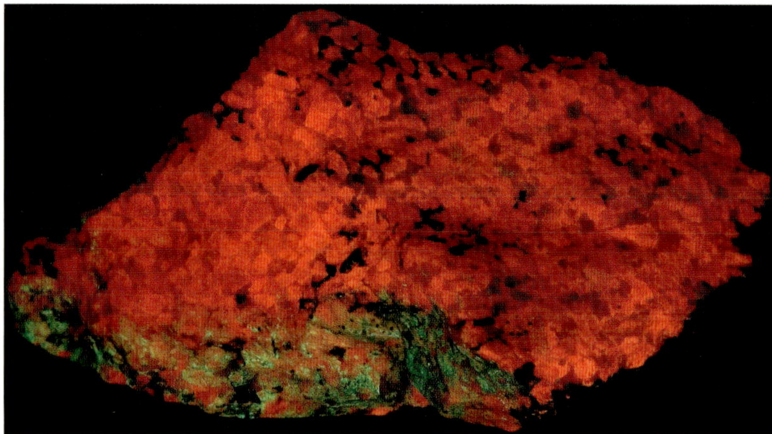

Same under LW

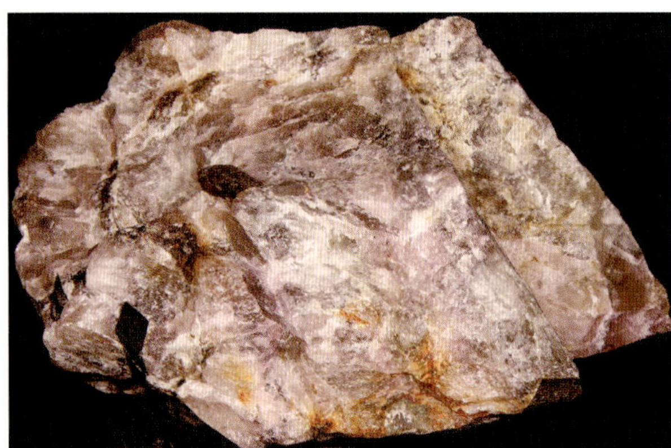

SODALITE. A very tenebrescent sodalite with arfvedsonite from Taseq slope, Ilimaussaq complex, Narsaq, Kitaa Province. Sodalite fluoresces orange LW. The piece is 3.5 x 3.0 x 2.0 inches. Courtesy of Mark Cole and MinerShop.

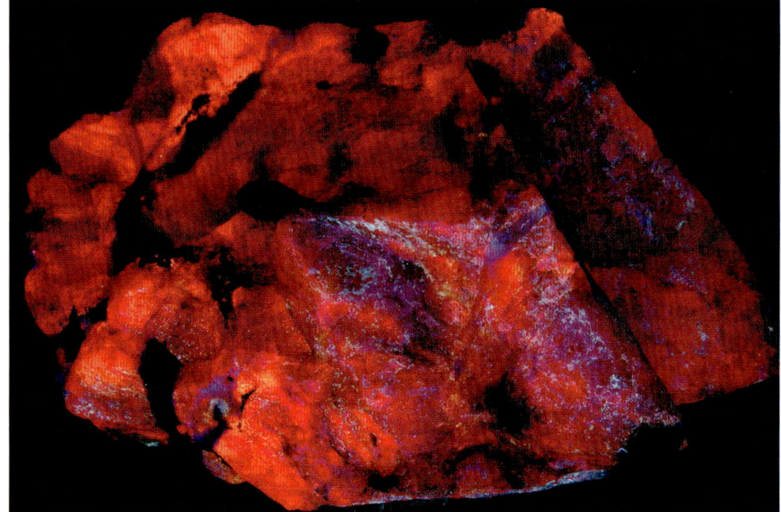

Same under LW

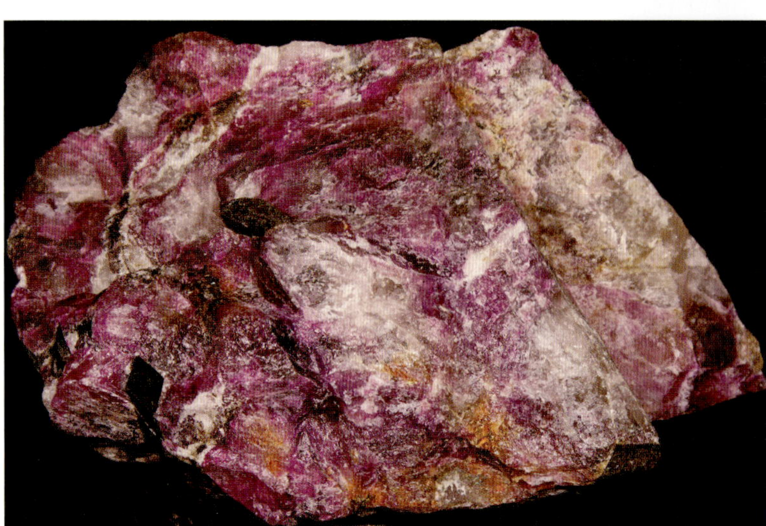

Same showing the tenebrescent effect.

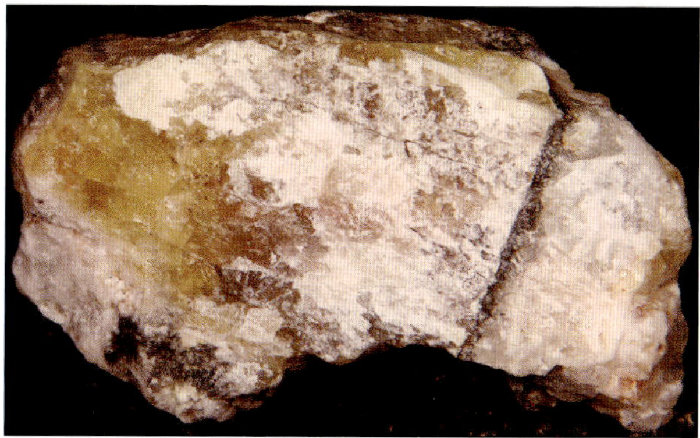

SODALITE. A very tenebrescent yellow sodalite from Taseq slopes, Ilimaussaq complex, Narsaq, Kitaa Province. Sodalite fluoresces orange LW. The piece is 4.0 x 2.5 x 2.0 inches. Courtesy of Mark Cole and MinerShop.

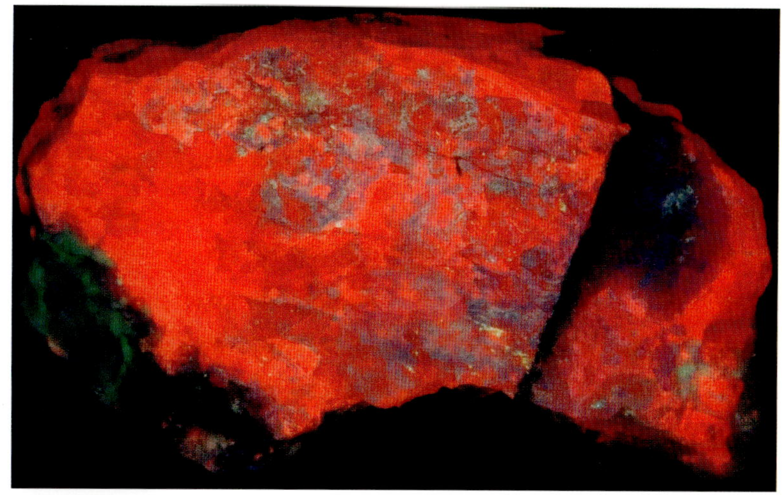

Same under combination of SW and LW

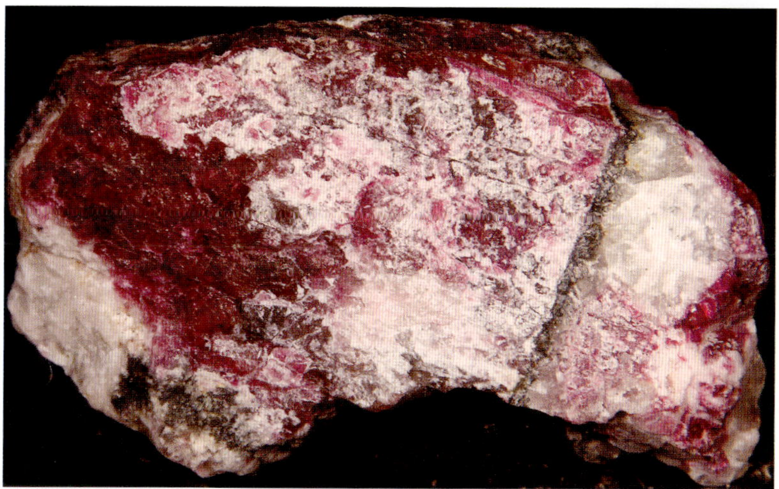

Same showing the tenebrescent effect.

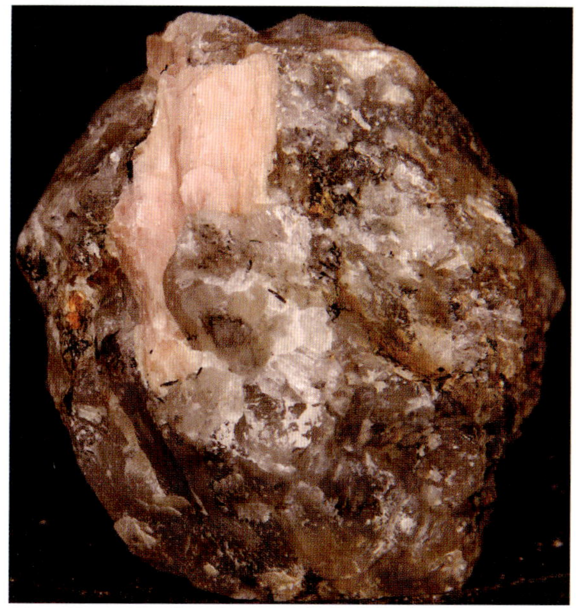

SORENSENITE. Rare sorensenite from Kvanefjeld, Ilimaussaq complex, Narsaq, Kitaa Province. Sorensenite fluoresces pale yellow SW. The piece is 4.0 x 4.0 x 3.0 inches. Courtesy of Mark Cole and MinerShop.

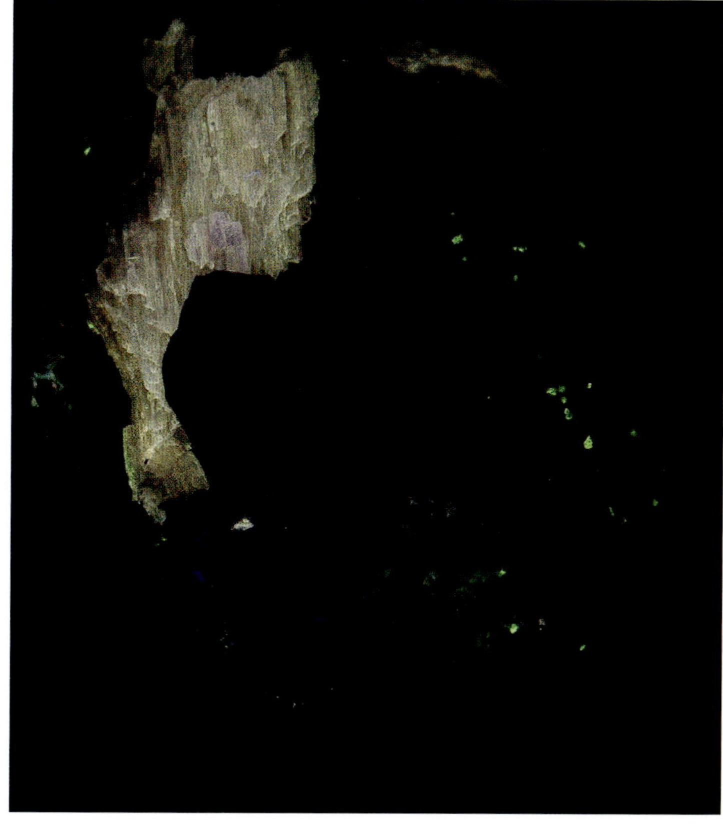

Same under SW

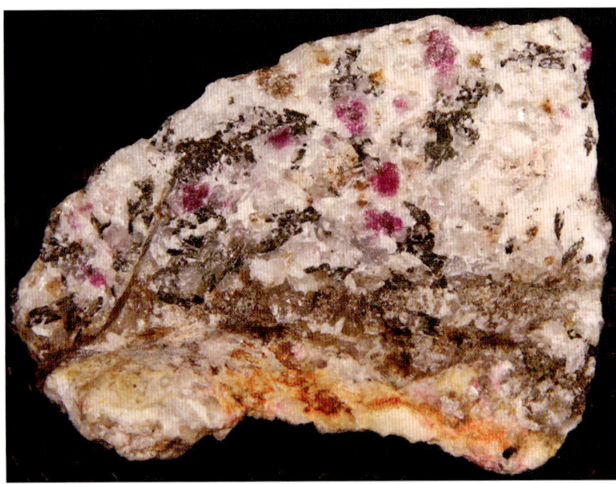

TUGTUPITE. A rare specimen with tugtupite, sodalite, and analcime from Taseq slope, Ilimaussaq complex, Narsaq, Kitaa Province. Sodalite fluoresces orange LW, tugtupite fluoresces red SW, and analcime fluoresces pale blue SW. The piece is 4.0 x 3.5 x 2.0 inches. Courtesy of Mark Cole and MinerShop.

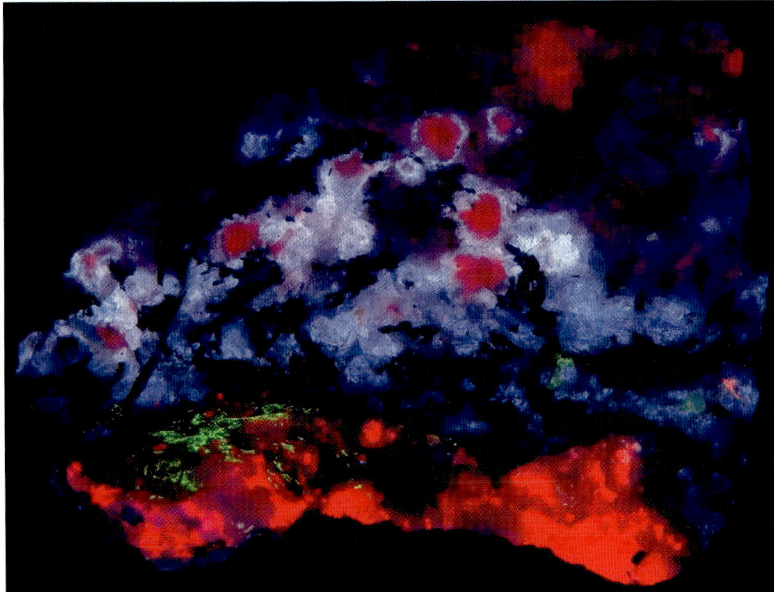

Same under a combination of SW and LW

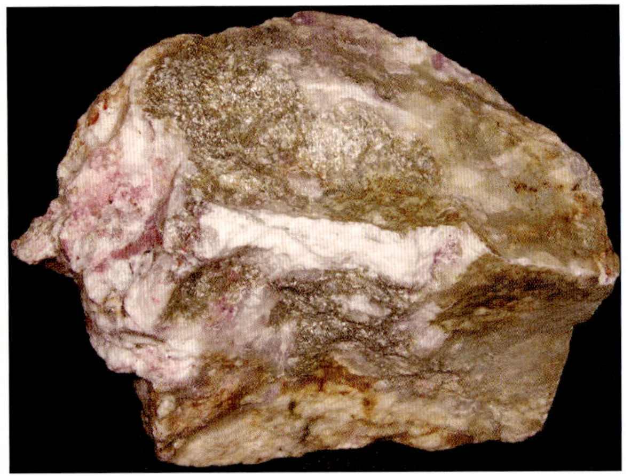

TUGTUPITE, etc. An attractive piece with tugtupite, chkalovite, natrolite, and polylithionite from Kangerlussaq, Ilimaussaq complex, Narsaq, Kitaa Province. Tugtupite fluoresces red SW, chkalovite fluoresces green SW, natrolite fluoresces green SW, and polylithionite fluoresces yellow SW. The piece is 4.0 x 3.5 x 2.0 inches. Courtesy of Mark Cole and MinerShop.

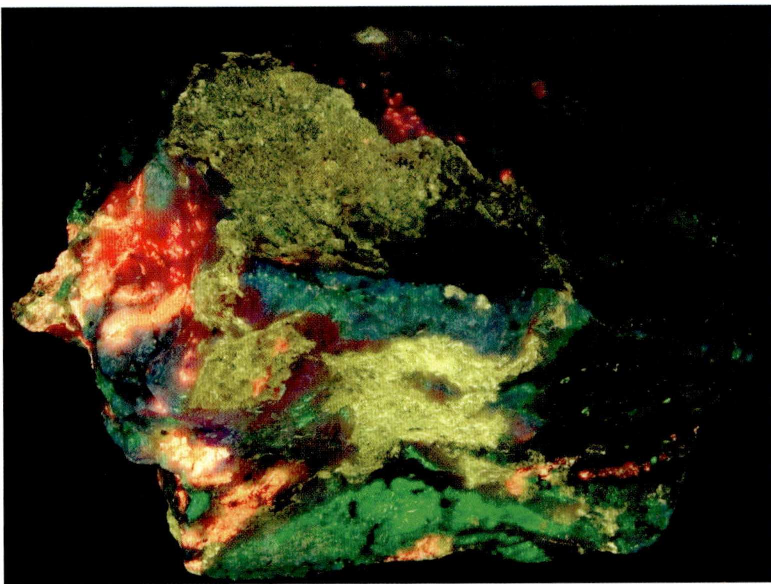

Same under SW

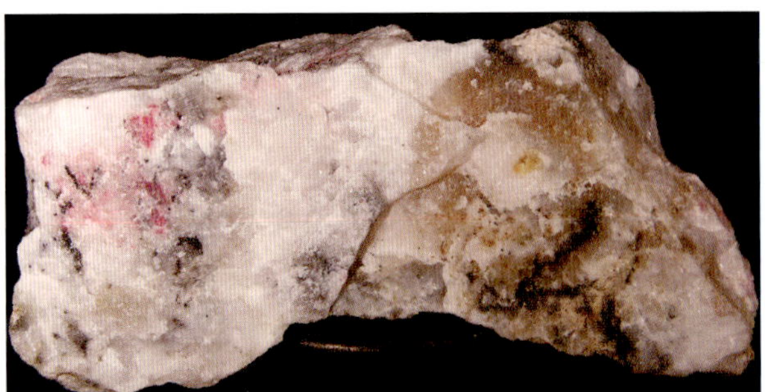

TUGTUPITE, etc. A beautiful specimen with tugtupite, analcime, beryllite, and polylithionite from Taseq slopes, Ilimaussaq complex, Narsaq, Kitaa Province. Tugtupite fluoresces red SW, analcime fluoresces pale blue SW, beryllite fluoresces pale gray SW, and polylithionite fluoresces yellow SW. The piece is 4.0 x 2.0 x 1.5 inches. Courtesy of Mark Cole and MinerShop.

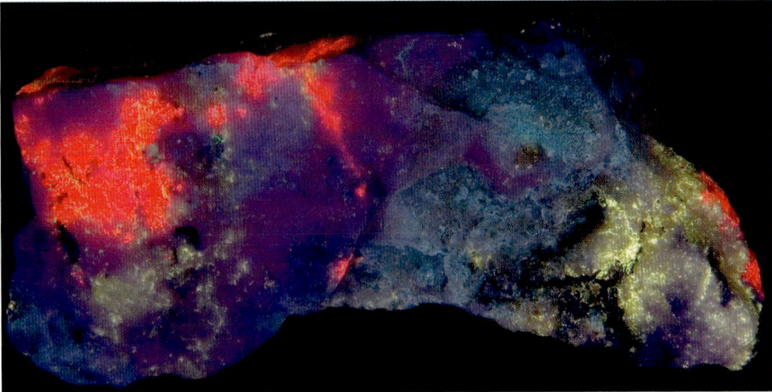

Same under SW

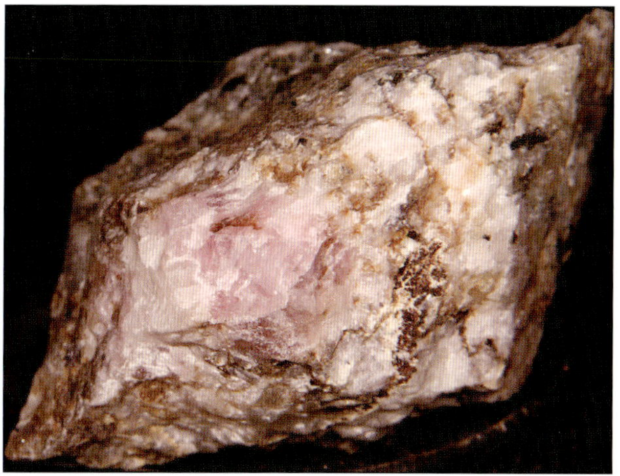

TUGTUPITE. A very tenebrescent lilac-colored tugtupite with other unidentified minerals from Kvanefjeld, Ilimaussaq complex, Narsaq, Kitaa Province. Tugtupite fluoresces red SW. The piece is 4.0 x 3.5 x 2.0 inches. Courtesy of Mark Cole and MinerShop.

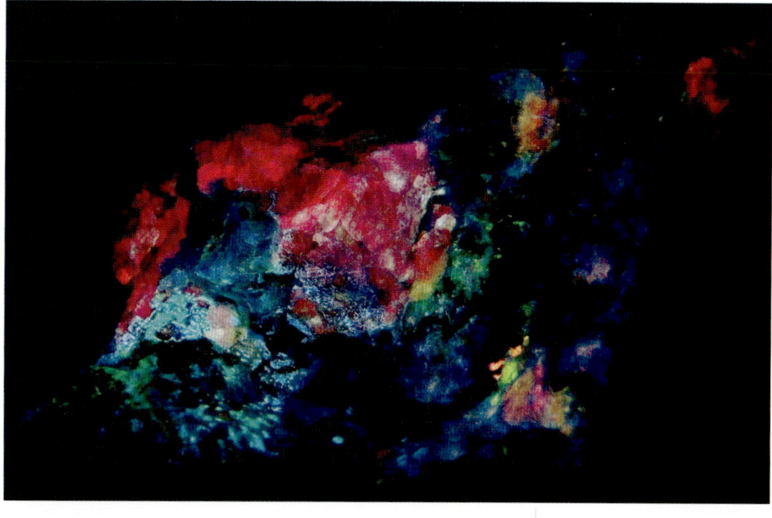

Top right:
Same under SW

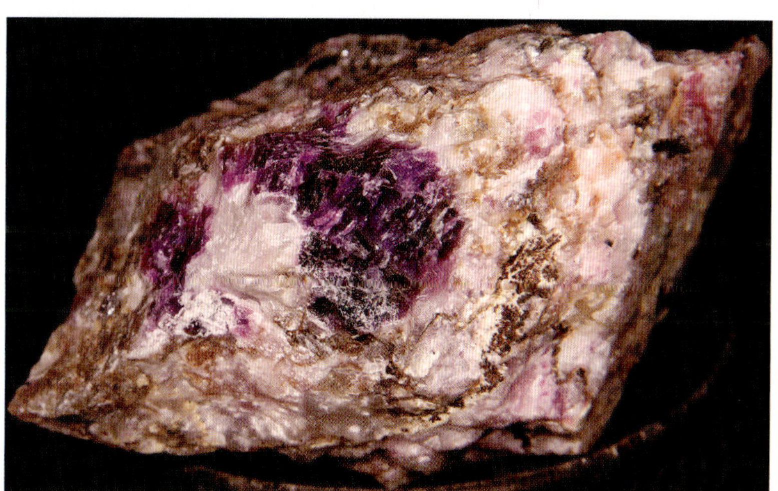

Right:
Same showing the tenebrescent effect.

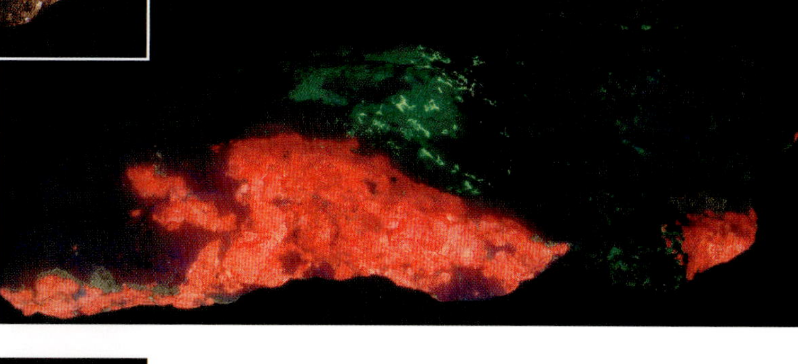

TUGTUPITE. An unusual pink tugtupite showing tenebrescence from Kangerlassaq, Ilimaussaq complex, Narsaq, Kitaa Province. Tugtupite fluoresces red SW. The piece is 4.0 x 2.0 x 1.5 inches. Courtesy of Mark Cole and MinerShop.

Same under SW

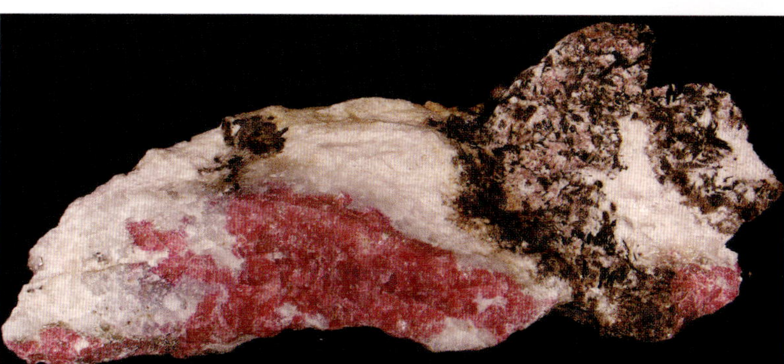

Same showing the tenebrescent effect.

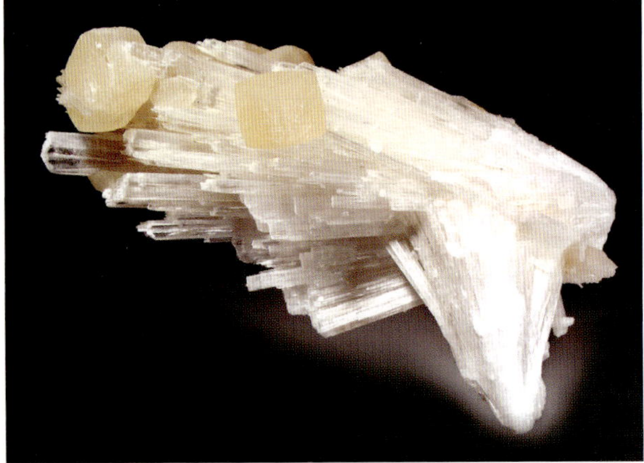

CALCITE. Beautiful pseudocubic crystals of calcite on feathery blades of scolecite from Nasik (also spelled "Nashik"), Pune (also spelled "Poona"), Maharashtra. The calcite fluoresces pale green SW and bright pale yellow LW. The piece weighs 4.0 oz. and is 3.8 x 2.0 x 1.5 inches. Value $90-110

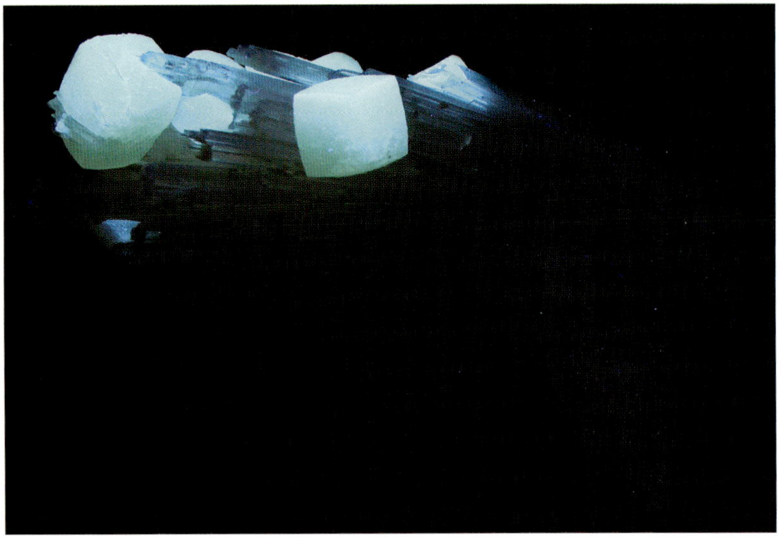

Same under SW

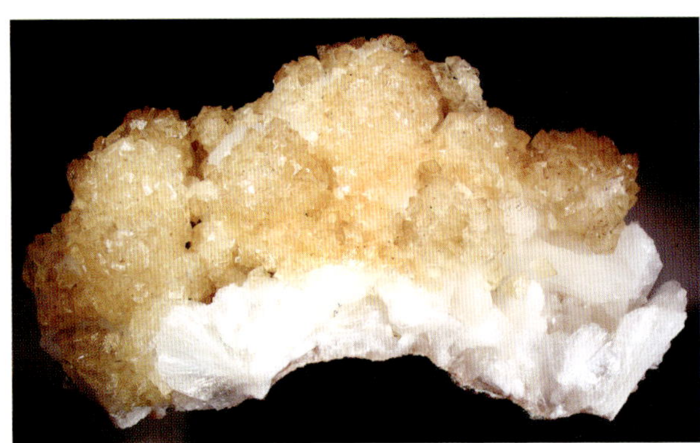

CALCITE. Calcite crystals on crystals of stilbite from Ahmadnagar, Pune, Maharashtra. The calcite fluoresces white LW and SW. The piece weighs 12.5 oz. and is 5.5 x 3.0 x 1.5 inches. Value $55-60

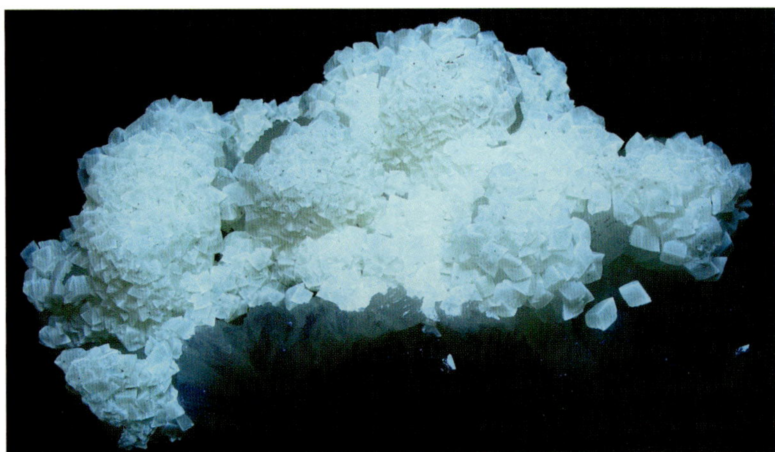

Same under SW

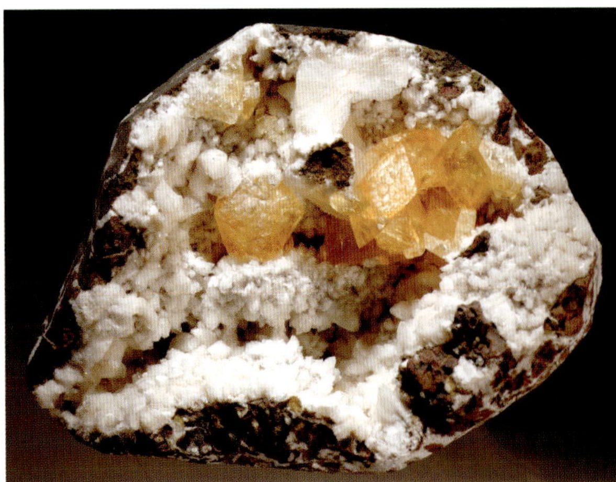

CALCITE. A geode containing amber spears of calcite on dolomite from Wagholi quarry, Wagholi, Pune, Maharashtra. The Wagholi quarry is better known for the bright blue cavansite that can be found there. Calcite from this locality is uncommon. The calcite fluoresces pale yellow SW and LW. The piece weighs 16.0 oz. and 4.5 x 3.5 x 2.3 inches. Value $150-175

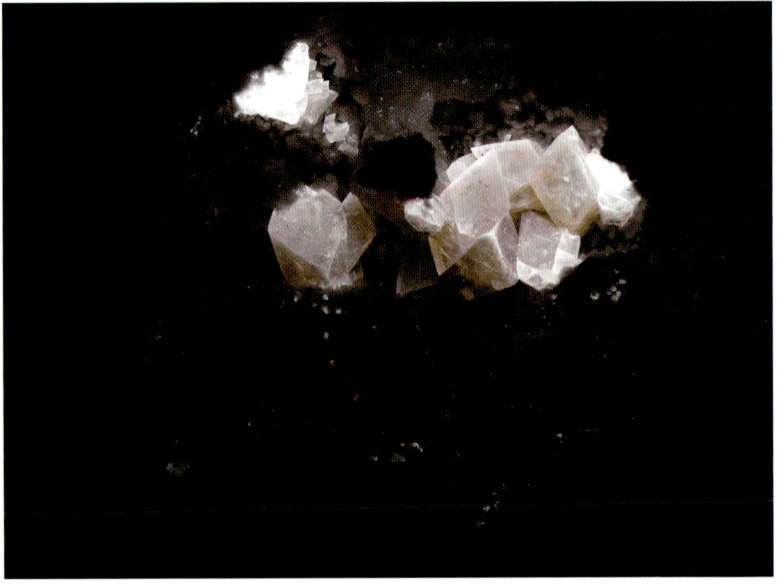

Same under SW

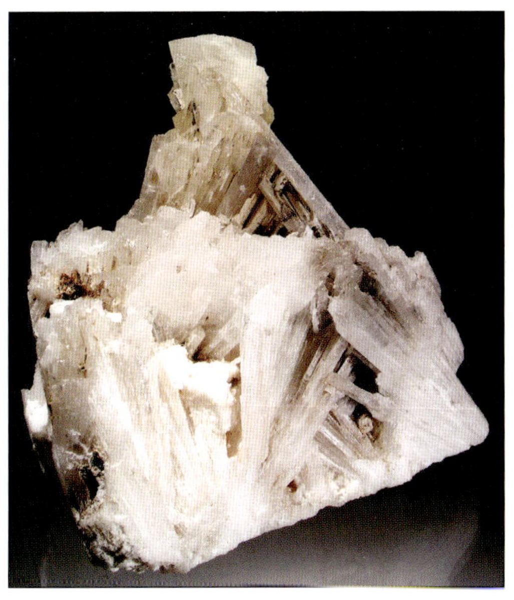

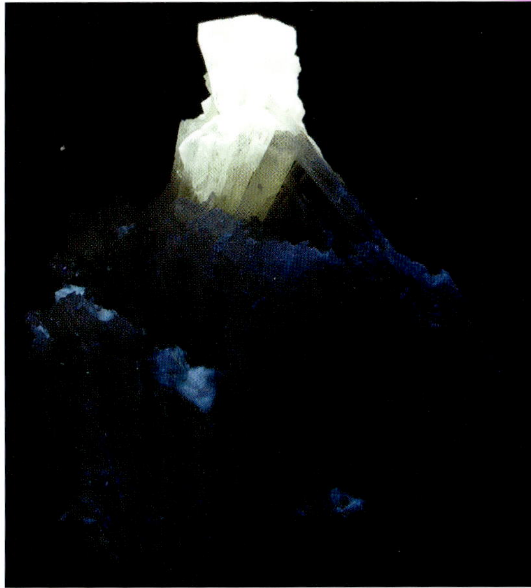

Far left:
POWELLITE. Powellite crystals on crystals of scolecite from Nasik, Pune, Maharashtra. Well-formed powellite crystals from India are expensive and rare. The powellite fluoresces pale yellow SW and pale white LW. The piece weighs 5.0 oz. and is 2.8 x 2.5 x 2.0 inches. Value $175-225.

Left:
Same under LW

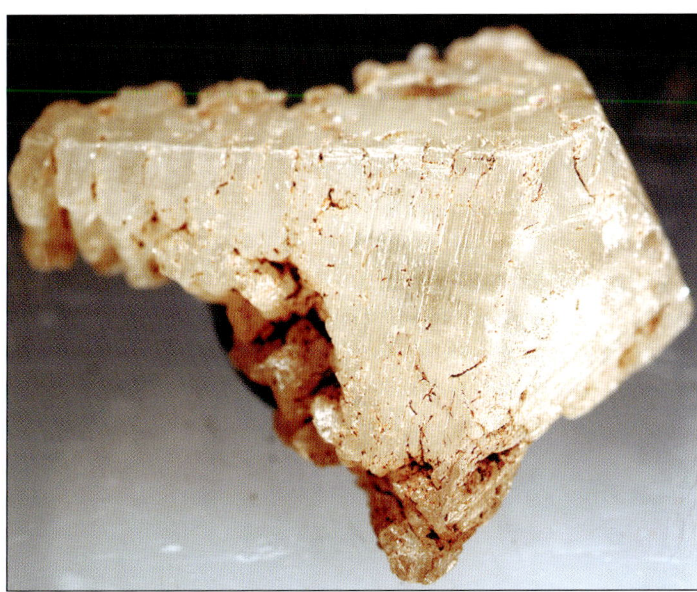

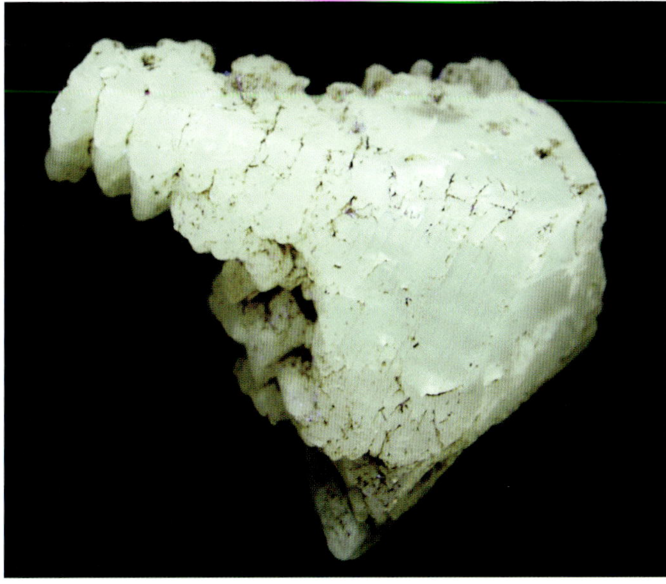

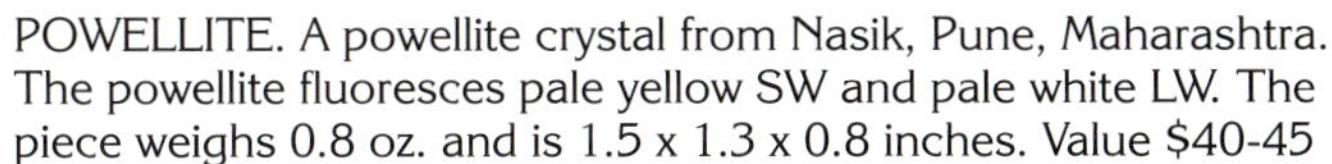
POWELLITE. A powellite crystal from Nasik, Pune, Maharashtra. The powellite fluoresces pale yellow SW and pale white LW. The piece weighs 0.8 oz. and is 1.5 x 1.3 x 0.8 inches. Value $40-45

Same under SW

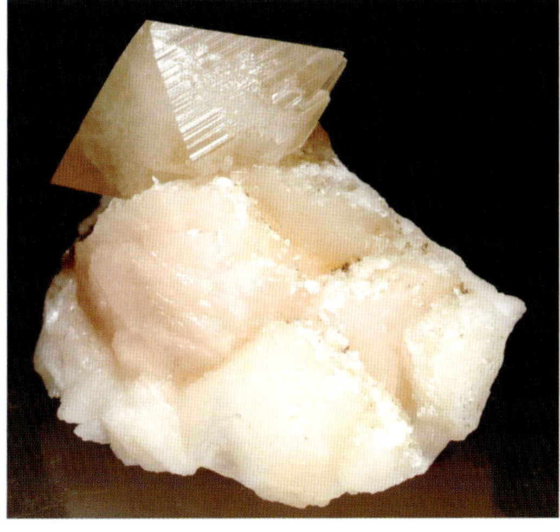

Far left:
POWELLITE. A superb powellite crystal (1.8 x 1.3 x 1.1 inches) on heulandite and stilbite crystals from Jalgaon, Maharashtra. This is a large perfect crystal. The powellite crystal fluoresces pale yellow SW and pale white LW. The piece weighs 5.8 oz. and is 3.0 x 2.0 x 2.0 inches. Value $1,200-1,300

Left:
Same under SW

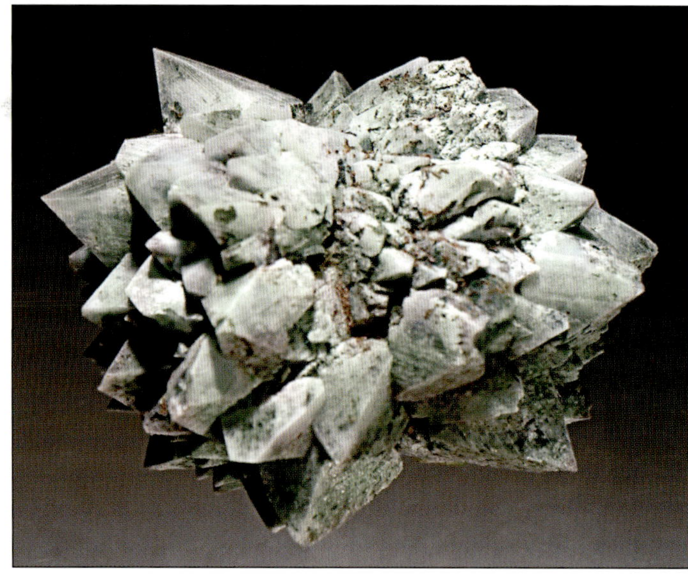

POWELLITE. A rare hedgehog-like cluster of powellite crystals from Jalgaon, Maharashtra. The powellite crystals fluoresce pale yellow SW and pale white LW. The piece weighs 5.5 oz. and is 2.3 x 1.8 x 1.5 inches. Value $500-600

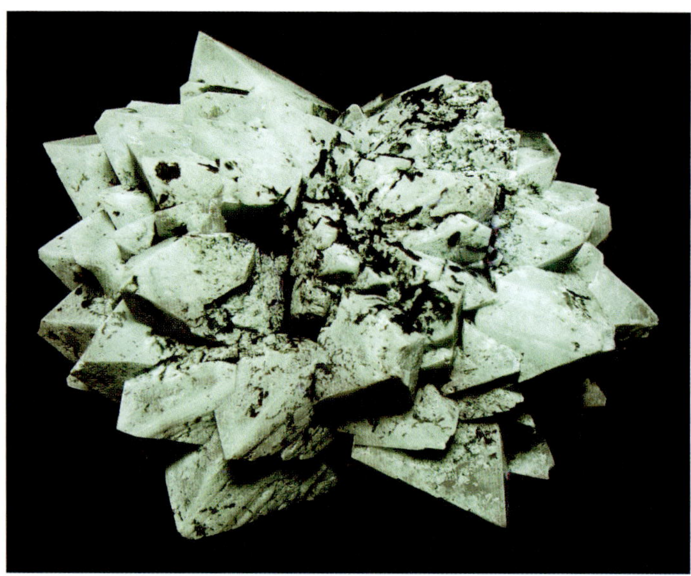

Same under SW

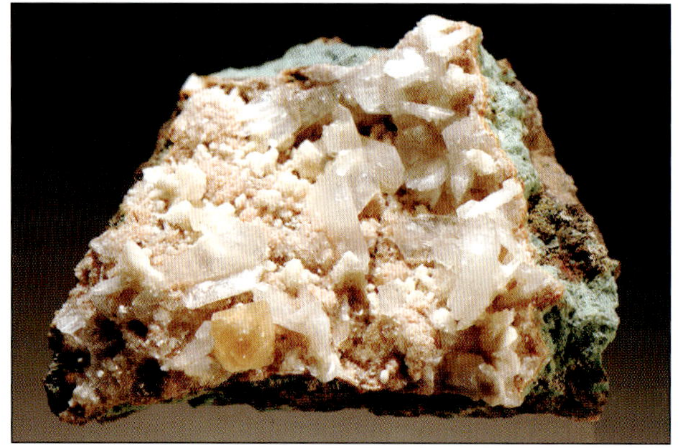

POWELLITE. A powellite crystal on heulandite/stilbite matrix from Nangaon, Aurangabad District, Maharashtra. The powellite fluoresces pale yellow SW and pale white LW. The piece weighs 5.5 oz. and is 2.8 x 2.0 x 1.5 inches. Value $50-60

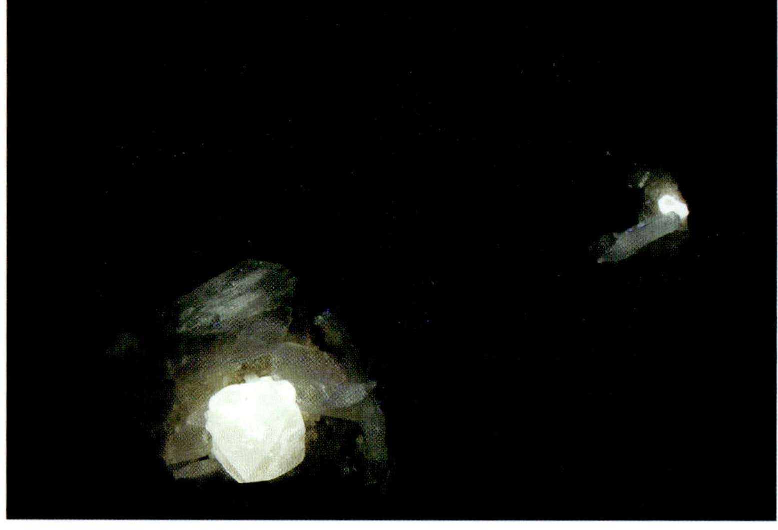

Same under LW

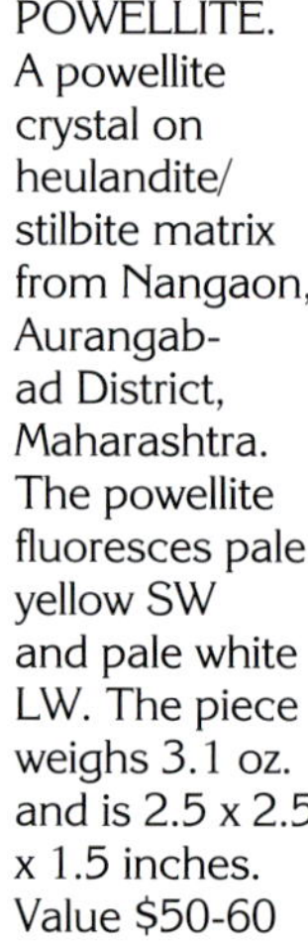

POWELLITE. A powellite crystal on heulandite/stilbite matrix from Nangaon, Aurangabad District, Maharashtra. The powellite fluoresces pale yellow SW and pale white LW. The piece weighs 3.1 oz. and is 2.5 x 2.5 x 1.5 inches. Value $50-60

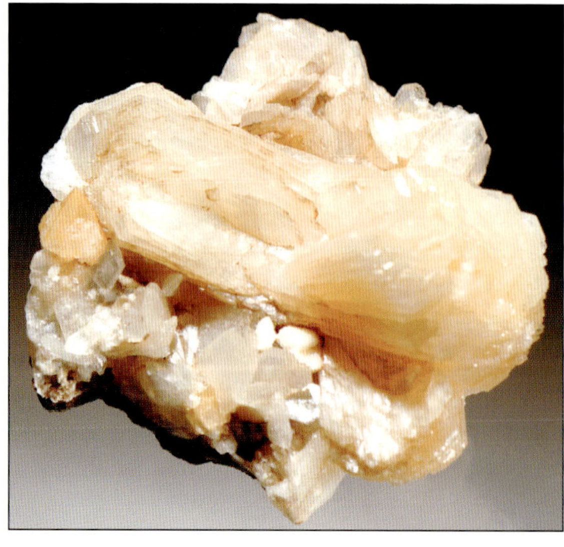

Same under LW.

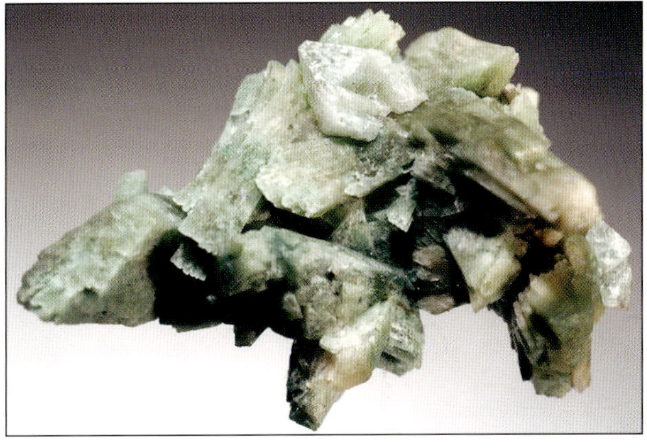

POWELLITE. A powellite crystal on fluorapophyllite crystals from Nangaon, Aurangabad District, Maharashtra. The powellite fluoresces pale yellow SW and pale white LW. The piece weighs 1.0 oz. and is 2.1 x 1.5 x 1.3 inches. Value $40-50

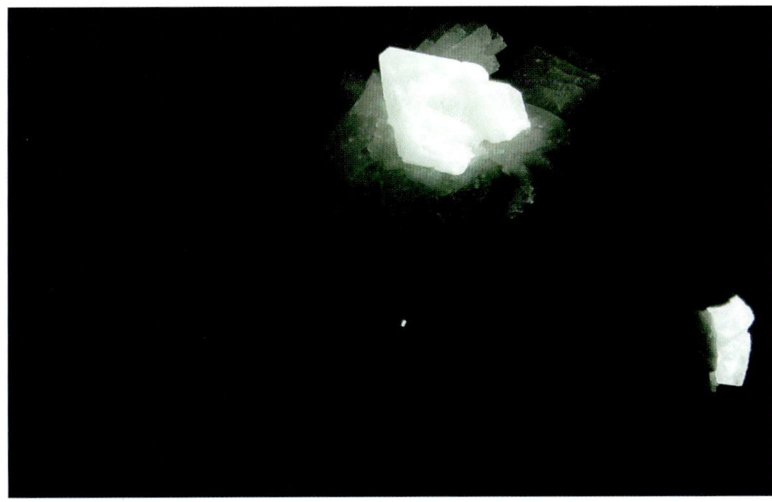

Same under LW

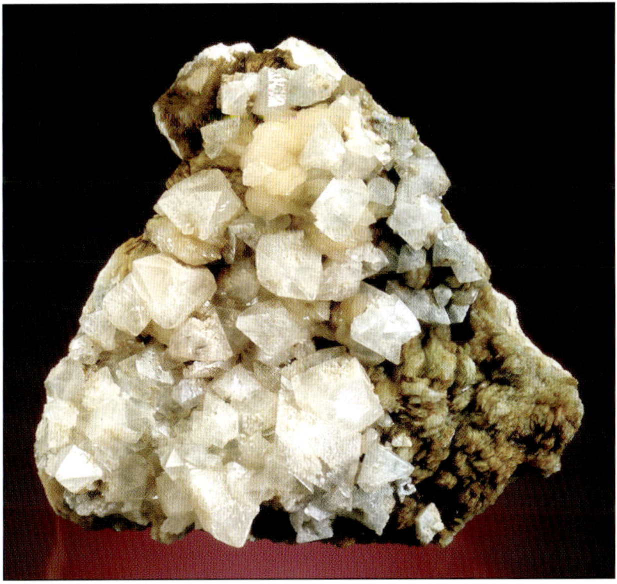

POWELLITE. Powellite crystals from Mahodari, Nasik, Pune, Maharashtra. The powellite fluoresces pale yellow SW and pale white LW. The piece weighs 14.8 oz. and is 1.5 x 3.8 x 4.0 inches. Courtesy of Don Newsome. Photo by Jeff Scovill.

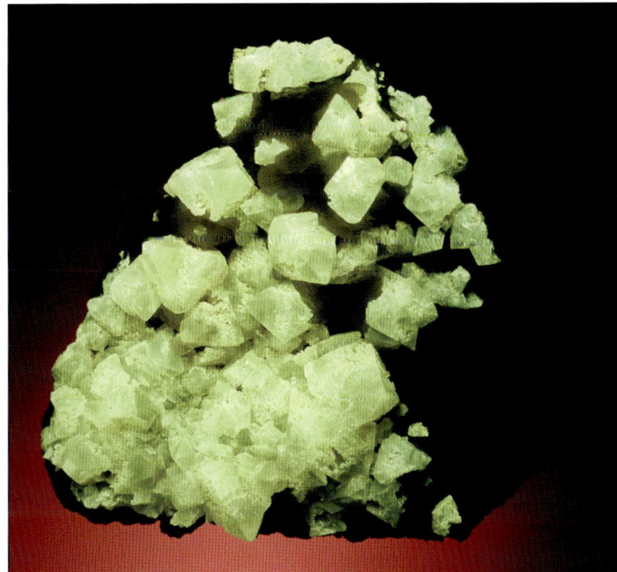

Same under SW

Iran

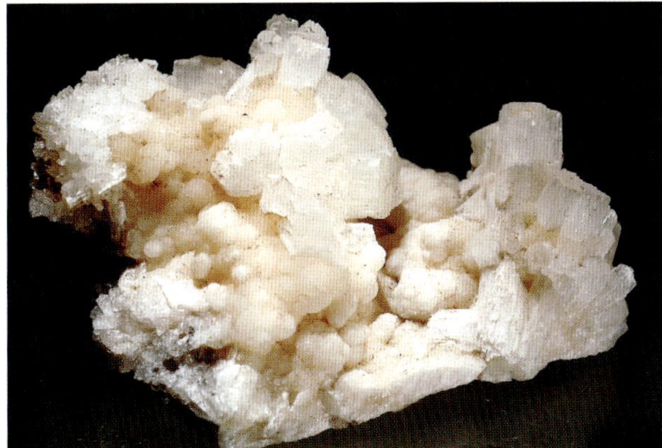

ARAGONITE. Superb crystals of aragonite on calcite from Dasht-e-Kavir, Qom Province. The aragonite fluoresces pale green SW. The piece weighs 7.5 oz. and is 4.0 x 2.8 x 1.9 inches. Value $35-40

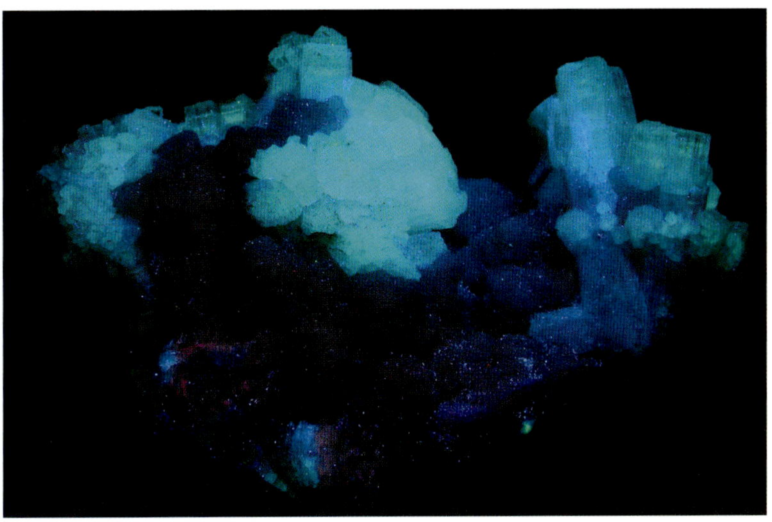

Same under SW

Ireland

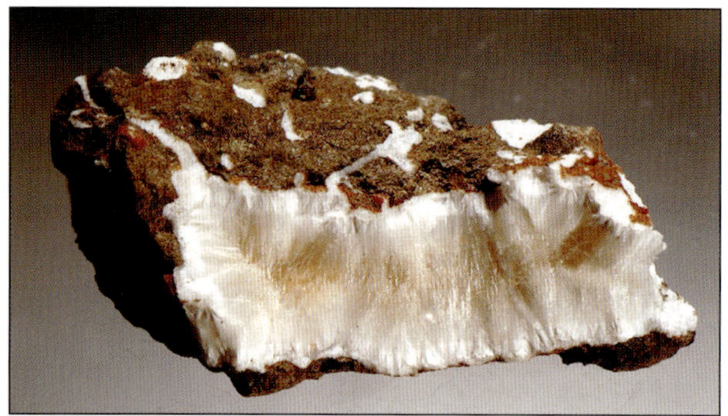

NATROLITE. Acicular crystals of natrolite from the Magheramourne quarry, Larne, County Antrim. The natrolite fluoresces (unevenly) pale blue SW. The piece weighs 2.5 oz. and is 3.0 x 1.5 x 1.3 inches. Value $20-25

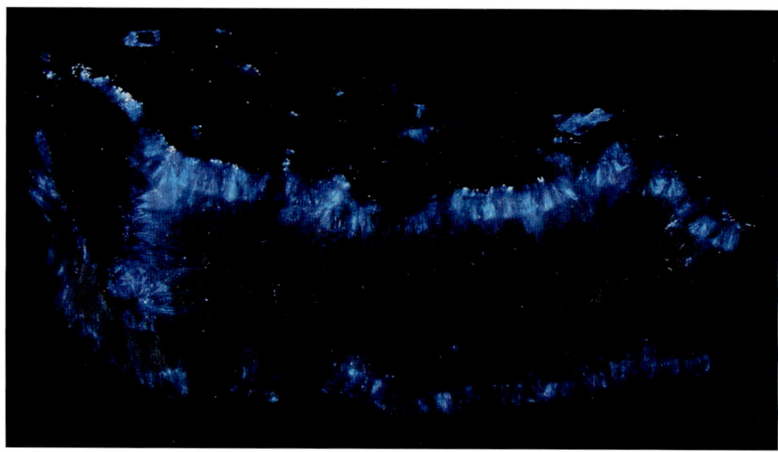

Same under SW

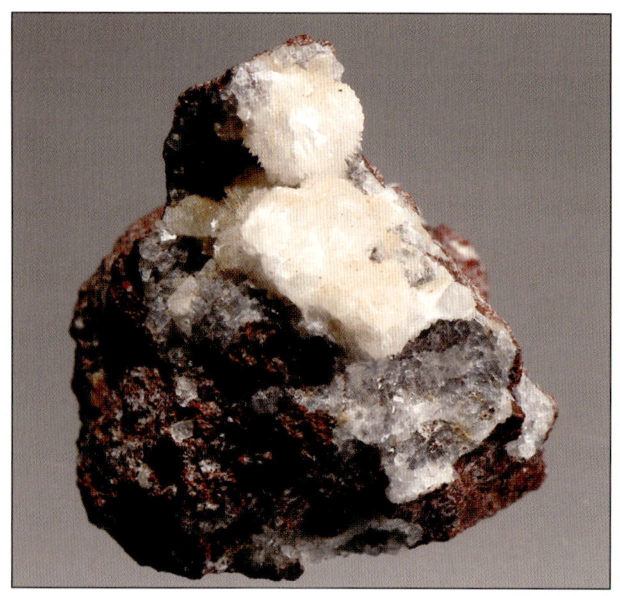

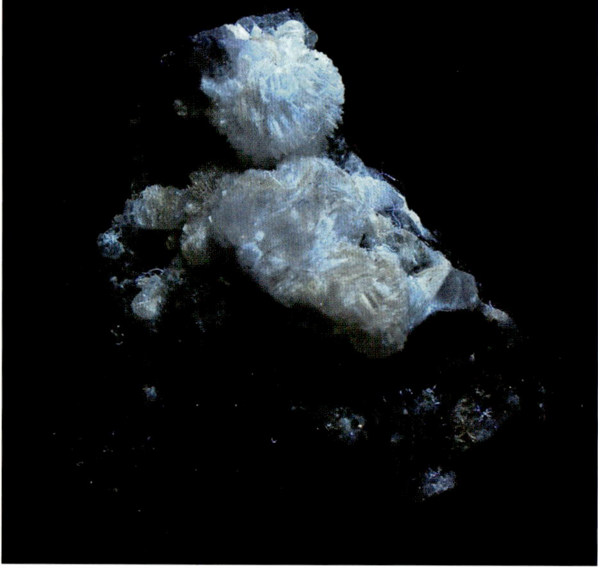

Far left:
THOMSONITE. Thomsonite crystals on matrix from the Parkate quarry, Larne, County Antrim. The thomsonite fluoresces pale tan SW. The piece weighs 1.0 oz. and is 1.3 x 1.1 x 1.1 inches. Value $15-20

Left:
Same under SW

Italy

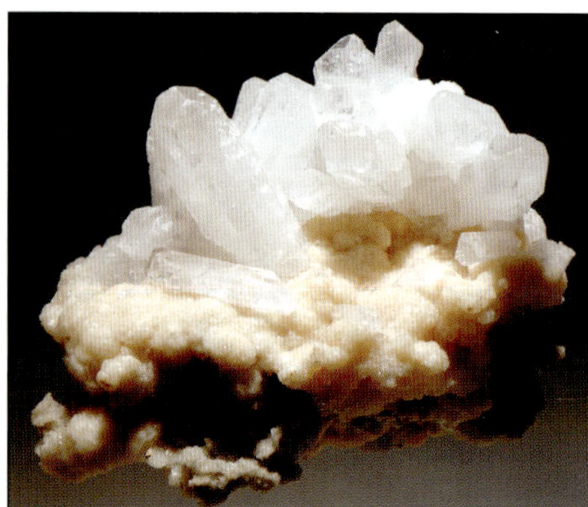

CELESTINE. Crystals of celestine, formerly known as celestite, on calcite from Girgenti, Agrigento Province, Sicily. The celestine fluoresces a pale blue SW and the calcite fluoresces a bright pale yellow SW and bright pink LW. The piece weighs 5.0 oz. and is 2.8 x 2.3 x 1.5 inches. Value $40-45

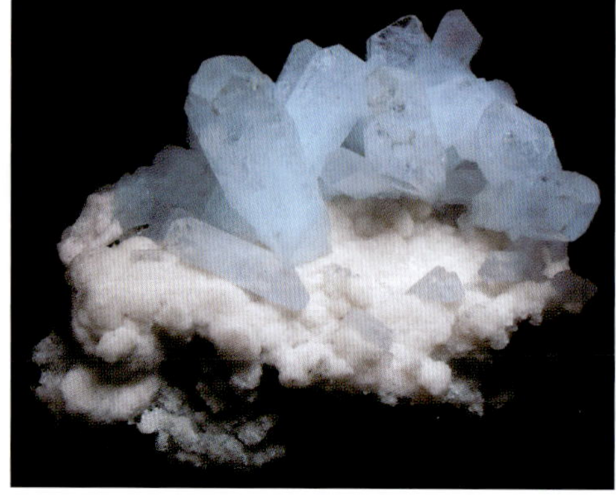

Same under SW

Same under LW

Madagascar

ORTHOCLASE. A yellow feldspar, var. orthoclase, crystal faceted as an octagonal-cut gem from Itrongay, Fianarantsoa Province. It fluoresces orange-red SW. It weighs 6.2 ct. and is 13.5 x 9.5 x 6.5 mm. Value $50-60

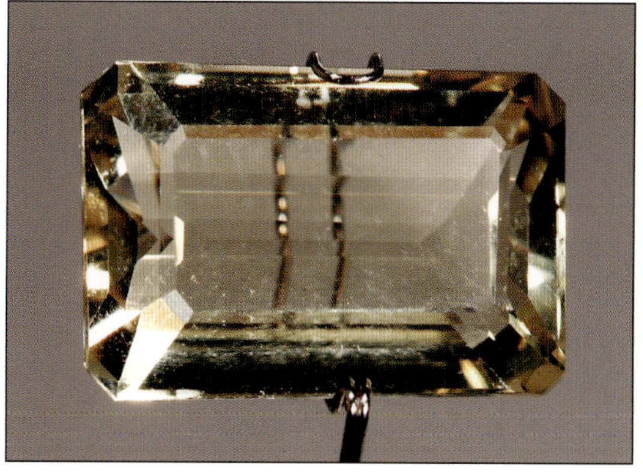

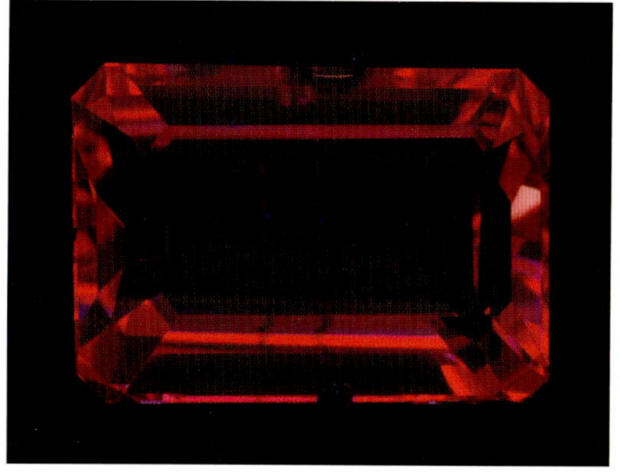

Same under SW

Malaysia

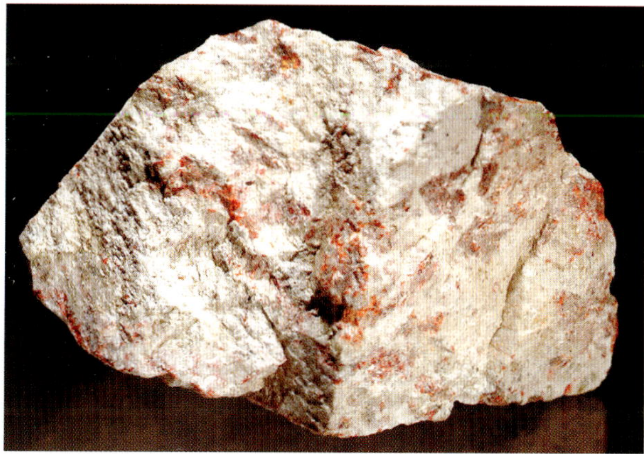

WOLLASTONITE. Cream to white radiating crystals of wollastonite, var. wollastonite-2M, (the collector name is parawollastonite) in calcite with Sarabauite from Sarawak Province, Borneo. The wollastonite-2M fluoresces pale pink SW, the calcite fluoresces weak red SW. The piece weighs 6.5 oz. and is 3.4 x 2.0 x 1.4 inches. Value $30-35

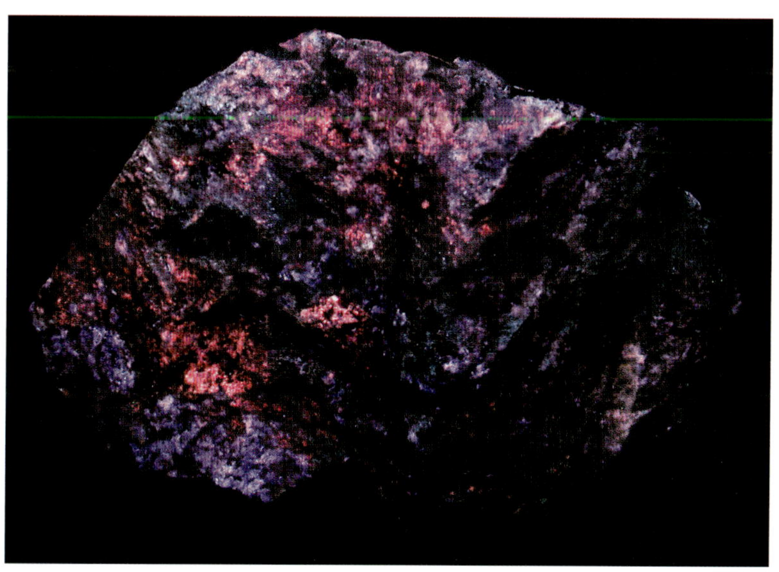

Same under SW

Malawi

FELDSPAR ZIRCON. Large crystals of feldspar with zircon crystals from Zomba. The feldspar fluoresces velvety red SW and the zircon fluoresces yellow SW. The piece weighs 2.8 oz. and is 1.8 x 1.5 x 1.5 inches. Value $40-45

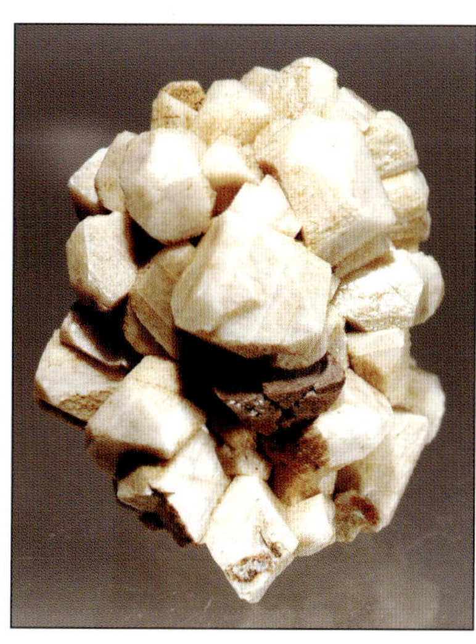

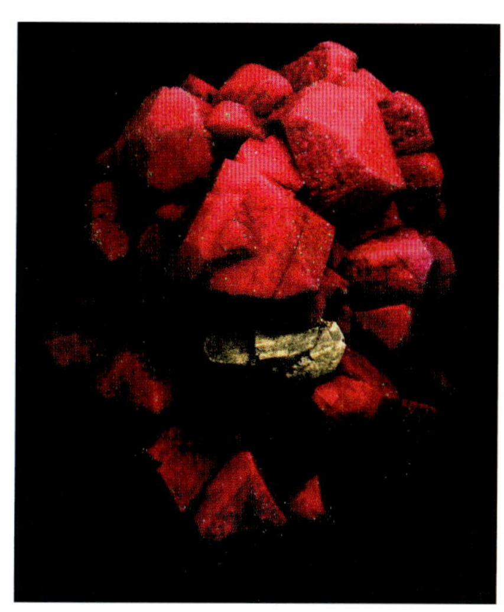

Same under SW

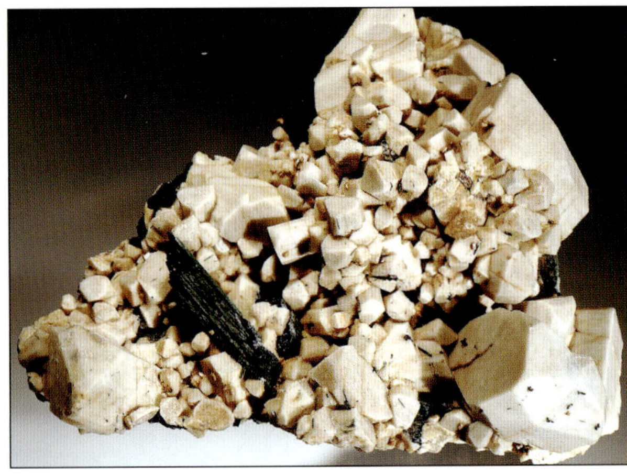

FELDSPAR ZIRCON. Zircon crystals on feldspar crystals with arfvedsonite crystals from Zomba. The zircon crystals fluoresce yellow SW and the feldspar fluoresces velvety red SW. The piece weighs 2 lb. 11.0 oz. and is 6 x 5 x 3 inches. Value $120-135

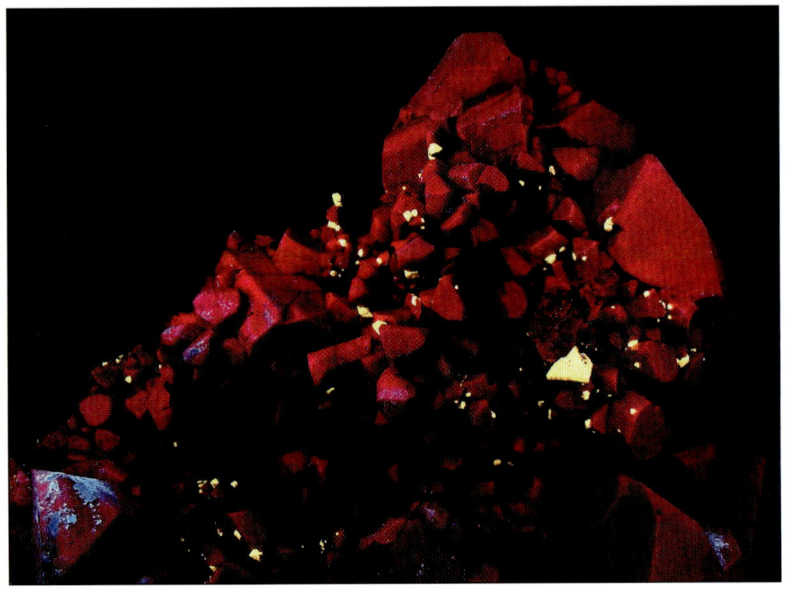

Same under SW

Morocco

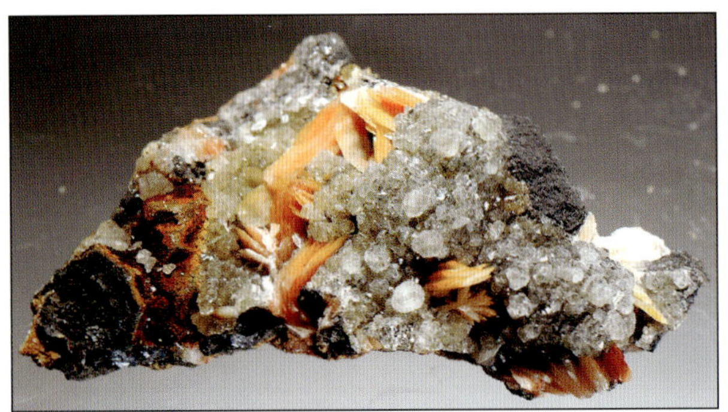

CERUSSITE. Gemmy cerussite crystals with pink barite from Mibladen, Khenifra. Cerussite fluoresces yellow LW. Cerussite is named from the Latin word "ceruse", a white lead pigment. The piece weighs 4.5 oz. and is 3.0 x 1.8 x 1.3 inches. Value $35-45

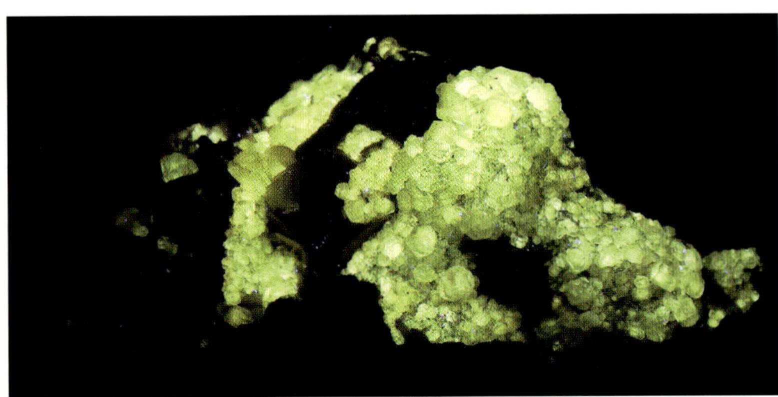

Same under LW

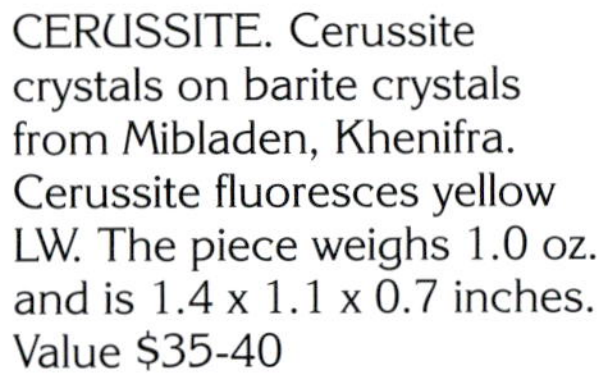

CERUSSITE. Cerussite crystals on barite crystals from Mibladen, Khenifra. Cerussite fluoresces yellow LW. The piece weighs 1.0 oz. and is 1.4 x 1.1 x 0.7 inches. Value $35-40

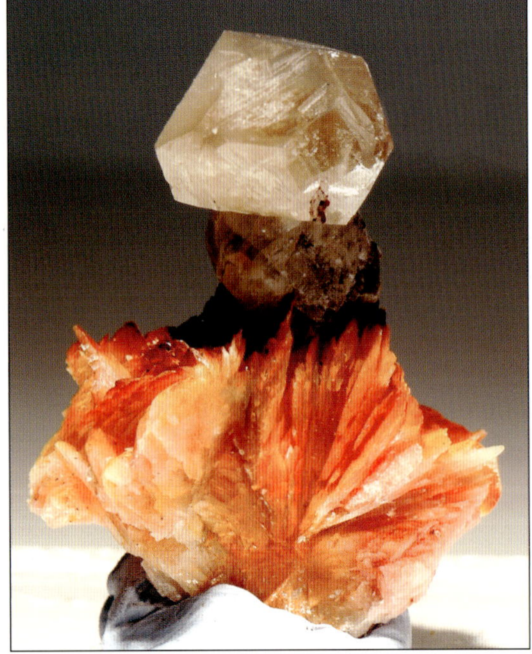

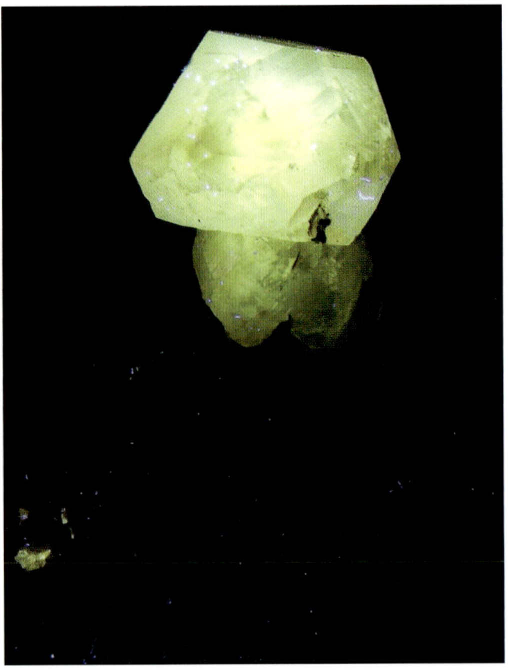

Same under LW

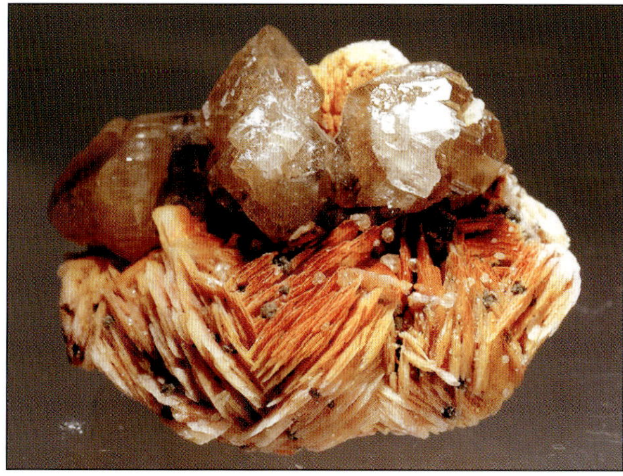

CERUSSITE. Cerussite crystals with pink barite from Mibladen, Khenifra. Cerussite fluoresces yellow LW. The piece weighs 4.0 oz. and is 2.3 x 1.5 x 1.1 inches. Value $30-35

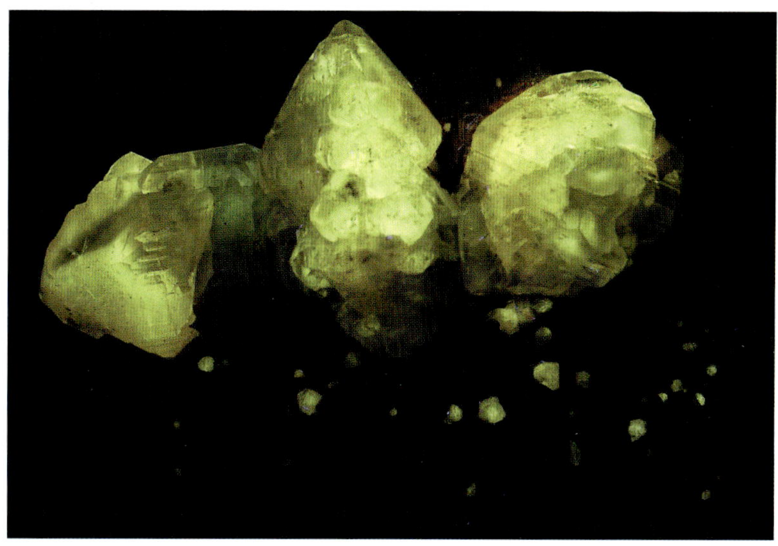

Same under LW

Namibia

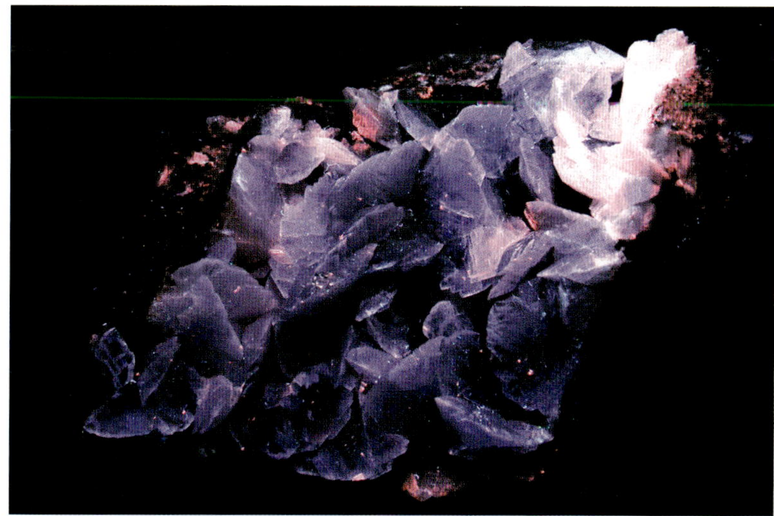

CALCITE. Calcite crystals from Tsumeb, Otjikoto Region. The calcite fluoresces two colors, pink-orange and weak violet SW. The piece weighs 6.5 oz. and is 3.1 x 2.3 x 1.8 inches. Value $35-40

Same under SW

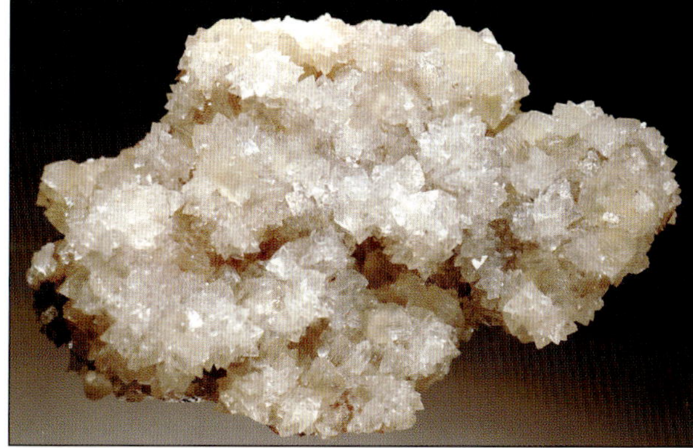

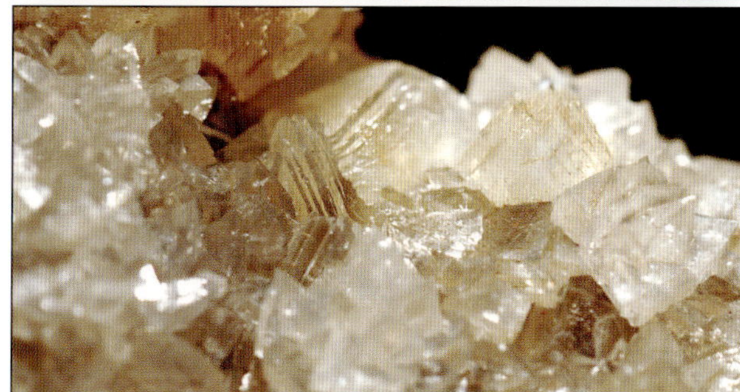

SMITHSONITE. Smithsonite crystals from Tsumeb, Otjikoto Region. The smithsonite fluoresces pale violet and pink SW on freshly broken surfaces. The piece weighs 11.8 oz. and is 3.7 x 2.3 x 1.8 inches. Value $55-60

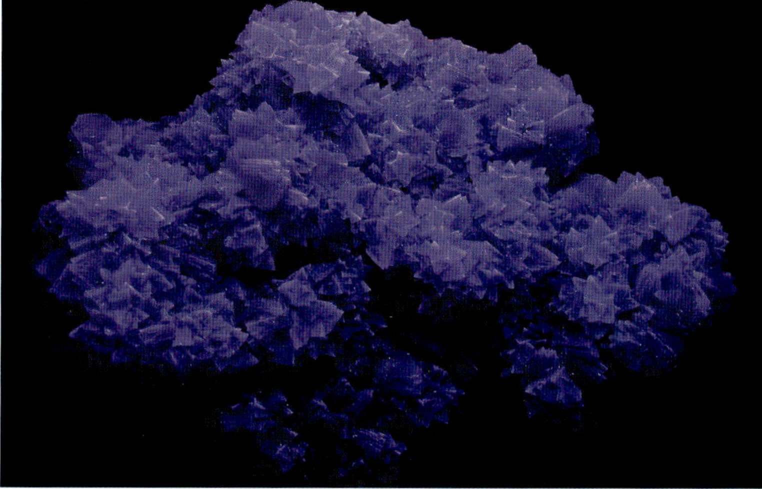

Same under SW

WILLEMITE. A willemite crystal from Tsumeb, Otjikoto Region that has been faceted as a gem. This willemite fluoresces pale blue-green SW. It is 4.5 mm wide. Value $55-65

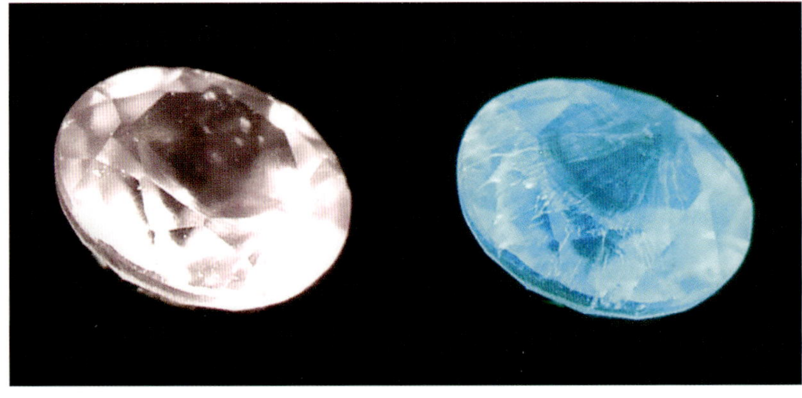

Norway

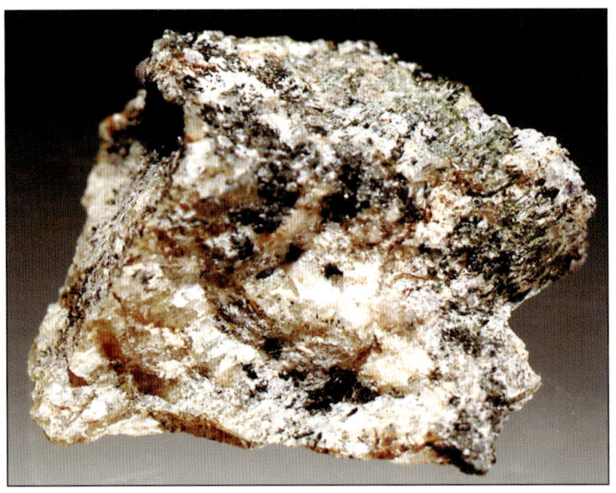

LEUCOPHANITE. Leucophanite from the Saga quarry, Strandåsen, Porsgrunn, Telemark. The leucophanite fluoresces pink SW. The piece weighs 0.4 oz. and is 0.8 x 0.8 inches. Value $22-25

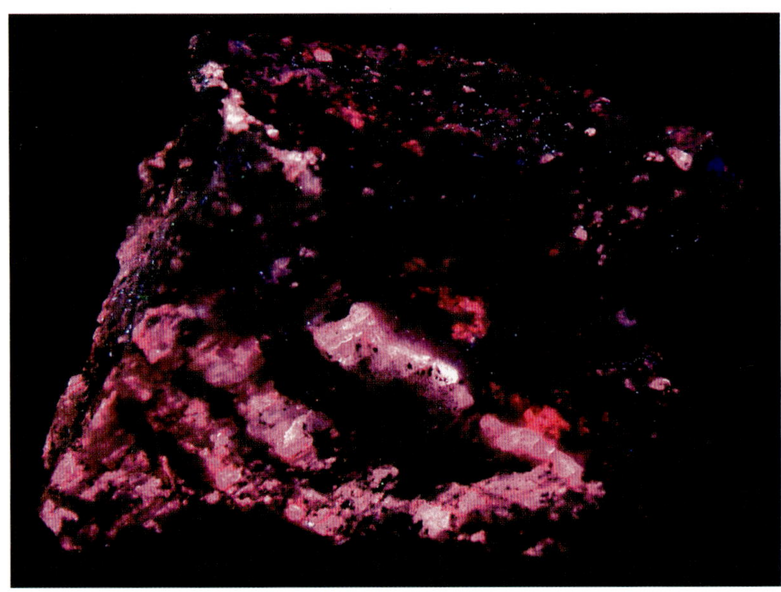

Same under SW

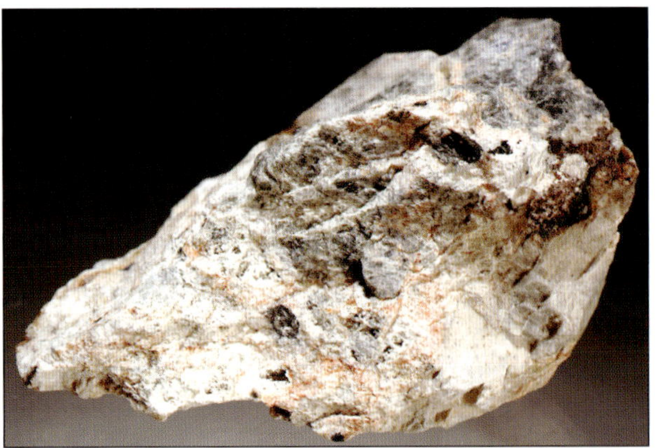

SODALITE, ZIRCON. Sodalite, feldspar, zircon, and parakeldyshite from Brevik, Langesundsfjorden, Porsgrunn, Telemark. Sodalite fluoresces orange LW, feldspar fluoresces velvety red SW, zircon fluoresces orange SW, and parakeldyshite fluoresces pale pale blue SW. The piece weighs 4.5 oz. and is 3.5 x 1.8 x 1.8 inches. Value $60-70

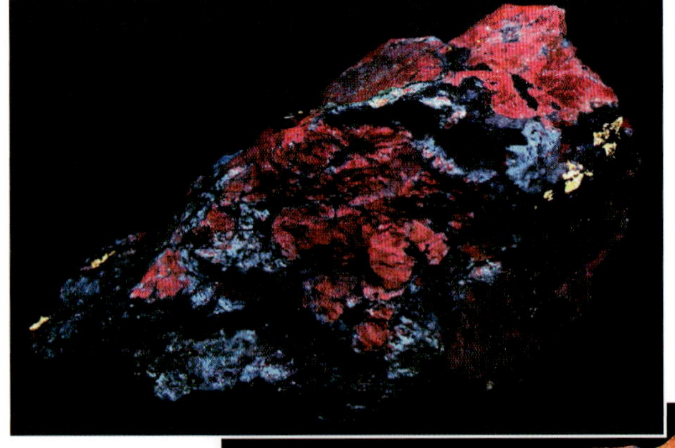

Same under SW

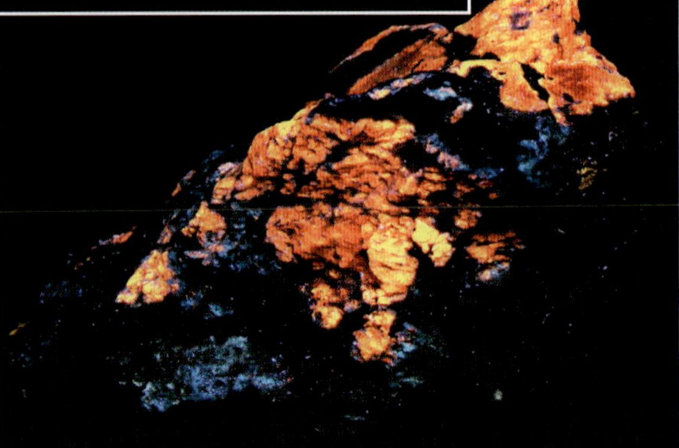

Same under LW

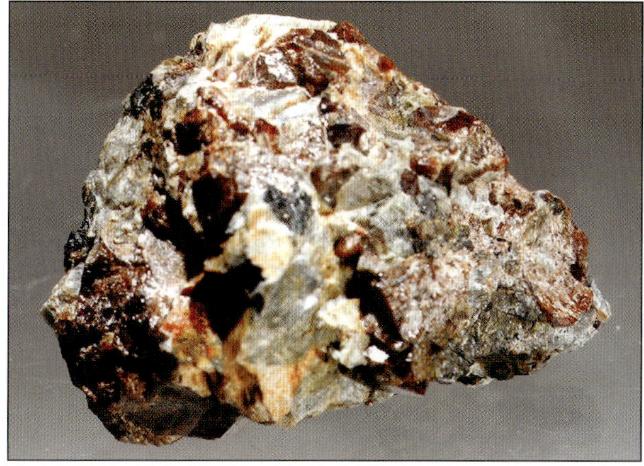

ZIRCON, Zircon crystals from Tvedalen, Larvik, Vestfold. Zircon fluoresces yellow-orange SW. The piece weighs 4.0 oz. and is 2.3 x 1.8 x 1.4 inches. Value $20-25

Same under SW

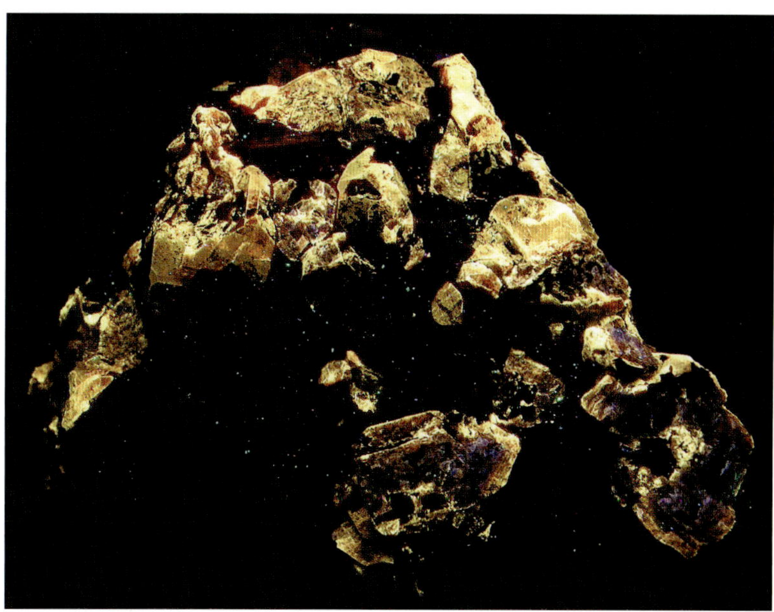

Pakistan

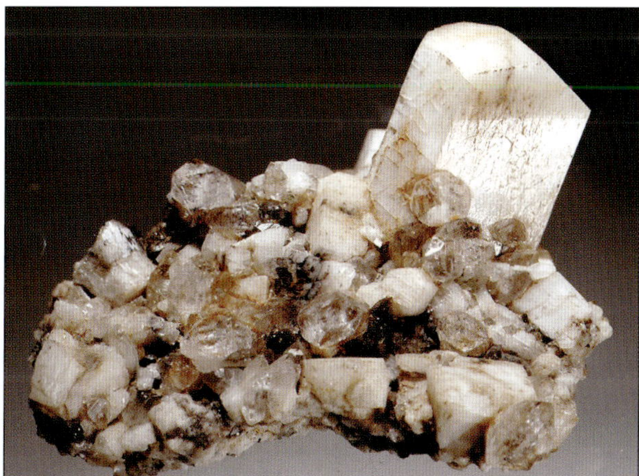

ALBITE. A Carlsbad twin crystal of albite with quartz crystals from Gilgit-Skardu Road, Gilgit district. Feldspars frequently exhibit phase changes with what appears to be a single crystal but actually is two or more individual crystals. The Carlsbad twin produces what appears to be two intergrown crystals growing in opposite directions. The albite fluoresces pale blue SW. The piece weighs 5.0 oz. and is 2.9 x 2.0 x 1.9 inches. Value $125-140

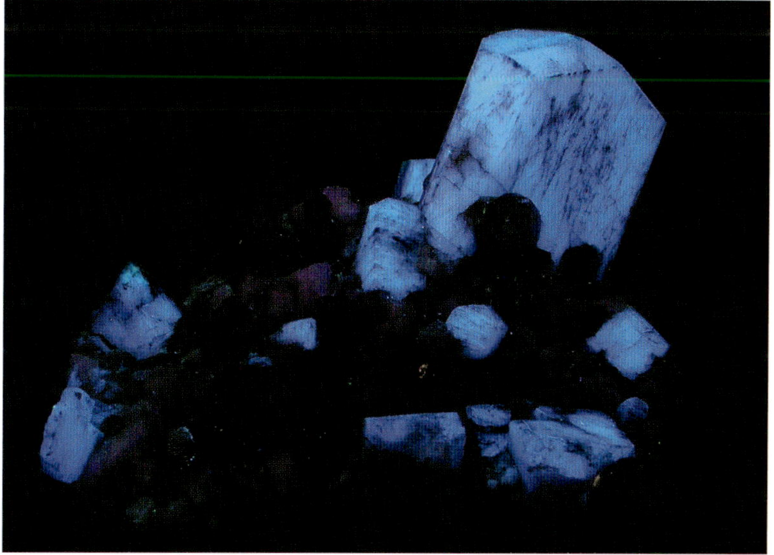

Same under SW

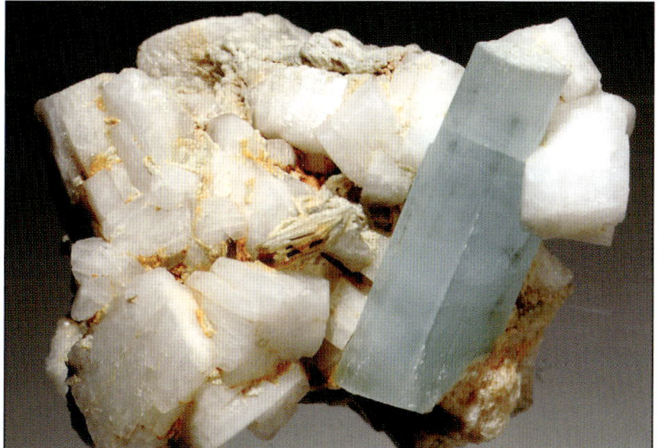

ALBITE. A beryl, var. aquamarine, crystal on albite with muscovite from Shigar Valley, Skardu district. The albite fluoresces dark pink SW. The piece weighs 3.5 oz. and is 2.5 x 1.8 x 1.5 inches. Value $140-150

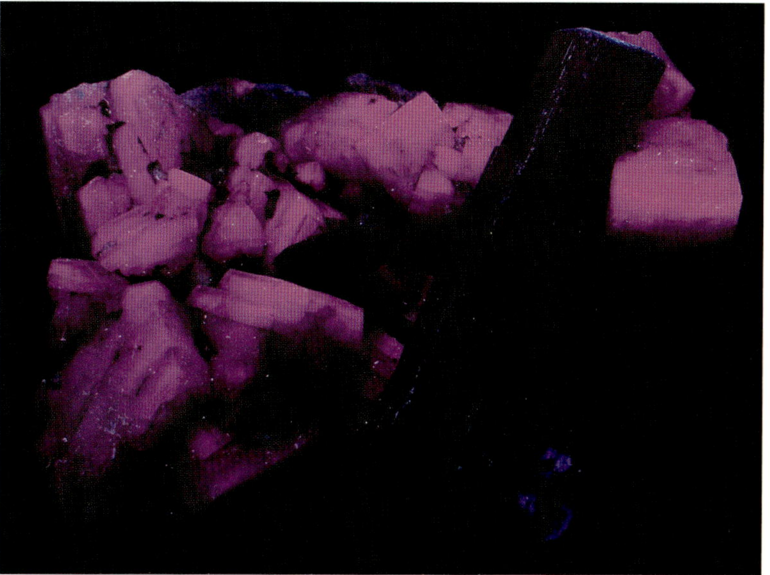

Same under SW

ALBITE. Albite crystals on a spessartine garnet from Shigar Valley, Skardu district. The albite fluoresces pink-blue SW. The piece weighs 1.4 oz. and is 1.4 x 1.3 x 1.1 inches. Value $40-45

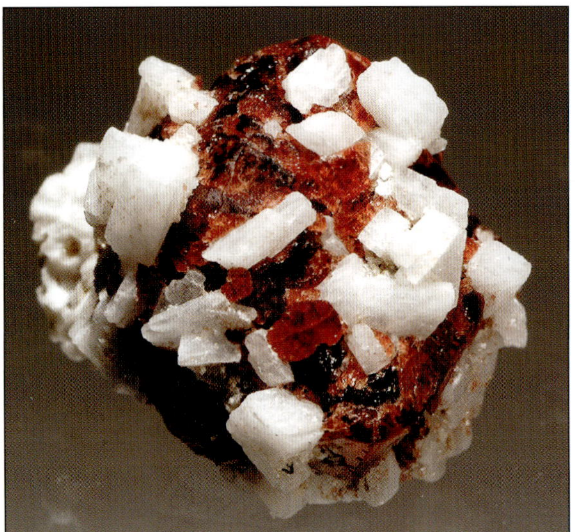

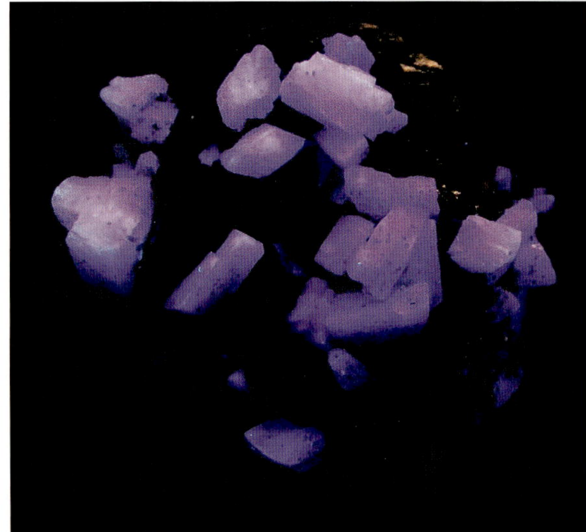

Same under SW

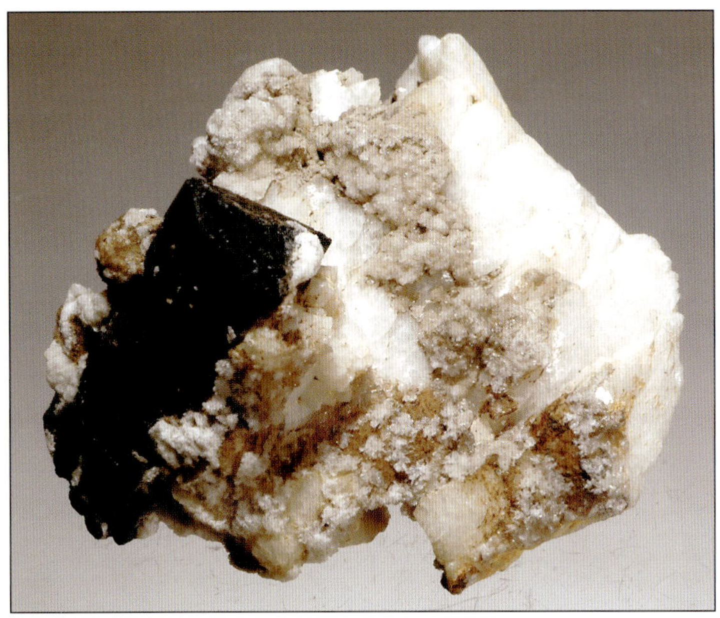

CALCITE. A black schorl crystal (tourmaline group) with calcite crystals from Shigar Valley, Skardu district. The calcite fluoresces green SW. The piece weighs 1.8 oz. and is 2.0 x 1.8 x 1.5 inches. Value $25-30

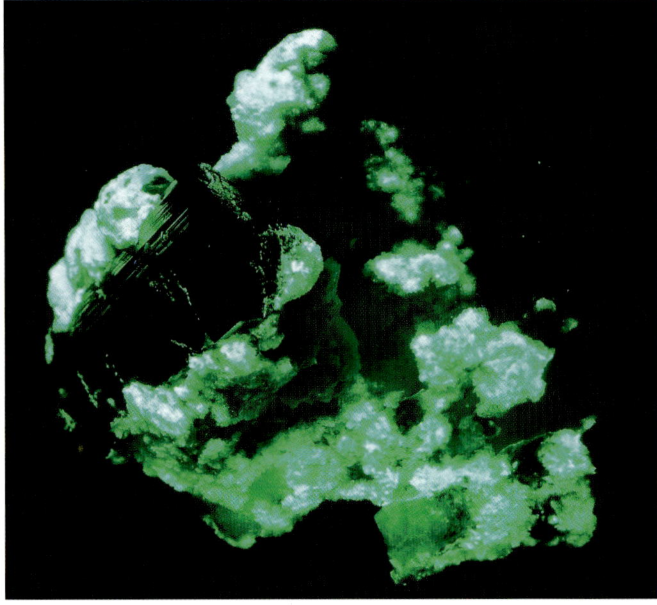

Same under SW

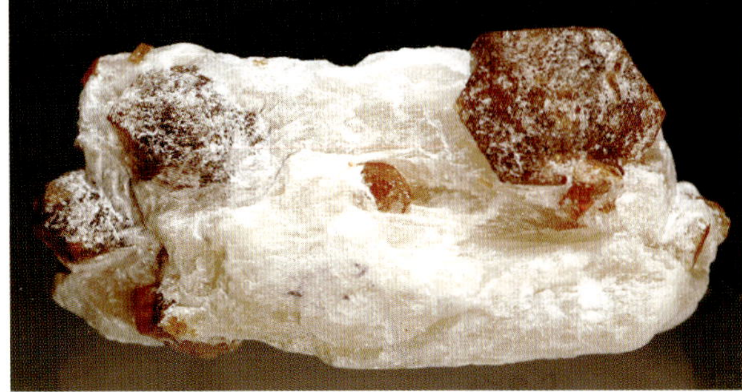

DRAVITE and TALC. Brown dravite crystals in talc from the Kohistan district. The dravite fluoresces mustard yellow SW and the talc fluoresces pale yellow SW. The piece weighs 1.5 oz. and is 2.1 x 1.3 x 1.0 inches. Value $40-50

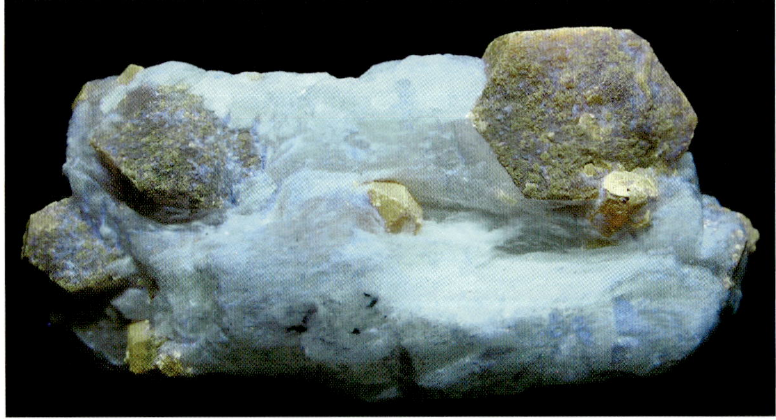

Same under SW

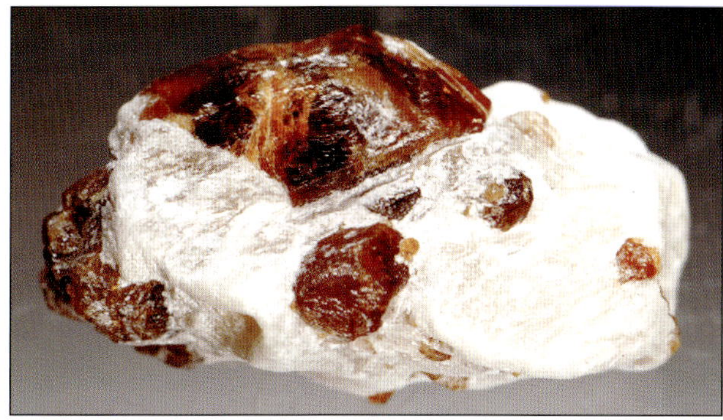

DRAVITE and TALC. Brown dravite crystals in talc from the Kohistan district. The dravite fluoresces mustard yellow SW and the talc fluoresces pale yellow SW. The piece weighs 1.0 oz. and is 1.5 x 0.8 x 0.8 inches. Value $35-45

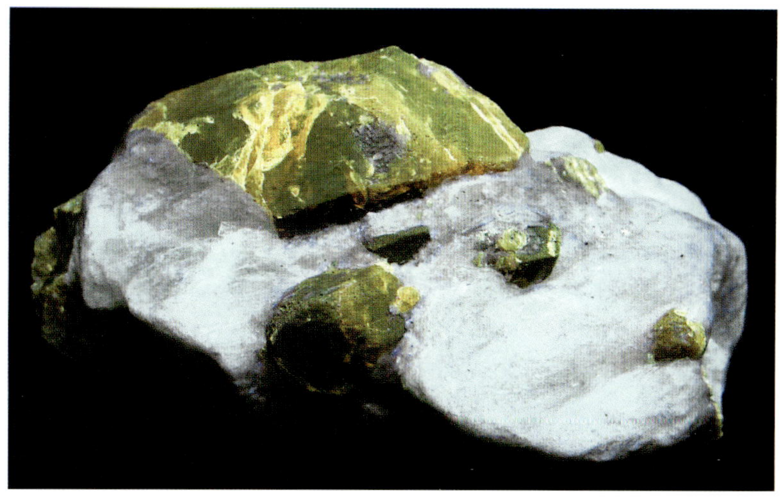

Same under SW

ELBAITE. Green to pink elbaite crystals (tourmaline group) from Stak Nala, Skardu district. The pink to colorless area fluoresces pale blue SW. The piece weighs 2.5 oz. and is 1.8 x 1.5 x 1.5 inches. Value $60-70

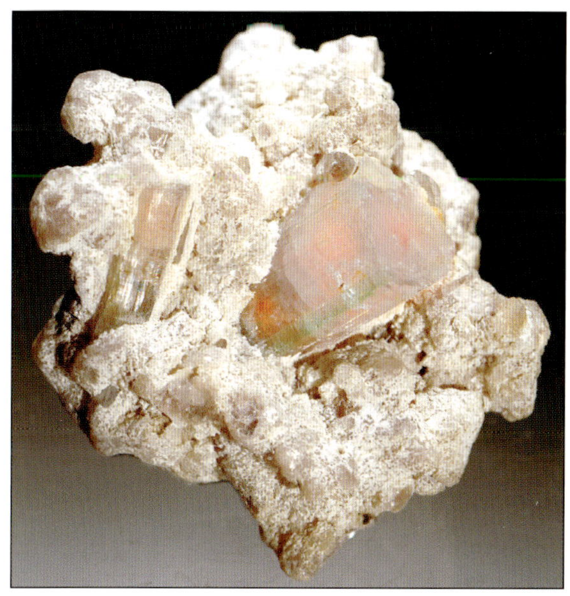

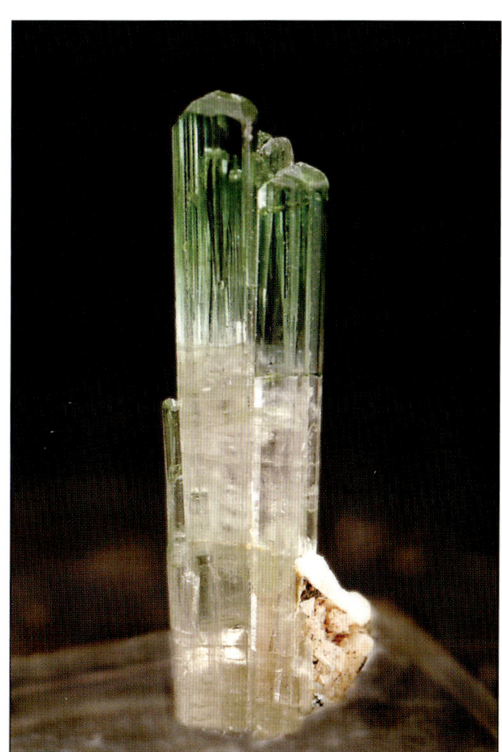

Same under SW

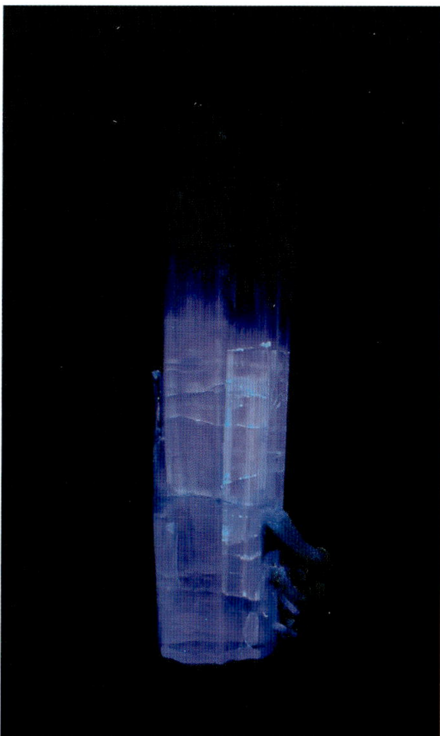

Far left:
ELBAITE. Green to colorless elbaite crystals (tourmaline group) from Stak Nala, Skardu district. The colorless area fluoresces pale blue SW. The piece weighs 0.3 oz. and is 1.5 x 0.4 x 0.3 inches. Value $80-85

Left:
Same under SW

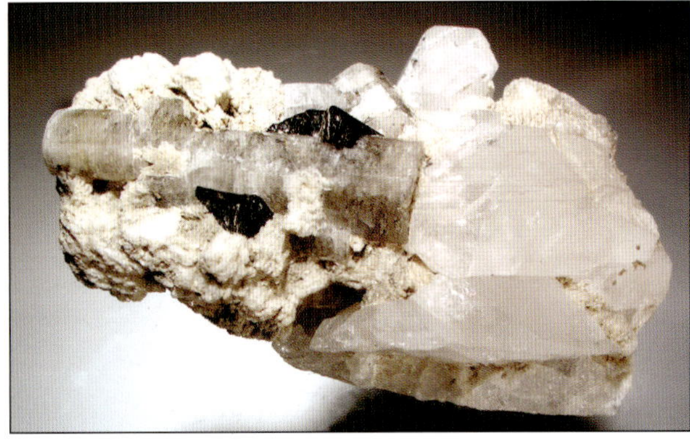

FLUORAPATITE. Fluorapatite on quartz from Shigar Valley, Skardu district. The fluorapatite fluoresces bright yellow under SW. The fluorapatite crystal is 2.0 inches long. The piece weighs 8.3 oz. and is 3.5 x 2.5 x 2.0 inches. Value $155-175

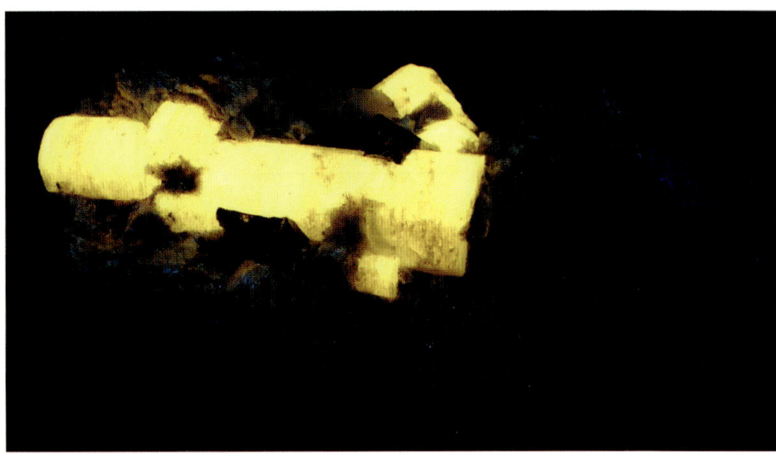

Same under SW

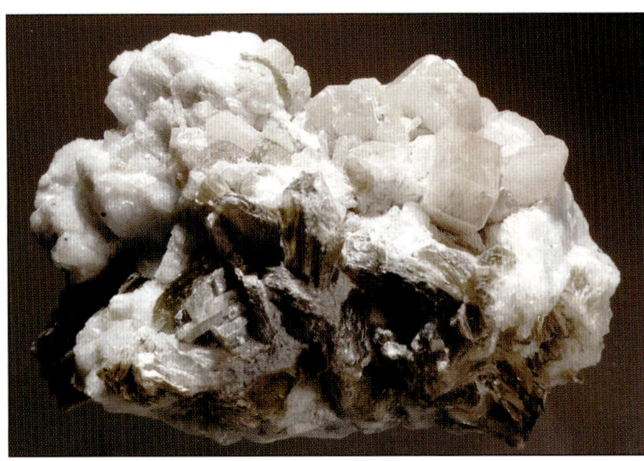

FLUORAPATITE. Fluorapatite crystals with muscovite from Shigar Valley, Skardu district. The fluorapatite crystals fluoresce bright yellow SW. The piece weighs 4.5 oz. and is 2.5 x 1.5 x 1.5 inches. Value $70-85

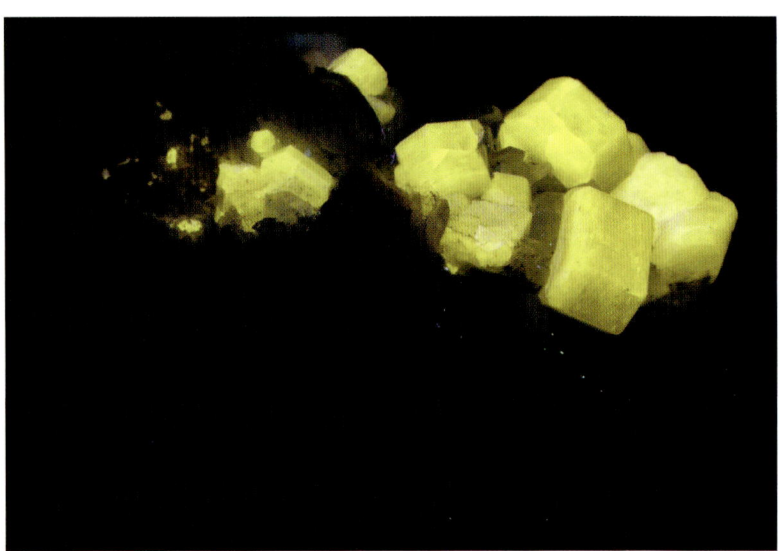

Same under SW

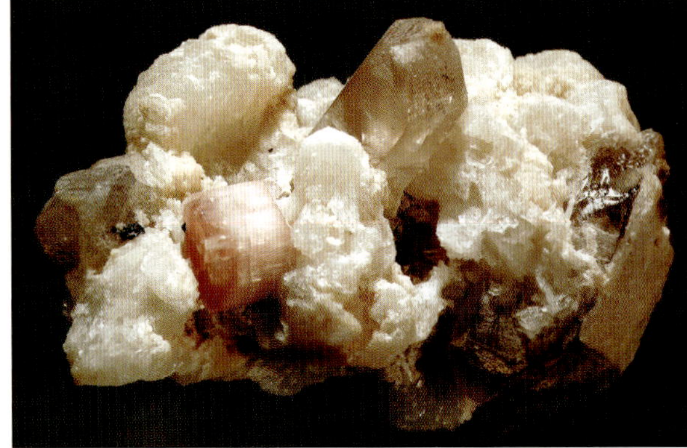

FLUORAPATITE. A fluorapatite crystal with quartz from Shigar Valley, Skardu district. The fluorapatite crystals fluoresce bright yellow SW. The piece weighs 4.8 oz. and is 2.8 x 1.8 x 1.8 inches. Value $60-70

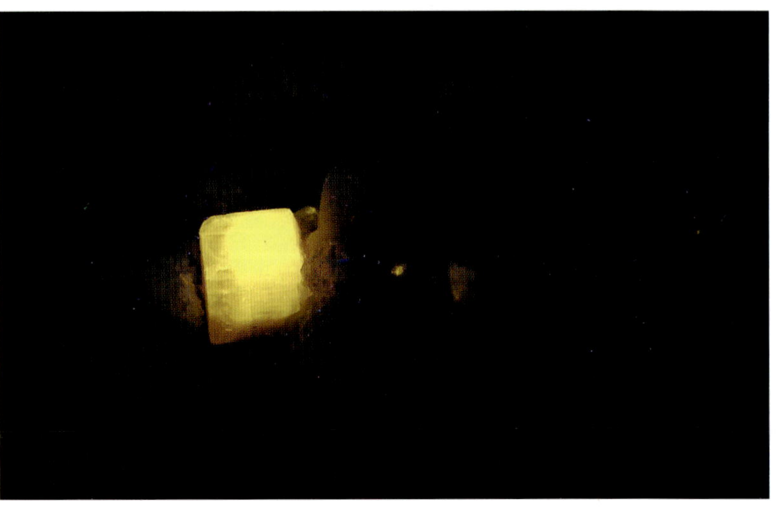

Same under SW

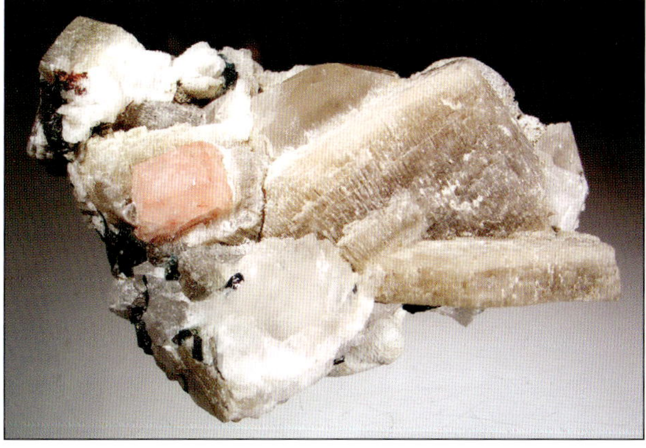

FLUORAPATITE. Fluorapatite in muscovite, quartz, and black schorl from Shigar Valley, Skardu district. The fluorapatite fluoresces bright yellow SW. The piece weighs 3.0 oz.and is 2.3 x 1.8 x 1.8 inches. Value $120-130

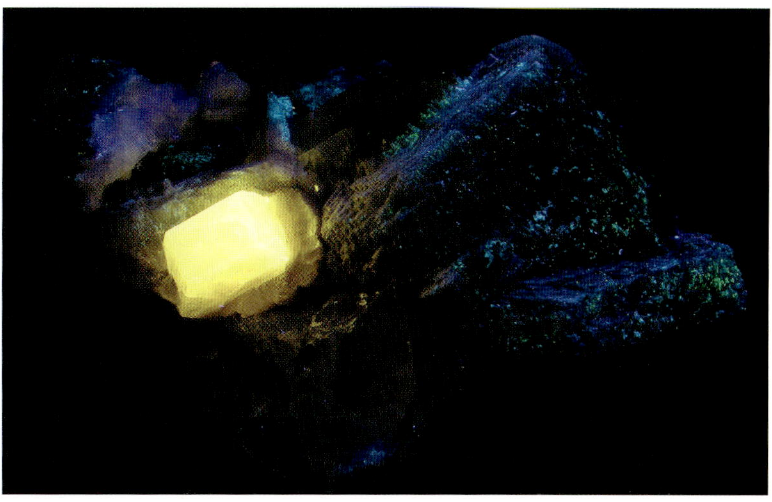

Same under SW

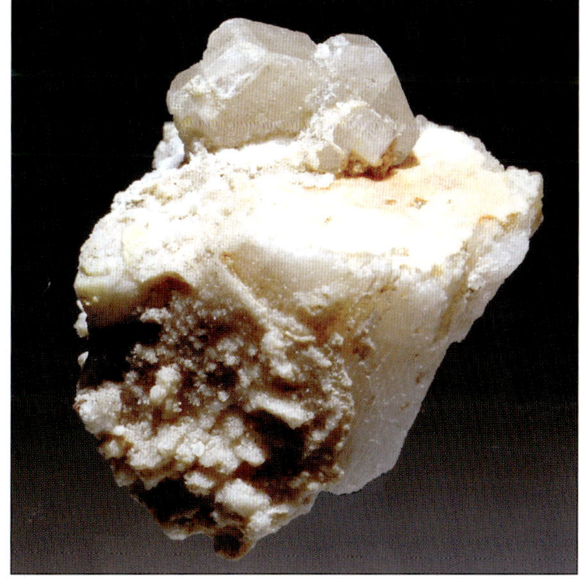

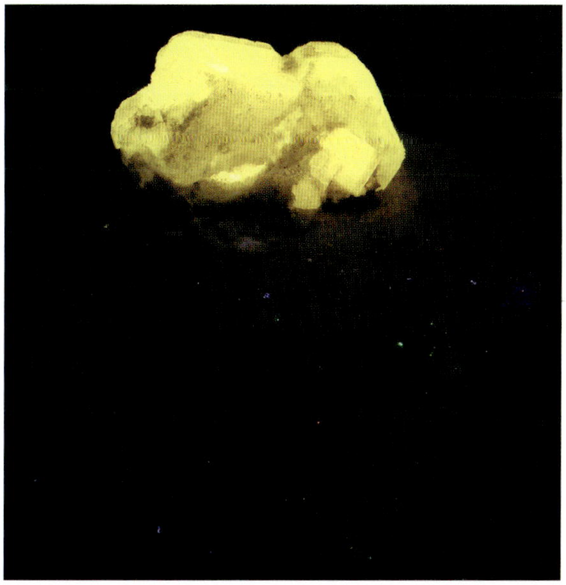

Far left: FLUORAPATITE. Crystals of fluorapatite in feldspar from Dassu, Gilgit district. The fluorapatite fluoresces yellow SW and the feldspar fluoresces pale blue SW. The piece weighs 3.0 oz. and is 2.3 x 1.5 x 1.5 inches. Value $60-70

Left: Same under SW

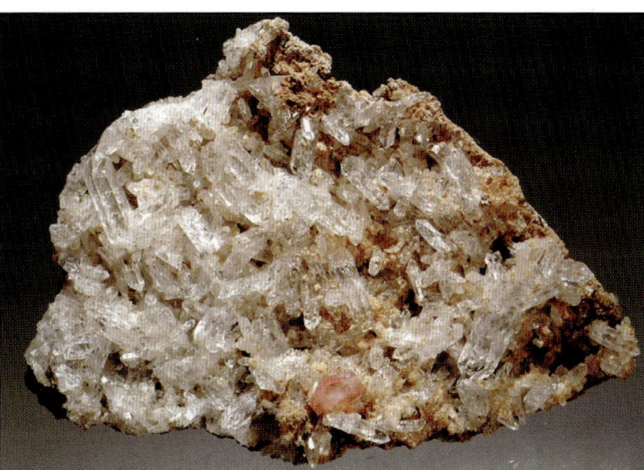

FLUORAPATITE. A fluorapatite crystal on quartz from Dalbandin, Balochistan. The fluorapatite crystals from this area are characterized by their thinness. The fluorapatite fluoresces dark pink SW. The piece weighs 2.5 oz. and is 3.5 x 2.5 x 0.5 inches. Value $80-90

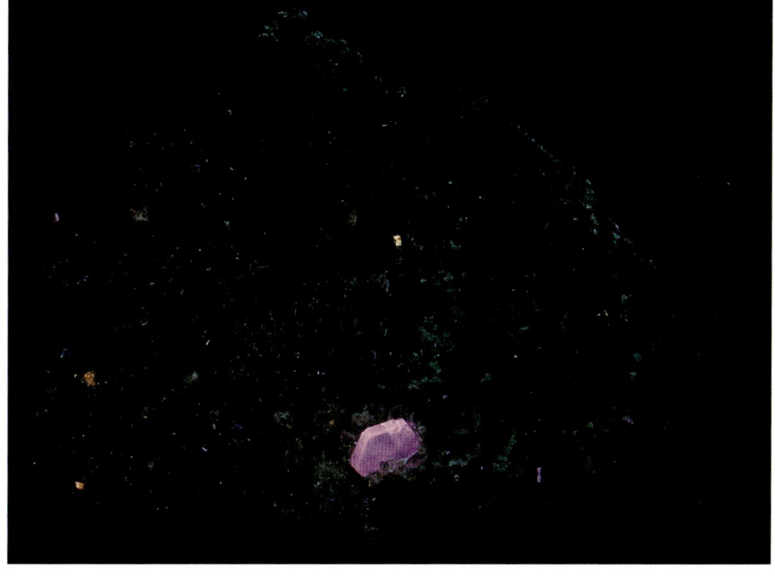

Same under SW

FLUORAPATITE. Crystals of fluorapatite on quartz from Tui, near Wana, Waziristan. This fluorapatite fluoresces dark pink SW. The piece weighs 1.0 oz. and is 1.8 x 1.3 x 0.8 inches. Value $70-85

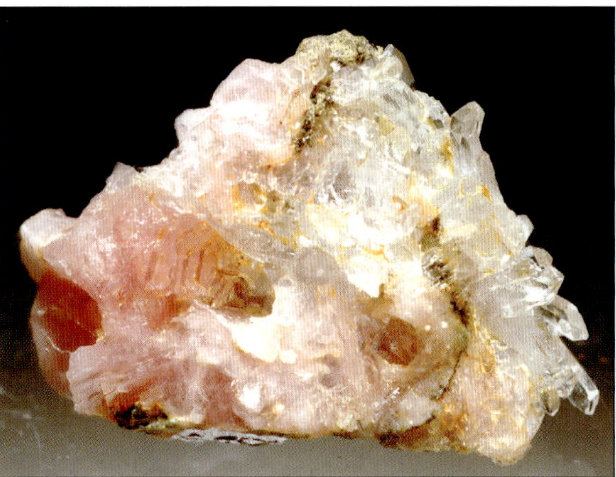

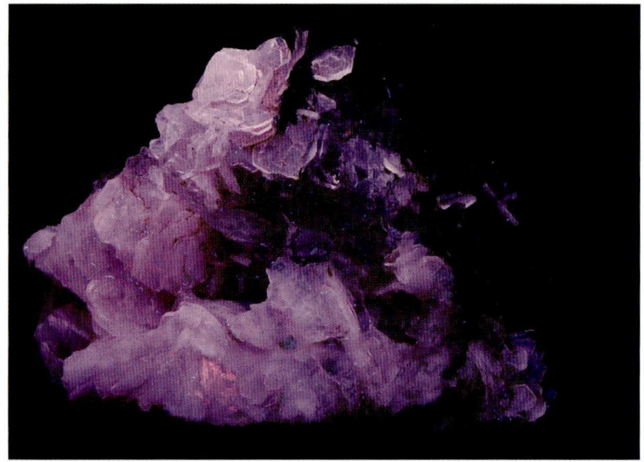

Same under SW

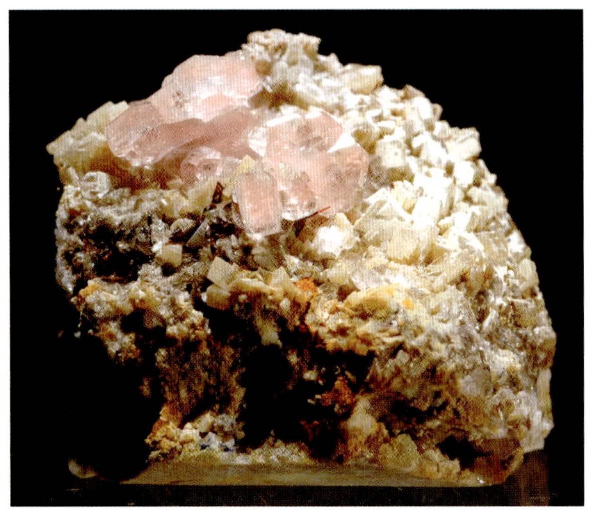

FLUORAPATITE. Hexagonal crystals of fluorapatite from Tui, near Wana, Waziristan. This fluorapatite fluoresces dark pink SW. The piece weighs 5.0 oz. and is 2.0 x 2.0 x 1.5 inches. Value $200-250

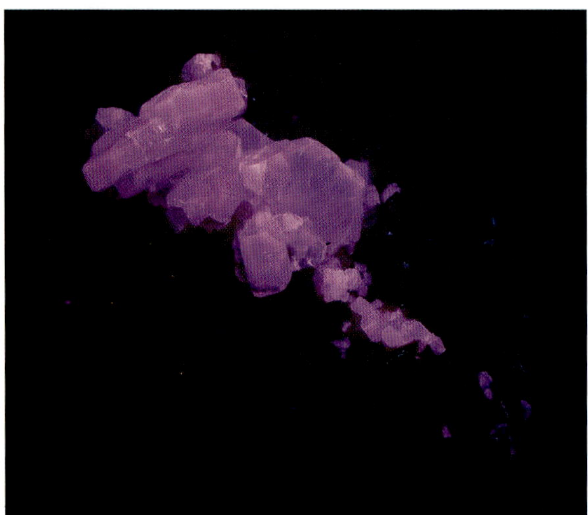

Same under SW

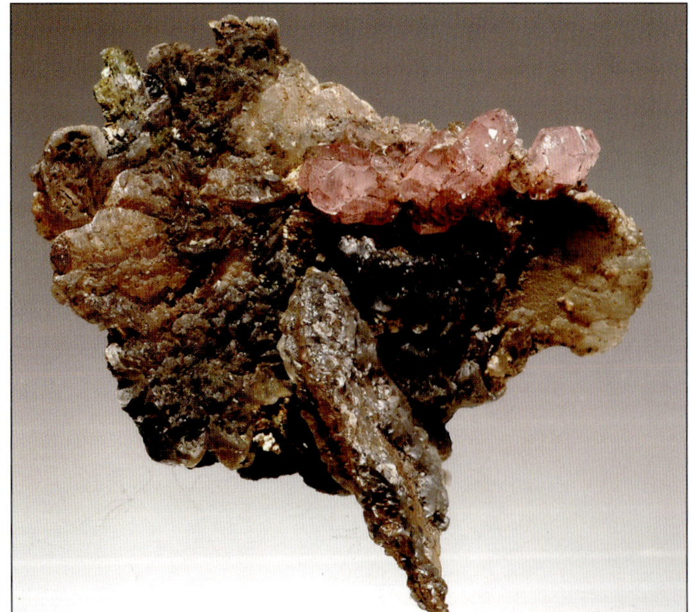

FLUORAPATITE. Hexagonal crystals of fluorapatite from Tui, near Wana, Waziristan. This fluorapatite fluoresces dark pink SW. The piece weighs 1.1 oz. and is 2.5 x 2.3 x 1.5 inches. Value $125-145

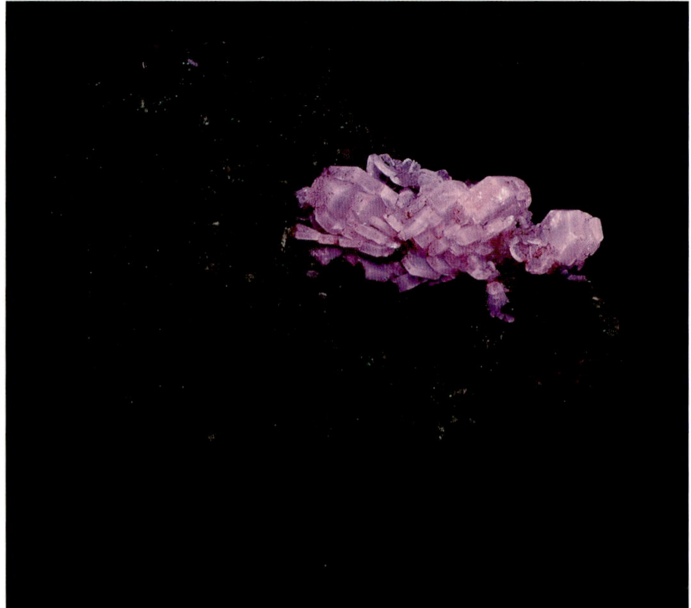

Same under SW

FLUORITE. Fluorite crystals from Hunza Valley, Gilgit district. The crystals fluoresce pale yellow SW and violet LW. The piece weighs 1.7 oz. and is 1.8 x 1.3 x 1 inches. Value $45-50

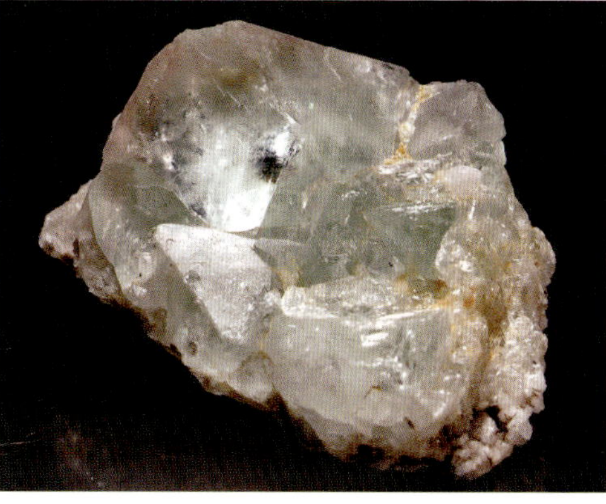

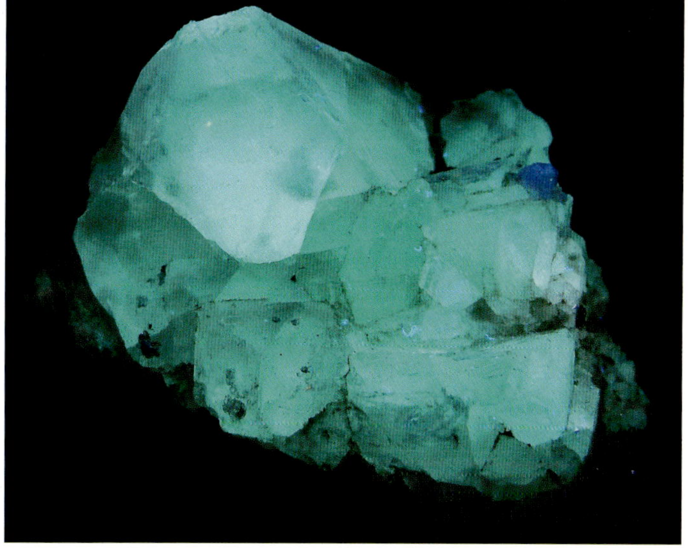

Same under SW

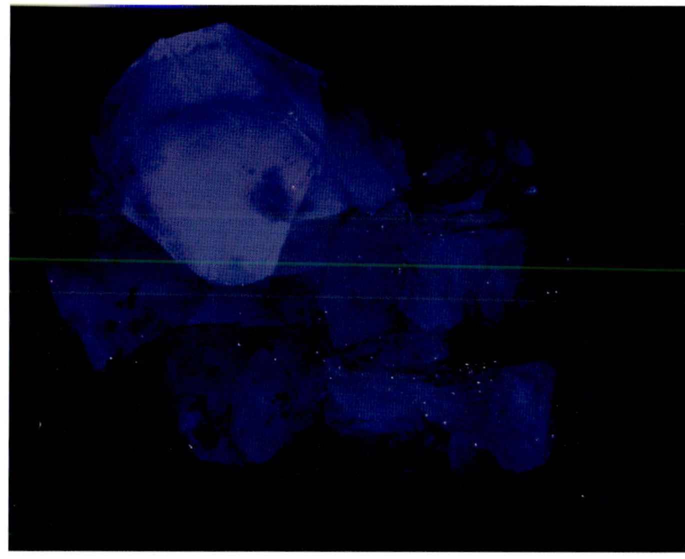

Same under LW

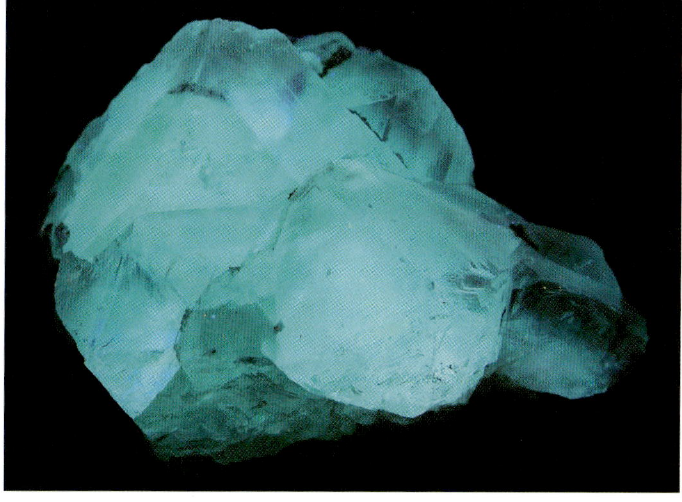

Same under SW

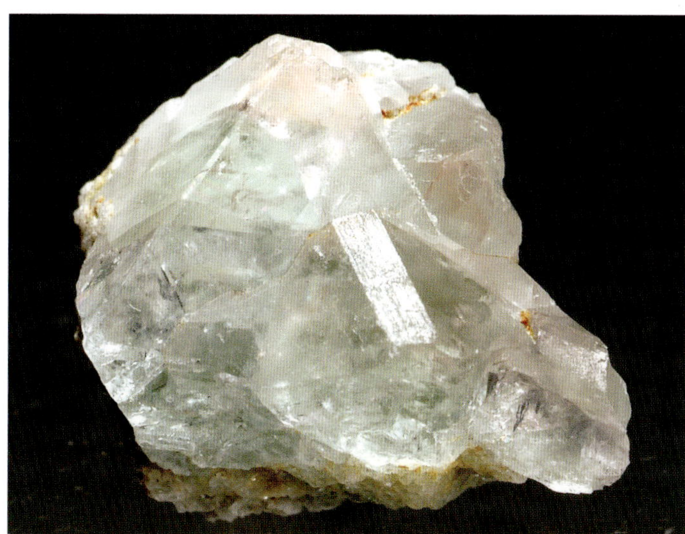

FLUORITE. FLuorite crystals from Hunza Valley, Gilgit district. The crystals fluoresce pale yellow SW and violet LW. The piece weighs 1.3 oz. and is 1.4 x 1.3 x 1.0 inches. Value $40-45

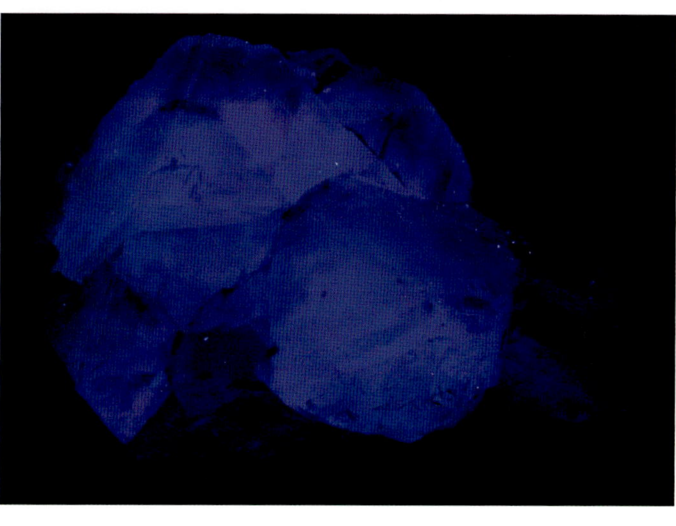

Same under LW

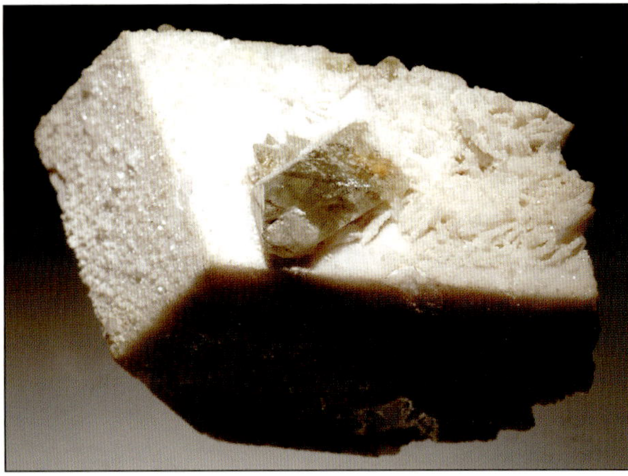

HERDERITE. A rare, gemmy crystal of herderite on feldspar from Shigar Valley, Skardu district. Herderite fluoresces purple SW and LW and feldspar fluoresces pale blue SW. Herderite was discovered in Freiberg, Saxony, Germany and named for Siegmund A. W. Von Herder (1776-1838), a mine official. The piece weighs 9.8 oz. and is 3.0 x 2.3 x 1.8 inches. Value $90-110

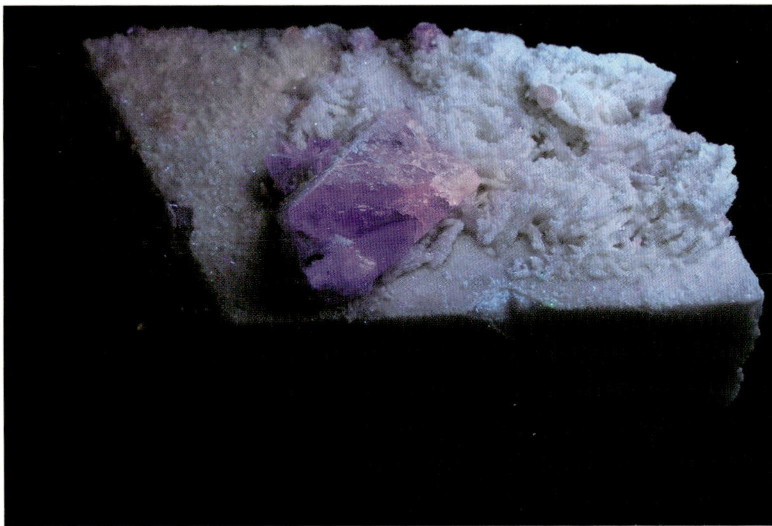

Same under SW. There is a coating of an unidentified crystalline mineral on the feldspar that is reducing its fluorescence on the side facing the camera.

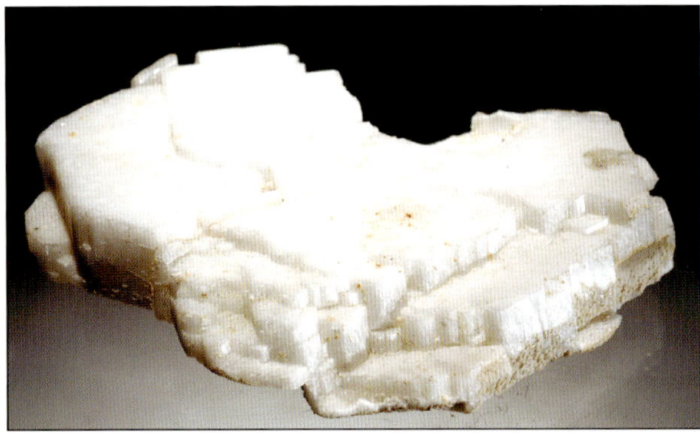

MICROCLINE. Large flat crystals of microcline from Dassu, Skardu district. The microcline fluoresces bright pale blue SW. The piece weighs 5.8 oz. and is 4.0 x 3.0 x 0.8 inches. Value $40-45

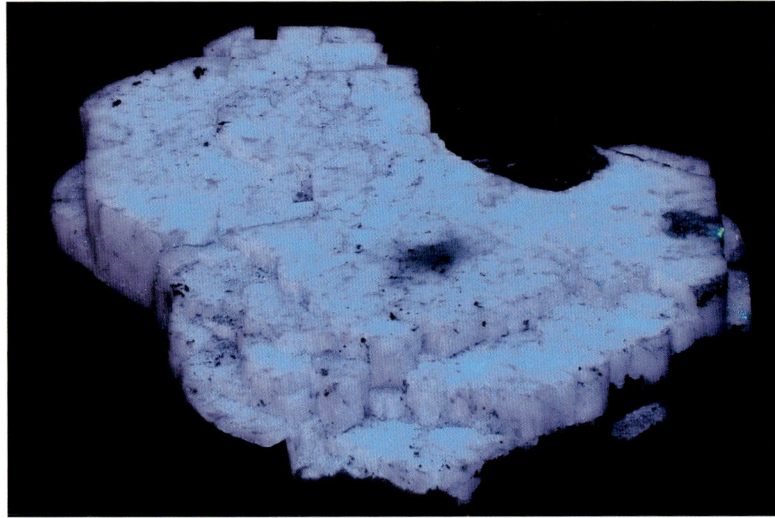

Same under SW

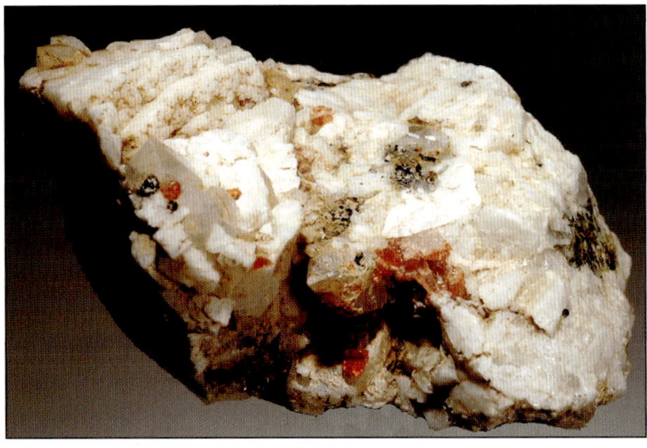

MICROCLINE. Spessartine garnet with quartz on microcline from Shigar Valley, Skardu district. The microcline fluoresces pale blue SW. The piece weighs 9.0 oz. and is 3.5 x 3 x 1.5 inches. Value $50-60

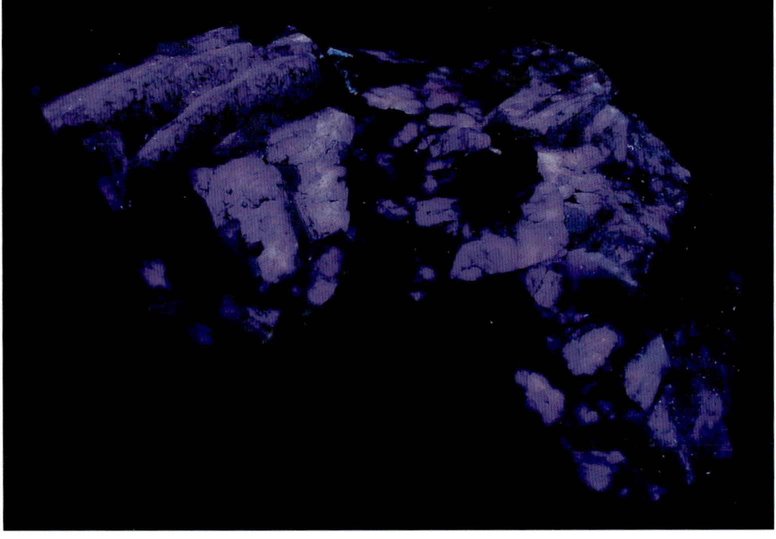

Same under SW

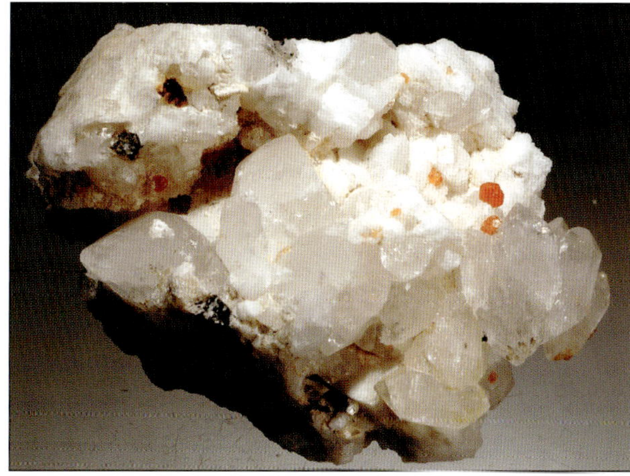

MICROCLINE. Spessartine garnet with quartz, titanite, and muscovite on microcline from Shigar Valley, Skardu district. The microcline fluoresces pale blue SW and the quartz fluoresces, unevenly, green SW. The piece weighs 9.5 oz. and is 3.8 x 2.5 x 2.0 inches. Value $75-85

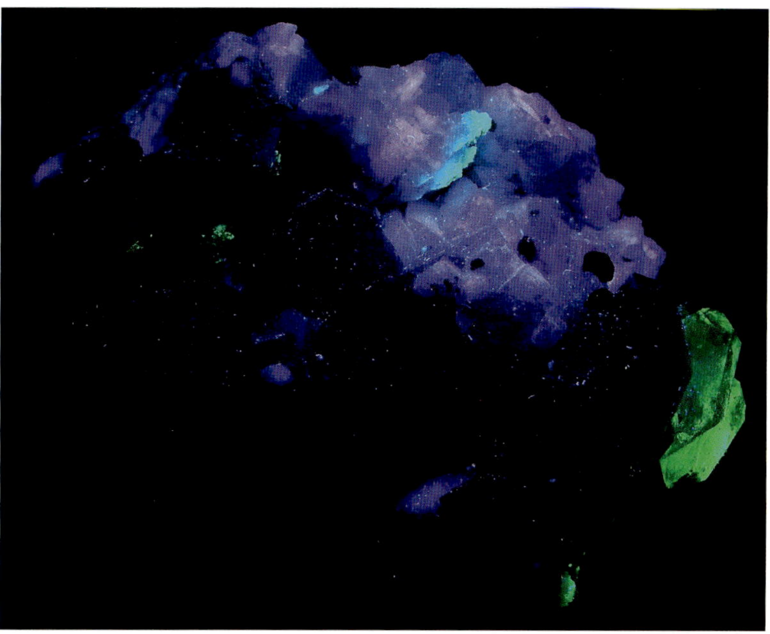

Same under SW

MICROCLINE. Large microcline crystals from Shigar Valley, Skardu district. The microcline fluoresces pale blue SW. The piece weighs 6.0 oz. and is 2.8 x 2.8 x 1.8 inches. Value $60-70

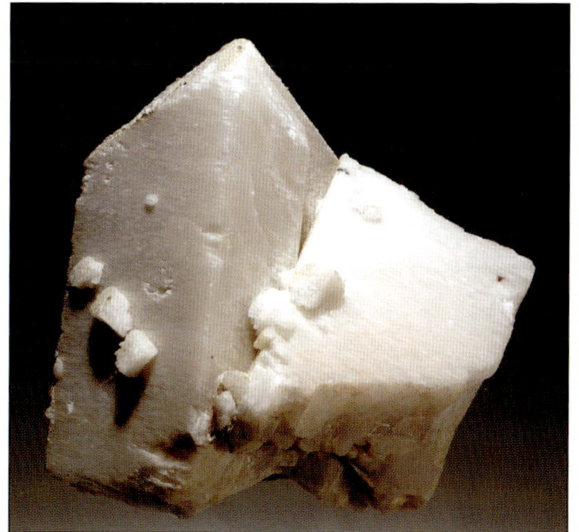

Same under SW

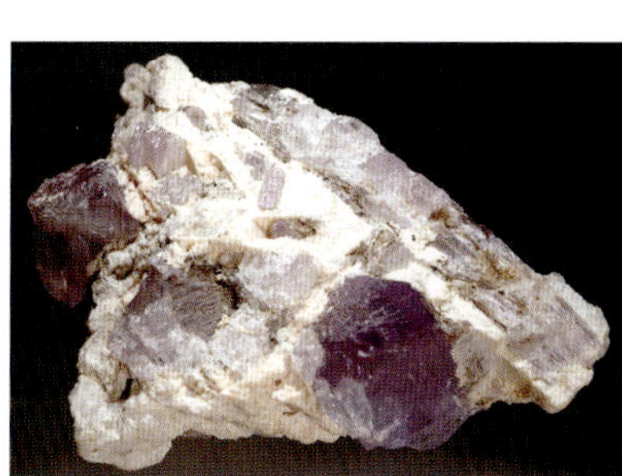

Same under SW

SPODUMENE. Gemmy crystals of spodumene, var. kunzite, from North Waziristan. The spodumene fluoresces violet SW and pink-violet LW. The piece weighs 8.5 oz. and is 3.5 x 2.5 x 1.8 inches. Value $225-275

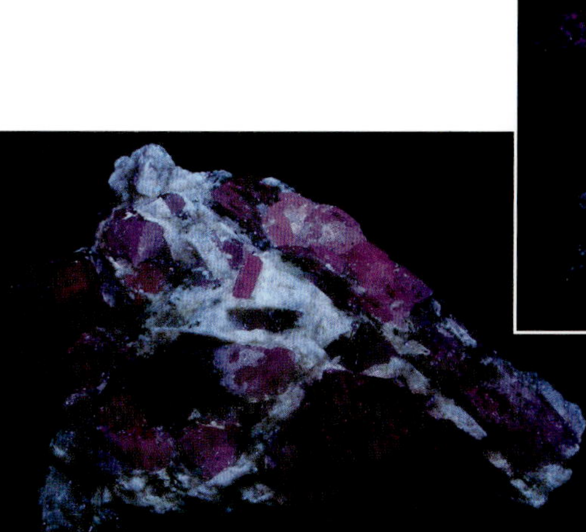

Same under LW

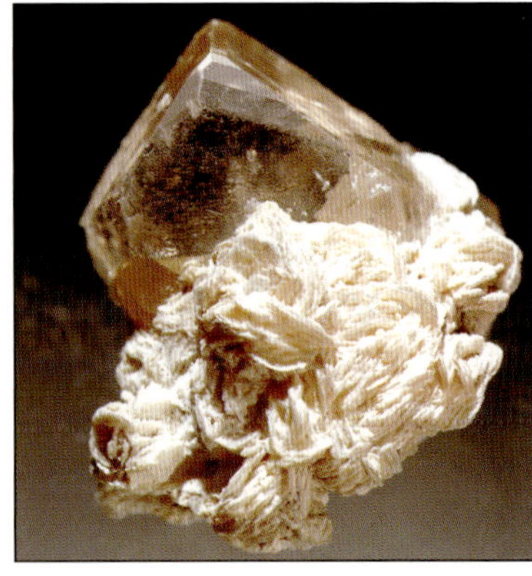

TOPAZ. A gemmy topaz crystal on muscovite from Shengus, Skardu district. The topaz has an uneven fluorescence. The topaz fluoresces pale yellow SW. The piece weighs 0.9 oz. and is 1.0 x 1.0 x 0.8 inches. Value $80-85

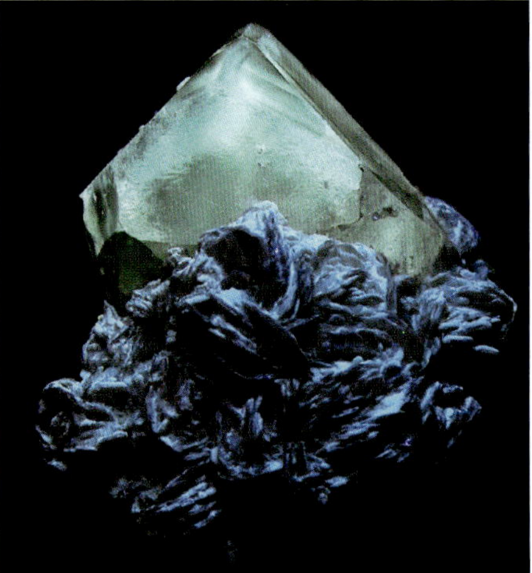

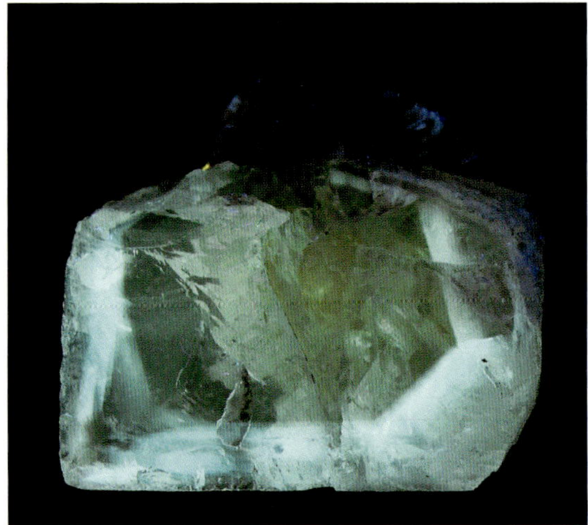

Same under SW. Two views of the crystal showing the areas that fluoresce unevenly.

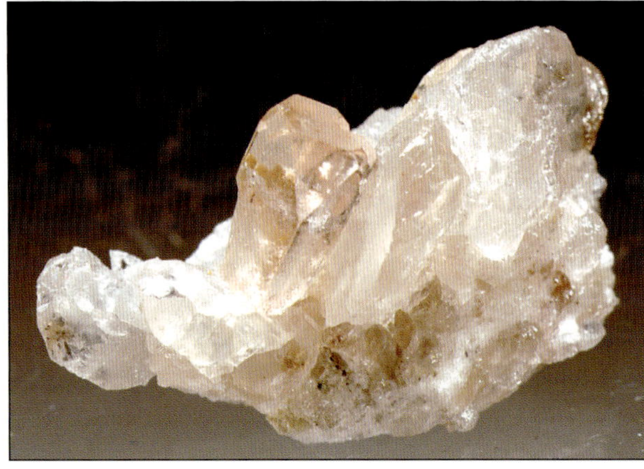

TOPAZ. Gemmy topaz crystals on quartz with grains of an unidentified mineral from Shengus, Skardu district. The topaz fluoresces pale yellow SW and the unidentified mineral (possibly zircon) fluoresces orange SW. The piece weighs 0.3 oz. and is 1.4 x 0.8 x 0.8 inches. Value $35-45

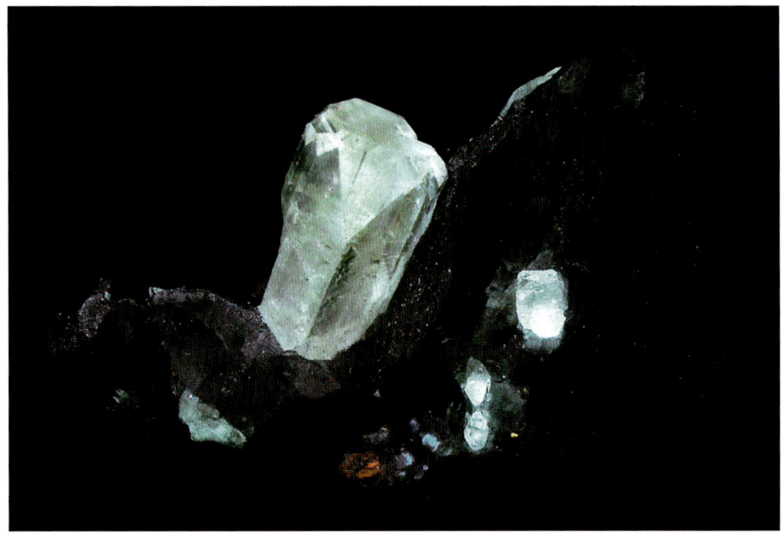

Same under SW

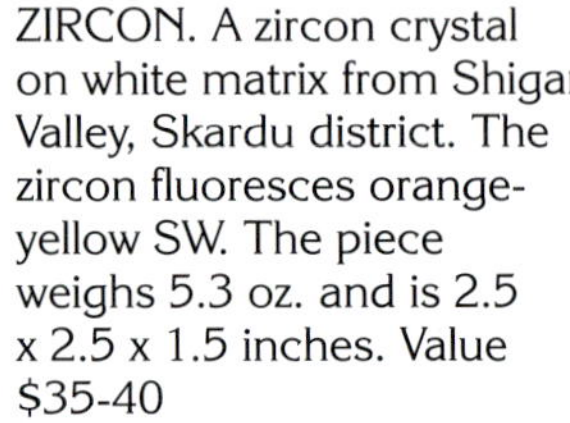

ZIRCON. A zircon crystal on white matrix from Shigar Valley, Skardu district. The zircon fluoresces orange-yellow SW. The piece weighs 5.3 oz. and is 2.5 x 2.5 x 1.5 inches. Value $35-40

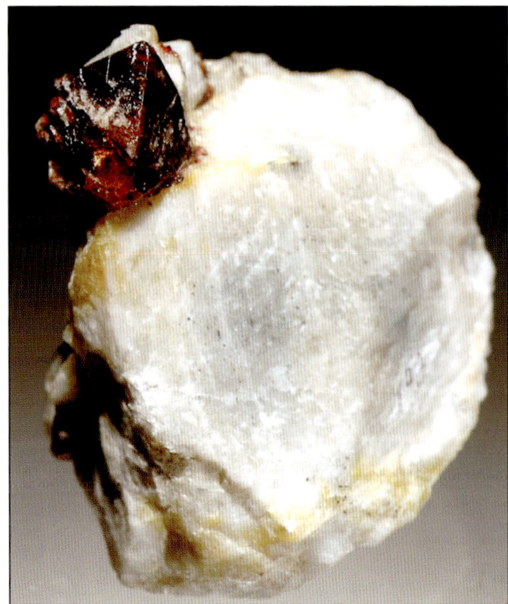

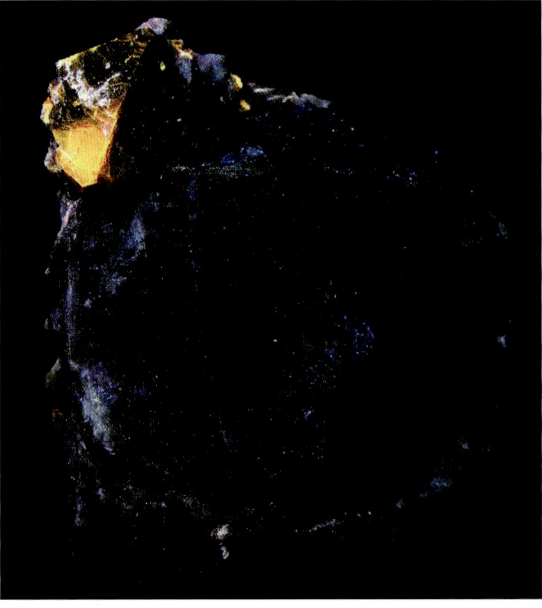

Same under SW

Peru

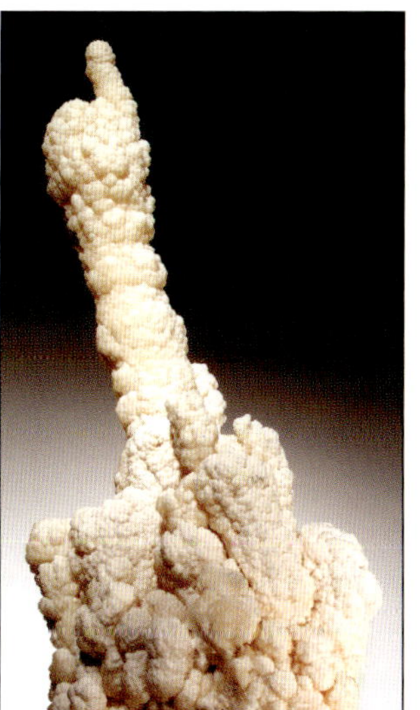

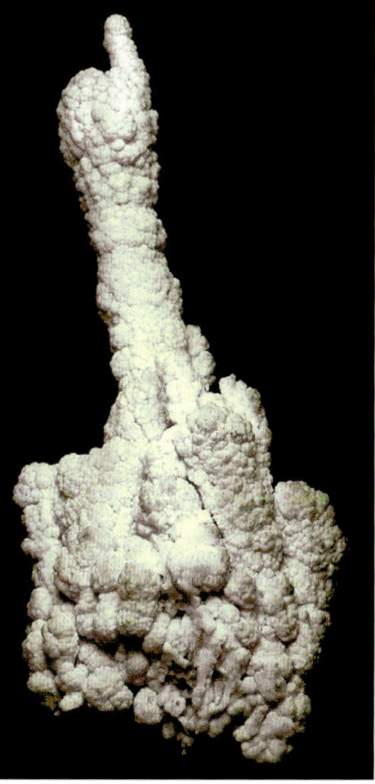

Far left:
ARAGONITE. An aragonite stalactite from La Oroyo, Junin Department. The aragonite fluoresces white SW with a long-lasting phosphorescence. The piece weighs 12.0 oz. and is 8.0 x 3.0 x 2.0 inches. Value $75-85

Left:
Same under SW

Right:
BARITE. Barite crystals from near Warihuayin, Huamalies Province, Huanuco Department. The crystals fluoresce pale yellow and have hour glass-shaped zones of fluorescence SW and LW. The piece weighs 5.5 oz. and is 2.8 x 2.4 x 1.5 inches. Value $45-50

Far right:
Same under SW

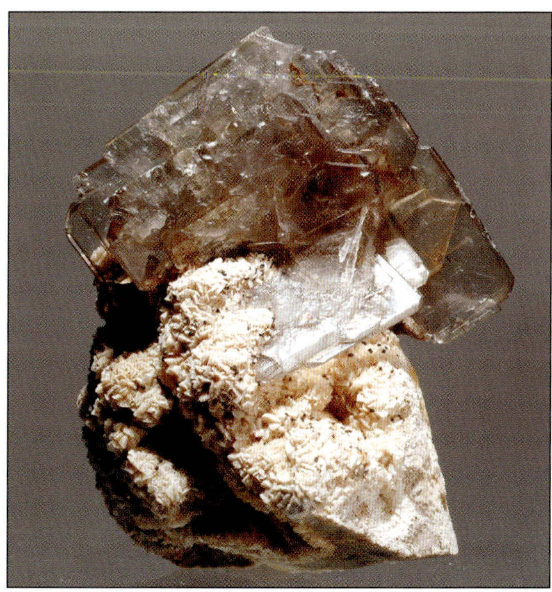

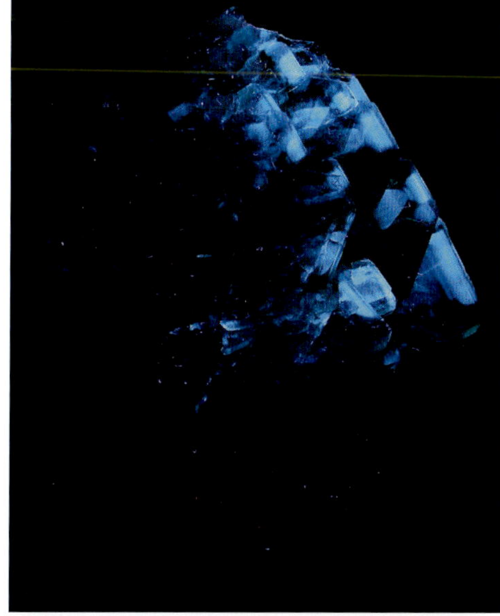

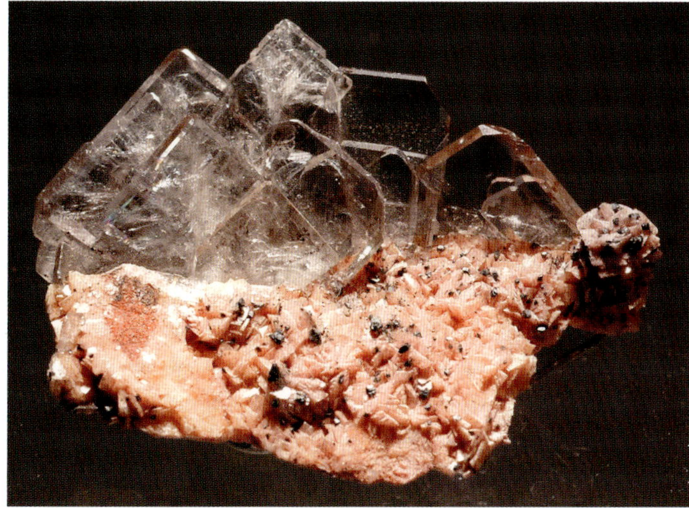

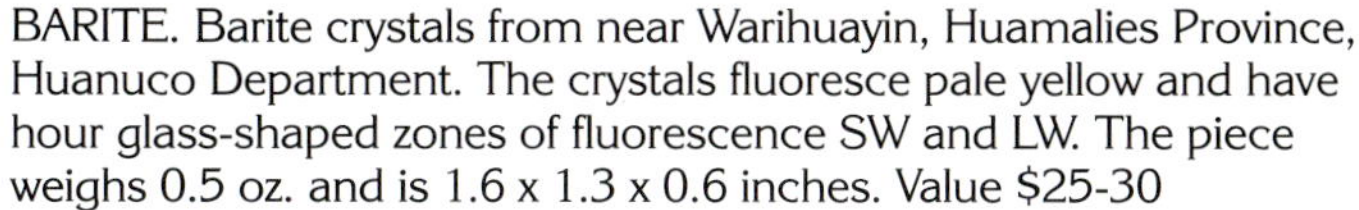

BARITE. Barite crystals from near Warihuayin, Huamalies Province, Huanuco Department. The crystals fluoresce pale yellow and have hour glass-shaped zones of fluorescence SW and LW. The piece weighs 0.5 oz. and is 1.6 x 1.3 x 0.6 inches. Value $25-30

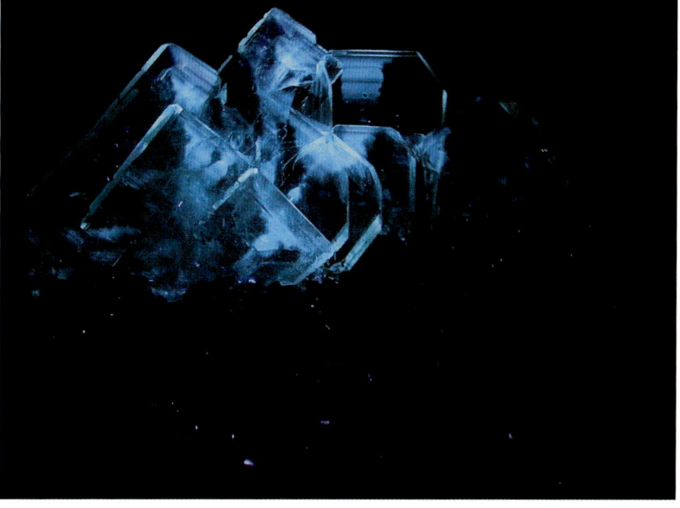

Same under SW

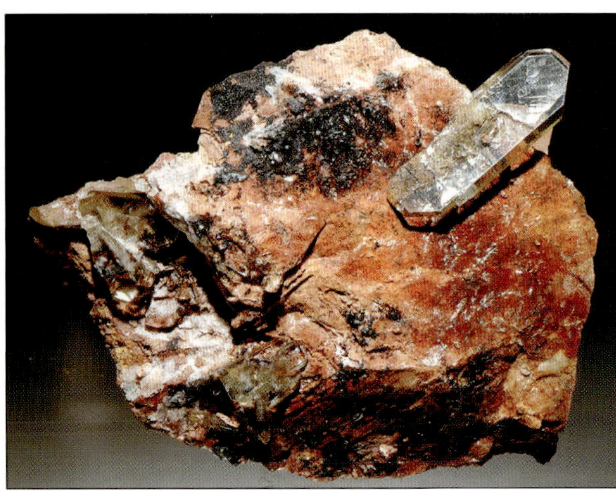

BARITE. Barite crystals from Dos de Mayo Province, Haunuco Department. The crystals fluoresce pale yellow and have hour glass-shaped zones of fluorescence SW and LW. The largest crystal is 1.8 inches long. The piece weighs 15.5 oz. and is 4.5 x 3.8 x 2.0 inches. Value $110-135

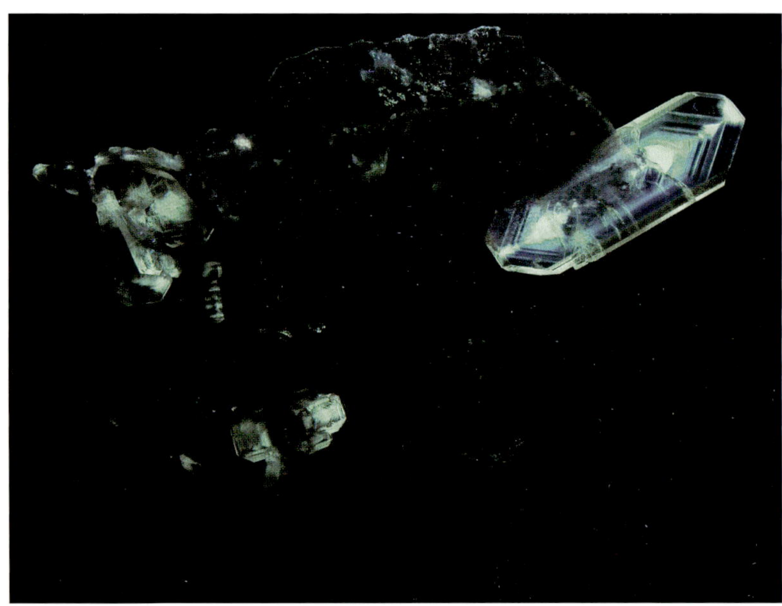

Same under SW

FLUORAPATITE. Fluorapatite on pyrite from the Huanzala mine, Huallanca district, Dos de Mayo Province, Huanuco Department. The fluorescent color (which is about the same under SW, MW, LW 350, and LW 370) is unique. It a hard-to-describe blue. The piece weighs 1 lb. 1.0 oz. and is 2.5 x 3.8 x 3.9 inches. Courtesy of Don Newsome. Photos by Jeff Scovil.

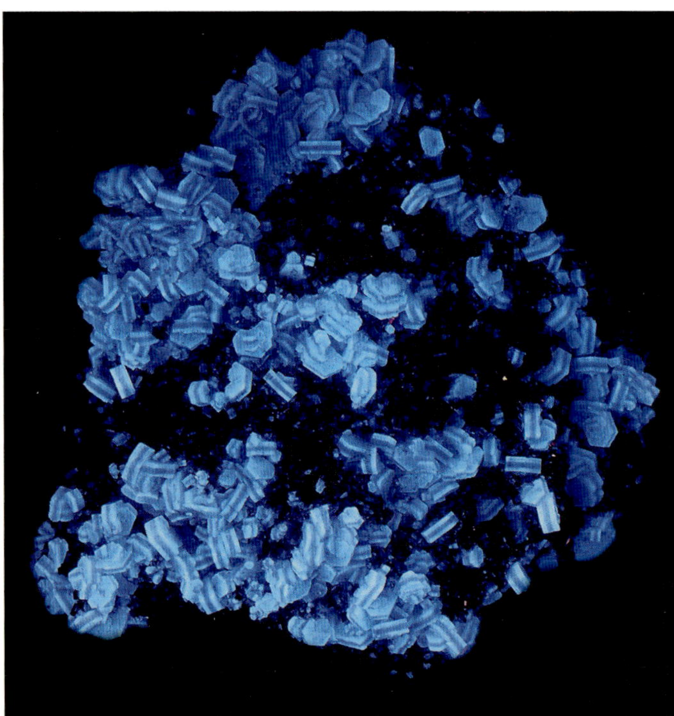

Same under SW

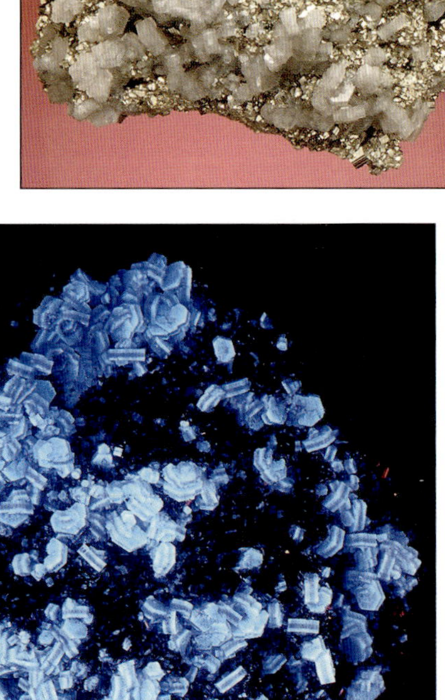

Same under 350 LW

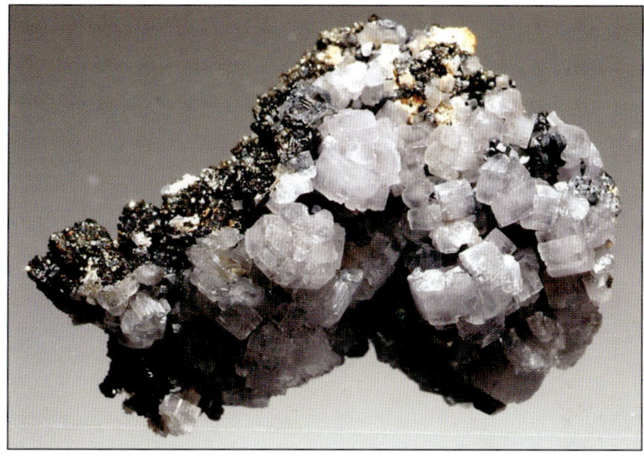

FLUORAPATITE. Fluorapatite crystals on pyrite crystals from the Huanzala mine, Huallanca district, Dos de Mayo Province, Huanuco Department. The fluorapatite fluoresces lavender SW. This piece is smaller than the following piece, but has more fluorapatite crystals. The piece weighs 2.3 oz. and is 2.1 x 1.4 x 1.3 inches. Value $100-125

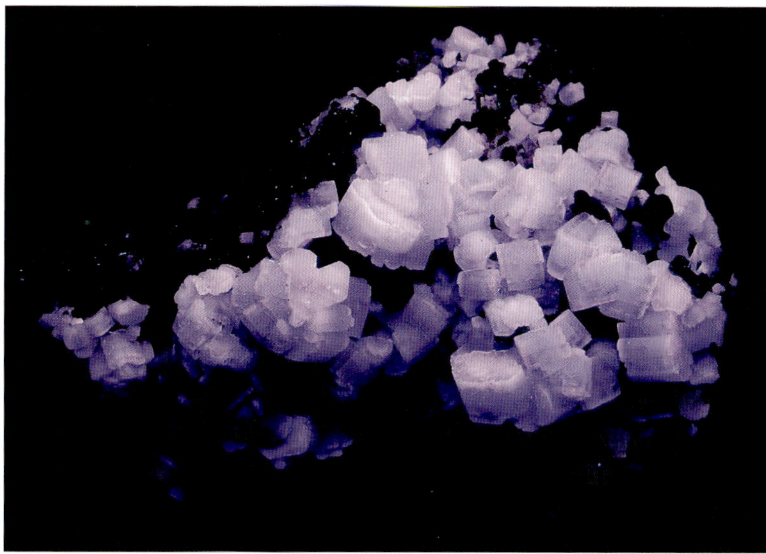

Same under SW

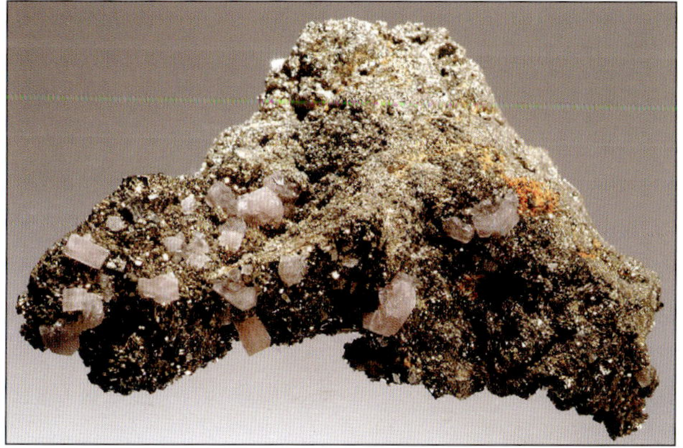

FLUORAPATITE. Fluorapatite crystals on pyrite crystals from the Huanzala mine, Huallanca district, Dos de Mayo Province, Huanuco Department. The fluorapatite fluoresces lavender SW. The piece weighs 6.8 oz. and is 3.3 x 2.0 x 1.5 inches. Value $100-125

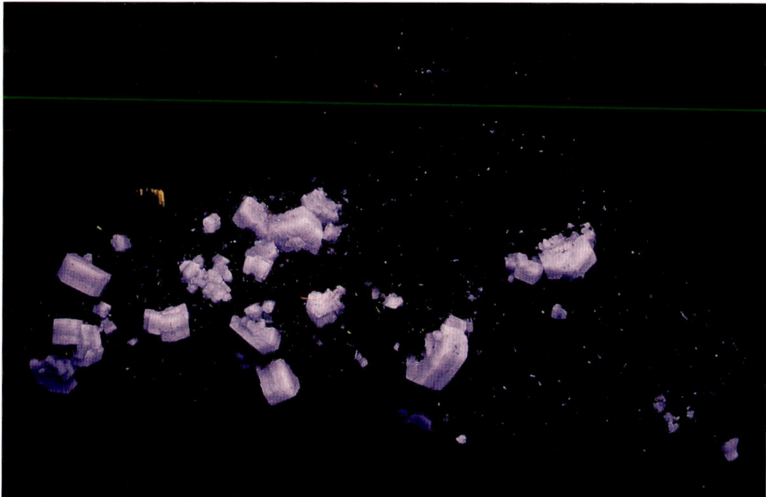

Same under SW

Poland

CELESTINE. Celestine crystal sprays with sulfur from the Machów mine, Tarnobrzeg. Celestine fluoresces pale blue SW. The piece weighs 4.5 oz. and is 3.0 x 2.0 x 1.8 inches. Value $30-40

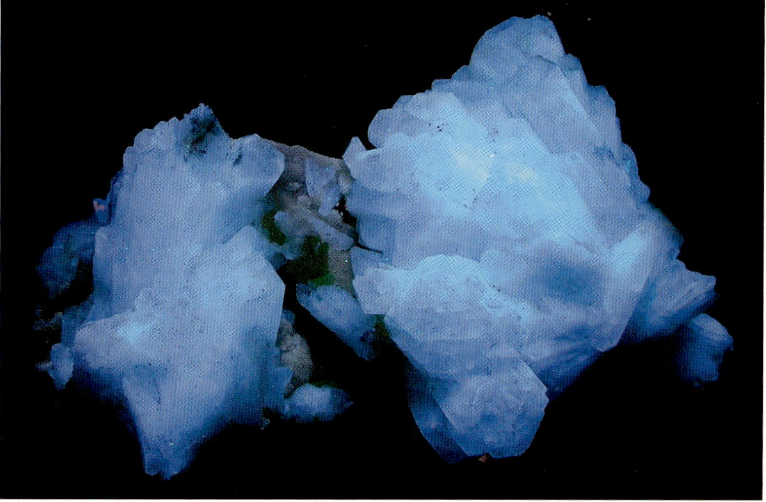

Same under SW

SELENITE. Gypsum, var. selenite, crystals from Poland (no further locality information). This is similar to the Winnipeg, Canada selenite clusters. The selenite fluoresces blue-gray SW and LW and has a long-lasting phosphorescence. The piece weighs 11.5 oz. and is 4.8 x 3.5 x 3.3 inches. Value $50-60

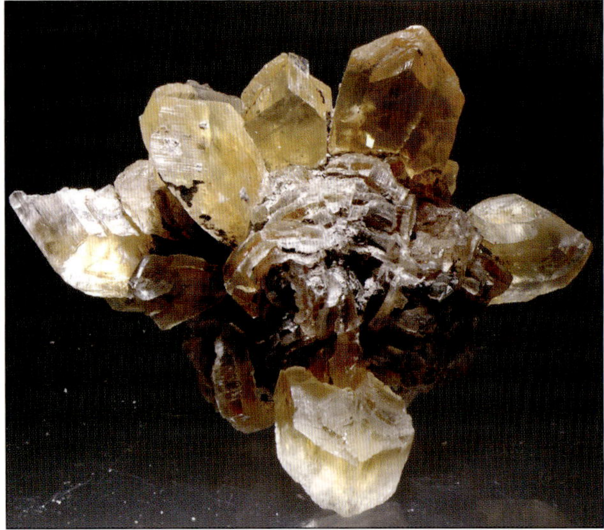

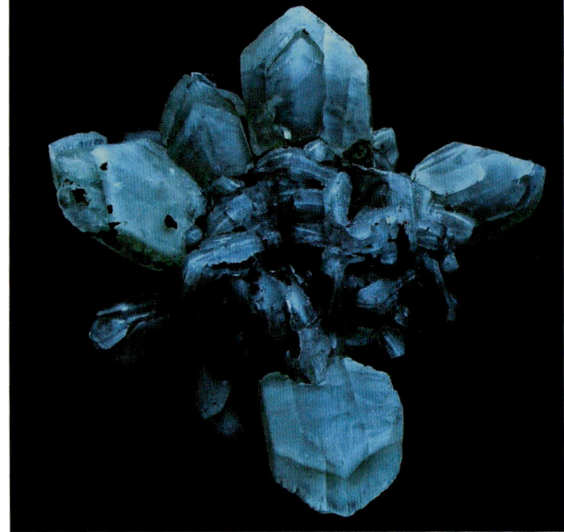

Same under SW

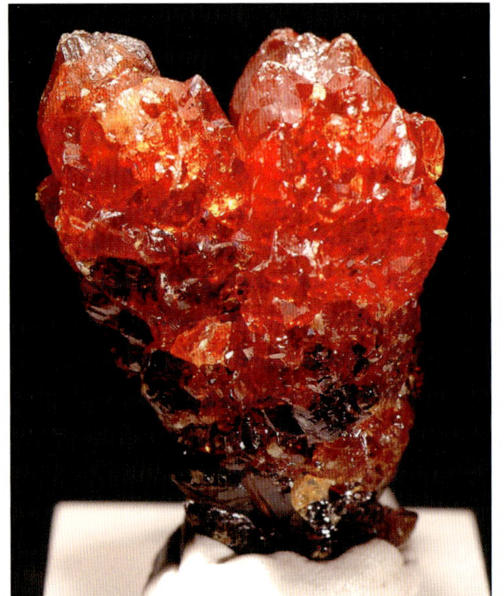

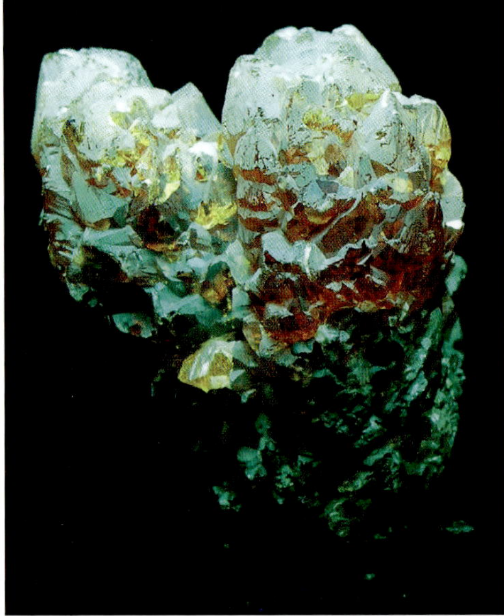

Far left:
ZINCITE. Man-made red zincite crystals from Lower Silesia, Poland. The zincite crystals fluoresce pale yellow SW and LW. It is claimed that when the old smelter was torn down, crystals of zincite were found in the chimneys. The piece weighs 2.3 oz. and is1.5 x 1.3 x 1.0 inches. Value $85-95

Left:
Same under SW

Portugal

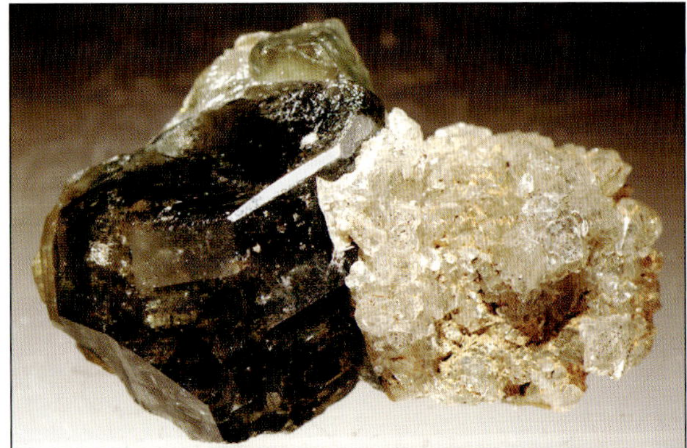

FLUORAPATITE. A large crystal of fluorapatite with siderite and dolomite crystals from Panasqueira, Castelo Branco district. This fluorapatite fluoresces yellow and pale blue SW. The piece weighs 1.1 oz. and is 2.0 x 1.3 x 0.8 inches. Value $150-175

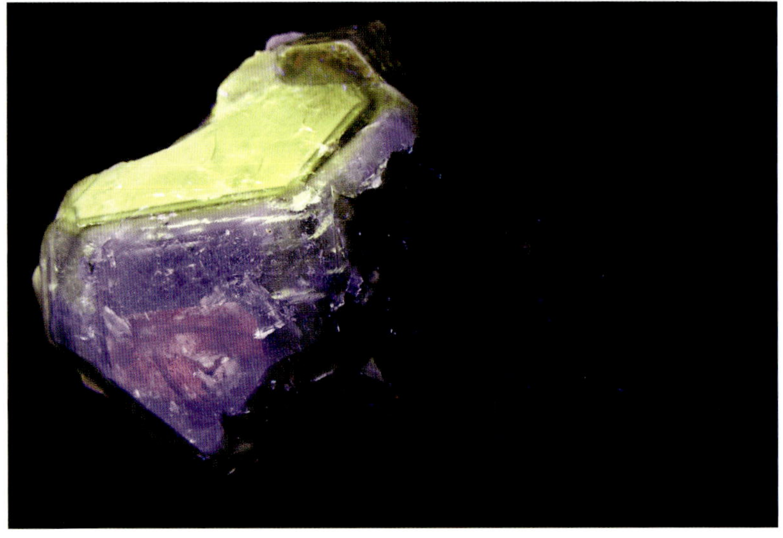

Same under SW

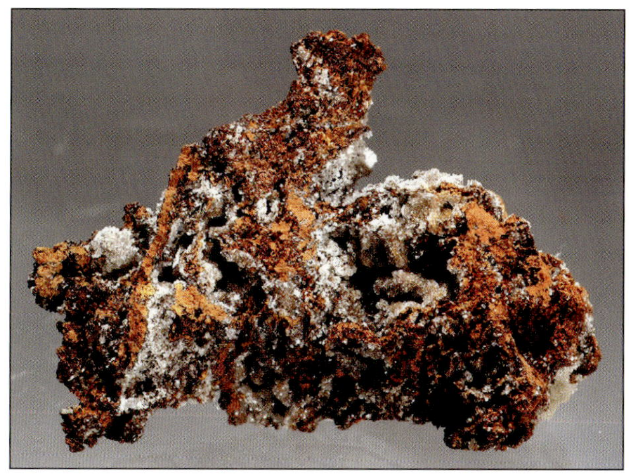

WILLEMITE. Glistening microcrystals of willemite from Sobral da Adiça, Moura, Beja district. Willemite fluoresces bluish green SW. The piece weighs 1.5 oz. and is 2.4 x 1.8 x 1.3 inches. Value $35-40

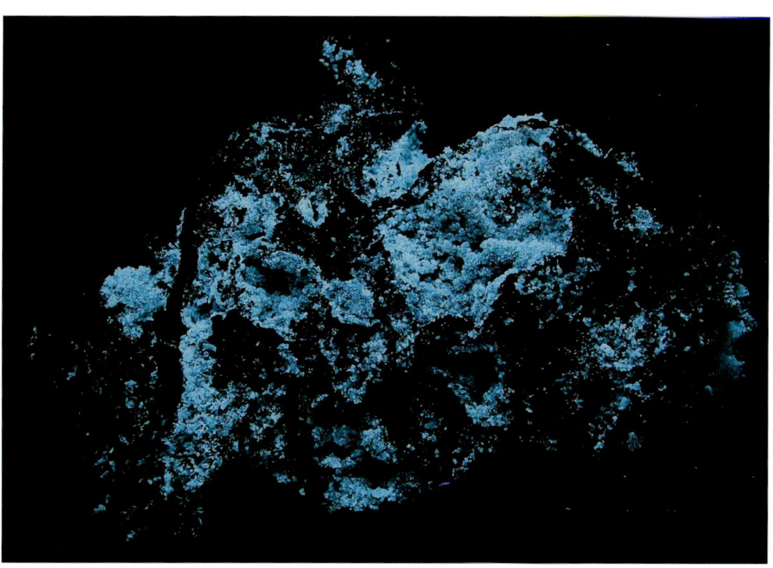

Same under SW.

Puerto Rico

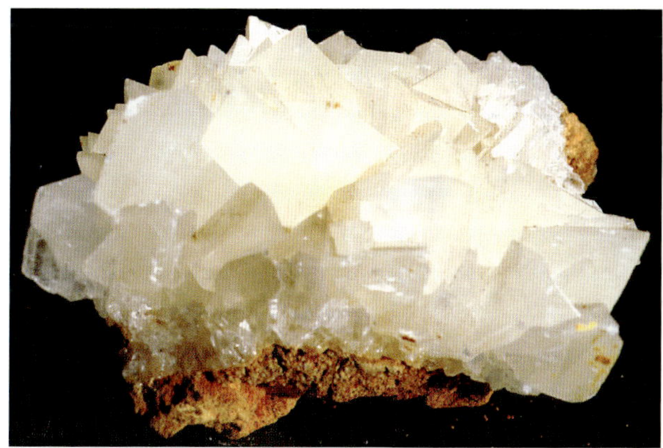

CALCITE. Calcite crystals from San Juan. The calcite fluoresces bright pale blue SW and LW. The piece weighs 4.0 oz. and is 2.5 x 2.0 x 1.5 inches. Value $20-25

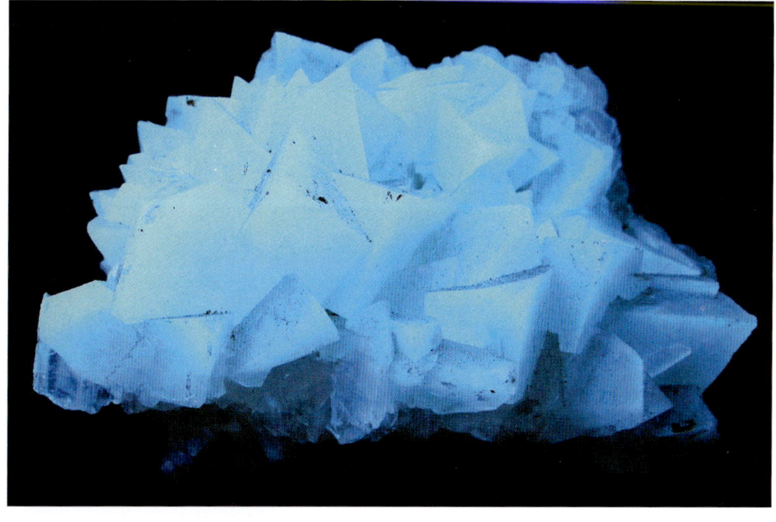

Same under SW

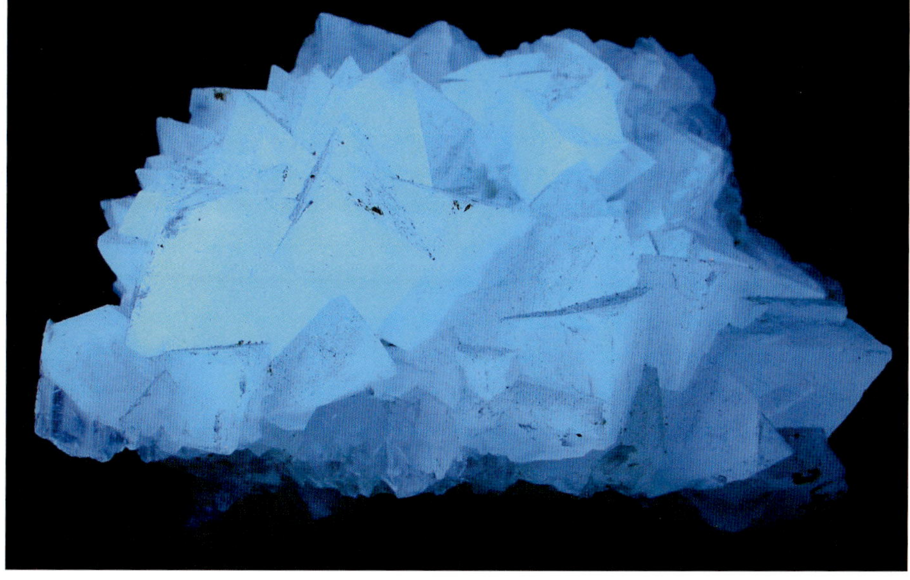

Same under LW

Romania

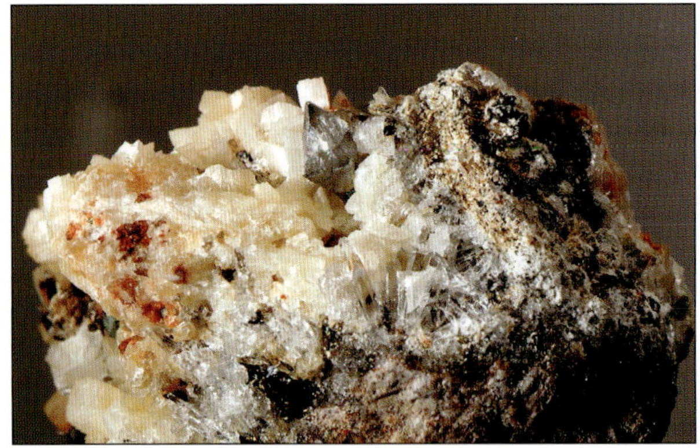

SCHEELITE. A gemmy scheelite crystal on dolomite from Baia Sprie, Maramures County. The scheelite fluoresces bright pale blue SW. The piece weighs 7.7 oz. and is 4.0 x 2.3 x 2.0 inches. Value $45-45

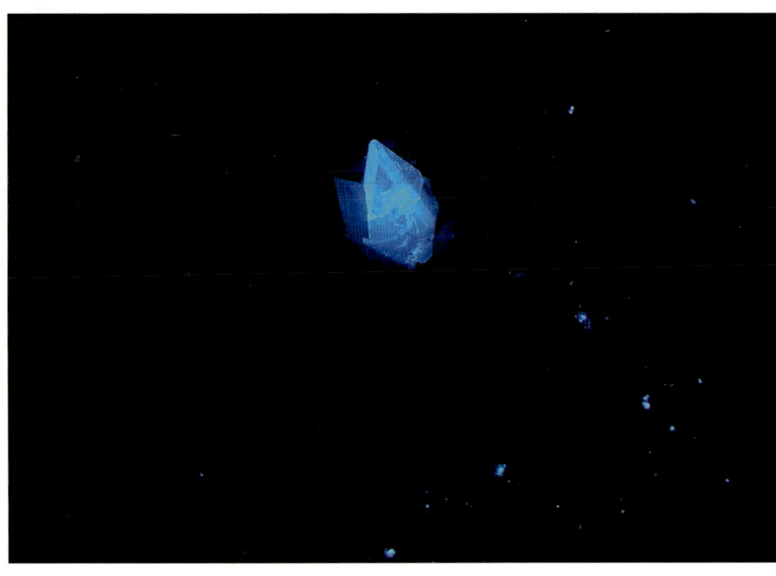

Same under LW

Russia

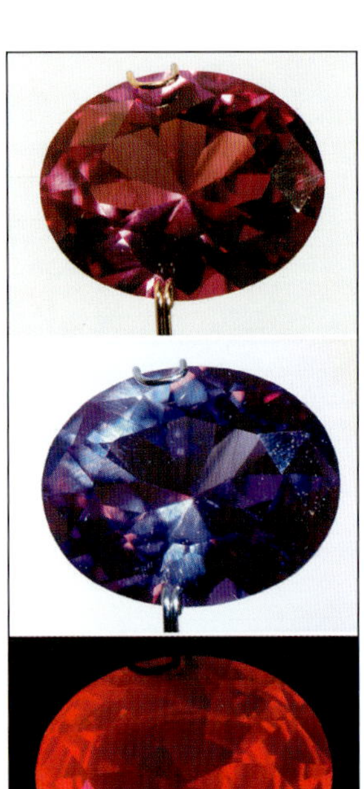

CORUNDUM. Lab-made, oval cut corundum gem (color-changing alexandrite). It fluoresces orange SW and LW. The piece weighs 3.9 ct. and is 11.0 x 9.0 mm. Value $25-30

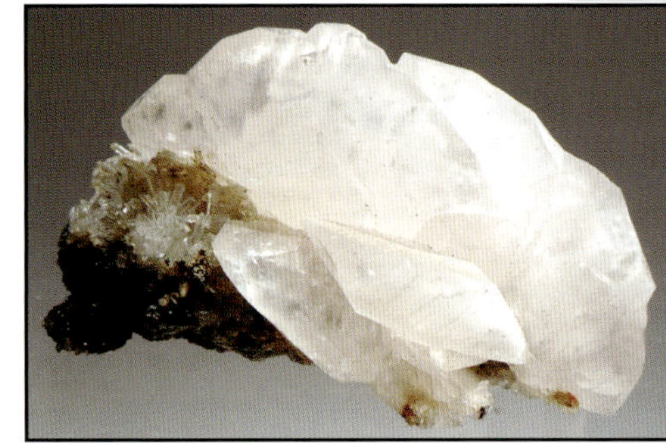

CALCITE. Calcite crystals with quartz from Dal'negorsk, Primorskiy Kray. The crystals fluoresce orange-red LW and (less brightly) SW. The piece weighs 3.0 oz. and is 2.5 x 1.5 x 1.5 inches. Value $22-27

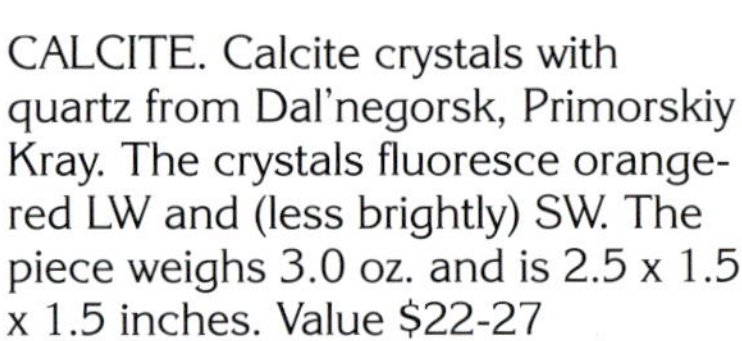

Same under LW

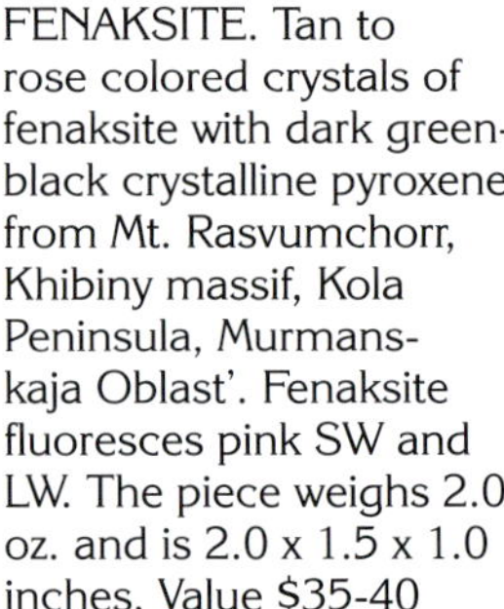

FENAKSITE. Tan to rose colored crystals of fenaksite with dark green-black crystalline pyroxene from Mt. Rasvumchorr, Khibiny massif, Kola Peninsula, Murmanskaja Oblast'. Fenaksite fluoresces pink SW and LW. The piece weighs 2.0 oz. and is 2.0 x 1.5 x 1.0 inches. Value $35-40

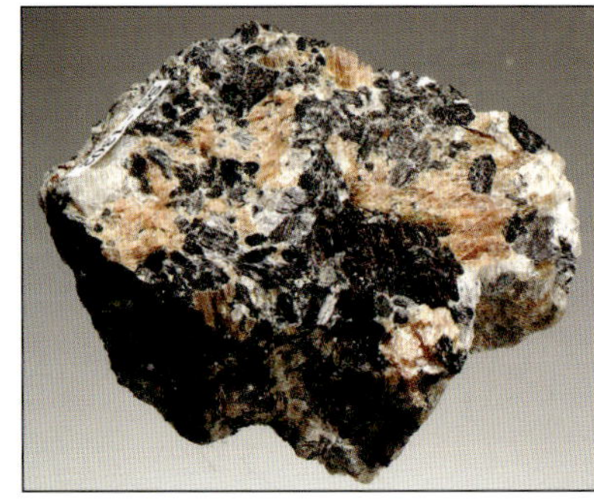

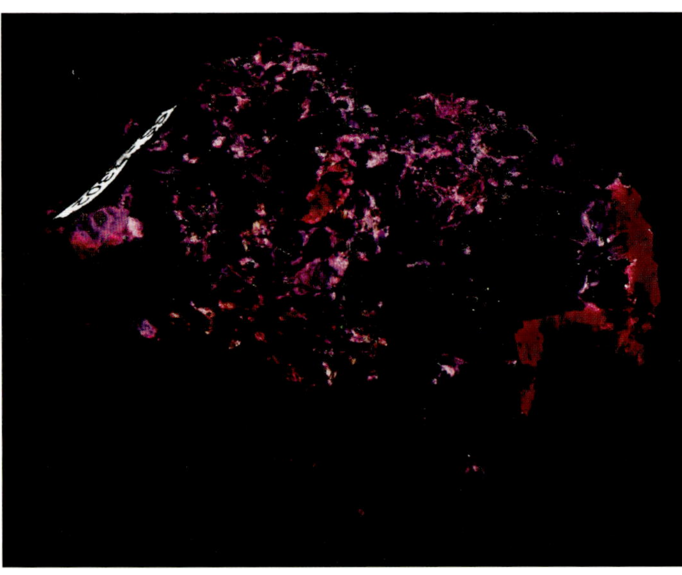

Same under SW

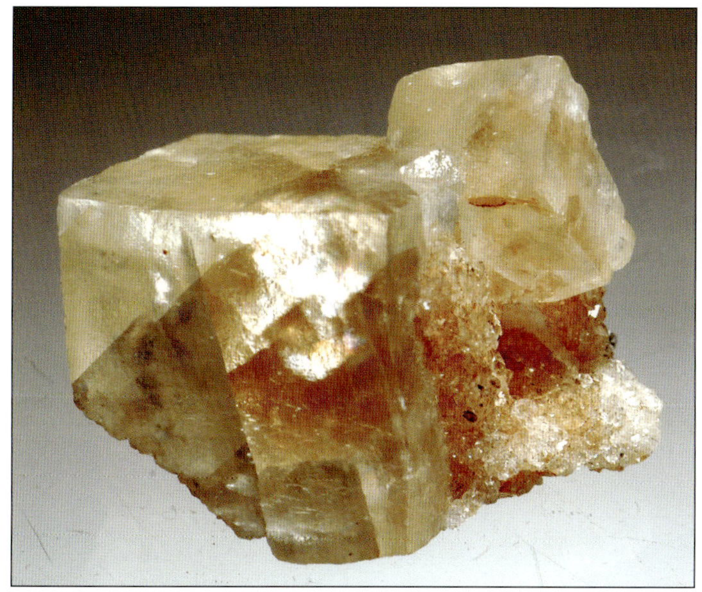

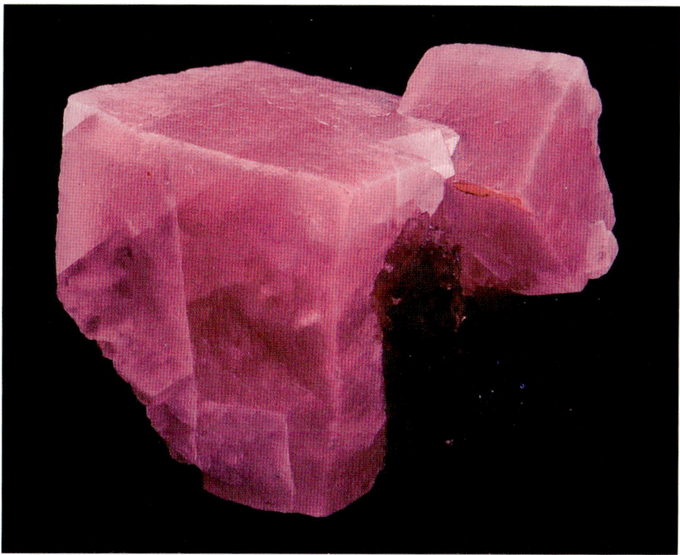

Same under SW

FLUORITE. Fluorite from Dal'negorsk, Primorskiy Kray. The fluorite fluoresces pink SW and pale violet LW. The piece weighs 0.5 oz. and is 1.3 x 1 x 0.8 inches. Value $25-30

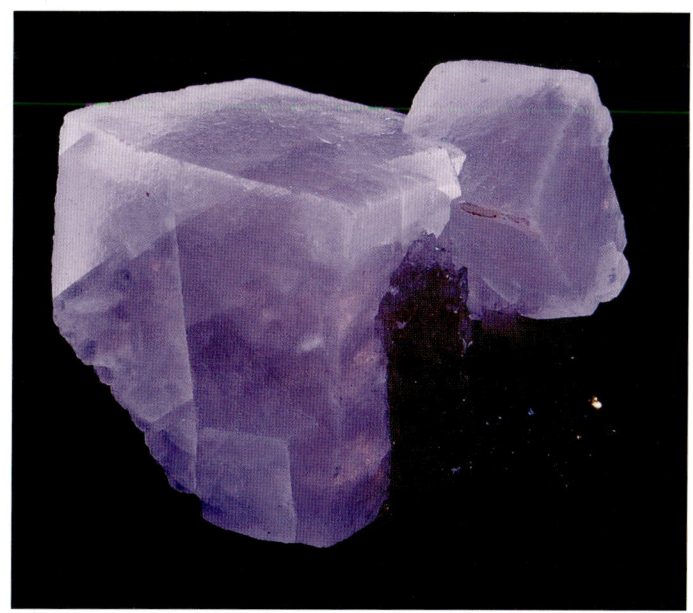

Same under LW

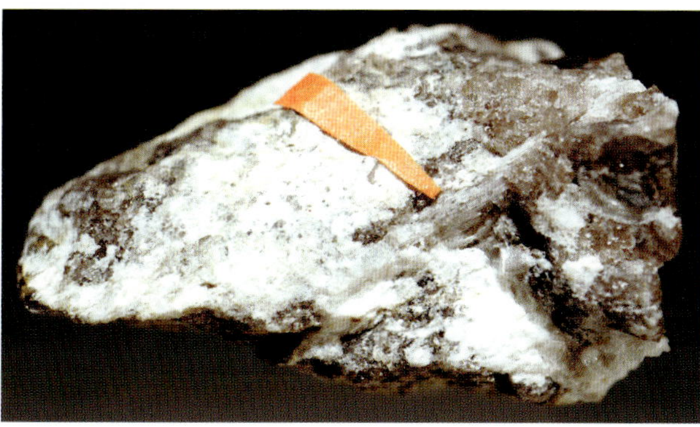

HOLTITE. Rare holtite with phlogopite from Vasin-Myl'k massif, Voron'i Tundry, Kola Peninsula, Murmanskaja Oblast'. The holtite fluoresces pale blue SW and the phlogopite fluoresces yellow SW. The piece weighs 0.8 oz. and is 1.3 x 1.0 inches. Value $22-25

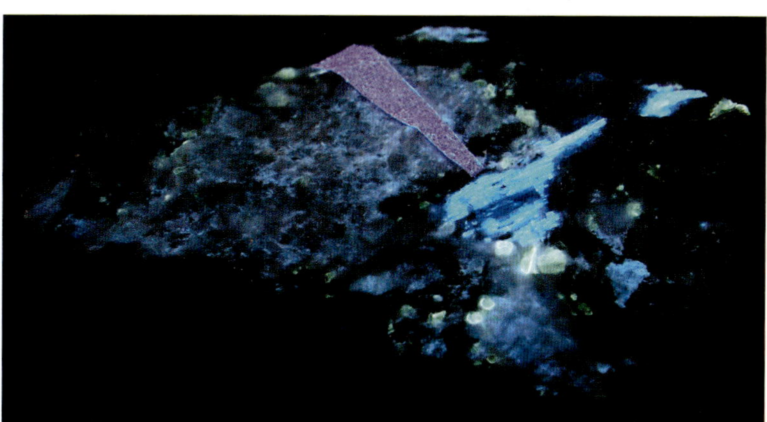

Same under SW

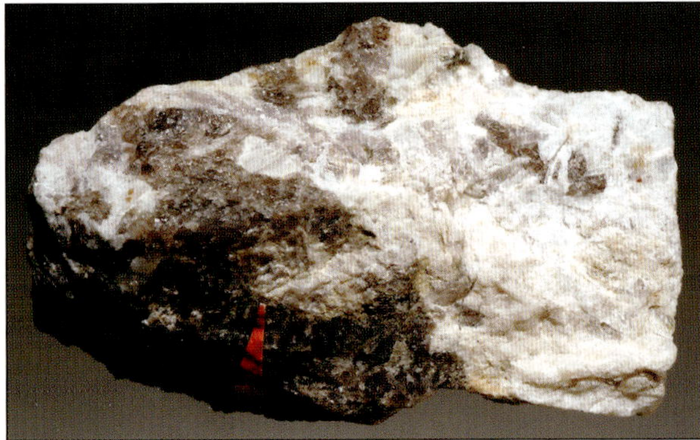

HOLTITE. Holtite with phlogopite from Vasin-Myl'k massif, Voron'i Tundry, Kola Peninsula, Murmanskaja Oblast'. The holtite fluoresces pale blue SW and the phlogopite fluoresces yellow SW. The piece weighs 2.3 oz. and is 2.1 x 1.3 x 1.0 inches. Value $40-45

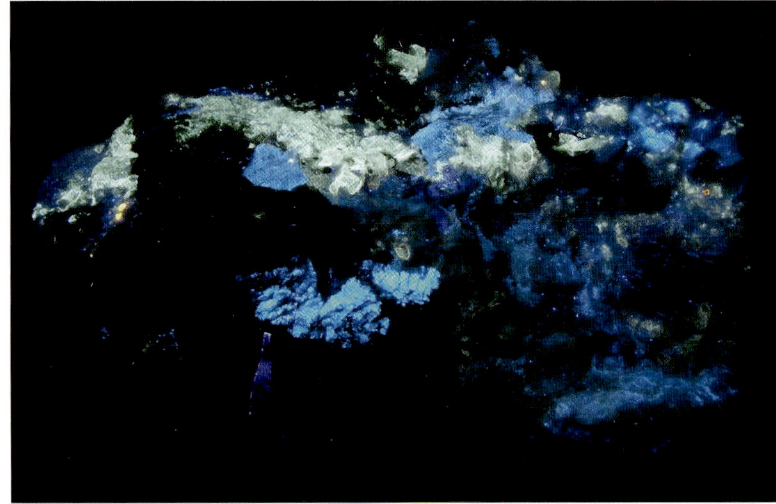

Same under SW

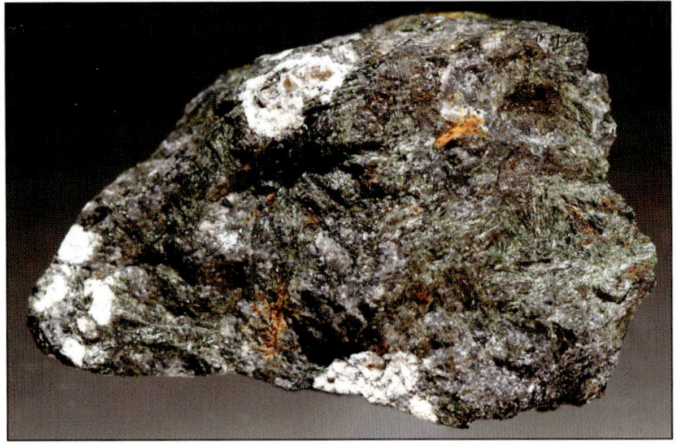

KELDYSHITE. Rare keldyshite and parakeldyshite crystals from Mt. Rasvumchorr, Khibiny massif, Kola Peninsula, Murmanskaja Oblast'. The keldyshite fluoresces yellow SW and the parakeldyshite fluoresces white SW. The piece weighs 0.5 oz. and is 1.3 x 1.0 inches. Value $20-22

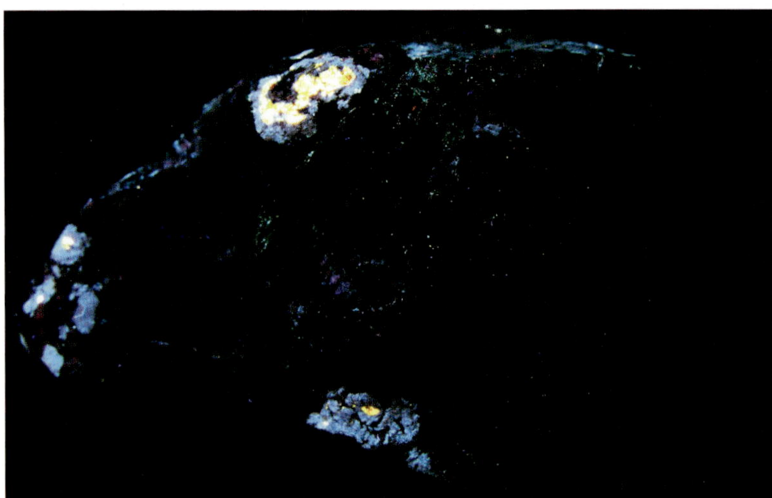

Same under SW

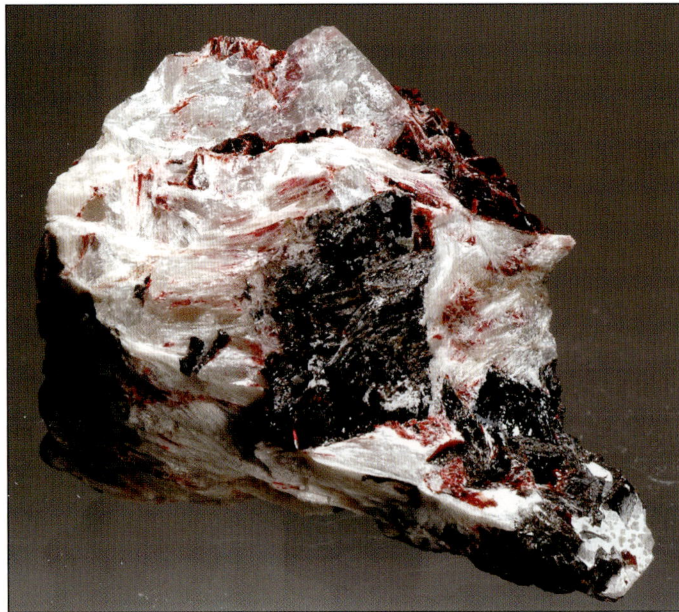

PECTOLITE. Pectolite crystals with villiaumite from Mt. Rasvumchorr, Khibiny massif, Kola Peninsula, Murmanskaja Oblast'. The pectolite fluoresces pink LW. The piece weighs 4.5 oz. and is 3.0 x 2.3 x 1.3 inches. Value $50-60

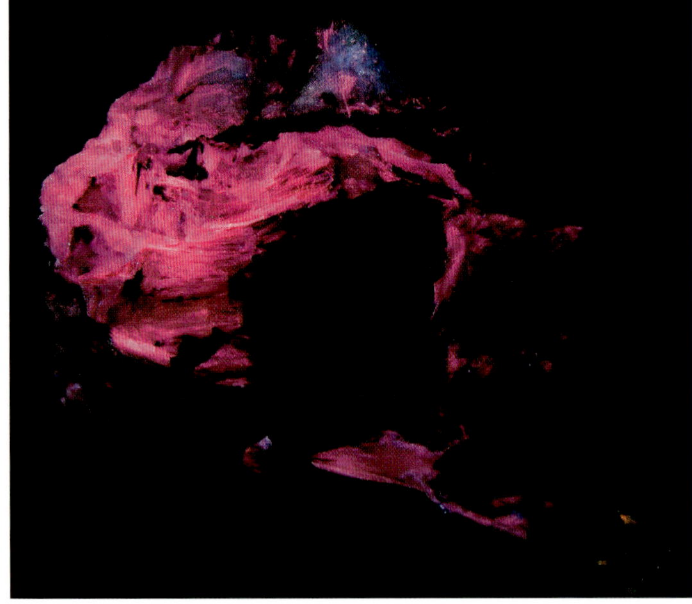

Same under LW

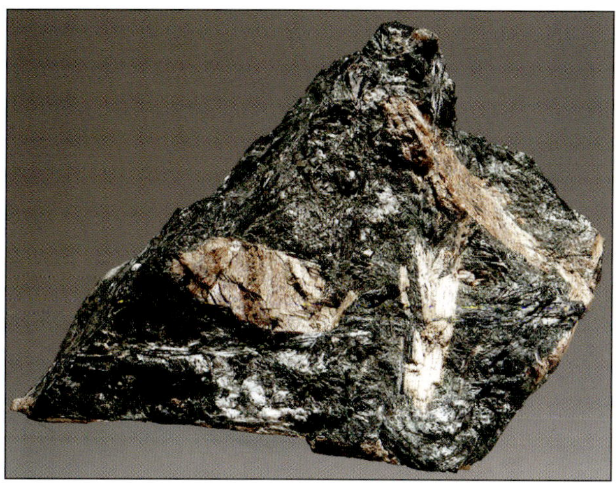

MURMANITE. Murmanite, a rare mineral from Lovozero massif, Kola Peninsula, Murmanskaja Oblast'. The murmanite fluoresces pale yellow SW. The piece weighs 3.3 oz. and is 2.8 x 2.0 x 0.8 inches. Value $35-40

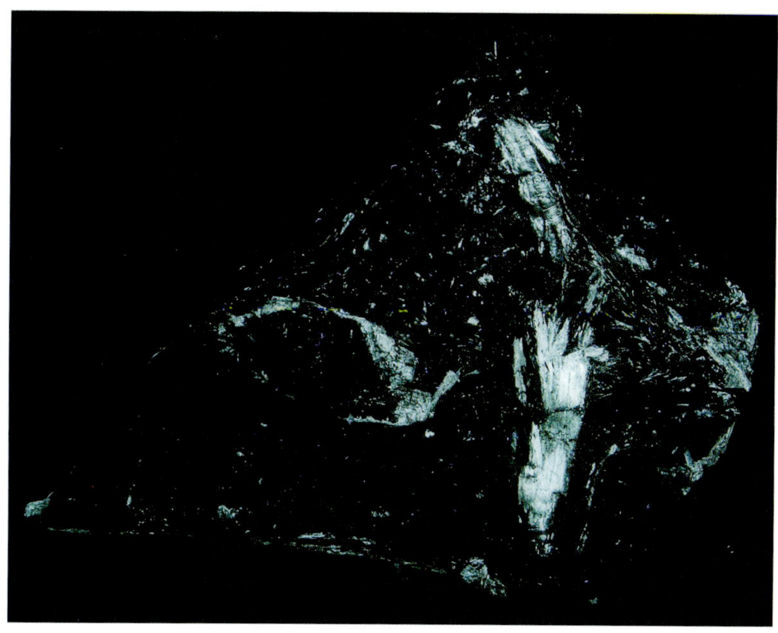

Same under SW

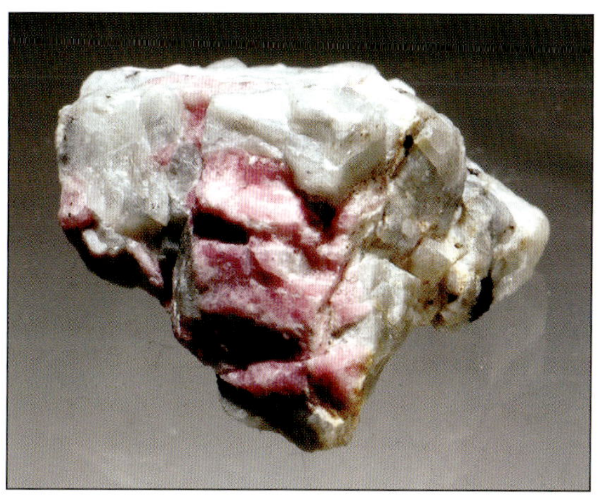

SODALITE. Sodalite, var. hackmanite, crystals on fluorite from Khibiny massif, Kola Peninsula, Murmanskaja Oblast'. The sodalite fluoresces orange LW and is tenebrescent. The fluorite fluoresces violet LW and SW. The piece weighs 0.5 oz. and is 1.3 x 1.0 x 0.9 inches. Value $12-15

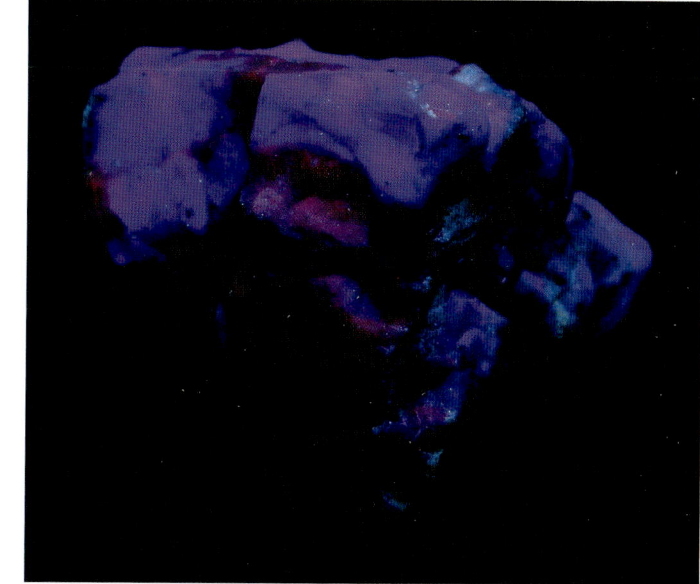

Same under SW

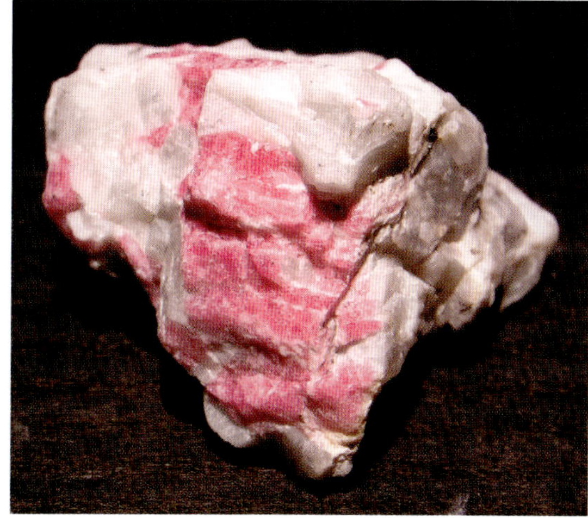

Same showing tenebrescence.

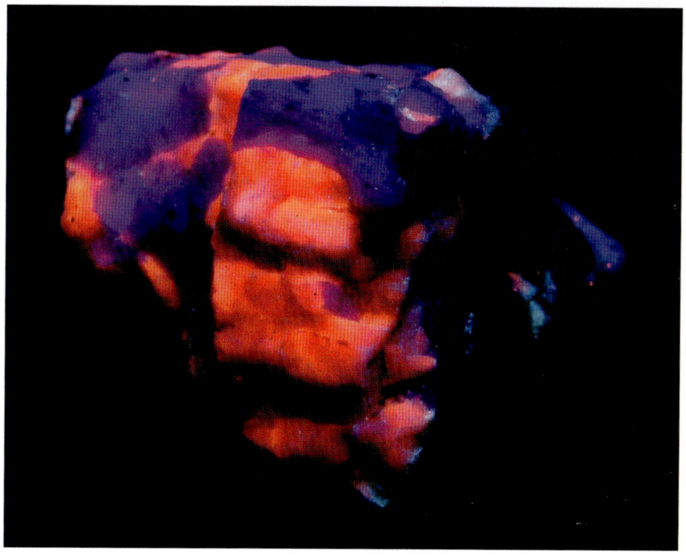

Same under LW

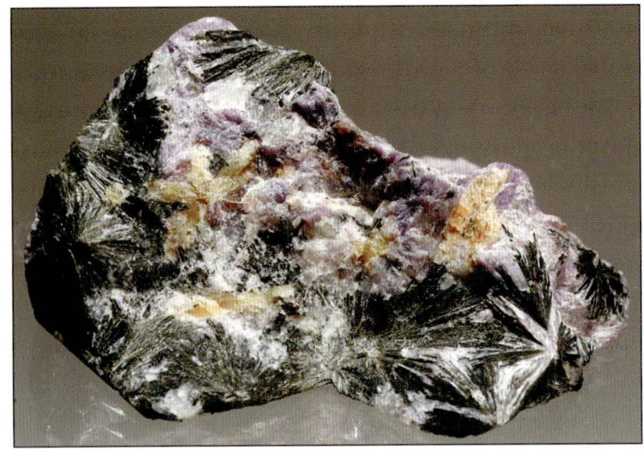

STEACYITE. Twinned crystals of steacyite with aegirine and albite in charoite from Aldan Shield, Sakha Republic (Yakutia). Steacyite was named after Harold R. Steacy, a mineralogist in Canada. The steacyite fluoresces green SW and white and green LW. The albite fluoresces cherry-red SW. The piece weighs 2.0 oz. and is 2.3 x 1.5 x 1.3 inches. Value $60-70

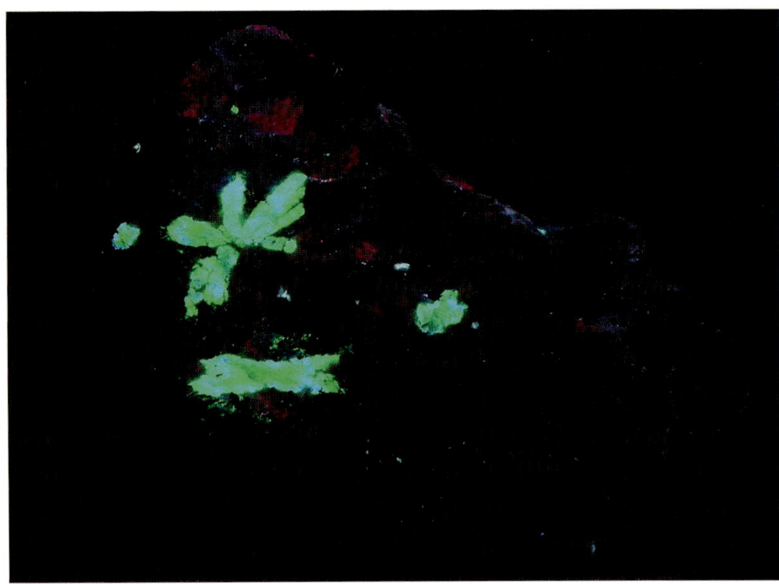

Same under SW

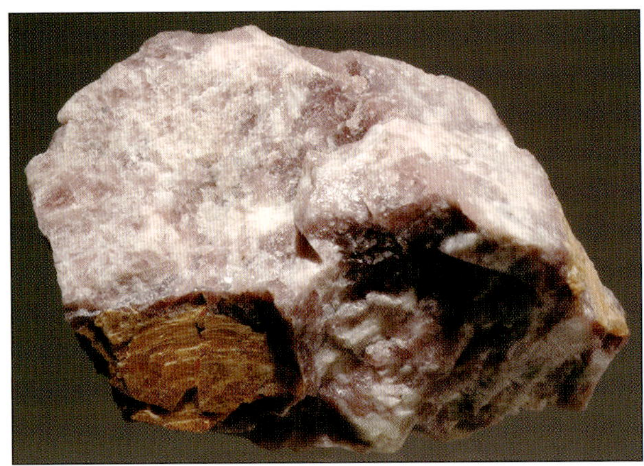

USSINGITE. Ussingite with serandite from Lovozero massif, Kola Peninsula, Murmanskaja Oblast'. The ussingite fluoresces green SW. The piece weighs 2.3 oz. and is 2.9 x 2.0 x 0.9 inches. Value $50-55

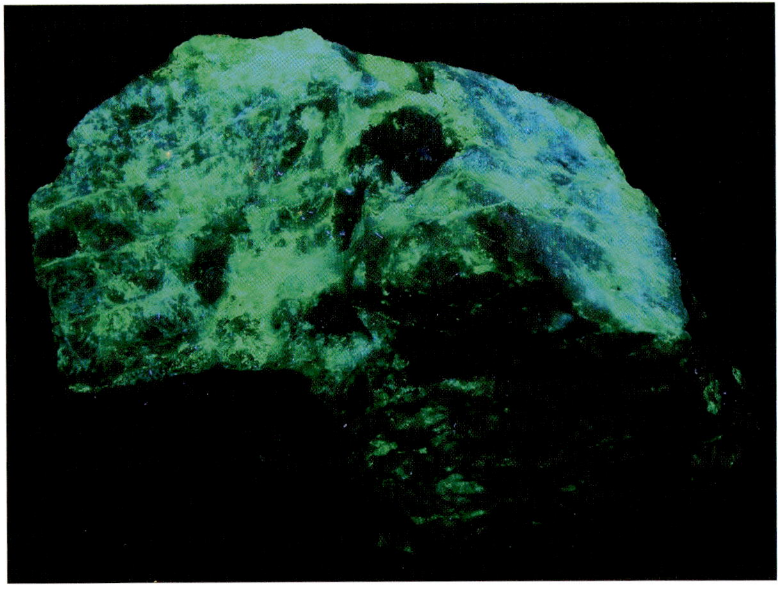

Same under SW

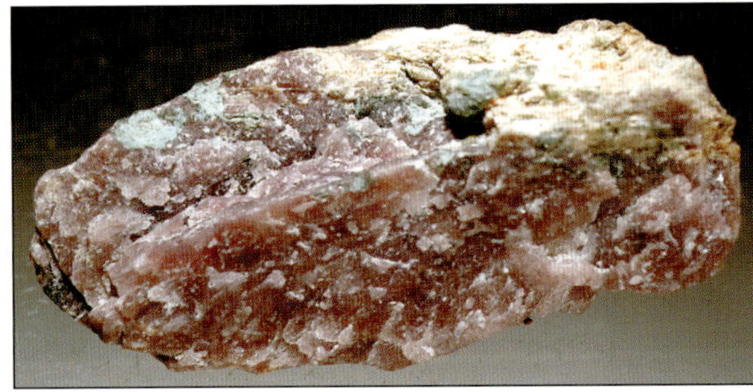

USSINGITE. Ussingite from Lovozero massif, Kola Peninsula, Murmanskaja Oblast'. Ussingite fluoresces green SW. The piece weighs 1.0 oz. and is 2.1 x 1.0 x 0.9 inches. Value $40-45

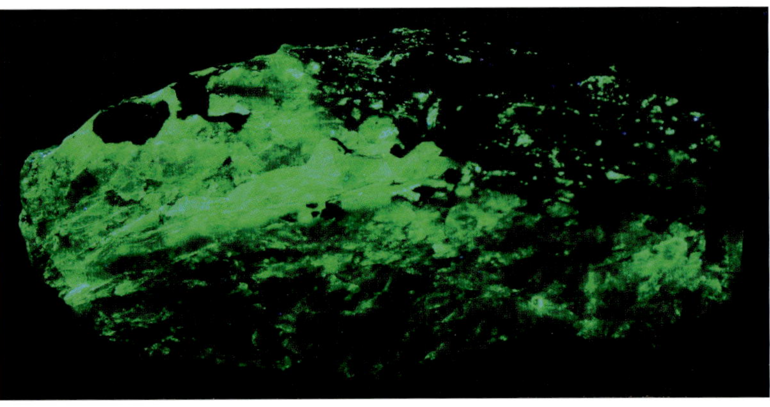

Same under SW

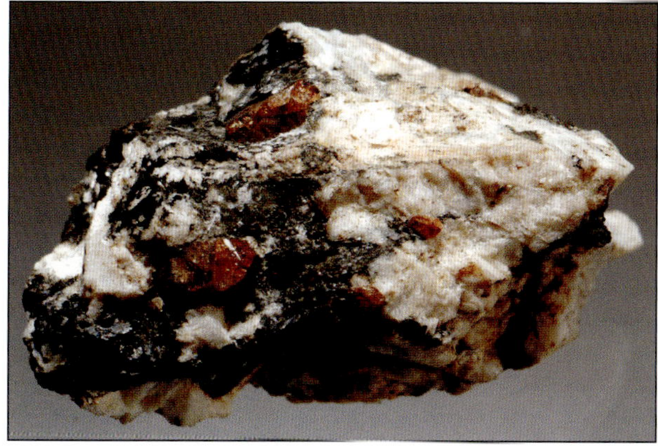

ZIRCON. Crystals of zircon from Lovozero massif, Kola Peninsula, Murmanskaja Oblast'. The zircon fluoresces yellow SW. The piece weighs 2.5 oz. and is 2.3 x 1.3 x 1.3 inches. Value $25-30

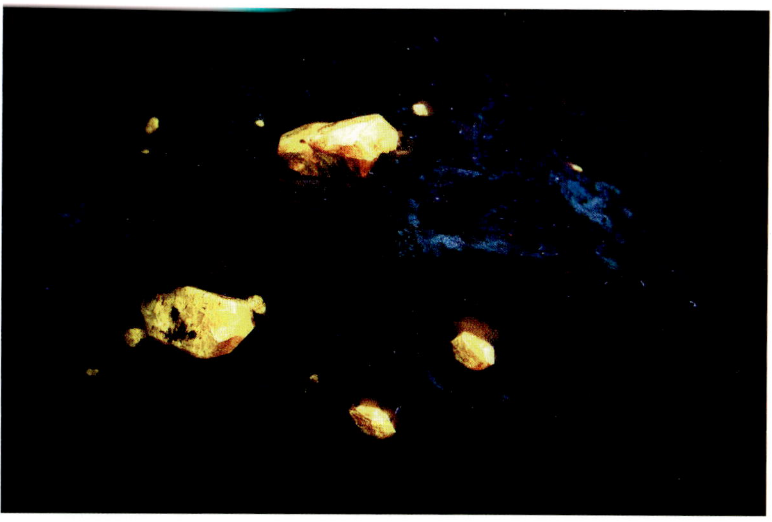

Same under SW

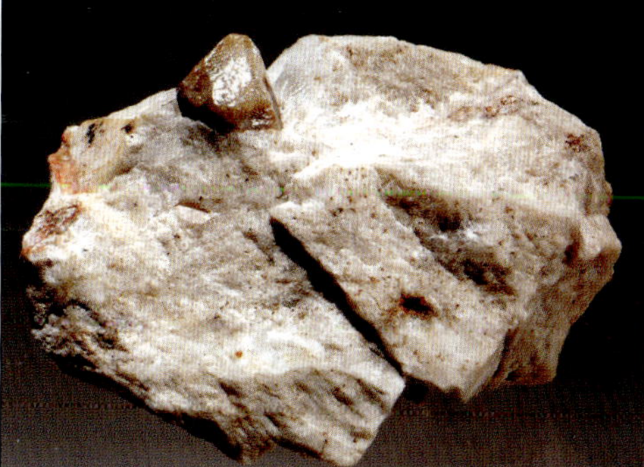

ZIRCON. A zircon crystal on matrix from Vishnevye massif, Chelyabinsk Oblast'. The zircon fluoresces yellow SW. The piece weighs 1.8 oz. and is 2.0 x 1.5 x 0.8 inches. Value $30-35

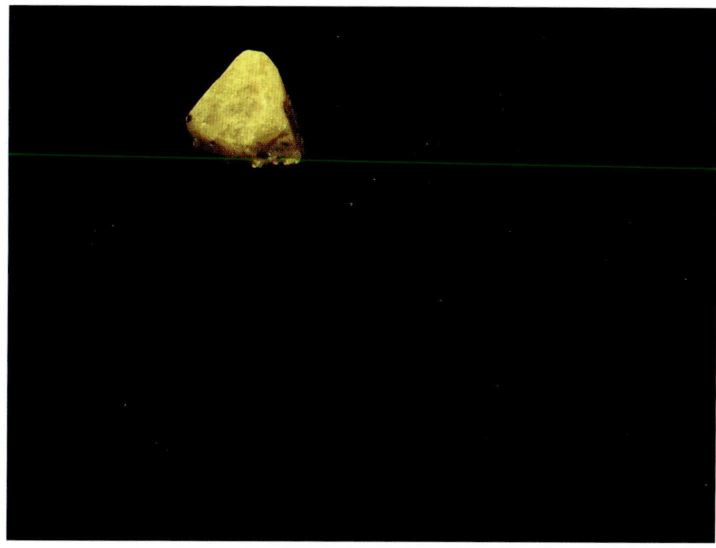

Same under SW

South Africa

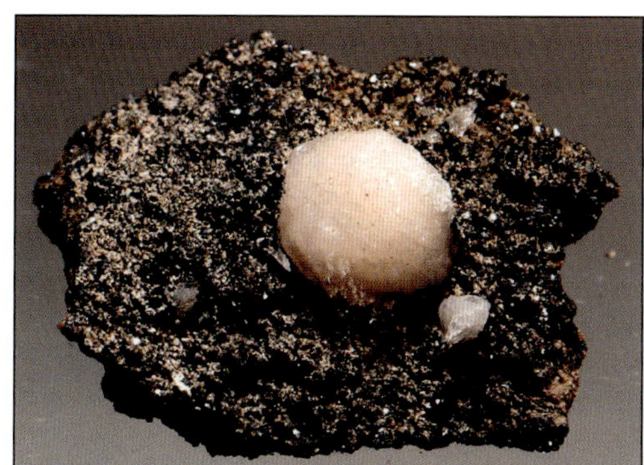

CALCITE. Ball-shaped calcite crystals from the Kalahari manganese fields (generally written as "KMF" on mineral labels), Kuruman. The calcite glows bright orange SW. The piece weighs 0.8 oz. and is 1.5 x 1.3 x 0.6 inches. Value $20-25

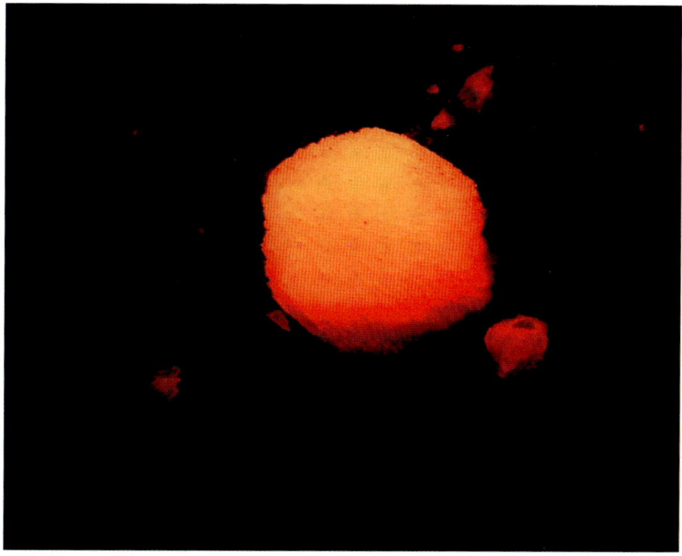

Same under SW

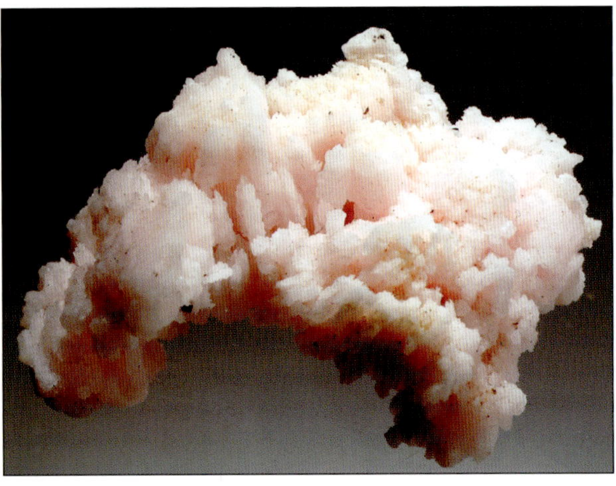

MANGANOCALCITE. Manganocalcite from the Wessels mine, Kalahari manganese fields, Kuruman. Manganocalcite looks similar to kutnohorite and sometimes can only be differentiated with an ultraviolet lamp. Kutnohorite does not fluoresce, whereas manganocalcite does. Manganocalcite fluoresces pink-orange SW and LW. The piece weighs 2.5 oz. and is 2.8 x 1.8 x 1.3 inches. Value $50-60

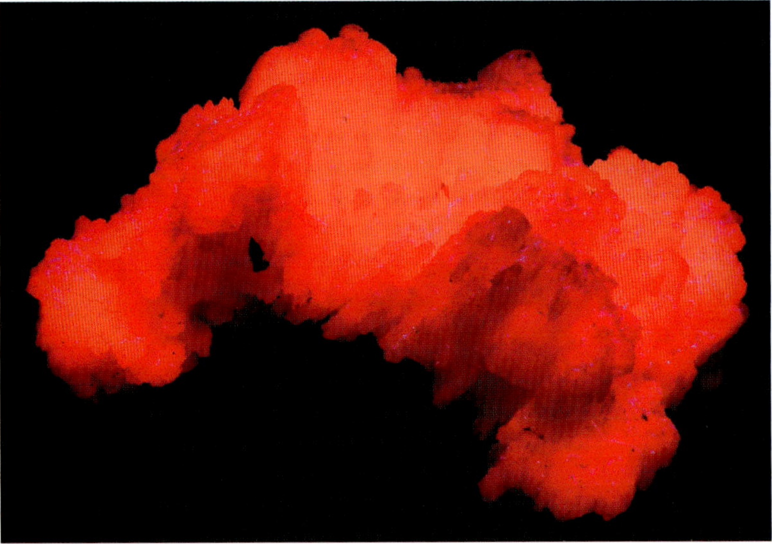

Same under SW

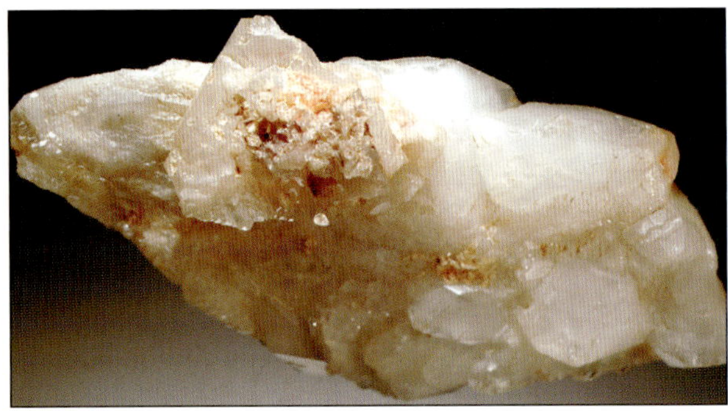

WILLEMITE. A quartz crystal on a plate of quartz with willemite from the Orange River, Northern Cape Province. Willemite fluoresces green SW. The piece weighs 3.8 oz. and is 4.3 x 2.0 x 1.0 inches. Value $20-25

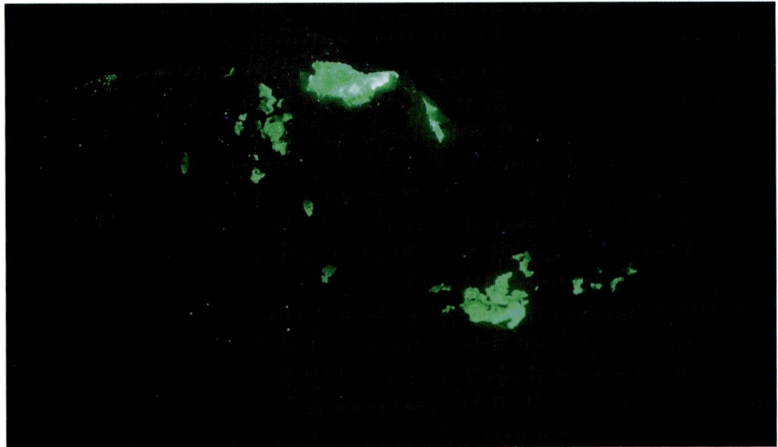

Same under SW

South Korea

Right:
SCHEELITE. Scheelite crystals on calcite from the Taewha mine, near Chungju. The scheelite fluoresces bight pale blue under SW. The piece weighs 5.0 oz. and is 2.3 x 2.3 x 1.5 inches. Value $60-70

Far right:
Same under SW

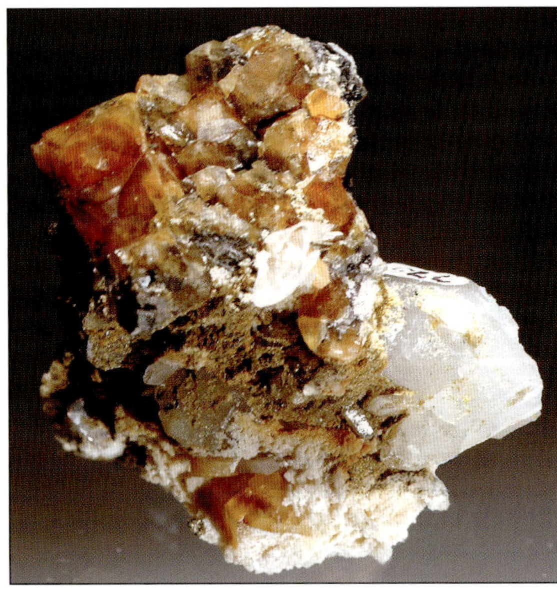

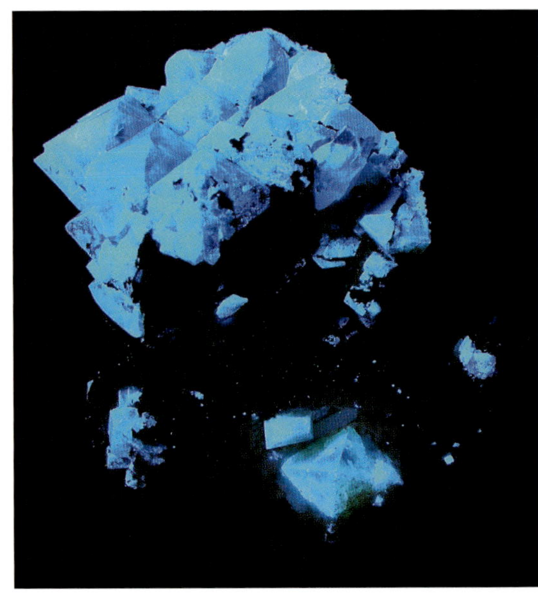

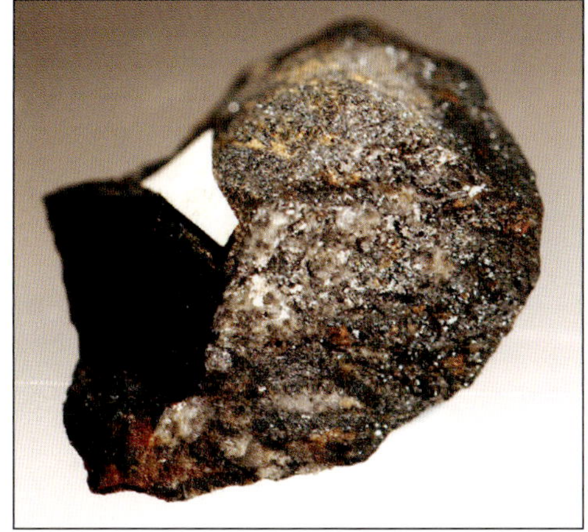

BERGSLAGITE. Rare bergslagite with calcite from Långban, Filipstad, Värmland. The thin coating of bergslagite fluoresces pale blue SW, the calcite fluoresces orange-red SW. The piece weighs 0.4 oz. and is 0.8 x 0.8 inches. Value $15-17

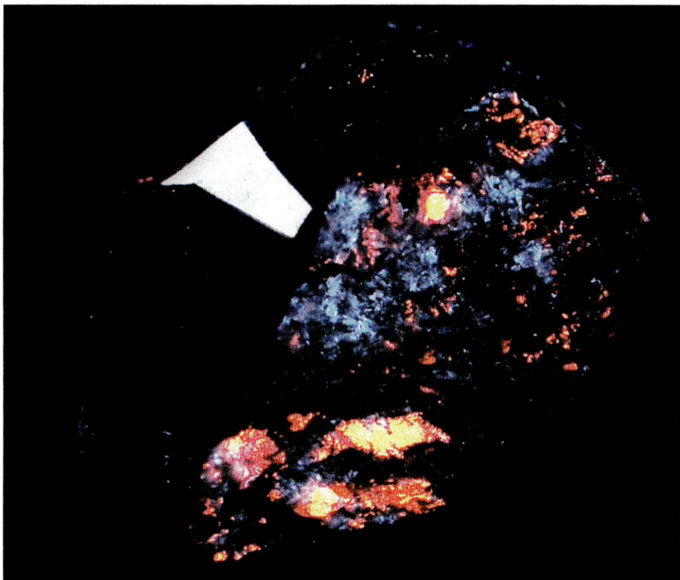

Same under SW

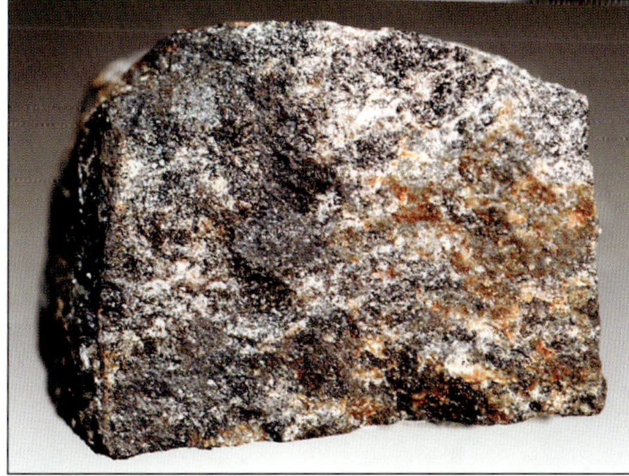

JOHNBAUMITE. Rare johnbaumite in calcite from the Jakobsberg mine, Nordmark, Filipstad, Värmland. The johnbaumite fluoresces pink-orange SW and the calcite fluoresces orange-red SW. The piece weighs 0.8 oz. and is 1.0 x 0.8 inches. Value $25-30

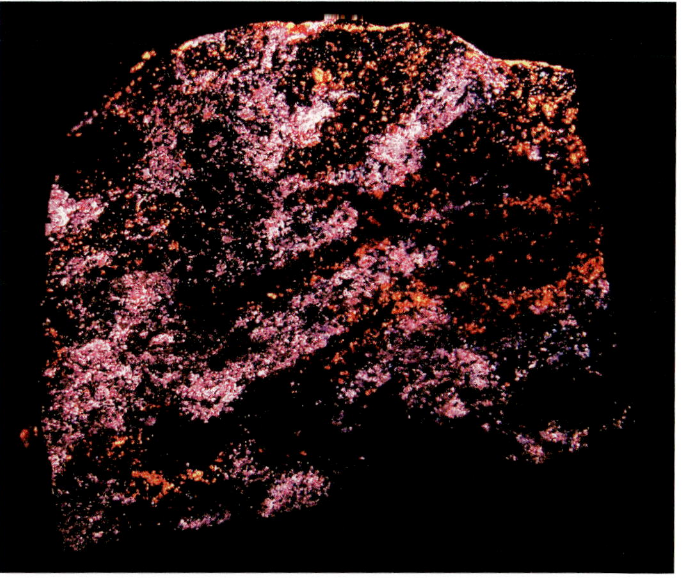

Same under SW

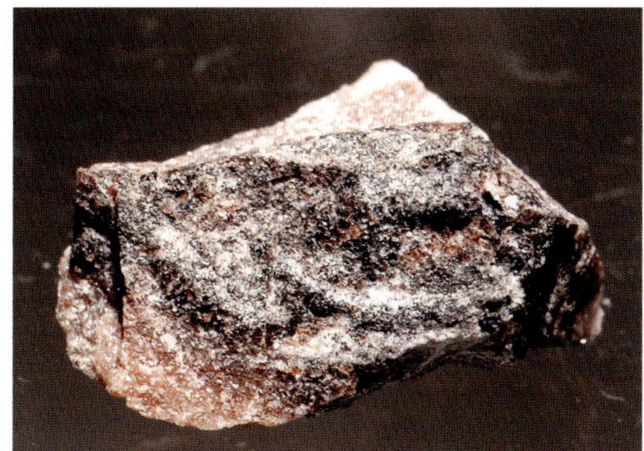

JOHNBAUMITE. Johnbaumite on calcite from the Jakobsberg mine, Nordmark, Filipstad, Värmland. The johnbaumite fluoresces pink-orange SW and the calcite fluoresces orange-red SW. The piece weighs 0.5 oz. and is 1.1 x 0.8 x 0.5 inches. Value $25-30

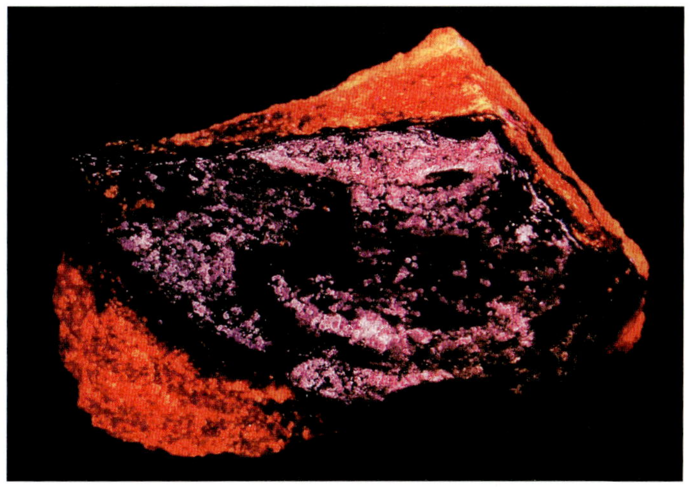

Same under SW

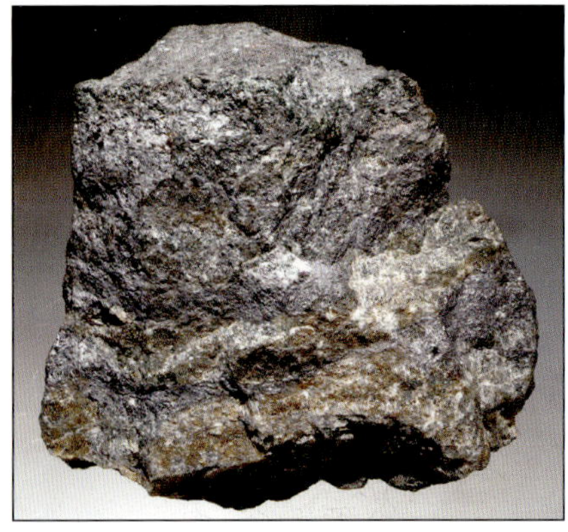

SPHALERITE. Calcite, willemite and sphalerite from Långban, Filipstad, Värmland. The sphalerite fluoresces bright orange SW and blue LW, the calcite fluoresces orange-red SW and the willemite fluoresces green SW. The piece weighs 11.3 oz. and is 3.0 x 2.8 x 1.5 inches. Value $45-55.

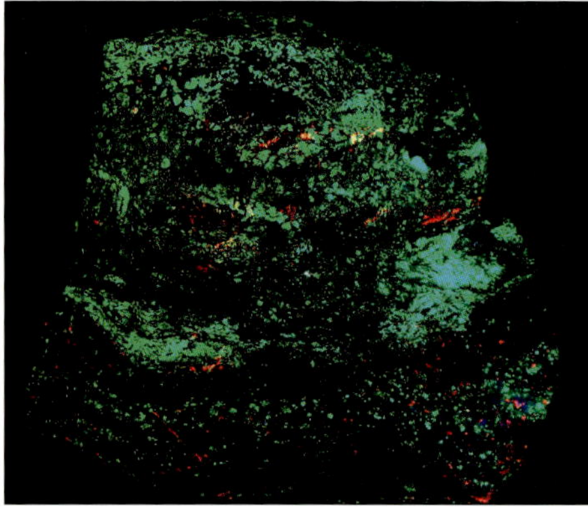

Same under SW

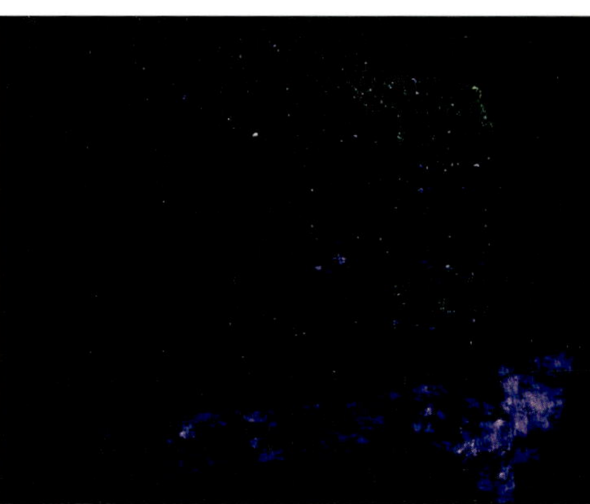

Same under LW

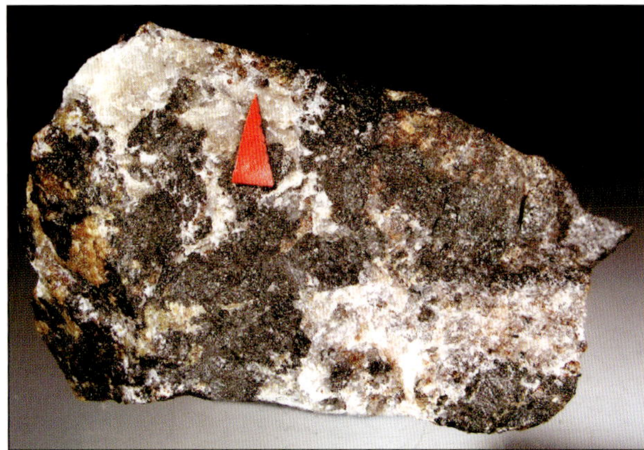

SVABITE. Svabite, calcite, romeite and grains of an unidentified mineral that fluoresces blue-green SW from Långban, Filipstad, Värmland. The svabite fluoresces orange SW and the calcite fluoresces orange-red SW. The piece weighs 3.3 oz. and is 2.0 x 1.0 x 1.0 inches. Value $20-25.

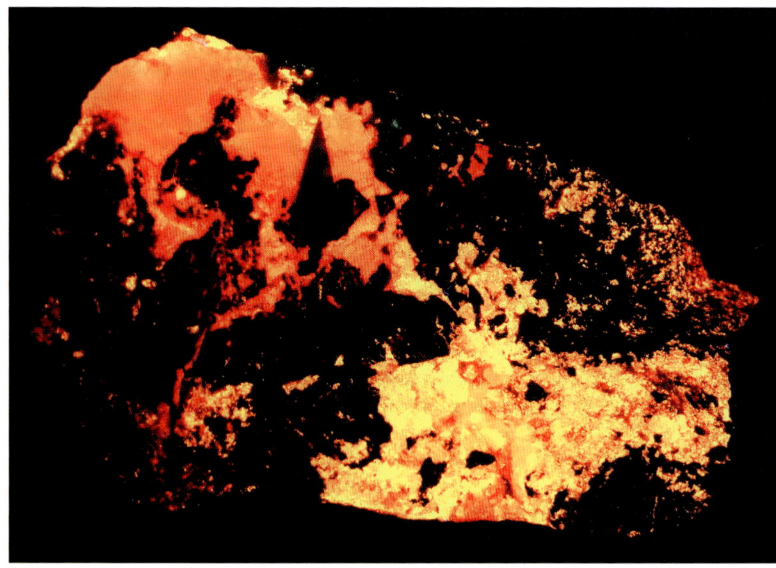

Same under SW

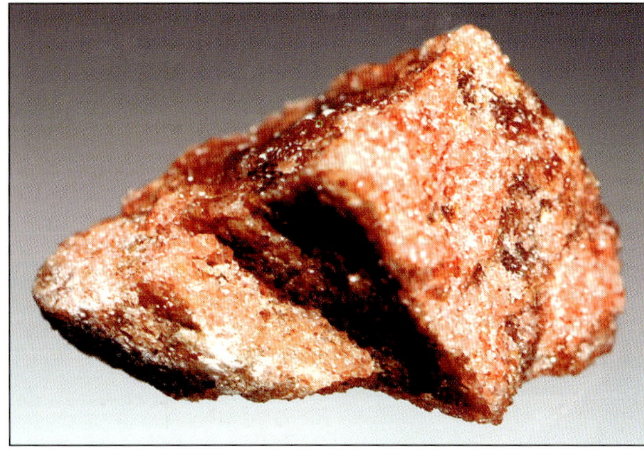

TILASITE. Rare tilasite from Långban, Filipstad, Värmland. The tilasite fluoresces orange SW. The piece weighs 0.3 oz. and is 0.9 x 0.8 inches. Value $15-17

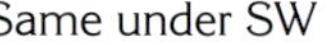

Same under SW

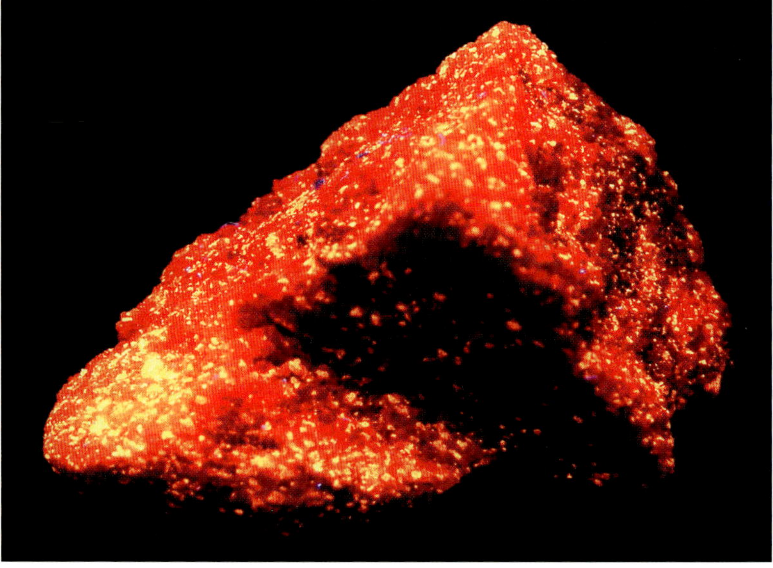

Switzerland

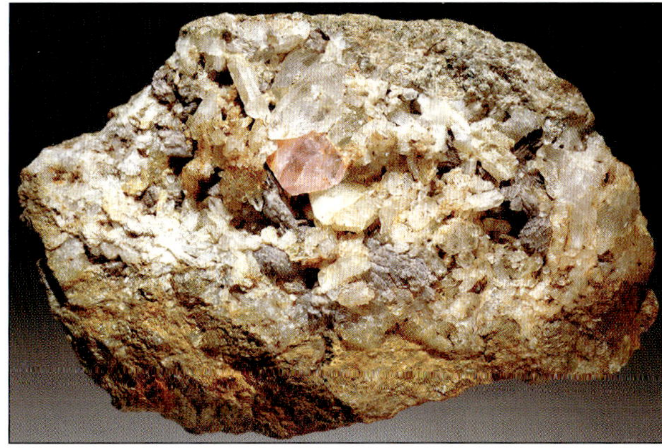

FLUORAPATITE. Crystals of fluorapatite with feldspar on quartz from Riedertobel, Amsteg, Canton Uri. The fluorapatite fluoresces violet SW and the feldspar fluoresces velvety red SW. The piece weighs 3.8 oz. and is 3.0 x 2.0 x 0.9 inches. Value $60-70

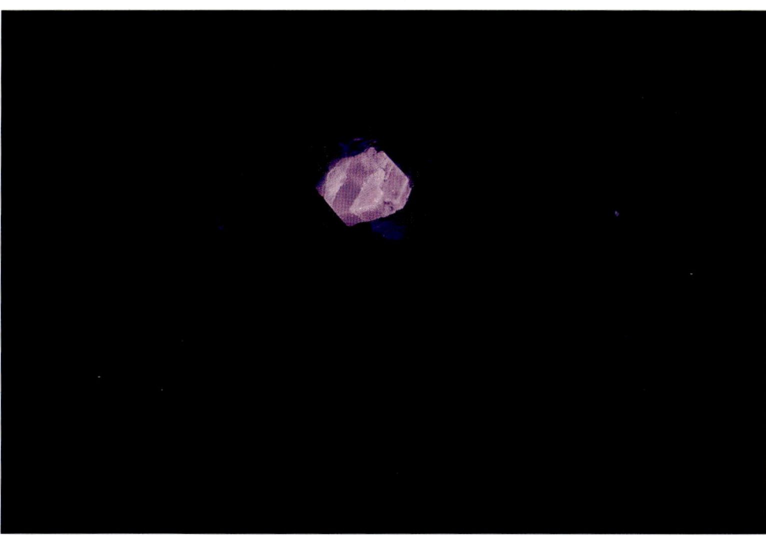

Same under SW

Tajikistan

(sometimes spelled "Tadzhikstan" or "Tadjikistan")

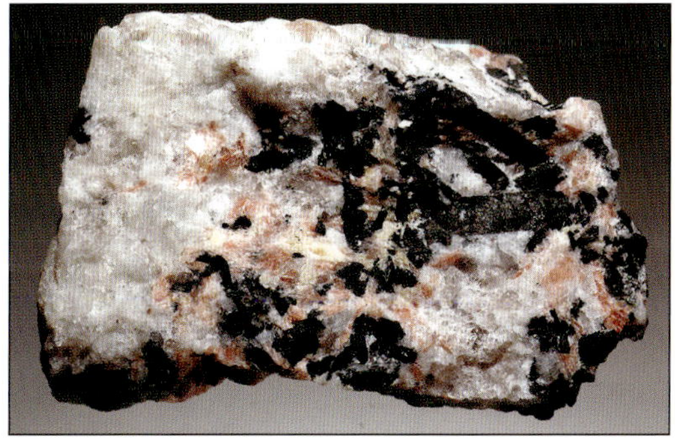

BARATOVITE. Calcite with pink, radiating crystals of baratovite from the Dara-i-Pioz glacier, Alayskiy Range, Tien Shan mountains, Districts of the Republican Subordination. The baratovite fluoresces a brilliant pale blue SW and the calcite fluoresces pale red SW. The piece weighs 2.8 oz. and is 2.3 x 1.5 x 1.0 inches. Value $60-70

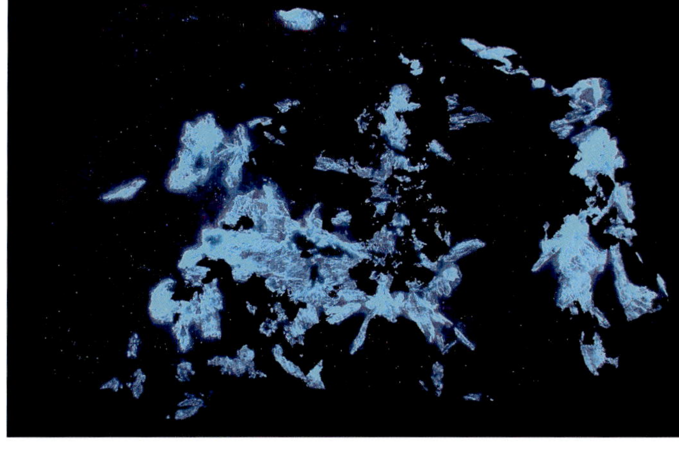

Same under SW

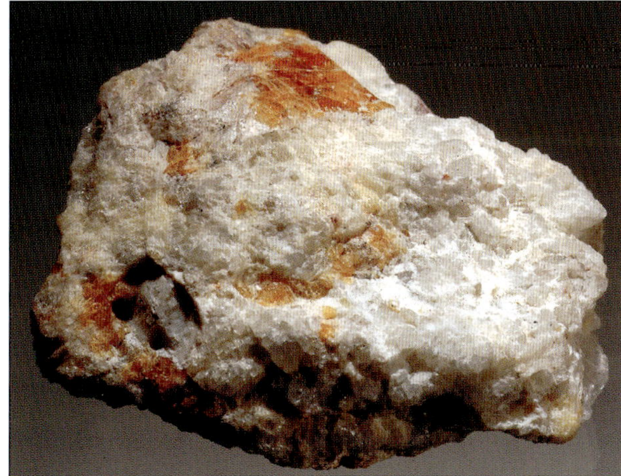

CLINOHUMITE. Orange-brown crystalline clinohumite from the Kukh-i-Lal gem spinel deposit, Pyandzh River valley, Pamir mountains, Viloyati Mukhtori Gorno-Badakhshan. The clinohumite fluoresces yellow SW. The piece weighs 3.3 oz. and is 2.0 x 1.5 x 1.3 inches. Value $45-50

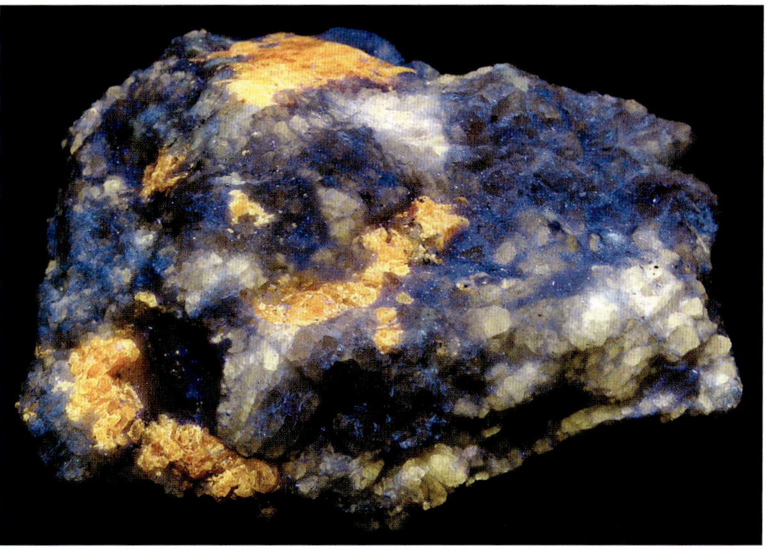

Same under SW

Right:
SOGDIANITE. Rare sogdianite from the Dara-i-Pioz glacier, Alayskiy Range, Tien Shan mountains, Districts of the Republican Subordination. The sogdianite fluoresces pale blue SW. The piece weighs 0.1 oz. and is 0.5 x 0.5 x .04 inches. Value $15-17

Far right:
Same under SW

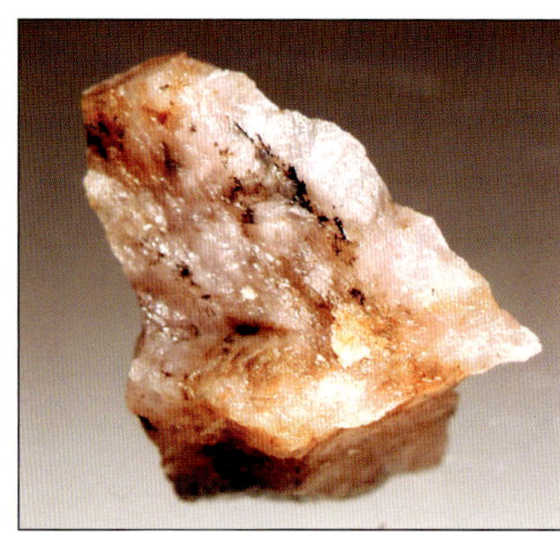

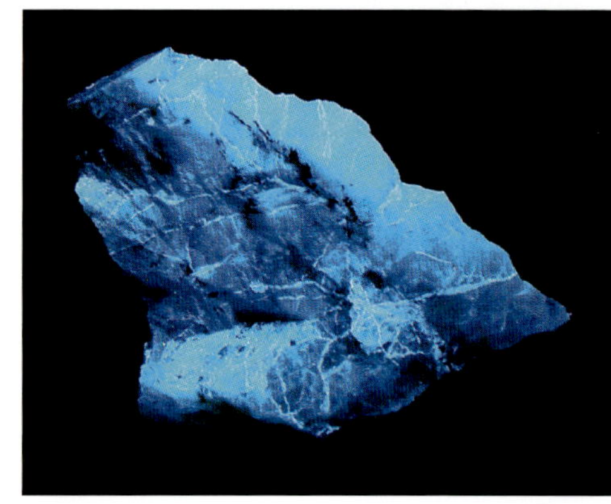

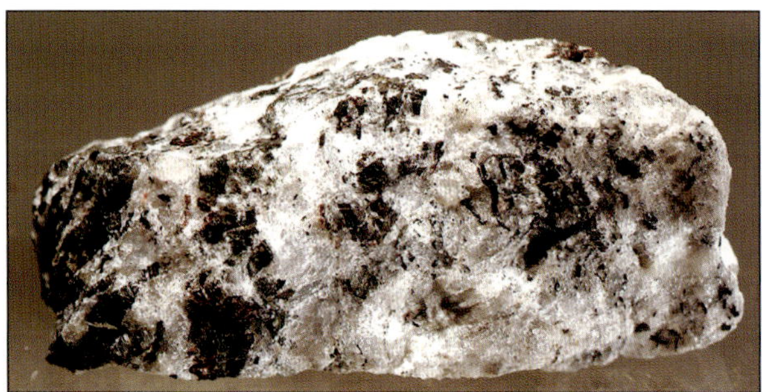

ZEKTZERITE. Zektzerite with neptunite in microcline from the Dara-i-Pioz glacier, Alayskiy Range, Tien Shan mountains, Districts of the Republican Subordination. Zektzerite fluoresces blue SW and microcline fluoresces red SW. The piece weighs 4.5 oz. and is 3.0 x 1.5 x 1.3 inches. Value $40-45

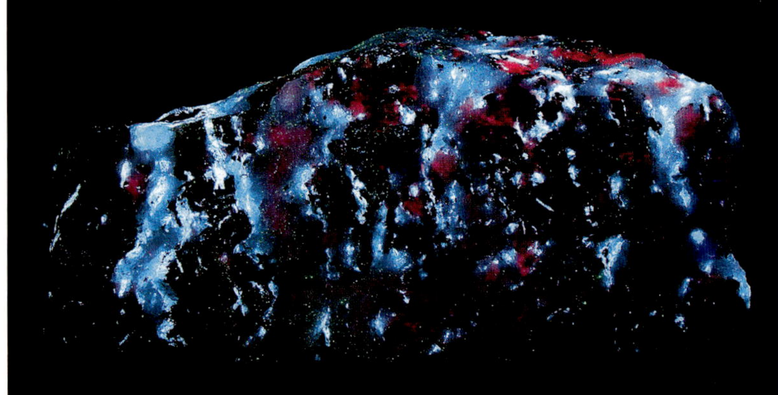

Same under SW

United Kingdom

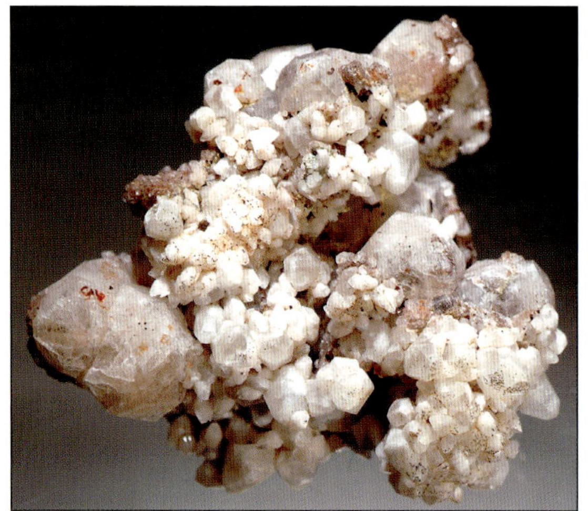

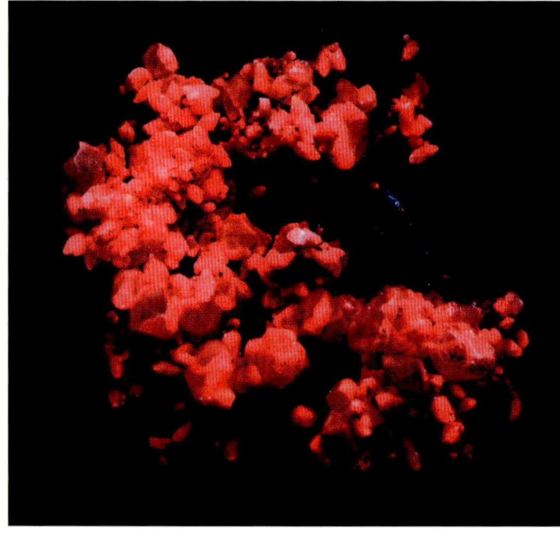

Far left:
CALCITE. Calcite crystals on analcime from the Croft quarry, Croft, Leicestershire. The calcite fluoresces orange-red SW. The piece weighs 2.0 oz. and is 2.3 x 2.3 x 1.0 inches. Value $25-30

Left:
Same under SW

Zambia

Right:
WILLEMITE. Crystals of willemite from Broken Hill. Willemite fluoresces green SW. The piece weighs 1.5 oz. and is 2.4 x 1.8 x 1.3 inches. Value $35-40

Far right:
Same under SW

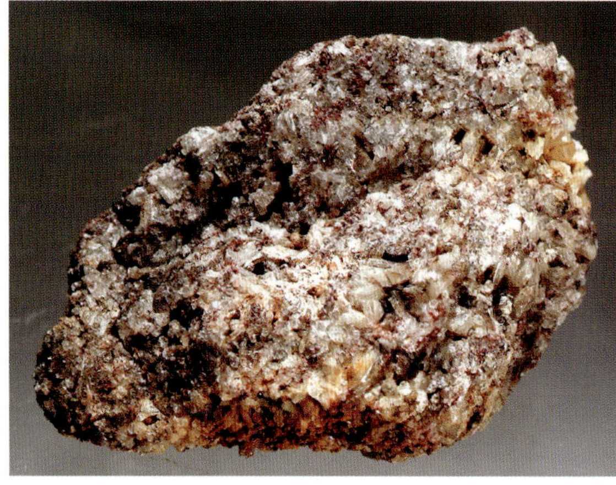

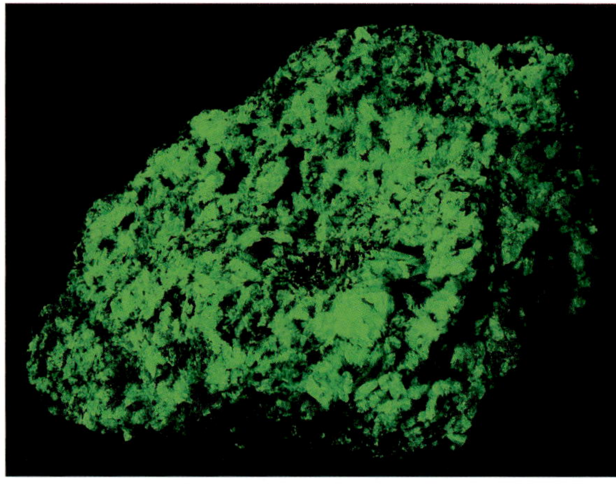

Miners Lamps

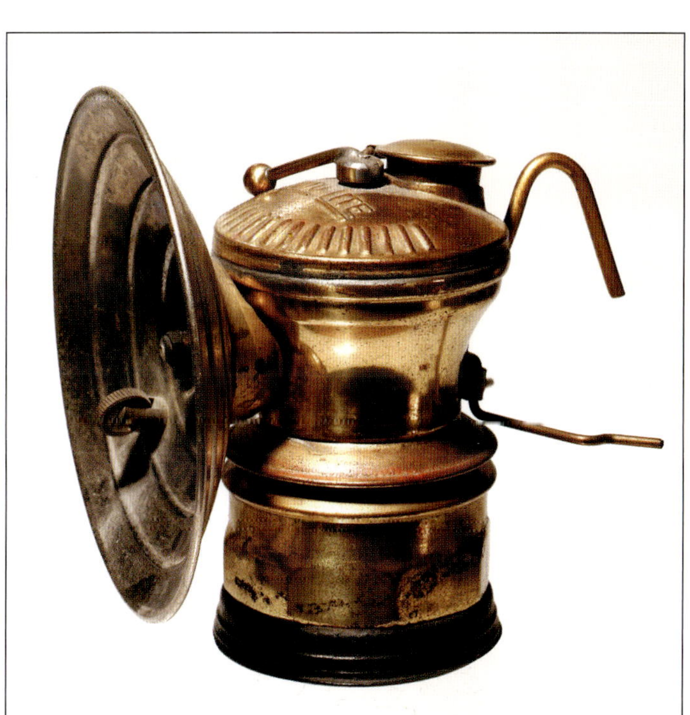

Autolite Miner's Lamp. A brass AutoLite Universal carbide miner's lamp with a 4 inch reflector. Many mineral collectors include miner's lamps and other mining memorabilia in their collections. Courtesy of Gar VanTassel.

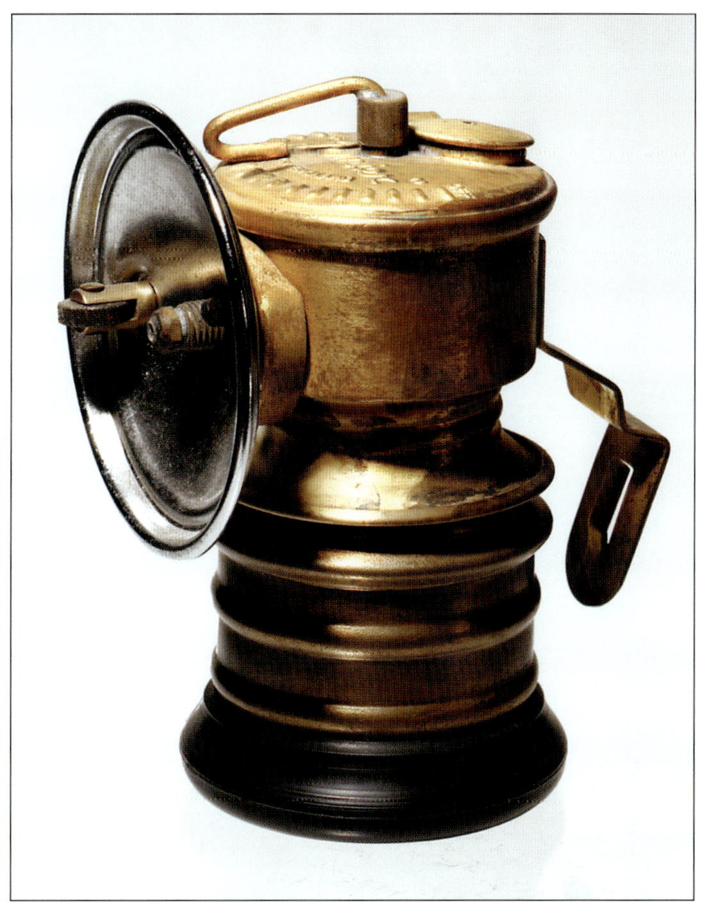

Premier Miner's Lamp. A brass Premier carbide miner's lamp with a 2.5 inch reflector. Courtesy of Gar VanTassel.

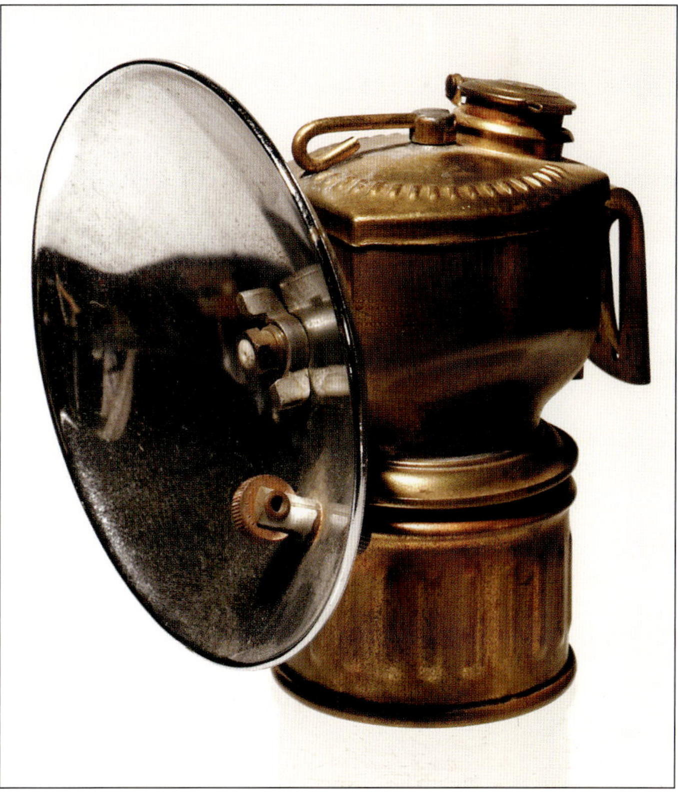

Justrite Miner's Lamp. A brass Justrite Universal carbide miner's lamp with a 4 inch reflector. Courtesy of Gar VanTassel.

Building a Small Battery Pack for Your Portable 12 Volt UV Lamp

by Mark Cole

I do most of my glow hounding in Greenland. Not nice and flat — mountains everywhere. The typical "honey holes" are found at 500 m above sea-level — and you start at sea-level! With about a 1 km climb, I think that's about a 50% grade. Carrying hammers, lights, BBQ grill covers (gives me the dark environment that I need), chisels, and a heavy lead-acid battery that's only going to last for 5 hours is not fun! And you've got to carry it back down, along with all the wonderful rocks you found (until the battery died).

I learned my lesson the first year, relearned it the second year, and solved it the third year. Adopting a technique used by remote control car enthusiasts, I built my own battery packs to power my light. I built two "2 hour" versions, each weighing only 11 ounces, and two "nine hour" versions, four pounds each (actually two "two pound packs" to spread the weight. More on that later).

NiMH batteries are quite powerful for their size. They make a "AA" 2300 mAh battery and a "D" cell 9000 mAh battery (and many others). I chose these sizes because at the time they were the highest capacity/lowest weight. Unlike alkaline and Ni-Cad batteries, NiMH batteries put out 1.2 volts (versus 1.5 volts). My SuperBright lamp needs 12 volts, thus I needed 10 batteries (10 x 1.2 v = 12 v). (Note: I now run my lights at 14.4 volts using twelve batteries for greater light output).

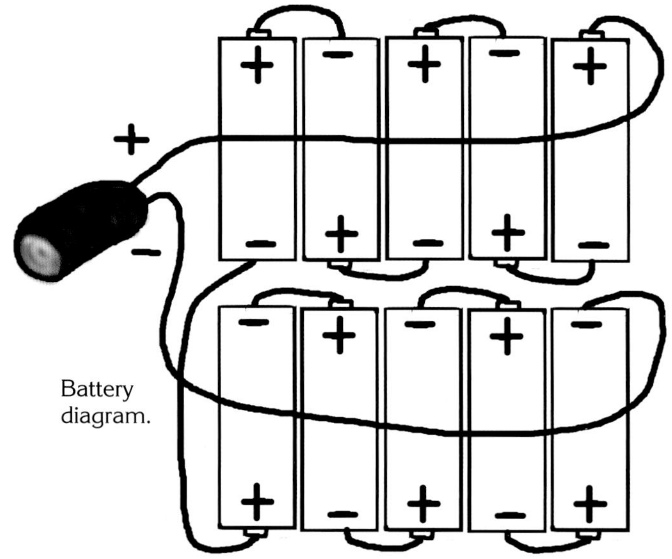

Battery diagram.

Perhaps a little discussions about "mAh" is in order. It stands for "Milliamp hours;" a 2,000 mAh battery would put out 2,000 ma (2,000 milliamps, or 2 amps) for one hour before being drained completely. Often, lead/acid batteries will just list them as 7 AH (7 amp hours, 1 amp for seven hours, or seven amps for one hour). It gets even better with NiMH batteries; they maintain their voltage rather consistently right to the end, then abruptly die. Conversely, a lead acid battery dies a slow death starting from the moment you start using it. The voltage (12 volts usually) starts out at 12 volts and rather linearly decreases over the length of time you use it. You may find less and less bright specimens towards the end of the night, until finally you realize your battery has died. An NiMH battery pack puts out 12 volts until the very end, then drops off quickly.

Having selected the batteries I had to figure out how to put them together, the tools needed, and the parts. This requires some basic (but good) soldering skills, a solder gun, some wire, and knowledge of the positive end (+) and negative end (-) of a battery. Luckily, there's an excellent web site that tells you how to do it all: http://www.rccentral.com/guides.asp?ATCL_ID=51. Battery pack supplies can be bought at many places. I used this one: http://www.dynamoelectrics.com/supplies.htm. Shop around on the Internet for the batteries and the best deal. They've gotten more powerful and cheaper since last year. The difficulty level for this project is: 1). Easy if you like to tinker and have not electrocuted yourself yet, 2) Hard if you find assembling your kid's bike on Xmas night a chore, 3) Impossible if you don't know how to use a soldering iron or soldering gun and would not recognize one if it fell in your lap.

The basic theory goes like this: To build a 12 volt battery you need to hook ten NiMH batteries in series. That means you hook up the positive end (the end with the little bump sticking out) of each battery to the negative end (the flat end) of the next battery, over and over again. You do this by soldering heavy-duty wire or braid to each battery. Short wires interconnect the batteries while longer wires at the end of the last two batteries are reserved for the connector that goes to the lamp.

Caution! Some batteries have a little hole around the bump. This hole allows the battery to vent when it is charging. Don't block it with solder. Also, be careful with heat — too much, too long, can ruin the battery.

I built my "AA" packs in two groups of five batteries (a couple of slightly longer wires interconnected the two groups), taped the result together, and covered with shrink tubing (a neat insulator that shrinks when heated and holds everything together very nicely; a hair dryer shrinks it). (Note from Stuart: I use a rubber tape that I found at the Ace hardware store, "Self Fusing Splicing Tape." You peel the backing off and it sticks to itself without a sticky surface. Once stuck, it bonds tightly.) The result was a small battery pack consisting of two rows of 5 "AA" batteries, in a pack measuring about 3" x 2" x 1". This pack provided 12 volt for 2 hours! The tricky part is figuring out how to hook it into your lamp. I have a SuperBright lamp where you can easily solder the output wires from your battery pack to one of those automotive 12 volt adapter plugs, but make sure you get the polarity right — positive goes to the middle! (Note from Stuart: I bought my connectors at Radio Shack. The best deal was one male plug connector that splits into two female connectors. I cut them apart and got two connectors for the plug from the light and one piece that I use with the battery charger. Radio Shack is also a good place to pick up some extra cords that go from the lamp to the battery pack).

Builder's note: make sure that you "strain-relief" the long wires coming out of the pack. Loop them back through the shrink tubing so that the solder connection does not receive any flexing. This is the weakest part of any system — the part that gets flexed.

If you are a little more electrically savvy (or know a friend) I would recommend losing the power connector on the SuperBright altogether and simply solder a 6-foot to 8-foot coiled wire into the innards of the SuperBright and strain-relief it by putting a Molex-style male connector (or another connector of your choice) on the other end, and a female matching connector on the wires sticking out of the battery pack (match polarities, plus to plus, negative to negative). The connector on my SuperBright always kept falling out. Plus this eliminates problematic fuses, which always seem to blow at the wrong time. My theory on fuses is that they are there to prevent fires, not protect the electronics. And I'm not worried about a fire on top of a pile of rocks. I worry more about lamp failure. Of course, all this fiddling will void your warranty.

Neat hint: while you're inside the SuperBright fiddling around, you'll find it rather easy to store a spare UV lamp behind the reflector by just removing a couple of screws. Then you'll always have a spare lamp, well-protected.

I built four of these "AA" packs and they were my primary source of power most of the time: light, easy to carry, and long-lasting. But for more power, simply use "C" or "D" cell batteries (again, ten of them) for four to nine hours of UV light (9,000 mAh, 9 hours at one amp). They're heavier but they last! I built these packs a little differently. Instead of shrink wrapping all ten batteries into one big pack I built two separate packs of five batteries each with enough wire between the two packs that I could put one in one pocket and one in another pocket for weight balancing. I bought a fly-fishing vest with lots of pockets and wore it with the batteries nicely tucked into each side – two pounds per side, nine hours of light! And the extra pockets were great for those delicate little rocks I kept finding.

A note on wire: use the heavy 13-gauge, high-strand-count wire. It is very flexible (it won't break from continuous flexing) and carries a lot of current. Use red for positive connections and black for negative connections.

Soldering: Make sure your solder connections are good. If you don't know how to solder find someone who does. A cold solder joint is a lousy connection and will result in a useless battery. The solder joint will separate just when you need the light. Also, too much heat on a battery can destroy it, and if you melt the plastic around the battery that'll cause more trouble (shorts, etc.) (Note from Stuart: I bought "C" batteries with tabs and it was easier to solder one tab to another and not have to use wire. Just remember to put some of the rubber electrical tape under each tab so that when the battery is getting banged around, the tab does not rub against the edge of the battery and short out.)

Final note on charging. NiMH batteries must be charged by a special charger: even more special because you will be charging ten of them at a time. Don't use the charger that came with your lead acid battery. Here's where a little investment will make a world of difference. I bought a "Maha MH-C777Plus-II" which easily charges ten batteries at a time (quickly), can be plugged into 120 vac or 220 vac without an adapter (important for Greenland), has a heat sensor for over-charging, and lots of other bells and whistles. Plus, you can charge all your other rechargeable batteries you have laying around the house. There are other NiMH chargers available from companies that sell remote-control supplies.

I know this all sounds pretty technical, but the instructions on the web sites really help, and if anyone has any questions feel free to e-mail me at info@minershop.com.

Lately some really neat pre-built battery packs have shown up on Ebay. They are 4,000 mAh packs, and come complete with chargers (make sure that you are buying NiMH batteries or battery packs and not Ni-Cad batteries). For Geo-Adventure 2004 most of the participants invested in several of these, modified the connectors to run their SuperBright lamps and they worked great. Personally, I still prefer my home-brew packs. The batteries seemed to be of higher quality and held a charge longer, and they were lighter (ounce per amp hour). But if you're not into home-brew soldering, this might be the way to go.

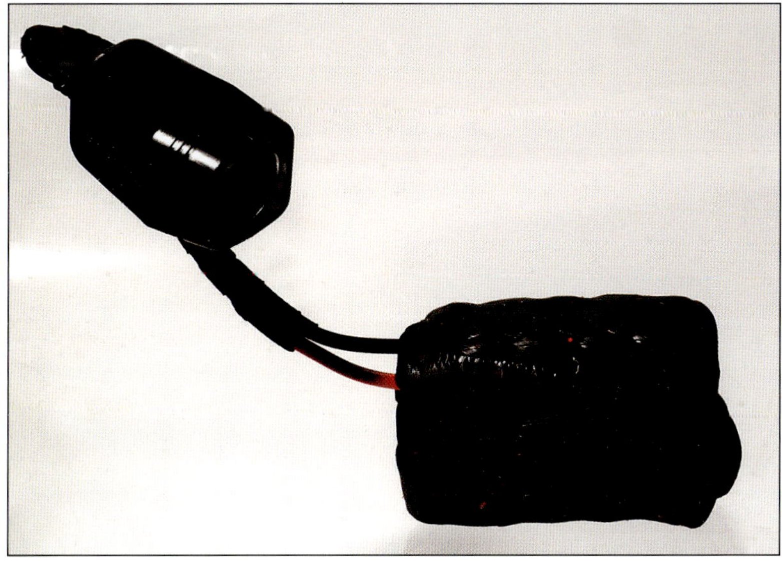

This is an AA NiMH battery pack with a receptacle for the connector to the ultraviolet lamp.

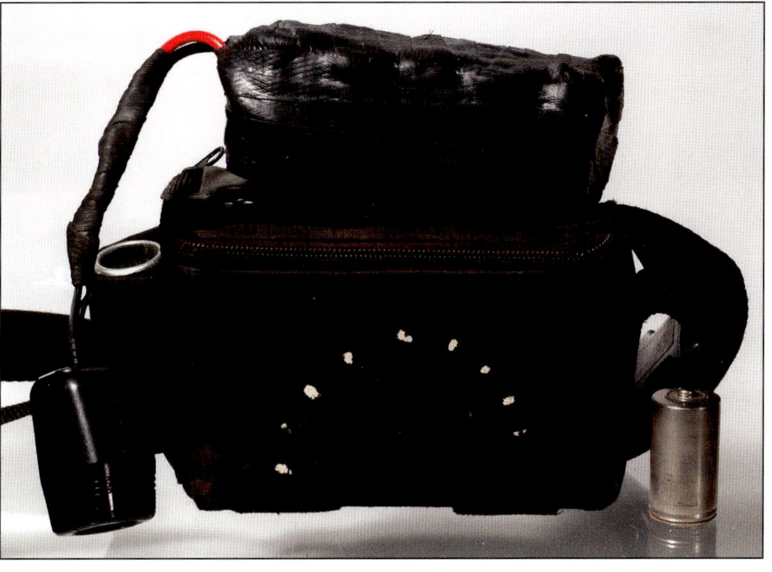

This is a C NiMH battery pack sitting on top of a typical lead-acid battery for size comparison.

Documenting Your Collection

Do you keep records about your collection? You should. All collectors should keep a database or log book that contains the information about the pieces in their collection. I number each new piece that I obtain. I used to keep a small bottle of "Wite Out" handy (available at office supply stores) to paint a white dab in an inconspicuous place on the specimen. Then I would write the number with a fine Sharpie pen. I now print numbers on the computer in size 8 type. Then the number is snipped out and glued to the bottom of each specimen. This helps where there are no flat surfaces on a specimen. Each piece gets an entry in my database with its number. I need only look at the number on a specimen and then check my printed-out sheet to know what it is and where it is from.

We are only the keepers of our collection during our lifetime. When we are no longer here, how will anyone know what it is, where it came from, and what we paid for it? I have attended several garage sales of rocks from deceased collectors. Some were in paper bags with a paper label, but most were in basements on shelves or in boxes or milk crates with no identifying information. I heard one recent horror story of a collector who always told his wife that each new specimen only cost a dollar or two, when in reality he was spending thousands. When he died, his wife threw out the collection. When other collectors inquired about the collection, they learned that she dumped it out since they only cost a few dollars apiece, and it was not worth her while to try to sell a "valueless" collection.

With computers it is really simple to keep a good database of your collection. If you are just starting now to enter 1000-plus pieces, just remember, "When eating an elephant take one bite at a time." Do not get overwhelmed by the size of the job. Just start now and enter the most recent pieces, and over time, keep entering information when you have an extra hour and nothing good is on television. It is worthwhile and it helps to let you appreciate what you have gathered. You will also be surprised to find things that you did not remember that you had. The benefits are that someone in the future will be able to go to the database or your print-out and see what you have. (When selling a piece, you should also provide the buyer with all the cataloged information for that piece). Databases can also print the tags that go with your specimens if you display or sell pieces. There are pre-made databases offered for sale in many mineral magazines. I use a program called "Filemaker Pro" that is available for all computers. It is simple to use, easy to make labels with, and easy to sort. Some people use Microsoft Excel, which is a spreadsheet that can act as a database. Any simple database program will work.

What should a database contain?

1) **Category:** This can be Fluorescent Minerals, Fossils, Sea Shells, etc.

2) **Item:** This is a simple description of the piece such as "scheelite," "scheelite from China," or "scheelite & beryl."

3) **Specific Item:** Here is where you enter a complete description such as "Amber-colored scheelite crystals with beryl crystals on muscovite from Ping Wu mine, Sichuan Province, China. Scheelite fluoresces bright pale blue (SW), beryl and muscovite are NF."

4) **Date purchased:** My database has this as "Date Purch."

5) **Number**: Here is where your item number goes.

6) **Cost**

7) **Size:** I enter the weight and dimensions.

8) **Place Purchased:** Sometimes this will help you find a piece once it is on your shelf, since you often remember where you found or purchased it and from whom you got it.

9) **Etc.:** This is where I enter the first few letters of the piece such as "scheel." This can be used to sort and find all the scheelites in your collection.

10) **Etc.2:** This is a catch-all category where you can add notes such as "Bob Jones gave me this piece at the Tucson show," or any other interesting information.

11) **Etc.3:** Here is where I add "The Schneider Collection, New Jersey." When I print labels this appears on the bottom of each label.

Earl Verbeek, the curator of the Sterling Hill Mining Museum, suggests that you also might add a category that says why you have this piece in your collection. For a collector, it could say something like, "This piece has crystals of xonotlite with crystals of prehnite. It is extremely rare and is my only example of this material." Earl Verbeek uses it to denote "significance." For example, as a geologist, he is aware that the association of wollastonite plus quartz in a marble host rock should not occur. If he finds a specimen that exhibits this association, it will be added to the collection. He will note the "significance" of this piece so that future curators of the collection, who may not realize the significance of the rare association, will understand why it is important to the collection.

If you sell minerals you might want to put a "Price Sold" and "Date Sold" category.

This suggested group of database categories appears on your "data entry" page. With a computer database you can create different ways to look at the information. I have a "data entry" view that gives me all of the information about a specific item. This is also where I enter information. I have a "list" view that just gives me just the "Item," "Specific Item," "Date purch," "Cost," and "Number." I have a "card" view that gives me "Specific Item," "Number," and "Etc.3" and is used to print cards. I have a "small print" view that gives me "Item," "Specific Item," "Size," "Date purch," and "Number" that is used when I print out my database that goes into a binder. You can create other views to suit your needs.

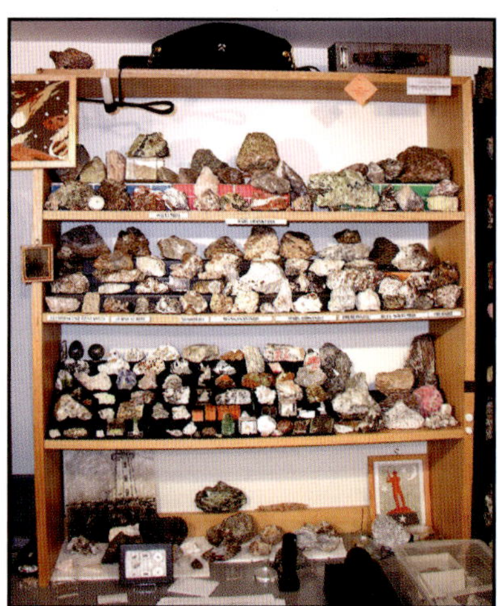

Before computers entered my life, I used a journal to record every purchase. When computers came along, I slowly transferred all that information to the database. My memory used to be much better years ago. I could remember every item I bought, where I bought it, what I paid, etc. As my collections grew, the memories became less clear. Now when I pick up a specimen in my display room and I can't recall what it is, I check its number against my printed database and know instantly what it is. If I need further information, I can check the computer for further details.

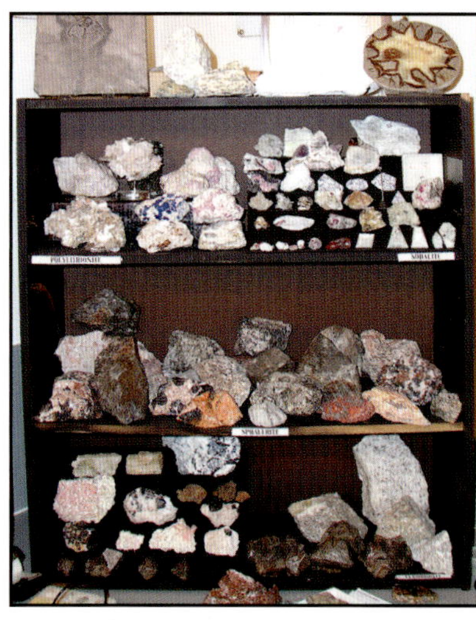

These are inexpensive book shelves that make good mineral display cases for my Franklin and Greenland collections.

Displaying Your Collection

The fluorescent-mineral collector and the nonfluorescent-mineral collector have similar goals in displaying their collections. They want to show off the pieces in their collections so that they look attractive, are protected from harm, and are somewhat easy to get to when showing individual pieces. The differences are that the nonfluorescent-mineral collector can put everything in glass display cases (which minimizes dust) and can light their specimens with regular lights. The fluorescent-mineral collector must either build a display case with built-in ultraviolet lamps (fairly expensive if done in a professional manner) or not use glass display cases, but some type of open-faced shelving, and light their pieces with hand-held lamps.

I have seen many fluorescent-mineral displays. Collectors have used ingenuity in building attractive showcases that do not cost a great deal of money. All collectors who share a home or apartment with others must take into consideration where they can put their collection. Some collectors want everything hidden so that visitors do not see the collection unless it is shown to them. Their collections reside in closets, corners of basements, portions of a spare bedroom, or even under a bed. Other collectors have a spouse that does not want the collection cluttering up the living area of the home. Again, their collection is usually tucked away somewhere out of the main living area. Some collectors live in smaller homes and just do not have very much room to display their collection.

I was fortunate when it came to displaying my collection. I was a photographer with a dedicated darkroom. When digital photography came along, I stopped using the darkroom and turned it into a fluorescent-mineral display room. With space devoted exclusively to my collection, I could have built permanent display cases with dedicated ultraviolet lighting. I would, however, rather spend my discretionary-use money on specimens rather than cases. I visited my local K-Mart and bought a few inexpensive book shelves. The shelves were spray painted black and I had my first display cases. I found some heavy-duty, cardboard, children's blocks at a garage sale. Once they were painted flat black, they gave me raised specimen stands so that pieces in the rear of the shelves were higher than those in the front. I later found some quality, tiered-metal stands at a gem and mineral show. These replaced the blocks and made a better display for the smaller pieces.

As my collection grew, I needed more display cases. My wife suggested VCR "bookcases." These cases are available at music and furniture stores or in mail order catalogs, and are used to hold video cassettes and DVDs. Their advantages are that they offer more shelf space for the area, they are not very deep, and they are built well enough, so that 400 pounds of rocks will not come crashing down. Bookshelves and VCR shelves allow you to store and show off your collection. When you want to illuminate your collection, a good, hand-held, ultraviolet lamp will, in an appropriately dark place, make everything fluoresce nicely.

As my collection grew larger, I realized that I could not display everything on shelves. I visited an office-supply store that sold second-hand furniture, and found some metal cases that resemble museum storage drawers. These were made for the storage of pre-printed forms. The one I selected had 48 drawers that were each about 3 inches tall, 9 inches wide, and 15 inches deep. They were made to hold heavy reams of paper and were perfect for rocks and minerals. The cost was a fraction of new museum storage drawers.

As I continued to spend my summers digging for minerals in Sussex County, New Jersey, I needed even more places to put my ever-growing collection. My wife was using plastic shoeboxes for her shoes, so I bought a few dozen of these for rocks. The plastic shoebox is a wonderful storage device. It is inexpensive, you are able to see what is inside each box, a marking pen lets you note what is in the box, and they are stackable (3 or 4 boxes high). If you want to go a bit more "industrial", there are mail order companies that make storage racks and boxes for the hardware and automotive industry. The boxes slide in and out of the racks and they are made to hold heavy items like nuts and bolts.

Whether you are setting up your display in a closet, taking over a room, or doing gem and mineral shows, an organized storage and display system is attractive, useful, and can be designed and purchased within your budget.

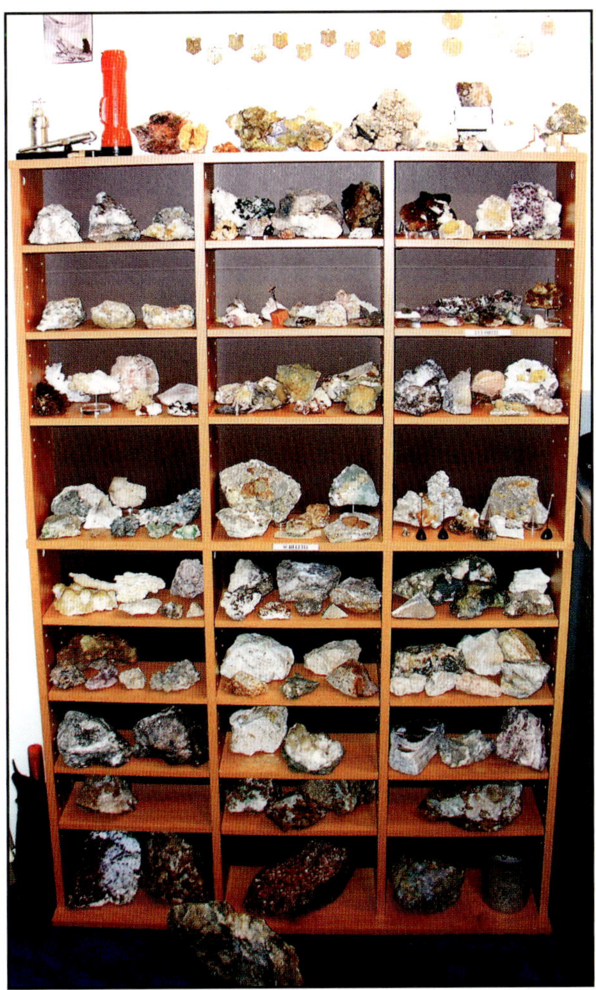

This is a VCR "bookcase" that displays my apatite, calcite, fluorite, and scheelite collections.

Things to Consider

1. **Buy the best.** Comparing fluorescent-mineral collecting to other collecting fields, it is expected that prices on items will continue to rise. Higher prices may be a blessing in disguise. Some people have no incentive to sell a piece for only a few dollars, but if they can get "a lot of money" they will sell a piece. You will pay more, but you will get a mineral that you may never have another chance to own. Collections can be put together for very little money, but, as has been proven true in every field of collecting, the best pieces in the best condition have held or increased in value at a greater rate than the more common pieces. Buy the best that you can afford. Upgrade your collection where possible. Remember, you will rarely regret having paid too much for an item, but you will always regret the good pieces that got away.

2. **Cost.** A fluorescent-mineral collection need not cost a great deal of money. Some people collect fluorescent minerals from around the world. Others have a collection of self-collected rocks. Joining a local mineral club can provide leads to items. If you use your resources wisely, a nice collection can be put together within your budget. Exhibiting your collection at a local library can generate new leads to the type of material that you collect. Tell friends what you collect. You may be amazed to find that they know some old-time collector who is no longer active and lives in your area.

3. **Buy books on minerals and join a club.** Learn from others' experiences. The more you know about minerals, the more likely you will be able to spot a good piece for your collection or to trade or sell. Join the Fluorescent Mineral Society and a local mineral club. Members of a local mineral club, even if they are not interested in fluorescent minerals, may help you learn more about the hobby and give you leads to local digs and other collectors.

4. **Go to local mineral shows.** eBay is a great boon to fluorescent-mineral collectors, but there is a lot to be said for having the piece in your hand for a careful examination. A local mineral show may turn up pieces from old collections that may not be available from other sources.

5. **Availability or "Will I ever find another?"** Some items are always available at mineral shows or on the Internet. Ask yourself if it is just a matter of dollars to acquire an example or is this a rare opportunity to own that item? With rare pieces, you may never get another chance to own one and another person may be waiting for you to put it down so that he or she can buy it. Don't hesitate to buy it.

6. **Investment.** Hopefully you are not "investing" in minerals. Collecting should be for fun and not just for profit.

7. **Think small.** Microminerals are getting more popular each year. They take up little space, often cost only a few dollars, and the variety can be incredible. They don't appeal to all collectors. A small piece of rock with a few clinohedrite crystals will delight a collector of microminerals and can easily be overlooked by other mineral collectors. Collectors of microminerals seem to have such fun on digs as they gaze through their magnifying lenses at tiny crystals that other collectors have passed without seeing.

Collector Names of Minerals

There are several names of minerals that have been discontinued, replaced, or otherwise no longer used by mineralogists. There are also mineral-group names that often appear on labels when the seller of the mineral has only a vague idea of what the actual mineral is.

You may come across a mineral with an unfamiliar name and pass it by. This could work against you when going through an old collection, especially when the minerals are in plastic boxes or plastic bags, since SW UV may not make it through the plastic. If you fail to turn your UV lamp on a particular mineral because you do not recognize the name, you might miss a good piece. Also, many dealers on eBay use the collector name rather than the official name. There is also the mineral's official name which may be known to mineralogists and not collectors. Here is a short list of names of minerals or mineral groups that you should always check.

Beta-willemite: A secondary willemite, properly called "yellow fluorescing willemite".

Calcium larsenite: An old name for esperite. Both were named for Esper Larsen, Jr. (1879-1961).

Caliche: A general term for any secondary calcium carbonate that forms in sediments or in voids in bedrock just below the surface in semiarid regions. It usually fluoresces pale yellow SW.

Chalcedony: A variety of quartz that usually fluoresces.

Cleiophane: A name for iron-free sphalerite that usually fluoresces blue SW. In the Franklin, NJ, area, this refers to blue-fluorescing sphalerite since most of the sphalerite from this area is iron-free.

Dravite: Brown tourmaline now called uvite. It often fluoresces yellow SW.

Elbaite: Tourmaline group. Most dealers refer to it as tourmaline.

Feldspar: The feldspar group includes hyalophane, microcline, albite, orthoclase, and amazonite that may all fluoresce.

Golden sphalerite: Golden sphalerite is a sphalerite which fluoresces a distinctive orange-yellow with blue highlights. It is rare and most of the time, the name is misused.

Gypsum: Selenite is a variety of gypsum

Humite: Humite group. This can be chondrodite, norbergite, humite, or clinohumite.

Jeffersonite: Augite. This does not fluoresce, but could be associated with fluorescing minerals.

Mahagony sphalerite: A fine-grained reddish-brown sphalerite.

Manganapatite: Now called manganoan fluorapatite or manganese-rich fluorapatite

Schefferite or zinc-schefferite: Aegirine-augite (pyroxene) group. If it is white to light brown it is probably diopside, dark brown is augite, and very dark brown or dark green probably aegirine. It does not fluoresce, but may be associated with fluorescent minerals.

Tirodite: Once called manganocummingtonite. Now known as Parvowinchite.

Wernerite: A mineral that fluoresces bright yellow LW. Actually scapolite var. meionite, or var. marialite. Most dealers sell it as wernerite.

Resources

Collecting Groups, Information, & Museums

The Fluorescent Mineral Society, Inc. This is a great organization that publishes newsletters and a journal and links you with a group of people who will not laugh when you tell them you collect fluorescing minerals. It was started in 1971. It is a must-join organization for fluorescent-mineral collectors. For information, write to Dr. Rodney Burroughs, P.O. Box 572694, Tarzana, CA 91357 or go to www.uvminerals.org. You can also order popular books on fluorescent-mineral collecting from the Fluorescent Mineral Society.

The Franklin-Ogdensburg Mineralogical Society, Inc. This is the local New Jersey collector's organization that puts on two great gem and mineral shows each year. They publish a newsletter and a journal, The Picking Table (which is easily worth more than the cost of membership). Membership is $20.00-yr. Write to Denise Kroth, Treasurer, 240 Union Ave., Wood-Ridge, NJ 07075

The Franklin Mineral Museum, 32 Evans Road, Franklin, NJ 07416. Open March (weekends only) then April 1 to December 1, Monday to Saturday 10 AM to 4 PM, Sunday 12:30 PM to 4:30 PM. www.franklinmineral-museum.com

Mountain Area Gem and Mineral Association (M.A.G.M.A.). This is a North Carolina digging group that runs trips to many different mines in North Carolina and other states. Website at: www.wncrocks.com/magma/magma.html or contact them at rick@wncrocks.com or stevenpenley@bellsouth.net or maitrimaitri@wncrocks.com.

The Sterling Hill Mining Museum and Thomas S. Warren Museum of Fluorescence, 30 Plant St., Ogdensburg, NJ 07439. Open April 1 through November 30, 7 days a week, 10 AM to 5 PM. Website at www.sterlinghill.org.

Stuart Schneider's Fluorescent Mineral Museum. Online at www.wordcraft.net. There are currently 16 pages.

The St. Lawrence Co. Rock & Mineral Club can be reached c/o Virginia Searles, 42 1/2 Maple Street, Potsdam, NY 13676, http://web.northnet.org/st.lawrence.co.mineral.club

The Alkali-Nuts. The Mont Saint-Hilaire, Canada Mineral Group's website is a wonderful source of information on Mont Saint-Hilaire minerals. It is at www.ssc.on.ca/mandm/fluo.html

Rock & Gem Magazine, 4880 Market St., Ventura, CA 93003-7783. A full-color gem and mineral magazine with articles and information on every aspect of mineral collecting.

Rocks & Minerals Magazine, 1319 Eighteenth St., NW, Washington, DC 20036 publishes a high-quality, full-color magazine.

Lapis International, P.O. Box 263, East Hampton, CT 06424 publishes an English edition of its magazine that has each issue concentrating on a particular mineral. The photography is first-class.

The Mineralogical Record, P.O. Box 35565, Tucson, AZ 85750 is another magazine with great photography, but the articles are more technical than those in most of the other magazines.

Mineral Dealers

Anderson Fluorescent Minerals is a dealer in fluorescent minerals. 1430 Vue Du Bay Ct., San Diego, CA 92109.
Website: www.globalpac.com/gems/afm.

Benitoite Mine was a source for benitoite, neptunite, etc. in natrolite. The mine was sold in 2005 and hopefully a new owner will take over the mining and continue selling the minerals.

Excalibur Mineral Corp. provides analytical service, rare minerals, books, supplies. 1000 North Division St., Peekskill, NY 10566
Website: www.excaliburmineral.com.

GSL Rocks is a dealer in fluorescent minerals and ultraviolet lamps. Write Greg at 4726 Porter Center Rd., Lewiston, NY 14905.
Email: gslrocks@aol.com.

MinerShop has information on traveling to Greenland and collecting your own fluorescent minerals. MinerShop, in conjunction with Jewel Stones of Greenland, offers guided Adventure Tours each summer season. Tours are limited to eight people per week.
Website: www.minershop.com
Email: sales@minershop.com

Polman Minerals is a dealer in fluorescent minerals. P.O. Box 93276, Phoenix, AZ 85070.
Website: www.polmanminerals.com
Email: polmans@compuserve.com.

Purple Passion mine, Bill Gardner, 4608 W. Bluefield Ave., Glendale, AZ 85308 sells fluorescent minerals from the mine and manufactures a quality line of ultraviolet lamps.
Email: wggardner@aol.com.

Rocko Minerals, Box 3A, Route 3, Margaretville, NY 12455 is a dealer in fluorescent minerals.
Email: rocko@catskill.net.

Stuart Schneider's Fluorescent Mineral Museum Online currently has 8 pages of minerals for sale.
Website: www.wordcraft.net

Simkev Micromounts/The Fine Mineral Company. They have examples of rare Mont Saint-Hilaire minerals.
Email: simkev@sympatico.ca.

Soenke Stolze's fluorescent minerals for sale (German dealer)
Website: www.systematic-minerals.com
Email: stolze@systematic-minerals.com.

Tom Jokela is a good source for Canadian fluorescent minerals.
Website: www.element51.com/fluorescents.htm.

Veronica Matthews Minerals, P.O. Box 588, Hammock Rd., Westbrook, CT 06498 is a dealer in fluorescent minerals.

Charles B. Ward is a dealer in fluorescent minerals. 4071 NC 80, Bakersville, NC 28705
Website: www.fluorescentminerals.com.

Ultraviolet Lamps and Supplies

Buy the best lamp that you can afford. They are expensive, but they are the most important tool that a fluorescent-mineral collector can own. Check out several sources here and ask questions.

GSL Rocks is a dealer in reasonably priced 9 to 13 watt "Way Too Cool" ultraviolet lamps. 4726 Porter Center Rd, Lewiston, NY 14905
Email: gslrocks@aol.com.

Mineralab. Supplies, lamps, and other great material. 2860 Live Oak Dr. #G, Prescott, AZ 86305
Website: www.mineralab.com.

Mineralogical Resource Company. A resource for ultraviolet lamps, supplies, etc. Mineralogical Resource Co., 15840 East Alta Vista Way, San Jose, CA 95127.
Website: www.minresco.com
Email: xtls@minresco.com.

Raytech Industries makes a good series of ultraviolet lamps, including a nice portable model. Contact them at 475 Smith Road, Middletown, CT 06457.
Website: www.raytech-ind.com
Email: info@raytech-ind.com.

The Rock Peddler has lapidary supplies at the best prices. Contact them at 771 Boston Post Road East, #180, Marlborough, MA 01752.
Website: www.rockpeddler.com
Email: rockpeddler@attbi.com.

Spectronics Corp., 956 Brush Hollow Road, P.O. Box 483, Westbury, NY 11590
Website: www.spectroline.com.

UV Systems. Makers of the SuperBright ultraviolet lamp (one of my favorites) now has an improved model that drives the lamp at a slightly higher output, UV Systems, 16605 127th SE, Renton, WA 98058
Website: www.uvsystems.com
Email: info@uvsystems.com.

UVP, Inc. makes a good series of ultraviolet lamps, including a nice portable model. Contact them at 2066 W. 11th St., Upland, CA 91786.
Website: www.uvp.com
email: uvp@uvp.com.

Remember, two is a coincidence, three is a collection. Happy hunting!

Bibliography

Abandoned Iron Mines of Sussex County, New Jersey, State of New Jersey, Dept. of Labor, Trenton, NJ, 1982

Arem, Joel E., *Color Encyclopedia of Gemstones, 2nd ed.*, Van Nostrand Reinhold Co., New York, NY, 1987

Bancroft & District Mineral Collecting Guide, Bancroft Chamber of Commerce, Bancroft, Ontario, 2001

The Borough of Andover New Jersey, Historical Society of Andover Borough, Andover, NJ, 1992

Bowersox, G., and B. Chamberlin, *Gemstones of Afghanistan*, Geoscience Press, Tucson, AZ, 1995

Ottens, Berthold, *Calcite from the Deccan Traps of India - Rocks & Minerals*, Volume 80, No. 2, March/April, 2005

Horvath, Lazlo and Robert Gault, *Mont Saint-Hilaire*, The Mineralogical Record, Vol. 21, #4, Tucson, AZ, 1990

Lapis International, LLC, *Pakistan, Minerals, Mountains & Majesty*, East Hampton, CT, 2004

Mandarino, J. and M. Back, *Fleischer's Glossary of Mineral Species 2004*, The Mineralogical Record, Tucson, AZ, 2004

Matlins, A., and A.C. Bonanno, *Gem Identification Made Easy: A Hands On Guide to More Confident Buying and Selling, 3rd Ed.*, Gemstone Press, Woodstock, VT, 2006

Palache, Charles, *The Minerals of Franklin and Sterling Hill, Sussex County, New Jersey*, U.S. Geological Survey, Washington, D.C., 1935, 1960

Pratt, M.E., *Report on the Diamond Joe Mining property*, Arizona Department of Mines and Mineral Resources, 1938

Robbins, Manuel, *Fluorescence: Gems and Minerals under Ultraviolet Light*, Geoscience Press, Inc., Phoenix, AZ, 1994

Robbins, Manuel, *The Collector's Book of Fluorescent Minerals*, Van Nostrand Reinhold Company, New York, NY, 1983

The First Hundred Years of the New Jersey Zinc Company, New Jersey Zinc Co., New York, NY, 1948

Warren, T., S. Gleason, R. Bostwick, and E. Verbeek, *Ultraviolet Light and Fluorescent Minerals*, Thomas S. Warren, available from The Sterling Hill Mining Museum, 1995

Franklin-Ogdensburg Mineralogical Society (FOMS), *The Picking Table*, The journal of the FOMS.

The Fluorescent Mineral Society, Inc. (FMS), *UV Waves*, The newsletter of the FMS.

Index